Boris M. Smirnov

Fundamentals of Ionized Gases

Related Titles

Ambaum, M.

Thermal Physics of the Atmosphere

2010
ISBN: 978-0-470-74515-1

LeBlanc, F.

An Introduction to Stellar Astrophysics

2010
ISBN: 978-0-470-69956-0

Stock, R. (ed.)

Encyclopedia of Applied High Energy and Particle Physics

2009
ISBN: 978-3-527-40691-3

Hippler, R., Kersten, H., Schmidt, M., Schoenbach, K. H. (eds.)

Low Temperature Plasmas
Fundamentals, Technologies and Techniques

2008
ISBN: 978-3-527-40673-9

Iliadis, C.

Nuclear Physics of Stars

2007
ISBN: 978-3-527-40602-9

Oks, E.

Plasma Cathode Electron Sources
Physics, Technology, Applications

2005
ISBN: 978-3-527-40634-0

Lieberman, M. A., Lichtenberg, A. J.

Principles of Plasma Discharges and Materials Processing

2005
ISBN: 978-0-471-72001-0

Woods, L. C.

Physics of Plasmas

2004
ISBN: 978-3-527-40461-2

Boris M. Smirnov

Fundamentals of Ionized Gases

Basic Topics in Plasma Physics

WILEY-VCH Verlag GmbH & Co. KGaA

The Author

Prof. Boris Smirnov
Joint Institute for High Temperatures
Russian Academy of Sciences
Moscow, Russian Federation
bmsmirnov@gmail.com

Cover Picture
Spiesz Design, Neu-Ulm

Library of Congress Card No.: applied for

British Library Cataloguing-in-Publication Data:
A catalogue record for this book is available from the British Library.

Bibliographic information published by the Deutsche Nationalbibliothek
The Deutsche Nationalbibliothek lists this publication in the Deutsche Nationalbibliografie; detailed bibliographic data are available on the Internet at http://dnb.d-nb.de.

Typesetting le-tex publishing services GmbH, Leipzig
Printing and Binding Fabulous Printers Pte Ltd, Singapore
Cover Design Grafik-Design Schulz, Fußgönheim

Printed in Singapore
Printed on acid-free paper

ISBN Print 978-3-527-41085-9

ISBN oBook 978-3-527-63710-2
ISBN ePub 978-3-527-63711-9
ISBN ePDF 978-3-527-63712-6
ISBN Mobi 978-3-527-63713-3

Contents

Preface

This book is based on the lecture courses on plasma physics which were given by the author at various education institutes in Russia and abroad over several decades. The courses were intended for undergraduate students and postgraduate students of physical and technical professions. The main goal of the book is to provide the reader with various concepts of plasma physics and related topics in physics. These problems are represented in a large number of plasma books, a list of which is given below, and therefore it is necessary to note the features of this book.

The form of this book is such that the author has tried, on one hand, to conserve the level of contemporary theoretical plasma physics, and, on other hand, to use a simple description of the problems. These requirements seem to be contradictory, but may be achieved partially by using limiting cases or simple models. These were accumulated by the author during his scientific activity, partially borrowed from colleagues, and represented in previous books by the author [1–5]. In this book, these limiting cases and simpler models are used and their number is increased, allowing a large number of concepts of plasma physics and related physics areas to be included in this book.

Since this book reckons on an active reader, it contains contemporary information, mostly in the text and tables. The plasma physics course in the subsequent text is divided into parts which include general plasma properties, elementary processes in plasmas, kinetics and transport processes in ionized gases, and waves, instabilities, and nonlinear processes in plasmas and excited gases. Certain plasma objects, such as a cluster plasma, a dusty plasma, an aerosol plasma, and an atmospheric plasma, are analyzed to demonstrate the connection between general plasma properties and constituent plasma particles. The author hopes that his long-term experience in plasma physics will be useful for the reader.

Moscow, June 2011 *Boris M. Smirnov*

References

1 Smirnov, B.M. (1972) *Physics of Weakly Ionized Gas*, 1st edn., Nauka, Moscow. 2nd edn, 1978, 3rd edn, 1985, in Russian.
2 Smirnov, B.M. (1975) *Introduction to Plasma Physics*, 1st edn, Nauka, Moscow. 2nd edn, 1982, in Russian.
3 Smirnov, B.M. (1977) *Introduction to Plasma Physics*, Mir, Moscow.
4 Smirnov, B.M. (1981) *Physics of Weakly Ionized Gases*, Mir, Moscow.
5 Smirnov, B.M. (2001) *Physics of Ionized Gases*, John Wiley & Sons, Inc., New York.

Books for Plasma Physics

Alexandrov, A.F., Bogdankevich, L.S., and Rukhadze, A.A. (1984) *Principles of Plasma Electrodynamics*, Springer, Berlin.

Baumjohann, W. and Treumann, R.A. (1997) *Basic Space Plasma Physics*, Imperial College Press, London.

Belan, P.M. (2006) *Fundamentals of Plasma Physics*, Cambridge University Press, Cambridge.

Bittencourt, J.A. (1986) *Fundamentals of Plasma Physics*, Pergamon Press, Oxford.

Bittencourt, J.A. (2004) *Fundamentals of Plasma Physics*, Springer, New York.

Bittencourt, J.A. (1982) *Fundamentals of Plasma Physics*, Pergamon Press, Oxford.

Boening, H.V. (1982) *Plasma Science and Technology*, Cornell University Press, Ithaca.

Boulos, M.I., Fauchais, P., and Pfender, E. (1994) *Thermal Plasmas*, Plenum, New York.

Brown, S.C. (1966) *Introduction to Electrical Discharges in Gases*, John Wiley & Sons, Inc., New York.

Brown, S.C. (1994) *Basic Data of Plasma Physics. The Fundamental Data on Electrical Discharges in Gases*, American Institute of Physics, New York.

Cairns, R.A. (1985) *Plasma Physics*, Blackie, Glasgow.

Chapman, B. (1980) *Glow Discharge Processes: Sputtering and Plasma Etching*, John Wiley & Sons, Inc., New York.

Chen, F.F. (1984) *Introduction to Plasma Physics and Controlled Fusion*, Plenum, New York.

Chen, F.F. and Chang, J.P. (2003) *Lectures Notes on Principles of Plasma Processing*, Kluwer, New York.

Cobine, J.D. (1958) *Gaseous Conductors*, Dover, New York.

D'Agostino, R. (1990) *Plasma Deposition, Treatment, and Etching of Polymers*, Academy Press, New York.

Davidson, R.C. (1974) *Theory of Nonneutral Plasmas*, Perseus Books, Reading.

Davidson, R.C. (1990) *Physics of Nonneutral Plasmas*, Perseus Books, Reading.

Dendy, R.O. (1995) *Plasma Physics*, Cambridge University Press, Cambridge.

Ecker, G. (1972) *Theory of Fully Ionized Plasmas*, Academic Press, New York.

Fortov, V.E. (ed.) (2000–2010) *Encyclopedia of Low Temperature Plasma*, Nauka, Moscow, in Russian.

Fortov, V.E. and Iakubov, I.T. (1990) *Physics of Nonideal Plasma*, Hemisphere, New York.

Fortov, V.E. and Iakubov, I.T. (2000) *The Physics of Non-ideal Plasma*, World Scientific, Singapore.

Fortov, V.E., Jakubov, I.I., and Khrapak, A.G. (2006) *Physics of Strongly Coupled Plasma*, Clarendon Press, Oxford.

Frank-Kamenetskii, D.A. (1972) *Plasma – the Fourth State of Matter*, Plenum Press, New York.

Freidberg, J.F. (2007) *Plasma Physics and Fusion Energy*, Cambridge University Press, Cambridge.

Fridman, A.A. and Kennedy, L.A. (2004) *Plasma Physics and Engineering*, Taylor and Francis, New York.

Fridman, A.A. (2008) *Plasma Chemistry*, Cambridge University Press, Cambridge.

Galeev, A.A. and Sudan, R.N. (eds) (1983) *Basic Plasma Physics*, North Holland, Amsterdam.

Golant, V.E., Zhilinskii, A.P., and Sakharov, I.E. (1980) *Fundamentals of Plasma Physics*, John Wiley & Sons, Inc., New York.

Goldston, R.J. and Rutherford, P.H. (1995) *Introduction to Plasma Physics*, Adam Hilger, Bristol.

Goossens, M. (2003) *An Introduction to Plasma Astrophysics and Magnetohydrodynamics*, Springer, New York.

Grill, A. (1993) *Cold Plasma in Material Fabrication*, IEEE Press, New York.

Grill, A. (1994) *Cold Plasma in Material Fabrication. From Fundamentals to Applications*, IEEE Press, New York.

Gurnett, D.A. and Bhattacharjee, A. (2005) *Introduction to Plasma Physics. With Space and Laboratory Applications*, Cambridge University Press, Cambridge.

Hasegawa, A. (1975) *Plasma Instabilities and Nonlinear Effects*, Springer, Berlin.

Hasetline, R.D. and Waelbroeck, F.L. (2004) *The Framework of Plasma Physics*, Westview, Boulder.

Hippler, R., Pfau, S., Schmidt, M., and Schoenbach, K.H. (eds) (2001) *Low Temperature Plasma Physics. Fundamental Aspects and Applications*, Wiley-VCH GmbH.

Howlett, S.R., S.R Timothy, and Vaughan, D.A. (1992) *Industrial Plasmas: Focusing UK Skills on Global Opportunities*, CEST, London.

Hoyaux, M.F. (1968) *Arc Physics*, Springer, New York.

Hutchinson, I.H. (2002) *Principles of Plasma Diagnostics*, Cambridge University Press, Cambridge.

Huxley, L.G.H. and Crompton, R.W. (1974) *The Diffusion and Drift of Electrons in Gases*, John Wiley & Sons, Inc., New York.

Ichimaru, S. (1973) *Basic Principles of Plasma Physics: A Statistical Approach*, Benjamin, London.

Ishimaru, S. (1985) *Plasma Physics. Introduction to Statistical Physics of Charged Particles*, Benjamin, Menlo Park.

Ichimaru, S. (1992) Basic principles, in: *Statistical Plasma Physics*, vol. 1, Perseus Books, Reading

Ishimaru, S. (1994) *Statistical Plasma Physics*, vol. 2. *Condensed Plasmas*, Addison-Wesley, Redwood City

Ichimaru, S. (2004) *Statistical Plasma Physics. vol. 1. Basic Principles. vol. 2. Condensed Plasmas*, Westview, Boulder.

Kadomtzev, B.B. (1965) *Plasma Turbulence*, Academy Press, New York.

Kettani, M.A. and Hoyaux, M.F. (1973) *Plasma Engineering*, John Wiley & Sons, Inc., New York.

Kobzev, G.A., Iakubov, I.T., and Popovich, M.M. (eds) (1995) *Transport and Optical Properties of Nonideal Plasma*, New York, Plenum Press.

Krall, N.A. and Trivelpiece, A.V. (1973) *Principles of Plasma Physics*, McGraw-Hill, New York.

Kruer, W. (1988) *The Physics of Laser Plasma Interactions*, Perseus Books, Reading.

Kuzelev, M.V. and Rukhadze, A.A. (1990) *Elecrtodynamics of Dense Electron Beam in Plasma*, Nauka, Moscow, in Russian.

Lagarkov, A.N. and Rutkevich, I.M. (1994) *Ionization Waves in Electrical Breakdown in Gases*, Springer, New York.

Lieberman, M.A. and Lichtenberger, A.J. (2005) *Principles of Plasma Discharge and Materials Processing*, John Wiley & Sons, Inc., Hoboken.

Lifshits, E.M. and Pitaevskii, L.P. (1981) *Physical Kinetics*, Pergamon Press, Oxford.

Lisitsa, V.S. (1994) *Atoms in Plasmas*, Springer, Berlin.

Loeb, L.B. (1955) *Basic Processes of Gaseous Electronics*, University California Press, Berkeley.

Manos, D.M. and Flamm, D.L. (1989) *Plasma Etching. An Introduction*, Academy Press, San Diego.

Massey, H.S.W. (1979) *Atomic and Molecular Collisions*, Taylor and Francis, New York.

Massey, H.S.W. and Burhop, E.H.S. (1969) *Electronic and Ionic Impact Phenomena*, Oxford University Press, Oxford.

McDaniel, E.W. and Mason, E.A. (1988) *Transport Properties of Ions in Gases*, John Wiley & Sons, Inc., New York.

Mesyats, G.A. and Proskurovsky, D.I. (1989) *Pulsed Electrical Discharge in Vacuum*, Springer, Berlin.

Mikhailovskii, A.B. (1974) *Theory of Plasma Instabilities*, Consultants Bureau, New York.

Mikhailovskii, A.B. (1992) *Electromagnetic Instabilities in an Inhomogeneous Plasma*, IOP Publishing, Bristol.

Nasser, E. (1971) *Fundamentals of Gaseous Ionization and Plasma Electronics*, John Wiley & Sons, Inc., New York.

Nezlin, M.V. (1993) *Physics of Intense Beams in Plasmas*, IOP Publishing, Bristol.
Nicholson, D.W. (1983) *Introduction to Plasma Theory*, John Wiley & Sons, Inc., New York.
Nishikawa, K. and Wakatani, M. (2000) *Plasma Physics*, Springer, Berlin.
Ochkin, V.N. (2009) *Spectroscopy of Low Temperature Plasma*, Wiley-VCH Verlag GmbH, Weinheim.
Ochkin, V.N., Preobrazhensky, N.G., and Shaparev, N.Ya. (1998) *Optogalvanic Effect in Ionized Gas*, Gordon and Breach, London.
Piel, A. (2010) *Plasma Physics. An Introduction to Laboratory Space and Fusion Plasmas*, Springer, Heidelberg.
Proud, J. *et al.* (1991) *Plasma Processes of Materials: Scientific Opportunities and Technological Challenges*, National Academy Press, Washington.
Raizer, Y.P. (1991) *Gas Discharge Physics*, Springer, Berlin.
Rickbery, D.S. and Mathews, A. (1991) *Surface Engineering: Processes, Characterization and Applications*, Blackie and Son, Glasgow
Sagdeev, R.Z. and Galeev, A.A. (1969) *Nonlinear Plasma Theory*, Benjamin, New York.
Salzmann, D. (1998) *Atomic Physics in Hot Plasmas*, Oxford University Press, Oxford.
Schevel'ko, V.P. and Vainstein, L.A. (1993) *Atomic Physics for Hot Plasmas*, Institute of Physics, Bristol.
Schram, P.P.J.M. (1991) *Kinetic Theory of Gases and Plasmas*, Kluwer, Dordrecht.
Seshardi, S.R. (1973) *Fundamentals of Plasma Physics*, Elsevier, New York.
Shukla, P.K. and Mamun, A.A. (2002) *Introduction to Dusty Plasma Physics*, Institute of Physics Publishing, Bristol.
Sitenko, A.G. (1982) *Fluctuations and Nonlinear Wave Interactions in Plasmas*, Pergamon Press, Oxford.
Sitenko, A. and Malnev, V. (1995) *Plasma Physics Theory*, Chapman and Hall, London.
Smirnov, B.M. (2007) *Plasma Processes and Plasma Kinetics*, Wiley-VCH GmbH.
Son, E.E. (1977) *Thermophysical Properties of High-temperature Gases and Plasma*, MFTI Press, Moscow, in Russian.
Son, E.E. (1978) *Stability of Low-temperature Plasma*, MFTI Press, Moscow, in Russian.
Son, E.E. (1978) *Kinetic Theory of Low-temperature plasma*, MFTI Press, Moscow, in Russian.
Son, E.E. (2006) *Lectures in Low Temperature Plasma Physics*, Massachusetts Institute of Technology, Plasma Science Fusion Center, Cambridge.
Spitzer, L. (1962) *Physics of Fully Ionized Gases*, John Wiley & Sons, Inc., New York.
Spitzer, L. (2006) *Physics of Fully Ionized Gases*, Dover, New York.
Stix, T.H. (1992) *Waves in Plasmas*, American Institute of Physics, New York.
Sturrock, P.A. (1994) *Plasma Physics*, Cambridge University Press, Cambridge.
Swanson, D.G. (2003) *Plasma Waves*, Institute of Physics Publishing, Bristol.
Toshoro, O. (1994) *Radiation Phenomena in Plasmas*, World Scientific, Singapore.
Tsytovich, V.N. (1970) *Nonlinear Effects in Plasmas*, Plenum Press, New York.
Woods, L.C. (2004) *Physics of Plasmas*, Wiley-VCH Verlag GmbH, Weinheim.
National Academy (1995) *Plasma Science: from Fundamental Research to Technological Applications* National Academies Press, Washington.
von Engel, A. (1983) *Electric Plasmas: Their Nature and Uses*, Taylor and Francis, London.
von Engel, A. (1997) *Ionized Gases*, American Institute of Physics, New York.)

1
General Concepts in Physics of Excited and Ionized Gases

1.1
Ideal Plasma

1.1.1
Plasma as a State of Matter

The word "plasma" was introduced into science by the Czech physiologist J.E. Purkinje in the middle of the nineteenth century to denote the uniform blood fluid that is released from particles and corpuscles. This term was suggested for a uniform ionized gas of the positive column of a gas discharge by Langmuir [1–3] and now this term denotes any system with electrons and ions where charged particles determine the properties of this system. The most widespread form of plasma is an ionized gas which consists of atoms or molecules with an admixture of charged particles, electrons and ions. Such a plasma is the subject of this book.

To understand the conditions required for the existence of such a plasma under equilibrium conditions, we compare it with an identical chemical system. Let us consider, for example, atmospheric air consisting basically of nitrogen and oxygen molecules. At high temperatures, along with the nitrogen and oxygen, nitrogen oxides can be formed. The following chemical equilibrium is maintained in air:

$$N_2 + O_2 \leftrightarrow 2NO - 41.5\,\text{kcal/mol} \tag{1.1}$$

Here and below, the sign $\leftrightarrow$ means that the process can proceed either in the forward direction or in the reverse direction. According to the Le Chatelier principle [4, 5], an increase in the temperature of the air leads to an increase in the concentration of the NO molecules.

A similar situation exists in the case of formation of charged particles in a gas, but this process requires a higher temperature. For example, the ionization equilibrium for nitrogen molecules has the form

$$N_2 \leftrightarrow N_2^+ + e - 360\,\text{kcal/mol} \tag{1.2}$$

Thus, the chemical and ionization equilibria are analogous, but ionization of atoms or molecules proceeds at higher temperatures than chemical transformations. To

Fundamentals of Ionized Gases, First Edition. Boris M. Smirnov.

illustrate this, Table 1.1 contains examples of chemical and ionization equilibria. This table gives the temperatures at which 0.1% of molecules are dissociated in the case of a chemical (dissociation) equilibrium or 0.1% of atoms are ionized for an ionization equilibrium at a gas pressure of 1 atm. Thus, a weakly ionized gas, which we shall call a plasma, has an analogy with a chemically active gas. Therefore, although a plasma has characteristic properties, which we shall describe, it is not really a new form or state of matter as is often stated. An ideal plasma is a form of a gas, whereas a dense nonideal plasma (a plasma with strong coupling) is an analogue of a condensed atomic system.

In most cases a plasma is a weakly ionized gas with a small degree of ionization. Table 1.2 gives some examples of real plasmas and their parameters – the number densities of electrons N_e and of atoms N_a, the temperature (or the average energy) of electrons T_e, and the gas temperature T. In addition, some types of plasma systems are given in Figures 1.1 and 1.2. We note also that an equilibrium between charged particles and atoms or molecules may be violated if the plasma is located in an external field. In particular, electric energy is introduced in a gas discharge plasma and it is transferred to electrons in a first stage, and then as a result of collisions it is transferred from electrons to atoms or molecules. This form of injection of energy may lead to a higher electron energy compared with the thermal energy of atoms or molecules, and the plasma becomes a nonequilibrium plasma. Figure 1.3 gives some examples of equilibrium and nonequilibrium plasmas.

Table 1.1 Temperatures corresponding to dissociation of 0.1% of molecules or ionization of 0.1% of atoms at a pressure of 1 atm.

Chemical equilibrium	T, K	Ionization equilibrium	T, K
$2CO_2 \leftrightarrow 2CO + O_2$	1 550	$H \leftrightarrow H^+ + e$	7 500
$H_2 \leftrightarrow 2H$	1 900	$He \leftrightarrow He^+ + e$	12 000
$O_2 \leftrightarrow 2O$	2 050	$Cs \leftrightarrow Cs^+ + e$	2 500
$N_2 \leftrightarrow 2N$	4 500		
$2H_2O \leftrightarrow 2H_2 + O_2$	1 800		

Table 1.2 Parameters of some plasmas. N_e and N are the number densities of electrons and neutral atomic particles, respectively, and T_e and T are their temperatures.

Type of plasma	N_e, cm^{-3}	N, cm^{-3}	T_e, K	T, K
Sun's photosphere	10^{13}	10^{17}	6000	6000
E layer of ionosphere	10^5	10^{13}	250	250
He-Ne laser	3×10^{11}	2×10^{16}	3×10^4	400
Ar laser	10^{13}	10^{14}	10^5	10^3

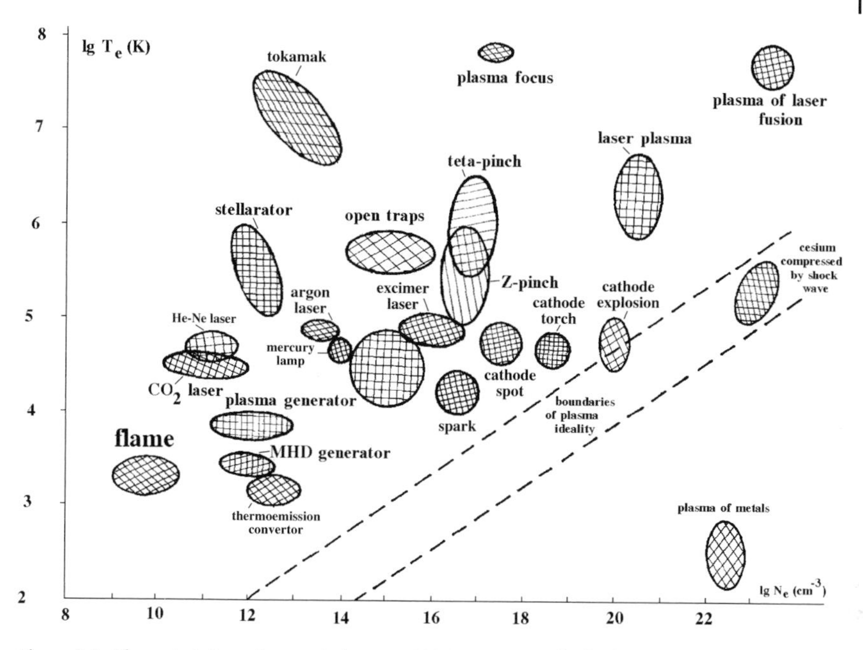

Figure 1.1 Characteristics of natural plasmas. MHD – magnetohydrodynamic.

It is seen that generation of an equilibrium plasma requires strong heating of a gas. One can create a conducting gas by heating the charged particles only. This

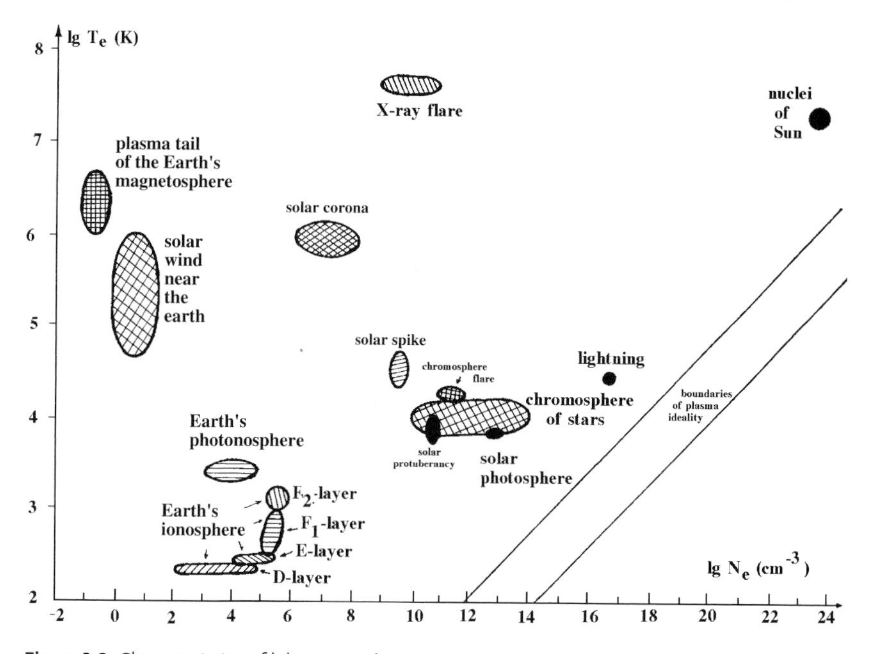

Figure 1.2 Characteristics of laboratory plasmas.

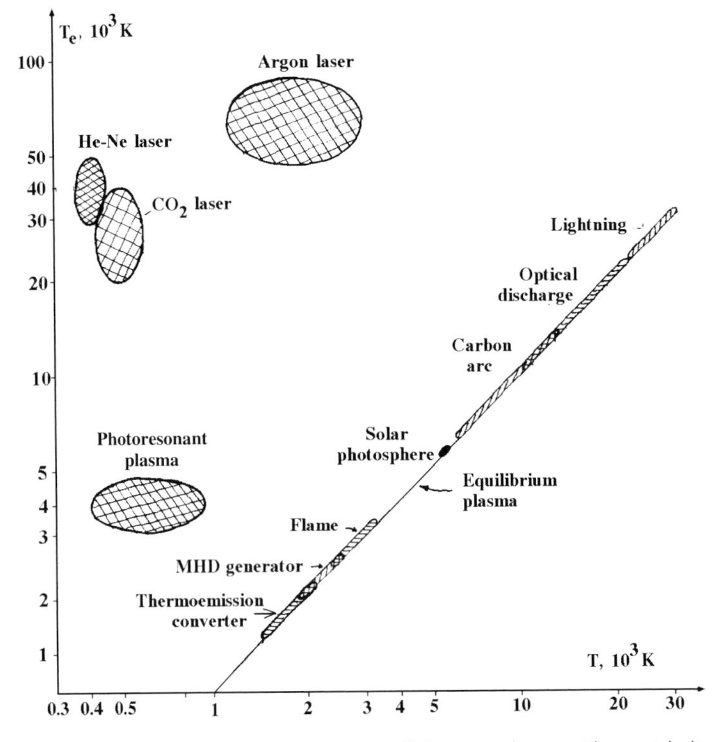

Figure 1.3 Electron and gas temperatures of laboratory plasmas. The straight line corresponds to an equilibrium plasma, whose electron and gas temperatures are equal.

takes place in gas discharges when an ionized gas is placed in an external electric field. Moving in this field, electrons acquire energy from the field and transfer it to the gas. As a result, the mean electron energy may exceed the thermal energy of neutral particles of the gas, and they can produce the ionization which is necessary for maintaining an electric current in the system. Thus, a gas discharge is an example of a plasma which is maintained by an external electric field. If the temperatures of electrons and neutral particles are identical, the plasma is called an equilibrium plasma; in the opposite case, we have a nonequilibrium plasma. Figure 1.3 gives some examples of equilibrium and nonequilibrium plasmas.

Thus, a plasma as a physical object has specific properties which characterize it. Because of the presence of charged particles, various types of interaction with external fields are possible and these lead to a special behavior of plasmas which is absent in ordinary gaseous systems. Furthermore, there are a variety of means for generation and application of plasmas, and these will be considered below.

1.1.2
The History of the Study of Electricity

Let us review briefly the history of plasma physics that is connected with the development of electrical techniques and electrical science [6]. One can date the start of the history of electric phenomena in gases to 1705, when the English scientist Francis Hauksbee made an electrostatic generator whose power allowed him to study luminous electric discharges in gases. In 1734, Charles Francois de Cisternay Dufay (France) discovered that air conducts electricity near hot bodies. In 1745, E.J. von Kleist (Germany) and P.V. Musschenbroek (Netherlands) constructed independently a type of electric capacitor named the Leyden jar. This made possible the study of electric breakdown in air. In 1752, the American scientist and statesman Benjamin Franklin created a theory of lightning on the basis of some experiments. He considered the lightning phenomenon as a flow of electricity through air that corresponds to contemporary understanding of this phenomenon as passage of an electric current through air.

The nineteenth century was a period in which electric processes and phenomena were studied intensely [7]. At that time the technique required to create a gas discharge was worked out and this paved the way for the subsequent development of the electrical sciences, including plasma physics. This included the creation of new gas discharge devices together with the development of sources of electricity and methods of electricity diagnostics. The sources of electricity were developed simultaneously in the creation and improvement of batteries, where electric energy resulted from chemical energy, and dynamos for transformation of mechanical energy into electric energy. In 1800, A. Volta (Italy) created a battery that was the prototype of the modern battery. It consisted of electrochemical cells joined in series, and each cell contained zinc and copper electrodes and an electrolyte between them, firstly H_2SO_4. Two types of ions, H^+ and SO_4^{2-}, are formed in this electrolyte, and protons capture electrons from the copper electrodes, forming bubbles of hydrogen (H_2), whereas negatively charged sulfate ions react with the zinc electrodes. This battery was improved many times and became a reliable source of electricity, and the general concept of this device was conserved in all subsequent devices of this type [8]. In the same period, the fuel element was also created (W.R. Grove, England, 1839). This generates electric energy as a result of the chemical reaction of H_2 and O_2.

On the path to the development of dynamos, J. Henry (USA) constructed an electric motor in 1831 that worked from a battery, that is, electric energy was transformed into mechanical energy, and M.H. Jacobi (Germany) showed that this dynamo could be used not only as a motor, but also as a generator of electricity, that is, it converts mechanical power into electric power. In 1870, Z.T. Gramme (France) constructed a magnetoelectric machine that was a prototype of contemporary generators of electricity in electric power plants. As a result of the creation and improvement of sources of electric energy, reliable equipment was made for various applications in manufacturing and the economy.

The creation and development of sources of electricity determined the development of the gas discharge technique. The first type of discharge was an arc discharge at atmospheric pressure. The subsequent development of the pumping technique allowed the creation of burned gas discharges of various types. Crookes tubes [9] were an important development, where on the basis of a mercury vacuum pump the pressure of the remaining gas in a sealed-off tube may be decreased to 10^{-7} atm. The gas discharge technique allowed study of various processes involving electrons and ions in gases, including the cathode rays discovered by J. Pluecker (Germany) in 1851. Firstly, they were discovered from the sharp straightforward boundary of fluorescence resulting from applying a high voltage to electrodes of the tube with a low gas pressure. Later, it was proved that the gas was excited by a beam of charged particles that came from the cathode and therefore they were named cathode rays (*Kathodenstrahlen*) [10, 11]. In reality, cathode rays are electron beams and they were used subsequently as a specific instrument which became important in the new physics at the beginning of the twentieth century. For this technique the creation of an inductive coil by Rumkopf in 1851 was important as it provided a possibility to obtain high voltages in a simple way. All these developments allowed various forms of gas discharges to be obtained, including a glow discharge, and allowed them to be used for various technical applications. Along with gas discharges, this technique allowed beams of charged particles to be generated.

The analysis of some results obtained by using the new technique led to a new understanding of the nature of ionized gases and related physical objects. The understanding of electricity carriers followed from the electrolysis laws of Michael Faraday (England) in 1833–1834. The electrolysis laws give the ratio of the charge of the carrier to its mass. Along with the establishment of the fundamental laws of electrolysis, Faraday introduced new terms such as “ion” – the electricity carrier – and “anode” (“way up” from Greek) and “cathode” (“way down” from Greek) for the electrodes involved in electricity transfer. In continuation of these studies, Stoney (England) suggested the term “electron” for the quantum of elementary charge and estimated its value in 1874.

The development of electricity sources and gas discharges was a basis for the creation of the new science branches, such as spectroscopy, atomic physics, and nuclear physics, and also led to quantum mechanics, the atomic quantum theory, and nuclear physics at the beginning of the twentieth century. In turn, this new physics stimulated the subsequent gas discharge development [12, 13]. Finally, understanding the gas discharge phenomenon as a self-consistent one due to the ionization balance in a gas allowed Townsend to describe the passage of current through a gas [14–16]. According to I. Langmuir, an ionized gas excited in a gas discharge is characterized by different spatial regions, and the region adjacent to the walls was named by him as a “sheath”, whereas a quasineutral ionized gas far from the walls was called a “plasma” [1–3] by analogy with blood plasma. These terms are used now [17, 18].

1.1.3
Methods of Plasma Generation

There are various methods to create a plasma as a result of action on matter, and we consider them briefly. A gas discharge plasma is the most widespread form of plasma and can have a variety of characteristics. It can be either stationary or pulsed, depending on the character of the external fields. An external electric field may cause electric breakdown of gas, which then generates different forms of plasma depending on the conditions of the process. In the first stage of breakdown, a uniform current of electrons and ions may arise. If the electric field is not uniform, an ionization wave can propagate in the form of an electron-avalanche streamer. In the next stage of the breakdown process, the electric current establishes a distribution of charged particles in space. This is one form of gas discharge.

After the external electric field is switched off, the plasma decays as a result of recombination processes of electrons with ions, and spatial diffusion of the plasma occurs. This plasma is called an afterglow plasma, and is used to study recombination and diffusion processes involving ions and excited atoms. A convenient way to generate plasma uses resonant radiation, that is, radiation whose wavelength corresponds to the energy of an atomic transition in the atoms constituting the excited gas. As a result of the excitation of the gas, a high density of excited atoms is attained, and collision of these atoms leads to formation of free electrons. Thus, the atomic excitation in the gas leads to its ionization and to plasma generation. This plasma is called a photoresonant plasma. The possibility of generating such a plasma has improved with the development of laser techniques. In contrast to a gas discharge plasma, a photoresonant plasma is formed as a result of excitations of atoms and therefore there are specific requirements for its formation. In particular, the temperature of the excited atoms can be somewhat in excess of the electron temperature. This plasma may be used for generation of multicharged ions, as a source of acoustic waves, and so on.

A laser plasma is created by laser irradiation of a surface and is described by some parameters such as the laser power and the duration of the process. In particular, if a short (nanosecond) laser pulse is focused onto a surface, material evaporates from the surface in the form of a plasma. If the number density of electrons exceeds the critical density (in the case of a neodymium laser, where the radiation wavelength is 1.06 μm, this value is 10^{21} cm^{-3}), the evolving plasma screens the radiation, and subsequent laser radiation heats this plasma. As a result, the temperature of the plasma reaches tens of electronvolts, and this plasma can be used as a source of X-ray radiation or as the source of an X-ray laser. Laser pulses can be compressed and shortened up to about 2×10^{-14} s. This makes possible the generation of a plasma in very short times, and it permits the study of fast plasma processes.

If the laser power is relatively low, the evaporating material is a weakly ionized vapor. Then, if the duration of the laser pulse is not too short (more than 10^{-6} s), there is a critical laser power (10^7–10^8 W/cm^2) beyond which laser radiation is absorbed by the plasma electrons, and laser breakdown of the plasma takes place. For values of the laser power smaller than the critical value, laser irradiation of a surface is a method for generating beams of weakly ionized vapor. This vapor can be used for formation and deposition of atomic clusters.

A widely used method of plasma generation is based on passage of electron beams through a gas. Secondary electrons can be used further for certain processes. For example, in excimer lasers, secondary electrons are accelerated by an external electric field for generation of excited molecules with short lifetimes. The electron beam as a source of ionization is convenient for excimer and chemical lasers because the ionization process lasts such a short time.

A chemical method of plasma generation occurs in flames. The chemical energy of reagents is expended on formation of radicals or excited particles, and chemionization processes with participation of active particles generate charged particles. Transformation of the chemical energy into the energy of ionized particles is not efficient, so the degree of ionization in flames is low. Electrons in a hot gas or vapor can be generated by small particles. Such a process takes place in the products of combustion of solid fuels.

Introduction of small particles and clusters into a weakly ionized gas can change its electric properties because these particle can absorb charged particles, that is, electrons and positive ions or negative ions and positive ions recombine on these particles by attachment to them. This process occurs in an aerosol plasma, that is, an atmospheric plasma which contains aerosols. In contrast, in hot gases small particles or clusters can generate electrons.

A plasma can be created under the action of fluxes of ions or neutrons when they pass through a gas. Ionization near the Earth's surface results from the decay of radioactive elements which are found in the Earth's crust. Ionization processes and formation of an ionized gas in the upper atmosphere of the Earth are caused by UV radiation from the Sun. There are various methods of plasma generation that lead to the formation of different types of plasmas. Some methods of plasma generation and the types of plasmas resulting from them are given in Table 1.3.

Table 1.3 Methods of plasma generation.

Character of action	Matter	Type of plasma
Electric field	Gas	Stationary gas discharge plasma
Electromagnetic wave	Gas	Alternative gas discharge plasma
Resonant radiation	Atomic vapor	Photoresonant plasma
Excitation from chemical reactions	Chemically active mixture	Chemical (flame) plasma
Laser	Surface or particles	Laser plasma
Injection of electrons or ions	Surface or ionized gas	Beam plasma
Injection of nucleating vapor	Ionized gas	Cluster plasma
Injection of dust particles	Ionized gas	Dusty plasma
Ionization by hard radiation	Gas (air) with aerosols	Aerosol plasma

1.1.4
Charged Particles in a Gas

We shall consider primarily a plasma whose properties are similar to those of a gas. As in a gas, each particle of the plasma will follow a straight trajectory as a free particle most of the time. These free-particle intervals will occasionally be punctuated by strong interactions with surrounding particles that will cause a change in energy and the direction of motion. This situation occurs if the mean interaction potential of the particle with its neighbors is small compared with the mean kinetic energy of the particle. This is the customary description of the gaseous state of a system, and a plasma that also satisfies this description is called an ideal plasma.

Let us formulate now a quantitative criterion for a plasma to be ideal. The Coulomb interaction potential between two charged particles has the absolute value $|U(R)| = e^2/R$, where e is the charge of an electron or a singly charged ion and R is the distance between interacting particles. Thus, the interaction potential at the mean distance between particles $R_0 \sim N_e^{-1/3}$ is $|U| \sim e^2 N_e^{1/3}$, and because the mean thermal energy of the particles is of the order of T (T is the plasma temperature expressed in energy units), the condition of the gaseousness of a plasma due to interaction of charged particles is characterized by the smallness of the plasma parameter γ, which is

$$\gamma = \frac{N_e e^6}{T^3} \ll 1 . \tag{1.3}$$

If the gaseousness condition (1.3) is fulfilled for a plasma, this plasma is ideal. In the following, we shall deal primarily with a plasma whose parameters satisfy (1.3).

We define a weakly ionized gas as a gas with a small concentration of charged particles. Nevertheless, some properties of the weakly ionized gas are governed by the charged particles. For example, the degree of ionization in power-discharge molecular lasers is 10^{-7}–10^{-5}. In these lasers, the energy is first transferred from an external source of energy to electrons, and then it is transformed to the energy of laser radiation. As we shall see, a relatively small concentration of electrons determines the operation of this system.

Some properties of a weakly ionized gas are determined by the interaction between charged and neutral particles, whereas other properties are created by charged particles only. Although the concentration of charged particles in a plasma is small, the long-range Coulomb interaction between them may be more important than the short-range interaction between neutral particles. We consider below the plasma properties that are associated with the presence of charged particles. The short-range interaction of neutral particles is not important for these properties.

We now study the penetration of an external electric field into a plasma. Since this field leads to a redistribution of the charged particles of a plasma, it creates an internal electric field that opposes the external field. This has the effect of screening the plasma from an external field. To analyze this effect, we consider the Poisson equation for an electric field in a plasma, which has the form

$$\operatorname{div} \mathbf{E} = -\Delta\varphi = 4\pi e(N_i - N_e) . \qquad (1.4)$$

Here $\mathbf{E} = -\nabla\varphi$ is the electric field strength, φ is the potential of the electric field, N_e and N_i are the number densities of electrons and ions, respectively, and ions are assumed to be singly charged. The effect of an electric field is to cause a redistribution of charged particles. According to the Boltzmann formula, the ion and electron number densities are given by

$$N_i = N_0 \exp\left(-\frac{e\varphi}{T}\right) , \quad N_e = N_0 \exp\left(\frac{e\varphi}{T}\right) , \qquad (1.5)$$

where N_0 is the average number density of charged particles in a plasma and T is the plasma temperature. Substitution of (1.5) into the Poisson equation (1.4) gives

$$\Delta\varphi = 8\pi N_0 e \sinh\left(\frac{e\varphi}{T}\right) .$$

Assuming that $e\varphi \ll T$, we can transform this equation to the form

$$\Delta\varphi = \frac{\varphi}{r_D^2} , \qquad (1.6)$$

where

$$r_D = \left(\frac{T}{8\pi N_e e^2}\right)^{1/2} \qquad (1.7)$$

is the so-called Debye–Hückel radius [19].

The solution of (1.6) describes an exponential decrease with distance from the plasma boundary. For example, if an external electric field penetrates a flat boundary of a uniform plasma, the solution of (1.6) has the form

$$\mathbf{E} = \mathbf{E}_0 \exp\left(-\frac{x}{r_D}\right) , \qquad (1.8)$$

where x is the distance from the plasma boundary in the normal direction. If the electron and ion temperatures are different, then (1.5) has the form

$$N_i = N_0 \exp\left(-\frac{e\varphi}{T_i}\right) , \quad N_e = N_0 \exp\left(\frac{e\varphi}{T_e}\right) ,$$

and we will obtain the same results as above, except that the Debye–Hückel radius takes the more general form

$$r_D = \left[4\pi N_0 e^2 \left(\frac{1}{T_e} + \frac{1}{T_i}\right)\right]^{-1/2} . \qquad (1.9)$$

Now let us calculate the field from a test charge placed in a plasma. In this case the equation for the potential due to the charge has the form

$$\Delta\varphi \equiv \frac{1}{r}\frac{d^2}{dr^2}(r\varphi) = \frac{\varphi}{r_D^2} ,$$

where r is the distance from the charge considered. If this charge is located in a vacuum, the right-hand side of this equation is zero, and the solution has the form $\varphi = q/r$, where q is the charge. Requiring the solution of the above equation to be coincident with this one at $r \to 0$, we obtain for the potential of a test charged particle in a plasma

$$\varphi = \frac{q}{r} \exp\left(-\frac{r}{r_D}\right) . \tag{1.10}$$

Hence, the interaction potential U of two charged particles with charges q_1 and q_2 is as follows:

$$U(r) = \frac{q_1 q_2}{r} \exp\left(-\frac{r}{r_D}\right) . \tag{1.11}$$

Thus, the Debye–Hückel radius is a typical distance at which fields in a plasma are shielded by the charged particles. The field of a charged particle is eliminated on this scale by fields of surrounding particles.

Now we shall check the validity of the condition $e\varphi \ll T$, which allowed us to simplify the equation for the electric field strength. Because this condition must work at distances of the order of r_D, it has the form

$$\frac{e^2}{r_D T} \sim \left(\frac{e^6 N_0}{T^3}\right)^{1/2} \ll 1 .$$

This condition according to criterion (1.3) characterizes an ideal plasma.

We can determine the number of charged particles that participate in shielding the field of a test particle. This value is of the order of magnitude of the number of charged particles located in a sphere of radius r_D. The number of charged particles is, to within a numerical factor,

$$r_D^3 N_0 \sim \sqrt{\frac{T^3}{N_0 e^6}} \gg 1 .$$

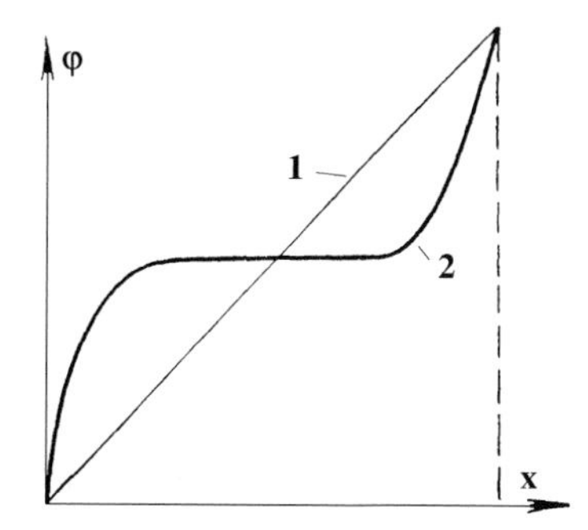

Figure 1.4 The distribution of the electric potential in a gap containing an ionized gas. 1 – the Debye–Hückel radius is large compared with the gap size; 2 – the Debye–Hückel radius is small compared with the gap size.

Hence, this value is large for an ideal plasma.

Thus, the Debye–Hückel radius is the fundamental parameter of an ideal quasineutral plasma. It is the distance over which an ensemble of charged plasma particles shields external electric fields or fields of individual plasma particles. Correspondingly, the character of the distribution of external fields in a plasma differs from that in a gas, as shown in Figure 1.4 for a gaseous or plasma gap.

1.1.5
Definition of a Plasma

Let us consider a gas-filled gap subjected to an external electric field. If the gas does not contain charged particles, the field is uniform in the gap. In the presence of charged particles in the gas, an external electric field is shielded near the edges of the gap (see Figure 1.4) at distances of the order of the Debye–Hückel radius. Thus, the character of the distribution of an electric field inside the gas is different in these two cases. Note that here we assume the plasma to be quasineutral up to its boundaries, an assumption that can be violated in real cases. On the basis of the above considerations, one can define a plasma as a weakly ionized gas whose Debye–Hückel radius is small compared with its size L, that is,

$$L \gg r_D \,. \tag{1.12}$$

For an electron temperature of $T_e \sim 1\,\text{eV}$ and a gap size of the order of 10 cm this criterion gives an electron number density $N_e \gg 10^4\,\text{cm}^{-3}$, which is a very small value (see Figures 1.1 and 1.2). For example, the electron number density of a glow discharge lies in the range 10^7–$10^{12}\,\text{cm}^{-3}$. Table 1.4 contains parameters of some plasma types in terms of a typical number density N_e and temperature T_e of electrons, and also important parameters of the plasma types under consideration, parameters such as the Debye–Hückel radius of the plasma r_D and its typical size L. These examples are a representative sampling of real plasmas.

Table 1.4 Parameters of some plasmas.

Type of plasma	N_e, cm^{-3}	T_e, K	r_D, cm	L, cm
E layer of ionosphere	10^5	300	0.3	10^6
He-Ne laser	10^{11}	3×10^4	3×10^{-3}	1
Mercury lamp	10^{14}	5×10^4	3×10^{-5}	0.1
Sun's chromosphere	5×10^8–5×10^9	3×10^4–3×10^5	0.01–0.1	10^9
Lightning	10^{17}	3×10^4	3×10^{-6}	100

1.1.6
Oscillations of Plasma Electrons

The Debye–Hückel radius is the parameter that characterizes an ideal quasineutral plasma. We can estimate a typical time for the response of a plasma to an external field. For this purpose we study the behavior of a uniform infinite plasma if all the plasma electrons are shifted at the initial time by a distance x_0 to the right starting from a plane $x = 0$. This creates an electric field whose strength corresponds to the Poisson equation (1.4):

$$\frac{dE}{dx} = 4\pi e(N_i - N_e) .$$

Assuming the electric field strength at $x < 0$ is zero, the Poisson equation gives an electric field strength for $x > x_0$ of $E = 4\pi e N_0 x_0$, where N_0 is the average number density of charged particles in the plasma. The movement of all the electrons under the influence of the electric field leads to a change in the position of the boundary. The equation of motion for each of the electrons can be written as

$$m_e \frac{d^2(x + x_0)}{dt^2} = -eE ,$$

where m_e is the electron mass and x is the distance of an electron from the boundary. Because x is a random value, not dependent on the phenomenon being considered, one can assume this value to be independent of time. Thus, the equation of motion of an electron is

$$\frac{d^2 x_0}{dt^2} = -\omega_p^2 x_0 ,$$

where [2, 20–23]

$$\omega_p = \left(4\pi N_0 \frac{e^2}{m_e}\right)^{1/2} \tag{1.13}$$

is called the plasma frequency, or Langmuir frequency.

The solution of the equation obtained predicts an oscillatory character for the electron motion. Accordingly, $1/\omega_p$ is a typical time for a plasma to respond to an external signal. Note that the value $r_D \omega_p = \sqrt{2T/m_e}$ is the thermal electron velocity. From this it follows that a typical time for a plasma to respond to an external signal is the time during which the electrons experience a displacement of the order of the Debye–Hückel radius. Thus, we have two fundamental parameters of an ideal quasineutral plasma: the Debye–Hückel radius r_D, which is a shielding distance for fields in a plasma, and the plasma frequency ω_p, so ω_p^{-1} is a typical time for the plasma to respond to external signals.

Thus, the Debye–Hückel radius and the plasma frequency are two fundamental parameters which reflect a long-range character of interaction of charged particles in an ionized gas. These parameters do not depend on a short-range interaction of neutral particles in an ionized gas, and therefore the character of reacting to external fields as well some plasma collective phenomena in ionized gases are determined by charged particles only and are independent of the presence of atoms or molecules in an ionized gas. As a result, collective phenomena in ionized gases have a universal character and may be identical for various ionized gases if the gas criterion holds true for neutral and charged particles.

1.1.7 Interaction of Charged Particles in an Ideal Plasma

We now calculate the average potential energy of a charged particle in an ideal plasma and the distribution function for the interaction potential of charged particles. From the interaction potential (1.11) for two charged particles, we have for its average value

$$\overline{U} = \int_0^\infty e\varphi \left[N_0 \exp\left(-\frac{e\varphi}{T}\right) - N_0 \exp\left(\frac{e\varphi}{T}\right)\right] d\mathbf{r} \,.$$

We assume the charges of electrons and ions in a plasma to be $\pm e$ and use (1.10) for the electric potential φ from an individual plasma particle of charge $+e$, that is,

$$\varphi = \frac{e}{r} \exp\left(-\frac{r}{r_D}\right) .$$

The above formula for the average interaction potential in an ideal plasma accounts for pairwise interactions of all the charged particles in a volume that have a Boltzmann distribution. In the case of an ideal plasma, the principal contribution to the integral occurs at small interactions $e\varphi \ll T$, which gives

$$\overline{U} = -\frac{2N_0}{T} \int_0^\infty (e\varphi)^2 4\pi r^2 dr = -\frac{4\pi N_0 r_D}{T} = -\frac{e^2}{2 r_D T} , \tag{1.14}$$

where we used expression (1.7) for the Debye–Hückel radius. Thus, the average energy of a charged particle in an equilibrium plasma is

$$\overline{\varepsilon} = \frac{3T}{2} - \frac{e^2}{2r_D} . \tag{1.15}$$

To estimate fluctuations ΔU of the interaction potential for a charged particle in a plasma, we assume this value to be determined by positions of other charged particles in a sphere of the Debye–Hückel radius centered on the test particle. The mean number of charged particles in this region is $n \sim N_0 r_D^3 \gg 1$, with fluctuations of the order of $\sqrt{n}$. Hence, the fluctuation of the interaction potential of the test charged particle in a plasma is

$$\Delta U \sim \sqrt{n}\frac{e^2}{r_D} \sim e^2 N_0^{1/2} r_D^{1/2} . \tag{1.16}$$

Since $\Delta U \gg U$, the distribution function for the interaction potentials yields $\overline{U} = 0$ and has the form

$$f(U)dU = (2\pi\Delta U^2)^{-1/2}\exp\left[\frac{-U^2}{2\Delta U^2}\right], \tag{1.17}$$

where $f(U)dU$ is the probability that the interaction potential lies in the interval from U to $U + dU$. The squared deviation of the distribution (1.17) is

$$\Delta U^2 = \overline{U^2} = 2\int_0^\infty (e\varphi)^2 N_0 4\pi r^2 dr = 4\pi N_0 e^4 r_D = \overline{U}T \gg (\overline{U})^2 ,$$

where the factor 2 takes into account the presence of charged particles of the opposite sign, and N_0 is the mean number density of charged particles of one sign. Thus, we have for an ideal plasma

$$\overline{U} \ll \Delta U \ll T .$$

1.1.8
Microfields in an Ideal Plasma

The charged plasma particles, electrons and ions, create electric fields in a plasma that act on atoms and molecules in a plasma. Although the average field at each point is zero, at this time electric fields are of importance. Guided by the action of these fields on atoms, for a corresponding time range one can average the fields from electrons, that is, the action of electrons on atoms as a result of their collision. Fields from ions at a given point are determined according to their spatial distribution. We will determine below the distribution function $P(E)$ for electric fields at a given point for a given time for a random spatial distribution of ions, so by definition $P(E)dE$ is the probability that the electric field strength at a given point ranges from E to $E + dE$, and we use an isotropic form of the distribution over the electric field strengths. We take into account the electric field strength from the ith ion at a distance $\mathbf{r}_i$ from it is

$$\mathbf{E}_i = \frac{e\mathbf{r}_i}{r_i^3} ,$$

and the total electric field of all the ions is

$$\mathbf{E} = \sum_i \mathbf{E}_i .$$

Evidently, we include in this sum ions located at a distance $r_i \ll r_D$ from a test point, that is, we assume that the main contribution to this sum is from ions located close to a test point. In addition, we take a random distribution of ions in space.

We first find the tail of the distribution function that corresponds to large fields. The probability of the nearest ion being located in a distance range from r to $r + dr$ is

$$P(E)dE = 4\pi r^2 dr N_i ,$$

where N_i is the number density of ions, and the electric field strength from an individual ion is $E = e/r^2$. From this we obtain

$$P(E)dE = \frac{2\pi e^{3/2} N_i dE}{E^{5/2}} , \quad \sqrt{\frac{e}{E}} \ll N_i^{-1/3} . \tag{1.18}$$

From this one can estimate a typical value E_0 of the electric field strength:

$$E_0 \sim e N_i^{2/3} . \tag{1.19}$$

This value corresponds to the electric field strength at the mean distance between ions.

We now determine the distribution function $P(\mathbf{E})$ on the basis of the standard method with the characteristic function $F(\mathbf{g})$, that is,

$$F(\mathbf{g}) = \int \exp(i\mathbf{E}\mathbf{g}) P(\mathbf{E}) d\mathbf{E} , \tag{1.20}$$

where $\mathbf{g}$ is a three-dimensional variable. Correspondingly, the inverse transformation gives

$$P(\mathbf{E}) = \frac{1}{(2\pi)^3} \int \exp(-i\mathbf{E}\mathbf{g}) F(\mathbf{g}) d\mathbf{g} . \tag{1.21}$$

Introducing the probability $p(\mathbf{E}_i)$ of a given electric field strength created by the ith ion, we have

$$P(\mathbf{E}) = \prod_i \int p(\mathbf{E}_i) d\mathbf{E}_i .$$

Evidently, the probability $p(\mathbf{E}_i)$ is identical for different ions, and the characteristic function is

$$F(\mathbf{g}) = \prod_i f_i(\mathbf{g}) , \quad f_i(\mathbf{g}) = \int \exp(i\mathbf{E}_i\mathbf{g}) p(\mathbf{E}_i) d\mathbf{E}_i . \tag{1.22}$$

Let n ions ($n \gg 1$) be located in a large volume Ω, so

$$p(\mathbf{E}_i) d\mathbf{E}_i = \frac{d\mathbf{r}_i}{\Omega} ,$$

and $d\mathbf{r}_i$ is the volume element where the ith ion is located with the origin of the frame of reference that is at a test point. For the partial characteristic function and using $p(\mathbf{E}_i) d\mathbf{E}_i = d\mathbf{r}_i/\Omega$ this gives

$$f_i(\mathbf{g}) = \frac{1}{\Omega} \int \exp(i\mathbf{E}_i\mathbf{g}) d\mathbf{r}_i = 1 + \frac{1}{\Omega} \int \left[\exp(i\mathbf{E}_i\mathbf{g}) - 1\right] d\mathbf{r}_i ,$$

and this gives

$$\ln f_i(\mathbf{g}) = \frac{1}{\Omega} \int \left[\exp(i\mathbf{E}_i(\mathbf{r}_i)\mathbf{g}) - 1\right] d\mathbf{r}_i .$$

From this, because of the identity of $f_\mathrm{i}(\mathbf{g})$, we obtain

$$\ln F(\mathbf{g}) = \sum_i \ln f_\mathrm{i}(\mathbf{g}) = n \ln f_\mathrm{i}(\mathbf{g}) = N_\mathrm{i} \int \left[\exp(i\mathbf{E}_i\mathbf{g}) - 1\right] d\mathbf{r}_i \,,$$

where $N_\mathrm{i} = n/\Omega$ is the number density of ions. In accordance with (1.21), from this we obtain for the distribution function $P(\mathbf{E})$ for the electric field strengths in a plasma

$$P(\mathbf{E}) = \frac{1}{(2\pi)^3} \int \exp\left\{-i\mathbf{E}\mathbf{g} + N_\mathrm{i} \int \left[\exp(i\mathbf{E}_i\mathbf{g}) - 1\right] d\mathbf{r}_i\right\} d\mathbf{g} \,.$$

Since the electric field strength from an individual ion is $\mathbf{E}_i = e\mathbf{r}_i/r_i^3$, we have

$$\int \left[\exp(i\mathbf{E}_i\mathbf{g}) - 1\right] d\mathbf{r}_i = \int \left[\exp\left(i\frac{eg\cos\theta_i}{r_i^2} - 1\right)\right] \cdot 2\pi d\cos\theta_i r_i^2 dr_i =$$

$$-4\pi(eg)^{3/2} \int_0^\infty \xi^2 d\xi \left(1 - \frac{\sin\xi}{\xi}\right) = -\frac{8\pi(eg)^{3/2}}{15} \,,$$

where $\xi = eg/r_i^2$. Hence, the distribution function for electric field strengths takes the form [24]

$$P(E) = \int \exp\left[-\frac{8\pi(eg)^{3/2} N_\mathrm{i}}{15} + i\mathbf{E}\mathbf{g}\right] \frac{d\mathbf{g}}{(2\pi)^3} = \frac{1}{E_0} H(z) \,,$$

$$z = \frac{E}{E_0} \,, \quad E_0 = 2\pi e \left(\frac{4N_\mathrm{i}}{15}\right)^{2/3} . \tag{1.23}$$

As is seen, the typical electric field strength E_0 corresponds to that of (1.19), and the Holtzmark function $H(z)$ is given by

$$H(z) = \frac{2}{\pi z} \int_0^\infty x \sin x \exp\left[-\left(\frac{x}{z}\right)^{3/2}\right] dx \,. \tag{1.24}$$

The Holtzmark function has the following expressions in the limiting cases

$$H(z) = \frac{4}{3\pi} z^2 \,, \quad z \ll 1 \,, \quad H(z) = \frac{15}{8}\sqrt{\frac{2}{\pi}} z^{-5/2} \,, \quad z \gg 1 \tag{1.25}$$

and the case of large electric field strengths $z \gg 1$ corresponds to (1.18). The Holtzmark function is represented in Figure 1.5, and its maximum corresponds to a typical electric field strength $E_\mathrm{max} \approx 1.6 E_0$ that is created at average distances between ions. The Holtzmark function may be constructed in a simple manner from its limiting expressions as

$$h(z) = \frac{4z^2}{3\pi\left[1 + \left(\frac{z}{z_0}\right)^{9/2}\right]} \,, \quad z_0 = 1.32 \,. \tag{1.26}$$

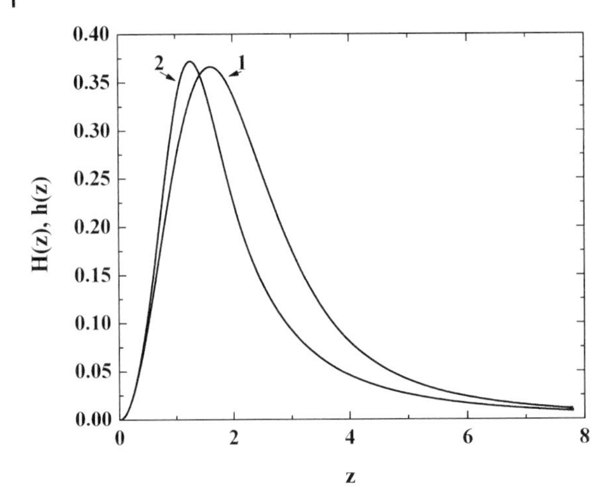

Figure 1.5 The Holzmark function $H(z)$ according to (1.24) (1) and its approximation (1.26) (2).

The approximation $h(z)$ is compared with the Holtzmark function in Figure 1.5.

One can expand these results for a plasma containing multicharged ions. If the ion charge is Z, it is necessary to change e to Ze, which for the typical electric strength E_0 in (1.23) gives

$$E_0 = 2\pi Z e \left(\frac{4N_\mathrm{i}}{15}\right)^{2/3} = 2.60 Z e N_\mathrm{i}^{2/3} . \tag{1.27}$$

Note that microfields in a plasma may be created by both electrons and ions. In this consideration we assume electrons to be distributed uniformly in space. This means that the observation time τ for the ion distribution exceeds significantly a typical time of approximately r/v_e for electron displacement at a distance r that determines this part of the distribution function (v_e is a typical electron velocity). For the maximum of the distribution function ($r \sim N_\mathrm{i}^{-1/3}$) this criterion gives

$$\tau \gg m_\mathrm{e}^{1/2} T_\mathrm{e}^{-1/2} N_\mathrm{i}^{-1/3} , \tag{1.28}$$

where m_e is the electron mass and T_e is a typical electron energy. Simultaneously, the observation time is small compared with a typical time for establishment of a uniform ion distribution, that is, for typical electric field strengths ($E \sim E_0$):

$$\tau \ll M^{1/2} T_\mathrm{i}^{-1/2} N_\mathrm{i}^{-1/3} , \tag{1.29}$$

where M is the ion mass and T_i is a typical electron energy. In particular, for a gas discharge helium plasma with typical parameters $N_\mathrm{i} \sim 10^{12}\,\mathrm{cm}^{-3}$, $T_\mathrm{e} \approx 1\,\mathrm{eV}$, and $T_\mathrm{i} = 400\,\mathrm{K}$ this formula gives $5 \times 10^{-10}\,\mathrm{s} \gg \tau \gg 10^{-12}\,\mathrm{s}$.

In addition, it is necessary to account for screening of the ion field in a plasma, which requires us to change the electric field strength E in (1.23) to $E \exp(-r/r_\mathrm{D})$

in accordance with (1.8). But in this case only electrons partake in screening of ion fields and hence the Debye–Hückel radius is given by

$$r_D = \sqrt{\frac{T_e}{4\pi N_0 e^2}}$$

instead of (1.9). For an ideal plasma (1.3) this effect is weak for the maximum of the distribution function $E \sim E_0$.

1.1.9
Beam Plasma

We considered above a quasineutral plasma as a widespread plasma type. In reality, plasma boundaries (called plasma sheaths, or double layers) contain a nonneutral plasma. Plasma properties in this region depend on processes that occur there. If charged particles are generated by a metallic surface or charged particles recombine on walls, an intermediate layer of nonneutral plasma arises between the plasma and the surface. A nonneutral plasma also occurs if charged particles are collected in certain regions or traps, and if they are transported through space by the action of external fields in the form of beams. Thus, a nonneutral plasma is a specific physical object [25, 26], so the primary characteristic of a nonneutral plasma arises from the strong fields created by the particle charge, and that restricts the plasma density. As an illustration, we shall consider a classic example of a nonneutral plasma formed near a hot cathode.

We start from a simple estimation. Let a beam of electrons have number density $N_e \sim 10^{10}\,\text{cm}^{-3}$ (the minimum boundary of the electron number density for a glow discharge) and beam radius $\rho_0 \sim 1\,\text{cm}$. Then from the Poisson equation (1.4) we have the following estimation for the electric voltage φ between the beam center and its boundary:

$$\varphi \sim 4\pi e N_e \rho_0^2 \sim 20\,\text{keV} .$$

This shows the necessity of using specific electric optics to conserve this plasma.

One can generate an electron beam in a simple way by heating a metallic surface in a vacuum; this will cause the surface to emit an electron flux as a result of thermoemission. Using electric fields allows us to accelerate electrons and remove them from the surface in the form of a beam. But the parameters of this beam can be limited by internal electric fields that arise from electron charges. We can find the properties of such a beam, created between two flat plates a distance L apart, with an electric potential U_0 between them.

The electron current density i is constant in the gap because electrons are not produced nor do they recombine in the gap. This gives

$$i = e N_e(x) v_e(x) = \text{const} ,$$

where x is the distance from the cathode, N_e is the electron number density, $v_e = \sqrt{2e\varphi(x)/m_e}$ is the electron velocity, and the electric potential is zero at the cathode surface, that is, $\varphi(0) = 0$. An electron charge creates an electric field that slows the electrons. We can analyze this relationship. The electric field strength $E = -d\varphi/dx$ satisfies the Poisson equation

$$\frac{dE}{dx} = -4\pi e N_e(x) = -4\pi i \sqrt{\frac{m_e}{2e\varphi}} .$$

Multiplication of this equation by $E = -d\varphi/dx$ provides an integrating factor that makes a simple integration possible. We obtain

$$E^2 = E_0^2 + 16\pi i \sqrt{\frac{m_e \varphi}{2e}} , \tag{1.30}$$

where $E_0 = E(0)$.

We need to establish the boundary condition on the cathode. We consider the regime where the current density of the beam is small compared with the electron current density of thermoemission. This means that most of the emitted electrons return to the metallic surface, and the external electric field does not significantly alter the equilibrium between the emitted electrons and the surface. Then the boundary condition on the cathode is the same as in the absence of the external electric field, so $E(0) = 0$. Formula (1.30) leads to the distribution of the electric potential in the gap, given by

$$\varphi(x) = \left(9\pi i \sqrt{\frac{m_e}{2e}}\right)^{2/3} x^{4/3} .$$

This can be inverted to obtain the connection between the electron current density and the parameters of the gap [27–30]:

$$i = \frac{2}{9\pi} \sqrt{\frac{e}{2m_e}} \frac{\varphi_0^{3/2}}{L^2} . \tag{1.31}$$

This dependence is known as the three-halves power law. This describes the behavior of a nonneutral plasma that is formed in the space between two plane electrodes with different voltages. This voltage difference in this case is determined by the charge that is created by charged atomic particles [29–31] and has a universal character; in particular, in a magnetron discharge [32, 33], where electrons are magnetized and hence reproduction of a charge in a magnetron discharge results from ion flux to the cathode with energy of hundreds of electronvolts that creates secondary electrons. Along with this, metal atoms are sputtered, which determines the applications of this gas discharge.

We consider one more example of transport of the flux of charged particles through a vacuum: an electron beam of radius a that is fixed by a longitudinal magnetic field inside a cylindrical metal tube of a radius ρ_0 [34]. According to the Gauss theorem, the electric field strength E at a distance ρ from the beam center outside the beam surface is

$$E = \frac{4\pi N_e}{2\pi\rho} = \frac{2I}{\rho e v_e} ,$$

where N_e is the number of electrons per unit beam length, I is the electron current, and v_e is the electron velocity in the beam. On the basis of the equation $E = -d\varphi/dx$ for the voltage difference between the beam and metal tube, this gives

$$\varphi = \frac{2I}{ev_e} \ln \frac{\rho_0}{a},$$

where we assume $\rho_0 \gg a$ and ignore the potential variation inside the electron current. Hence, when the electron beam enters a space inside the tube through a grid or when it intersects a surface with the electric potential of the walls, each electron loses energy $e\varphi$. Let the initial energy of electrons in the beam be E, so the energy becomes $E - e\varphi$ after entry into the tube. This corresponds to the electron velocity

$$v_e = \sqrt{\frac{2(E - e\varphi)}{m_e}}$$

or to the electron current

$$I = \frac{ev_e\varphi}{2\ln(\rho/a)} = \frac{\varphi\sqrt{(2e/m_e)(E - e\varphi)}}{2\ln(\rho/a)} .$$

From this we find the maximum current that is possible under given conditions,

$$I_B = \frac{1}{3}\sqrt{\frac{2}{3m_e}} \cdot \frac{E^{3/2}}{\ln(\rho/a)} , \tag{1.32}$$

which is called the Bursian current [35] and corresponds to the beam voltage with respect to the walls:

$$\varphi_{max} = \frac{2E}{3e} .$$

The formula for the maximum electric current has the form of the three-halves power law and may be written in the form

$$I_B = I_0 \frac{E^{3/2}}{\ln(\rho/a)} , \tag{1.33}$$

and if the electron energy is expressed in electronvolts, we have [34] $I_0 = 12.7\ \mu A$.

In this case of propagation of the electron beam through a vacuum, the electric potential brakes the electron beam because of the noncompensated charge in the electron beam [34, 36, 37], which is similar to the previous case of electron thermoemission and propagation between plane electrodes. One can expect that neutralization of the beam by introduction of the ion component in the beam region will remove the current limit. Nevertheless, in the latter case of beam propagation inside a cylindrical tube the limiting current is conserved, but its value increases and is $3\sqrt{3}I_B$ [34, 37–39].

1.2
Statistics of Atomic Particles in Excited and Weakly Ionized Gases

1.2.1
Distribution Function of a System of Identical Particles

The subject of this book is a weakly ionized gas that consists of a large number of atoms or molecules and a small admixture of electrons and ions. Each of its components is a system of many weakly interacting identical particles, and our first task is to represent the distribution of particles in these systems for various parameters, and apply to a system of many identical particles the laws of statistical physics. In this case we deal with the distribution function of free particles $f(x)$ over a parameter x, so for a macroscopic uniform system of particles $f(x)dx$ is the number density of particles with the value of this parameter between x and $x+dx$. Within the framework of statistical physics [40, 41], we expound the distribution function for each parameter twofold using the ergodic theorem [42, 43]. In the first case, we deal with an ensemble of a large number of particles, and the distribution function determines the average number particles with a value of this parameter in an indicated range. In the second case, one test particle is observed, and a given parameter of this particle varies in time as a result of interaction of this particle with other particles. Then the distribution function characterizes the relative time when a given parameter is found in an indicated range.

We start from the energy distribution for weakly interacting particles in a closed system. In an ensemble of a large number of particles, each of the particles is in one of a set of states described by the quantum numbers i. The goal is to find the average number of particles that are found in one of these states. For example, if we have a gas of molecules, the problem is to find the molecular distribution of its vibrational and rotational states. We shall consider problems of this type below. Consider a closed system of N particles, so this number does not change with time, as well as the total energy of particles. Denoting the number of particles in the ith state by n_i, we have the condition of conservation of the total number of particles in the form

$$N = \sum_i n_i \,. \tag{1.34}$$

Since the ensemble of particles is closed, that is, it does not exchange by energy with other systems, we require conservation of the total energy E of the particles, where

$$E = \sum_i \varepsilon_i n_i \,, \tag{1.35}$$

where ε_i is the energy of a particle in the ith state. In the course of evolution of the system, an individual particle can change its state, but the average number of particles in each state stays essentially the same. Such a state of a closed system is called a state of thermodynamic equilibrium.

Transitions of individual particles between states result from collisions with other particles. We denote by $W(n_1, n_2, \ldots, n_i \ldots)$ the probability that n_1 particles are found in the first state, n_2 particles are found in the second state, n_i particles are found in the ith state, and so on. We wish to calculate the number of possible realizations of this distribution. First, we take the n_1 particles for the first state from the total number of particles N. There are

$$C_N^{n_1} = \frac{N!}{(N-n_1)!n_1!}$$

ways to do this. Next, we select n_2 particles corresponding to the second state from the remaining $N - n_1$ particles. This can be done in $C_{N-n_1}^{n_2}$ ways. Continuing this operation, we find the probability distribution to be

$$W(n_1, n_2, \ldots n_i, \ldots) = \text{const} \cdot N! \Big/ \prod_i (n_i!) \,, \tag{1.36}$$

where const is a normalization constant. By this formula, Boltzmann introduced the equipartition law [44–47], which is the basis of statistical mechanics. We note the contradiction between the description of an ensemble of identical particles within the framework of statistical mechanics and the dynamical description [48, 49]. The principle of detailed balance is valid in the dynamical description of this ensemble, which means that reverse of time $t \to -t$ requires the development of this particle ensemble in the inverse direction, that is, particles move strictly along the same trajectories in the inverse direction. The equipartition law requires the development of the ensemble to its equilibrium state. This contradiction will be removed if the trajectories of particles are became slightly uncertain.

1.2.2 The Boltzmann Distribution

Let us determine the most probable number of particles $\overline{n}_i$ that are to be found in a state i assuming a large number $\overline{n}_i \gg 1$ of particles in each state and requiring that the probability W as well as its logarithm have maxima at $n_i = \overline{n}_i$. We then introduce $dn_i = n_i - \overline{n}_i$, assume that $\overline{n}_i \gg dn_i \gg 1$, and expand the value $\ln W$ over the interval dn_i near the maximum of this value. Using the relation

$$\ln n! = \ln\left(\prod_{m=1}^{n} m\right) \approx \int_1^n dx \ln x \,,$$

we have $(d/dn)(\ln n!) \approx \ln n$.

On the basis of this relation, we obtain from (1.36)

$$\ln W(n_1, n_2, \ldots, n_i \ldots) = \ln W(\overline{n}_1, \overline{n}_2, \ldots \overline{n}_i, \ldots) - \sum_i \ln \overline{n}_i dn_i - \sum_i dn_i^2/(2\overline{n}_i) \,. \tag{1.37}$$

The condition that this quantity is maximal is

$$\sum_i \ln \overline{n}_i dn_i = 0 \,. \tag{1.38}$$

In addition to this equation, we take into account the relations following from (1.34) and (1.35) which account for the conservation of the total number of particles and their total energy:

$$\sum_i dn_i = 0 \,, \tag{1.39}$$

$$\sum_i \varepsilon_i dn_i = 0 \,. \tag{1.40}$$

Equations (1.38), (1.39), and (1.40) allow us to determine the average number of particles in a given state. Multiplying (1.39) by $-\ln C$ and (1.40) by $1/T$, where C and T are characteristic parameters of this system, and summing the resulting equations, we have

$$\sum_i \left(\ln \overline{n}_i - \ln C + \frac{\varepsilon_i}{T}\right) dn_i = 0 \,.$$

Because this equation is fulfilled for any dn_i, one can require that the expression in the parentheses is equal to zero. This leads to the following expression for the most probable number of particles in a given state:

$$\overline{n}_i = C \exp\left(-\frac{\varepsilon_i}{T}\right) \,. \tag{1.41}$$

This formula is the Boltzmann distribution.

We now determine the physical nature of C and T in (1.41) that follows from the additional equations (1.34) and (1.35). From (1.34) we have $C \sum_i \exp(-\varepsilon_i/T) = N$. This means that the value C is the normalization constant. The energy parameter T is the particle temperature and characterizes the average energy of a particle. Below we express this parameter in energy units and hence we will not use the dimensioned proportionality factor – the Boltzmann constant $k = 1.38 \times 10^{-16}$ erg/K – as is often done. Thus, the Kelvin is the energy unit, equal to 1.38×10^{-16} erg (see Appendix B).

We can prove that at large $\overline{n}_i$ the probability of observing a significant deviation from $\overline{n}_i$ is small. According to (1.37) and (1.38) the probability W near its maxima is

$$W(n_1, n_2, \ldots n_i \ldots) = W(\overline{n}_1, \overline{n}_2, \ldots \overline{n}_i, \ldots) \exp\left[-\sum_i \frac{(n_i - \overline{n}_i)^2}{2\overline{n}_i}\right] .$$

From this it follows that a significant shift of n_i from its average value $\overline{n}_i$ is $|n_i - \overline{n}_i| \sim 1/(\overline{n}_i)^{1/2}$. Since $\overline{n}_i \gg 1$, the relative shift of the number of particles in one state is small $(|n_i - \overline{n}_i|/\overline{n}_i \sim (\overline{n}_i)^{3/2}$. Thus, the observed number of particles in a given state differs little from its average value.

1.2.3
Statistical Weight of a State and Distributions of Particles in Gases

Above we used subscript i to refer to one particle state, whereas below we consider a general case where i characterizes a set of degenerate states. We introduce the statistical weight g_i of a state that is a number of degenerate states i. For example, a diatomic molecule in a rotational state with the rotational quantum number J has the statistical weight $g_i = 2J + 1$, equal to the number of momentum projections on the molecular axis. Including accounting for the statistical weight, formula (1.41) takes the form

$$\overline{n_j} = C g_j \exp\left(-\frac{\varepsilon_j}{T}\right) , \tag{1.42}$$

where C is the normalization factor and the subscript j refers to a group of states. In particular, this formula gives the relation between the number densities N_0 and N_j of particles in the ground and excited states, respectively:

$$N_j = N_0 \frac{g_j}{g_0} \exp\left(-\frac{\varepsilon_j}{T}\right) , \tag{1.43}$$

where ε_j is the excitation energy and g_0 and g_j are the statistical weights of the ground and excited states.

We now determine the statistical weight of states in a continuous spectrum. The wave function of a free particle with momentum p_x moving along the x-axis is given by $\exp(i p_x x/\hbar)$ to within an arbitrary factor if the particle is moving in the positive direction and by $\exp(-i p_x x/\hbar)$ if the particle is moving in the negative direction. (The quantity $\hbar = 1.054 \times 10^{-27}$ erg s is the Planck constant h divided by 2π.) Suppose that the particle is in a potential well with infinitely high walls. The particle can move freely in the region $0 < x < L$ and the wave function at the walls goes to zero. To construct a wave function that corresponds to free motion inside the well and goes to zero at the walls, we superpose the basic free-particle solutions, so $\psi = C_1 \exp(i p_x x/\hbar) + C_2 \exp(-i p_x x/\hbar)$. From the boundary condition $\psi(0) = 0$ it follows that $\psi = C \sin(p_x x/\hbar)$, and from the second boundary condition $\psi(L) = 0$ we obtain $p_x L/\hbar = \pi n$, where n is an integer. This procedure thus yields the allowed quantum energies for a particle moving in a rectangular well with infinitely high walls.

From this it follows that the number of states for a particle with momentum in the range from p_x to $p_x + dp_x$ is given by $dn = L dp_x/(2\pi\hbar)$, where we take into account the two directions of the particle momentum. For a spatial interval dx, the number of particle states is

$$dn = \frac{dp_x dx}{2\pi\hbar} .$$

Generalizing to the three-dimensional case, we obtain

$$dn = \frac{dp_x dx}{2\pi\hbar} \frac{dp_y dy}{2\pi\hbar} \frac{dp_z dz}{2\pi\hbar} = \frac{d\mathbf{p}d\mathbf{r}}{(2\pi\hbar)^3} . \tag{1.44}$$

Here and below we use the notation $d\mathbf{p} = dp_x dp_y dp_z$ and $d\mathbf{r} = dxdydz$. The quantity $d\mathbf{p}d\mathbf{r}$ is called a differential element of the phase space, and the number of states in (1.44) is the statistical weight of the continuous spectrum of states because it is the number of states for an element of the phase space.

Let us consider now some cases of the Boltzmann distribution of particles. First we analyze the distribution of diatomic molecules in vibrational and rotational states. The excitation energy for the vth vibrational level of the molecule is given by $\hbar\omega v$ if the molecule is modeled by a harmonic oscillator. Here $\hbar\omega$ is the energy difference between neighboring vibrational levels. On the basis of (1.42), we have

$$N_v = N_0 \exp\left(-\frac{\hbar\omega v}{T}\right), \tag{1.45}$$

where N_0 is the number density of molecules in the ground vibrational state. Because the total number density of molecules is

$$N = \sum_{v=0}^{\infty} N_v = N_0 \sum_{v=0}^{\infty} \exp\left(-\frac{\hbar\omega v}{T}\right) = \frac{N_0}{\left[1 - \exp\left(-\frac{\hbar\omega}{T}\right)\right]},$$

the number density of excited molecules is

$$N_v = N \frac{\exp(-\hbar\omega v/T)}{[1 - \exp(-\hbar\omega/T)]}. \tag{1.46}$$

The excitation energy of the rotational state with angular momentum J is given by $BJ(J+1)$, where B is the rotational constant of the molecule, and the statistical weight of this state is $2J + 1$. On the basis of the normalization condition $\sum_J N_{vJ} = N_v$ and assuming $B \ll T$ (as is usually done), the number density of molecules in a given vibrational–rotational state is

$$N_{vJ} = N_v (2J+1) \frac{B}{T} \exp\left[-\frac{BJ(J+1)}{T}\right]. \tag{1.47}$$

As an example of the particle distribution in an external field, let us consider the distribution of particles in the gravitational field. In this case (1.42) gives $N(x) \sim \exp(-U/T)$, where U is the potential energy of the particle in the external field. For the gravitational field we have $U = mgz$, where m is the mass of the molecule, g is the free-fall acceleration, and z is the altitude above the Earth's surface. Formula (1.42) takes the form of the barometric distribution, and then has the form

$$N(z) = N(0) \exp\left(-\frac{mgh}{T}\right), \quad l = \frac{T}{mg}, \tag{1.48}$$

where $N(z)$ is the number density of molecules at altitude z. For atmospheric air near the Earth's surface at room temperature we have $l \approx 9\,\text{km}$, that is, atmospheric pressure falls noticeably at altitudes of a few kilometers.

1.2.4
The Maxwell Distribution

We now consider the velocity distribution of free particles. This distribution is the end result of energy-changing collisions of the particles. The Boltzmann formula (1.42) provides the necessary information. In the one-dimensional case, the particle energy is $mv_x^2/2$, and the statistical weight of this state is proportional to dv_x, that is, to the number of particles $n(v_x)$ whose velocity is in the interval from v_x to $v_x + dv_x$. Formula (1.42) then yields

$$N(v_x)dv_x = C\exp\left(-\frac{mv_x^2}{2T}\right)dv_x\,,$$

where C is the normalization factor. Correspondingly, in the three-dimensional case we have

$$N(v)d\mathbf{v} = C\left(-\frac{mv^2}{2T}\right)d\mathbf{v}\,,$$

where the vector $\mathbf{v}$ has components v_x, v_y, and v_z, $d\mathbf{v} = dv_x dv_y dv_z$, and the kinetic energy of the particle $mv^2/2$ is the sum of the kinetic energies for all the directions of motion. In particular, for the number density of particles $N(\mathbf{v})$ we have, after using the normalization condition,

$$N(\mathbf{v}) = N\left(\frac{m}{2\pi T}\right)^{3/2}\exp\left(-\frac{mv^2}{2T}\right)\,, \tag{1.49}$$

where N is the total number density of particles. When we introduce the function $\varphi(v_x) \sim N(v_x)$, normalized such that

$$\int_{-\infty}^{\infty}\varphi(v_x)dv_x = 1\,, \quad \varphi(v_x) = \left(\frac{m}{2\pi T}\right)^{1/2}\exp\left(-\frac{mv_x^2}{2T}\right)\,, \tag{1.50}$$

then

$$N(\mathbf{v}) = N\varphi(v_x)\varphi(v_y)\varphi(v_z) = N\left(\frac{m}{2\pi T}\right)^{3/2}\exp\left(-\frac{mv^2}{2T}\right)\,. \tag{1.51}$$

These particle velocity distributions are called Maxwell distributions [50–52]. The average kinetic energy of particles following from (1.49) is

$$\frac{1}{2}m\overline{v^2} = \frac{1}{2}m\overline{v_x^2} + \frac{1}{2}m\overline{v_y^2} + \frac{1}{2}m\overline{v_z^2} = \frac{3}{2}m\overline{v_x^2}\,.$$

This is to be combined with the result

$$\begin{aligned}\frac{1}{2}m\overline{v_x^2} &= \frac{\int_{-\infty}^{\infty}(mv_x^2/2)\exp\left[-mv_x^2/(2T)\right]dv_x}{\int_{-\infty}^{\infty}\exp\left[-mv_x^2/(2T)\right]dv_x} \\ &= -\frac{d}{d(1/T)}\ln\left\{\int_{-\infty}^{\infty}\exp\left[-mv_x^2/(2T)\right]dv_x\right\} \\ &= -\frac{d}{d(1/T)}\ln(aT^{1/2}) = \frac{T}{2}\,,\end{aligned}$$

where the constant a does not depend on the temperature. Thus, the particle kinetic energy per degree of freedom is $T/2$, and correspondingly the average particle kinetic energy in the three-dimensional space is $m\overline{v^2}/2 = 3T/2$. These relations can be used as the definition of the temperature.

1.2.5
The Saha Distribution

We considered above the distributions of gas particles in bound or free states. Now we analyze the specific distribution for plasma systems that contain both bound and free electron states. We must examine the equilibrium between continuous and discrete electron states. This equilibrium is maintained by the processes

$$A^+ + e \leftrightarrow A\,,$$

where e is the electron, A^+ is the ion, and A is the atom. We consider a quasineutral plasma in which the electron and ion number densities are the same.

Consider an ionized gas in a volume Ω and denote the average number of electrons, ions, and atoms in this volume by n_e, n_i, and n_a, respectively (note that $n_e = n_i$ here). Relation (1.43) gives the ratio between the number density of free and bound states of electrons as

$$\frac{n_i}{n_a} = \frac{g_e g_i}{g_a}\frac{1}{(2\pi\hbar)^3}\int \exp\left(-\frac{J + p^2/2m_e}{T}\right) d\mathbf{p}d\mathbf{r}\,.$$

Here $g_e = 2$, g_i, and g_a are the statistical weights of electrons, ions, and atoms corresponding to their electronic states, J is the atomic ionization potential, and p is the free electron momentum, so $J + p^2/(2m_e)$ is the energy for transition from the ground state of the atom to a given state of a free electron. It is assumed that atoms are found only in the ground state.

Integration of this expression over electron momenta yields

$$\frac{n_i}{n_a} = \frac{g_e g_i}{g_a}\left(\frac{m_e T}{2\pi\hbar^2}\right)^{3/2} \exp\left(-\frac{J}{T}\right)\int d\mathbf{r}\,.$$

Integrating over the volume, we take into account that transposition of the states of any pair of electrons does not change the state of the electron system. Therefore $\int d\mathbf{r} = \Omega/n_e$. Introducing the number densities of electrons $N_e = n_e/\Omega$, ions $N_i = n_i/\Omega$, and atoms $N_a = n_a/\Omega$, we deduce that

$$\frac{N_e N_i}{N_a} = \frac{g_e g_i}{g_a}\left(\frac{m_e T}{2\pi\hbar^2}\right)^{3/2} \exp\left(-\frac{J}{T}\right)\,. \tag{1.52}$$

This result is called the Saha distribution [53].

One can write the Saha distribution in the form of the Boltzmann distribution (1.42) as

$$\frac{N_i}{N_a} = \frac{g_c}{g_a}\exp\left(-\frac{J}{T}\right)\,, \quad g_c = \frac{g_e g_i}{N_e}\cdot\left(\frac{m_e T}{2\pi\hbar^2}\right)^{3/2}\,, \tag{1.53}$$

where g_c is the effective statistical weight of the electron continuous spectrum. For an ideal plasma, this statistical weight is rather large because the electron number density N_e is small compared with a typical atom number density. This leads to the conclusion that a remarkable degree of ionization takes place at relatively small temperatures, $T \ll J$. However, the probability of atomic excitation is very small at these temperatures; that is, the number density of excited atoms is small. Hence, at these temperatures, atoms are either in the ground state or ionized.

1.2.6 Dissociative Equilibrium in Molecular Gases

Equilibrium between atoms and molecules in a molecular gas is maintained by the processes

$$X + Y \leftrightarrow XY .$$

This equilibrium bears analogy to the equilibrium between discrete and continuous atomic states corresponding to bound and free states of atoms. We can find the relation between the equilibrium number densities of atoms and molecules in this case by analogy with the Saha distribution. On the basis of (1.52), one can express the relationship between the number densities of atoms and molecules in the ground state as [41]

$$\frac{N_X N_Y}{N_{XY}(v = 0, J = 0)} = \frac{g_X g_Y}{g_{XY}} \left(\frac{\mu T}{2\pi\hbar^2}\right)^{3/2} \exp\left(-\frac{D}{T}\right) , \qquad (1.54)$$

where g_X, g_Y, and g_{XY} are the statistical weights of atoms and molecules with respect to their electron states, μ is the reduced mass of atoms X and Y, and D is the dissociation energy of the molecule.

In contrast to the ionization equilibrium of atoms, in this case most molecules are found in excited states. Using (1.46) and (1.47), which connect the number density of molecules in the ground state to their total number density, we can transform (1.54) to the form

$$\frac{N_X N_Y}{N_{XY}} = \frac{g_X g_Y}{g_{XY}} \left(\frac{\mu T}{2\pi\hbar^2}\right)^{3/2} \frac{B}{T} \left[1 - \exp\left(-\frac{\hbar\omega}{T}\right)\right] \exp\left(-\frac{D}{T}\right) , \qquad (1.55)$$

where N_{XY} is the total molecular number density.

1.2.7 Laws of Blackbody Radiation

An ionized gas contains excited atoms or molecules that emit radiation, so it is necessary to examine how the gas interacts with radiation. If the interaction of radiation with a gas is strong, the distance that an individual photon travels before being absorbed is relatively small. Then we deal with so-called equilibrium radiation [54]. Radiation in a vessel whose walls are at a temperature T will be absorbed and emitted by the walls, and these processes establish the equilibrium between the radiation and the walls of the vessel. This radiation is called blackbody radiation.

To calculate the average number of photons in a particular state, we use the fact that photons obey Bose–Einstein statistics. Therefore, the presence of a photon in a given state does not depend on whether other photons with this energy are also in this state. Then according to the Boltzmann formula (1.43), the relative probability that n photons with energy $\hbar\omega$ are found in a given state is $\exp(-\hbar\omega n/T)$. Thus, the mean number of photons in this state is

$$\overline{n_\omega} = \frac{\sum_n n \exp(-\hbar\omega n/T)}{\sum_n \exp(-\hbar\omega n/T)} = \frac{1}{\left[\exp(\hbar\omega/T) - 1\right]} . \qquad (1.56)$$

This formula is called the Planck distribution [55].

We introduce the spectral radiation density U_ω as the radiation energy per unit time and volume in a unit frequency range. We shall show below how this quantity can be determined. The radiation energy in the frequency interval from ω to $\omega + d\omega$ is $\Omega\, U_\omega d\omega$, where Ω is the volume of a region where the radiation is located. Alternatively, this quantity can be expressed as $2\hbar\omega\, n_\omega\, \Omega\, d\boldsymbol{k}/(2\pi)^3$, where the factor 2 accounts for the two polarizations of an electromagnetic wave, $\boldsymbol{k}$ is the photon wave number, n_ω is the number of photons in a single state, and $\Omega\, d\boldsymbol{k}/(2\pi)^3$ is the number of states in an element of the phase space. Using the dispersion relation $\omega = ck$ between the frequency ω and wave vector $\boldsymbol{k}$ of the photon (c is light velocity), the equivalence of these two aspects of the same quantity yields

$$U_\omega = \frac{\hbar\omega^3}{\pi^2 c^3} n_\omega . \qquad (1.57)$$

When the Planck distribution (1.56) is inserted into (1.57), we obtain the Planck radiation formula:

$$U_\omega = \frac{\hbar\omega^3}{\pi^2 c^3 \left[\exp(\hbar\omega/T) - 1\right]} . \qquad (1.58)$$

In the limiting case $\hbar\omega \ll T$, this result transforms to the Rayleigh–Jeans formula [56, 57]:

$$U_\omega = \frac{\omega^2 T}{\pi^2 c^3} , \quad \hbar\omega \ll T . \qquad (1.59)$$

This expression corresponds to the classical limit, and hence does not contain the Planck constant. The opposite limit yields the Wien formula [58]:

$$U_\omega = \frac{\hbar\omega^3}{\pi^2 c^3} \exp\left(-\frac{\hbar\omega}{T}\right) , \quad \hbar\omega \gg T . \qquad (1.60)$$

We shall now apply (1.58) to find the radiative flux emitted by a blackbody surface. It may be defined as the flux of radiation coming from a hole in a cavity with perfectly absorbing walls when this cavity contains the blackbody radiation. The blackbody surface emits an isotropic flux $c U_\omega$, and hence the photon flux at a frequency ω outside the blackbody boundary is

$$j_\omega = \int \frac{d\Theta}{4\pi} c U_\omega = \frac{c U_\omega}{4} = \frac{\hbar \omega^3}{4\pi^2 c^2 \left[\exp(\hbar\omega/T) - 1\right]} , \tag{1.61}$$

where the element of solid angle is $d\Theta = d\varphi d(\cos\theta)$ and we assumed that the photon flux is directed perpendicular to the blackbody surface. From this we obtain the Stefan–Boltzmann formula for the total radiative flux leaving the emitting surface:

$$J = \int_0^\infty j_\omega d\omega = \frac{c}{4} \int_0^\infty U_\omega d\omega = \sigma T^4 , \tag{1.62}$$

where the Stefan–Boltzmann constant σ is given by

$$\sigma = \frac{1}{4\pi^2 c^2 \hbar^3} \int_0^\infty \frac{1}{e^x - 1} x^3 dx = \frac{\pi^2}{60 c^2 \hbar^3} = 5.67 \times 10^{-12} \frac{\mathrm{W}}{\mathrm{cm^2\,K^4}} .$$

One can evaluate the functional dependence of the radiation flux (1.62) in a simple way on the basis of the dimensional analysis. The result must depend on the following parameters: T (the radiation temperature), $\hbar$ (the Planck constant), and c (the light velocity). From these parameters one can compose only one combination that has the dimension of a flux; it is $J \sim T^4 \hbar^{-3} c^{-2}$, consistent with (1.62).

1.2.8
Ionization Equilibrium in a Plasma with Particles

Plasma properties can be influenced by the presence in the plasma of small particles on a variety of size scales, including atomic and molecular clusters. We shall refer to all such small particles as aerosols, and the plasma containing them as an aerosol plasma, although the size of the particles can have an influence on the plasma properties. One such plasma property is the ionization equilibrium. The presence of small particles in a hot gas may alter the ionization equilibrium because the energy for an electron to bind to a surface is smaller than the ionization potential of atoms constituting this surface. For example, the copper ionization potential is 7.73 eV, whereas the copper work function, the energy for an electron to bind to a copper surface, is 4.40 eV. The corresponding values are 7.58 and 4.3 eV for silver, and 3.89 and 1.81 eV for cesium. Thus, the presence of such particles in a hot vapor alters the equilibrium density of charged particles. We assume in the following that electrons in a hot gas or vapor result from small particles only. Our goal is to determine the equilibrium charge of these particles and the number density of electrons in a plasma.

For simplicity, we assume all particles to be spherical and to have the same radius r_0. This radius is taken to be sufficiently large such that

$$r_0 \gg \frac{e^2}{T} \,. \tag{1.63}$$

This criterion allows us to consider a particle as bulk matter, rather than needing to describe its microscopic properties. The addition of a single electron to the aerosol particle makes only a slight difference to the electric potential of the particle. We use this property to write the relationship between the number densities of particles n_Z and n_{Z+1} that possess charges Z and $Z+1$, respectively. By analogy with the Saha distribution (1.52), we have

$$\frac{n_Z N_e}{n_{Z+1}} = 2\left(\frac{m_e T}{2\pi\hbar^2}\right)^{3/2} \exp\left(-\frac{W_Z}{T}\right) , \tag{1.64}$$

where W_Z is the electron binding energy for the particle with charge Z, N_e is the electron number density, and the factor 2 accounts for the electron statistical weight (two spin projections). The electron binding energy of a charged particle W_Z is the sum of the electron binding energy for the neutral particle W_0 of a given material and the potential energy of the charged particle. Using the electric potential for a particle of charge $Z+1/2$ (the average between Z and $Z+1$), we have

$$W_Z = W_0 + \left(Z + \frac{1}{2}\right)\frac{e^2}{r_0} \,.$$

Substituting this into (1.64) transforms it to the form

$$\frac{n_Z N_e}{n_{Z+1}} = 2\left(\frac{m_e T}{2\pi\hbar^2}\right)^{3/2} \exp\left(-\frac{W_0}{T} - \frac{\left(Z+\frac{1}{2}\right)e^2}{r_0 T}\right) . \tag{1.65}$$

This relation gives the charge distribution of the particles. If the average charge is large, this distribution is sharp. Specifically, introducing n_0, the number density of neutral particles, (1.65) leads to

$$n_Z = n_{Z-1} A \exp\left(-\frac{Ze^2}{r_0 T}\right) = n_0 A^Z \exp\left(-\frac{Z^2 e^2}{2 r_0 T}\right) ,$$

where $A = (2/N_e)\cdot[m_e T/(2\pi\hbar^2)]^{3/2}\exp(-W_0/T)$. For charges that are close to the average, this relationship is conveniently written in the form

$$n_Z = n_{\overline{Z}} \exp\left(-\frac{(Z-\overline{Z})^2}{2\Delta Z^2}\right) ,$$

where $\Delta Z^2 = r_0 T/e^2 \gg 1$ because of (1.63). The average charge of the particles follows from the relation $\overline{Z}e^2/(r_0 T) = \ln A$, which gives

$$\overline{Z} = \frac{r_0 T}{e^2}\ln\left[\frac{2}{N_e}\left(\frac{m_e T}{2\pi\hbar^2}\right)^{3/2}\exp\left(-\frac{W_0}{T}\right)\right] . \tag{1.66}$$

This result must be combined with the condition for plasma quasineutrality

$$N_e = \overline{Z} n ,$$

where n is the total number density of particles. Combining these equations to remove the electron number density, we find the average charge of the particles to be

$$\overline{Z} = \frac{r_0 T}{e^2} \left\{ \ln \left[\frac{2}{\overline{Z} n} \left(\frac{m_e T}{2\pi \hbar^2} \right)^{3/2} \right] - \frac{W_0}{T} \right\} . \tag{1.67}$$

1.2.9
Thermoemission of Electrons

For high temperatures or large particle size, $Ze^2/(r_0 T)$ becomes small. For example, at $r_0 = 10\,\mu\text{m}$ and $T = 2000\,\text{K}$, the very large particle charge of $Z = 1200$ would be required to make this parameter become unity. If the parameter is small, then it follows from (1.66) that

$$N_e = 2 \left(\frac{m_e T}{2\pi \hbar^2} \right)^{3/2} \exp\left(-\frac{W_0}{T} \right) . \tag{1.68}$$

This formula describes the equilibrium density of electrons above a flat surface if the electric potential of the particle is small compared with a typical thermal energy. Therefore, the conditions near the particle and far from it are identical. Then the average particle charge is determined from the relation $N_e = \overline{Z} n$, where the number densities of electrons and of particles are both known.

Equation (1.68) allows us to obtain a simple expression for the electron current from the surface of a hot cathode. In the case of equilibrium between electrons and a hot surface, the electron current from the surface is equal to the current toward it. Assuming unit probability for electron attachment to the surface upon their contact, we obtain that the electron current density i toward the surface is equal to the electron current density from the surface:

$$i = e \left(\frac{T}{2\pi m_e} \right)^{1/2} N_e = \frac{e m_e T^2}{4\pi^2 \hbar^3} \exp\left(-\frac{W_0}{T} \right) . \tag{1.69}$$

This result is known as the Richardson–Dushman formula [61–65] describing the electron current density for electron emission by a hot surface. This type of emission is called thermoemission of electrons. For the analysis of gas discharge problems, it is convenient to rewrite the Richardson–Dushman formula (1.69) for the thermoemission current density in the form

$$i = A_R T^2 \exp\left(-\frac{W}{T} \right) , \tag{1.70}$$

wherein the Richardson parameter A_R, according to (1.69), has the value [66] $60\,\text{A}/(\text{cm}^2\,\text{K}^2)$. Table 1.5 contains the parameters in (1.69) for real metals. In this table, W is the metal work function, T_b is the metal boiling point, and i_b is the current density at the boiling point.

Table 1.5 Parameters in (1.70) for electron thermoemission from metals [59, 60], A_R is expressed in square meters per square centimeter per square Kelvin.

Material	A_R	W, eV	Material	A_R	W, eV	Material	A_R	W, eV
Ba	60	2.11	Feα	26	4.5	Pt	33	5.32
Be	300	3.75	Feγ	1.5	4.21	Si	8	3.6
Cs	160	1.81	Mo	55	4.15	Ta	120	4.25
C	15	4.5	Ni	30	4.61	Th	70	3.38
Cr	120	3.90	Nb	120	4.19	Ti	60	3.86
Co	41	4.41	Os	1100	5.93	U	60	3.27
Cu	120	4.57	Pa	60	4.9	W	60	4.54
Hf	22	3.60	Pd	60	4.99	Y	100	3.27
Ir	120	5.27	Re	720	4.7	Zr	330	4.12

1.2.10
The Treanor Effect

A weakly ionized gas can be regarded as a system of weakly interacting atomic particles. This system can be divided into subsystems, and in the first approximation each subsystem can be considered as an independent closed system. The next approximation, taking into account a weak interaction between subsystems, makes it possible to establish connections among subsystem parameters. There are a variety of ways in which this decomposition can be done, with the selection depending on the nature of the problem. Often it is convenient to divide an ionized gas into atomic and electronic subsystems. Energy exchange in electron–atom collisions is slight owing to the large difference in their masses, so equilibrium within the atomic and electronic subsystems is established separately. If a weakly ionized gas is located in an external electric or electromagnetic field, which acts on electrons mostly, the electron and atom temperatures may be different. This means that both atoms and electrons can be characterized by Maxwell distributions for their translational energies, but with different mean energies.

Another example of weakly interacting subsystems that we shall consider below relates to a molecular gas in which exchanges of vibrational energy between colliding molecules have a resonant character. This is a more effective process than collisions of molecules with transitions of the vibrational energy to excitations of rotational and translational degrees of freedom. If vibrational and translational degrees of freedom are excited or are cooled in a different way, then different vibrational and translational temperatures will exist in such a molecular gas. This situation occurs in gas discharge molecular lasers, where vibrational degrees of freedom are excited selectively, and also in gas dynamical lasers, where a rapid cooling of translational degrees of freedom occurs as a result of gas expansion. The same effect occurs in shock waves and as a result of gas expansion after a nozzle. There is thus a wide variety of situations where a molecular gas is characterized by different vibrational and translational temperatures. However, the resonant character of exchange of vibrational excitation takes place only for weakly excited molecules. At moderate excitations, the resonant character is lost because of molecular anharmonicity. This leads to a particular type of distribution of molecular states that we shall now analyze.

We consider a nonequilibrium gas consisting of diatomic molecules where the translational temperature T differs from the vibrational temperature T_ν. The equilibrium between vibrational states is maintained by resonant exchange of vibrational excitations in collisions of molecules, as expressed by

$$M(v_1) + M(v_2) \leftrightarrow M(v_1') + M(v_2') , \tag{1.71}$$

where the quantities in the parentheses are vibrational quantum numbers. Assuming molecules to be harmonic oscillators, we obtain from this the condition

$$v_1 + v_2 = v_1' + v_2' .$$

The excitation energy of a vibrational level is

$$E_v = \hbar\omega(v + 1/2) - \hbar\omega x_e(v + 1/2)^2 ,$$

where ω is the harmonic oscillator frequency and x_e is the anharmonicity parameter. The second term of this expression is related to the establishment of equilibrium in the case being considered, where translational and vibrational temperatures are different. Specifically, the equilibrium condition leads to the relation

$$N(v_1)N(v_2)k\left(v_1, v_2 \to v_1', v_2'\right) = N(v_1')N(v_2')k(v_1', v_2' \to v_1, v_2) ,$$

where $N(v)$ is the number density of molecules in a given vibrational state and $k(v_1, v_2 \to v_1', v_2')$ is the rate constant for a given transition. Because these transitions are governed by the translational temperature, the equilibrium condition gives

$$k\left(v_1, v_2 \to v_1', v_2'\right) = k\left(v_1', v_2' \to v_1, v_2\right)\exp(\Delta E/T) ,$$

where $\Delta E = \Delta E(v_1) + \Delta E(v_2) - \Delta E(v_1') - \Delta E(v_2')$ is the difference of the energies for a given transition, where $\Delta E(v) = -\hbar\omega x_e(v + 1/2)^2$. From this, one finds the number density of excited molecules to be [67, 68]

$$N(v) = N_0 \exp\left[-\frac{\hbar\omega v}{T_\nu} + \frac{\hbar\omega x_e v(v+1)}{T}\right] , \tag{1.72}$$

where N_0 is the number density of molecules in the ground vibrational state. This formula is often called the Treanor distribution.

Formula (1.72) gives a nonmonotonic population of vibrational levels as a function of the vibrational quantum number. Assuming the minimum of this function to correspond to large vibrational numbers, we have for the position of the minimum

$$v_{\min} = \frac{1}{2x_e}\frac{T}{T_\nu} \gg 1 , \tag{1.73}$$

Table 1.6 Parameter $1/(2x_e)$ in (1.73).

Molecule	H_2	OH	CO	N_2	NO	O_2
$1/(2x_e)$	18	22	82	84	68	66

and the minimum number density of excited molecules is given by

$$N(v_{\min}) = N_0 \exp\left(-\frac{\hbar\omega v_{\min}}{2T_v}\right).$$

Table 1.6 contains the values of the first factor in (1.73) for some molecules. The effect considered is remarkable at $v \sim 10$ in terms of the distinction between vibrational and translational temperatures.

Thus, the special feature of the Treanor effect is that at high vibrational excitations, collisions of molecules with transfer of vibrational excitation energy to translational energy become effective. This causes a mixing of vibrational and translational subsystems, and invalidates the Boltzmann distribution for excited states as a function of vibrational temperature. Note that the model employed is not valid for very large excitations because of the vibrational relaxation processes.

1.2.11
Normal Distribution

A commonly encountered case in plasma physics as well as in other types of physics is one in which a variable changes by small increments, each change occurring randomly, and the distribution of the variable after many steps is studied. Examples of this are the diffusive motion of a particle, or the energy distribution of electrons in a gas. This energy distribution as it occurs in a plasma results from elastic collisions of electrons with atoms, with each collision between an electron and an atom leading to an energy exchange between them that is small because of the large difference in their masses. Thus, in a general statement of this problem, we seek the probability that some variable z has a given value after $n \gg 1$ steps, if the distribution for each step is random and its parameters are given.

Let the function $f(z,n)$ be the probability that the variable has a given value after n steps, with $\varphi(z_k)dz_k$ the probability that after the kth step the change of the variable lies in the interval between z_k and $z_k + dz_k$. Since the functions $f(z)$ and $\varphi(z)$ are probabilities, they are normalized by the condition

$$\int_{-\infty}^{\infty} f(z,n)dz = \int_{-\infty}^{\infty} \varphi(z)dz = 1 .$$

By definition of the above functions we have

$$f(z,n) = \int_{-\infty}^{\infty} dz_1 \cdots \int_{-\infty}^{\infty} dz_n \prod_{k=1}^{n} \varphi(z_k) , \quad z = \sum_{k=1}^{n} z_k . \tag{1.74}$$

We introduce the Fourier transforms

$$G(p) = \int_{-\infty}^{\infty} f(z) \exp(-ipz) dz \,, \quad g(p) = \int_{-\infty}^{\infty} \varphi(z) \exp(-ipz) dz \,, \quad (1.75)$$

which can be inverted to give

$$f(z) = \frac{1}{2\pi} \int_{-\infty}^{\infty} G(p) \exp(ipz) dp, \quad \varphi(z) = \frac{1}{2\pi} \int_{-\infty}^{\infty} g(p) \exp(ipz) dp \,.$$

Equation (1.75) yields

$$g(0) = \int_{-\infty}^{\infty} \varphi(z) dz = 1 \,, \quad g'(0) = -i \int_{-\infty}^{\infty} z\varphi(z) dz = -i\overline{z_k} \,, \quad g''(0) = -\overline{z_k^2} \,, \quad (1.76)$$

where $\overline{z_k}$ and $\overline{z_k^2}$ are the mean shift and the mean square shift of the variable after one step. From (1.74) and (1.76) it follows that

$$G(p) = \int_{-\infty}^{\infty} \exp\left(-ip \sum_{k=1}^{n} z_k\right) \prod_{k=1}^{n} \varphi(z_k) dz_k = g^n(p) \,,$$

and hence

$$f(z) = \frac{1}{2\pi} \int_{-\infty}^{\infty} g^n(p) \exp(ipz) dp = \frac{1}{2\pi} \int_{-\infty}^{\infty} \exp(n \ln g + ipz) dp \,.$$

Since $n \gg 1$, the integral converges at small p. Expanding $\ln g$ in a power series of p, we have

$$\ln g = \ln\left(1 - i\overline{z_k} p - \overline{z_k^2} p^2/2\right) = -i\overline{z_k} p - \overline{z_k^2} p^2/2 + \left(\overline{z_k}\right)^2 p^2/2 \,,$$

which gives

$$f(z) = \frac{1}{2\pi} \int_{-\infty}^{\infty} \exp\left[ip(n\overline{z_k} - z) - n\overline{z_k^2} p^2/2\right] dp$$

$$= \frac{1}{\sqrt{2\pi\Delta^2}} \exp\left[-\frac{(z - \overline{z})^2}{2\Delta^2}\right] . \quad (1.77)$$

In this expression, $\overline{z} = n\overline{z_k}$ is the mean shift of the variable for n steps, $\overline{z^2} = n\overline{z_k^2}$, and $\Delta^2 = \overline{z^2} - (\overline{z})^2$ is the root-mean-square deviation of this value. Formula (1.77) is called the normal distribution, or the Gaussian distribution. It is valid if the principal contribution to the integral in (1.76) comes from small values of p, that is, if $\overline{z_k} p \ll 1$ and $\overline{z_k^2} p^2 \ll 1$. Because this integral is determined by a range $n\overline{z_k^2} p^2 \sim 1$, the Gaussian distribution is valid for a large number of steps $n \gg 1$.

1.3
Rarefied and Dense Plasmas

1.3.1
Criteria for an Ideal Plasma

Many varieties of plasmas are known, with most of them composed of weakly interacting particles (see Figures 1.1–1.3). These plasma systems are analogues of gaseous systems of neutral particles. When they experience weak interactions, the system of charged particles is an ideal plasma, with the criterion for plasma ideality expressed by (1.3) between plasma parameters. The small parameter of the theory, the plasma parameter, has the form [69]

$$\gamma = \frac{N_e e^6}{T_e^3} , \tag{1.78}$$

and the condition to have an ideal plasma is $\gamma \ll 1$. Specific values for γ can be estimated on the basis of fundamental physical considerations.

The first criterion for an ideal plasma refers to the condition that the mean interaction energy of a plasma particle with its neighbors must be small compared with its kinetic energy $3T/2$. (We assume in the following that electrons and other plasma particles have the same temperature.) The electric potential arising from a charged plasma particle is given by (1.10), and close to the nucleus it is

$$\varphi = \frac{q}{r} - \frac{q}{r_D} , \quad r \ll r_D ,$$

where q is the particle charge, r is the distance from this particle, and r_D is the Debye–Hückel radius. The first term in this expression is the particle potential in a vacuum, and the second term is the electric potential that is created by the neighboring plasma particles. This means that the average interaction energy of a particle of charge e with other plasma particles is e^2/r_D, and the criterion for an ideal plasma ($e^2/r_D \ll 3T/2$) leads to the value

$$\gamma \ll \frac{9}{32\pi} = 0.09 . \tag{1.79}$$

A second criterion follows from the condition that many charged particles are located in the sphere of radius r_D, that is, $4\pi r_D^3 N_e/3 \gg 1$. This gives the value

$$\gamma \ll \frac{1}{96\pi} = 0.003 . \tag{1.80}$$

Criteria (1.79) and (1.80) both contain identical combinations of parameters, but yield different numerical values. Criterion (1.79) is preferred, because it encompasses a larger region in which the ratio of the potential energy of a particle to its kinetic energy can be a small parameter, and thus is useful for expansion of plasma parameters. Nevertheless, Figures 1.1 and 1.2 display criteria (1.79) and (1.80) as boundaries between an ideal plasma and a dense plasma.

The quantity selected to characterize a plasma may not be γ, but may instead may be some function of γ. Often one uses the coupling constant Γ of the plasma, introduced as the ratio of the Coulomb interaction potential of a charged particle with its nearest neighbors to the thermal energy,

$$\Gamma = \frac{e^2}{r_W T} = \left(\frac{4\pi\gamma}{3}\right)^{1/3} , \tag{1.81}$$

where $r_W = (4\pi N_e/3)^{-1/3}$ is the Wigner–Seitz radius. A dense plasma is one with $\Gamma \gg 1$, and is called a strongly coupled plasma. The condition for the plasma to be ideal is $\Gamma \ll 1$, which implies that

$$\gamma \ll \frac{3}{4\pi} = 0.2 . \tag{1.82}$$

We can express plasma properties as a function of the plasma coupling constant Γ. The ratio of the average interaction energy of a charged plasma particle with other particles to its mean kinetic energy has the form

$$\frac{e^2}{r_D} \cdot \frac{2}{3T} = \frac{4}{3}\sqrt{2\pi\gamma} = \frac{(2\Gamma)^{3/2}}{2^{1/2}} .$$

The number of electrons in a sphere of radius r_D is

$$N_D = \frac{4\pi r_D^3 N_e}{3} = \frac{1}{6\sqrt{8\pi\gamma}} = \frac{1}{(6\Gamma)^{3/2}} = \frac{0.07}{\Gamma^{3/2}} ,$$

and the criterion $N_D \gg 1$ of an ideal plasma gives $\Gamma \ll 0.2$.

Note that the plasma under consideration consists of electrons and ions with a single charge. For a plasma with multicharged ions, the ideality criterion (1.5) is violated at low ion density. In particular, for interaction between ions after changing in criterion (1.5) e to Ze, where Z is the ion charge, we obtain the plasma parameter (1.78) in the form

$$\gamma = \frac{Z^6 e^6 N_i}{T^3} .$$

In particular, for a dusty plasma with typical parameters [70] $Z = 10^3$, the number density of particles $N_p = 10^3\ \mathrm{cm}^{-3}$, and $T = 400\ \mathrm{K}$ we have for the plasma parameter $\gamma \sim 10^5$, that is, in this example one can consider the plasma to have strong coupling.

1.3.2
Conditions for Ideal Equilibrium Plasmas

We expect a plasma to fail to be an ideal plasma as the plasma density increases, but if a plasma is prepared from a gas and ionization equilibrium is supported between electrons and atoms, this plasma is ideal. Below we convince ourselves of this with a simple example. For this goal we take a weakly ionized gas at a low temperature and find the dependence of the plasma parameter γ defined by (1.78) on the plasma density when there is equilibrium between charged and neutral plasma particles (ionization equilibrium). From the Saha distribution (1.52) for the electron number density N_e and the atom number density N_a in a quasineutral plasma, we obtain the relation

$$\frac{N_e^2}{N_a} = g \left(\frac{m_e T_e}{2\pi\hbar^2} \right)^{3/2} \exp\left(-\frac{J}{T_e} \right) ,$$

where $g = g_e g_i / g_a$, where g_e, g_i, and g_a are the statistical weights of electrons, ions, and atoms, respectively, T_e is the electron temperature, and J is the atomic ionization potential. We take the total number density of nuclei, $N = N_e + N_a (N_e = N_i)$, as a parameter, and determine the dependence $\gamma(N)$.

Let us write the Saha distribution in the form

$$\gamma^2 = \left(N - \frac{\gamma T_e^3}{e^6} \right) \frac{C}{T_e^{9/2}} \cdot \exp\left(-\frac{J}{T_e} \right) ,$$

where $C = g m_e^{3/2} e^{12} / [\hbar^3 (2\pi)^{3/2}]$. Concentrating our attention on plasmas with the maximum departure from ideal plasma conditions, we choose the plasma temperature at a given N such that γ is maximal. The condition $d\gamma / dT_e = 0$ leads to the expressions

$$N = g \frac{3T(J - 3T_e/2)}{(J - 9T_e/2)^2} \left(\frac{m_e T_e}{2\pi\hbar^2} \right)^{3/2} \exp\left(-\frac{J}{T_e} \right) , \tag{1.83}$$

$$N_e = N \left(\frac{J - 9T_e/2}{J - 3T_e/2} \right) . \tag{1.84}$$

The maximum values of γ are located at $T_e \leq 2J/9$, and, in the limit of large N, the temperature approaches $2J/9$. In this limit the degree of plasma ionization goes to zero as N increases, and the plasma parameter γ increases with the increase of N. On the basis of the above equations, we analyze the limiting case where there is basic violation of the conditions for an ideal plasma. We take into account that (1.83) and (1.84) are valid for an ideal plasma. The maximum values of γ at a given large N lead to the expressions for the temperature T_e and the degree of ionization N_e/N:

$$\frac{(T_e - T_0)^2}{T_e^2} = \frac{4g}{9N} \left(\frac{m_e T_0}{2\pi\hbar^2} \right)^{3/2} e^{-9/2} , \tag{1.85}$$

$$\frac{N_e}{N} = \frac{3}{2} \frac{(T_e - T_0)}{T_0} , \tag{1.86}$$

where T_0 is defined as $T_0 = 2J/9$. It is evident that the plasma parameter behaves as $\gamma \sim (N a_0^3)^{1/2}$ in the limit of large N, and the coupling constant of the plasma is estimated to be $\Gamma \sim (N a_0^3)^{1/6}$, where a_0 is the Bohr radius. In particular, in the limit of large N, these expressions for a hydrogen plasma become

$$\frac{N_e}{N} = \frac{5.1 \times 10^{-3}}{(N a_0^3)^{1/2}} , \quad \gamma = 3.7 \left(N a_0^3 \right)^{1/2} , \quad \Gamma = 2.5 (N a_0^3)^{1/6} , \tag{1.87}$$

with validity constrained by $Na_0^3 \ll 1$. From this, it follows for the hydrogen plasma that $\Gamma = 1$ at $Na_0^3 = 0.004$.

One can conclude from these results that the degree of plasma ionization decreases with an increase of the plasma density. This means that departure from the ideal nature of the plasma is accompanied by an increase in the number density of neutral particles. Thus, interactions involving neutral atomic particles are of importance for the properties of a strongly coupled plasma. One can consider the above results from another standpoint. We have an equilibrium plasma that is characterized by the electron temperature T_e and the electron number density N_e, and the relation between these parameters is determined by the Saha relation (1.52). In addition to this, one can obtain a given electron temperature T_e from the energy balance because the specific energy of a plasma is $(3T_e + J)N_e$, where we assume identical temperatures of electrons and ions. From this it is seen that one can obtain an equilibrium plasma with different values of the plasma parameter γ depending on the plasma energy.

1.3.3 Instability of Two-Component Strongly Coupled Plasmas

From the above analysis it follows that a plasma prepared from a gas is ideal if ionization equilibrium between electrons and atoms is supported and external fields are absent. However, a dense plasma without neutral particles can be created for a short time under nonequilibrium conditions. For example, let us ionize an atomic gas by laser radiation as a result of atom photoionization if the photon energy is close to the atomic ionization potential. For example, if an atomic gas at room temperature under a pressure of 6 Torr is ionized fully such that the electron temperature of a forming plasma is 1000 K, the plasma parameter (1.78) for this plasma $\gamma = 1$. Our goal is to estimate the lifetime of such a plasma and compare it with actual times for plasma generation.

Under these conditions the energy $(3T_e + J)N_e$ related to unit volume of this plasma will be conserved in the course of plasma evolution, but the number density of electrons will be decreased as a result of electron–ion recombination, and released energy will go into electron heating. This leads to a decrease of the plasma parameter (1.78), that is, this plasma becomes ideal. In evaluating the rate of this process, we take into account the conservation of the specific energy $(3T_e + J)N_e$ of this plasma, and assuming the thermal energy of electrons to be small compared with the atomic ionization potential J $(T_e \ll J)$, we obtain from this the following connection between the rates of change of the electron temperature and number density in the course of plasma relaxation:

$$\frac{dT_e}{dt} = -\frac{J}{3}\frac{d\ln N_e}{dt} .$$

This leads to the following equation for evolution of the plasma parameter (1.78):

$$\frac{d\gamma}{dt} = \gamma \cdot \frac{J}{T_e}\frac{d\ln N_e}{dt} . \tag{1.88}$$

We now analyze the character of plasma relaxation to ionization equilibrium in the first stage when the gas is fully ionized, and one can ignore ionization of atoms. Then relaxation results from recombination of electrons and ions according to the scheme

$$2e + A^{+} \rightarrow e + A^{*} \rightarrow e + A\,, \tag{1.89}$$

and the balance equation for the electron number density is

$$\frac{dN_e}{dt} = -K N_e^2 N_i\,, \tag{1.90}$$

where for a quasineutral plasma the number densities of electrons and ions are identical $N_e = N_i$, and K is the recombination coefficient for the three body process (1.89). Subsequently, we analyze this process in detail, but now we find the rate constant K from the dimension considerations, according to which it has dimensions centimeters to the sixth power per second and includes the parameters e^2, m_e, and T_e ($T_e \ll J$). Then the recombination coefficient is

$$K = \frac{a e^{10}}{T_e^{9/2} m_e^{1/2}}\,, \tag{1.91}$$

where the numerical coefficient $a \sim 1$.

Taking $a = 1$, we obtain the balance equation (1.88) in the form

$$\frac{d\gamma}{dt} = -\gamma \cdot \frac{J}{T_e} K N_e^2 = -\gamma^3 \cdot \frac{J\sqrt{T_e}}{e^2\sqrt{m_e}} = -\frac{\gamma^3}{\tau}\,, \quad \tau = \frac{e^2}{J}\sqrt{\frac{m_e}{T_e}}\,. \tag{1.92}$$

As is seen, although a typical relaxation time τ exceeds the atomic time, the difference between these times is not great. For example, for a hydrogen plasma and $T_e = 1000$ K, τ is about 10^{-15} s, whereas the minimum time for plasma generation is approximately 10^{-14} s for short-pulse lasers, and the minimum time of a laser pulse exceeds a typical ionization relaxation time for this plasma. This means that it is impossible to create a strongly coupled plasma of low temperature in a two-component plasma consisting of electrons and ions only. Hence, the neutral component of a plasma or external fields are of importance to support a plasma with strong coupling, in contrast to an ideal plasma, where many of its properties (such as plasma oscillations and interaction with electromagnetic waves) are independent of its neutral components. Thus, the word "plasma" in its general sense meaning a system of charged particles is not appropriate for a dense plasma of low temperature, whereas it is a suitable description for ideal plasmas that are weakly ionized gases at low temperature.

1.3.4
Special Features of Strongly Coupled Plasmas

The above analysis of a strongly coupled plasma makes it possible to identify the distinguishing features of such an object. We first present a contrast to a weakly coupled plasma. When we apply the term "plasma" to a weakly ionized gas, we concentrate on properties that are determined by charged particles. These properties of an ideal plasma do not depend on the presence of neutral particles even though the density of neutral particles is greatly in excess of the density of charged particles. For example, the presence of neutral atoms in the Earth's ionosphere has no effect on the character of the propagation of electromagnetic waves through it. This explicitly plasma property is qualitatively distinct from such ionosphere properties as thermal capacity and thermal conductivity that are determined by the neutral component only. We can thus use the term "plasma" as a universal description of a wide variety of systems of weakly ionized gases whose electric or electromagnetic properties are determined by charged particles only. This allows us to analyze in a general way the properties of systems that may be very diverse in terms of the properties arising from their neutral components.

The explanation for this situation is to be found in the character of the interactions in a weakly ionized gas. There is a short-range interaction between neutral particles – atoms or molecules – and a long-range Coulomb interaction between charged particles. These interactions produce effects that can be treated independently in a gaseous system. Therefore, if a particular property of this system is determined by a long-range interaction between charged particles, one can ignore the short-range interactions in this system, That is, the presence of neutral particles does not affect these properties of the system. It follows from the nature of this deduction that such a conclusion can be valid only for systems with weak interactions among the constituents; that is, it can apply only to gaseous systems. On the other hand, a strongly coupled plasma is a system with strong interactions among the particles, and these strong interactions make it impossible to divide interactions into independent short-range and long-range types. Therefore, the term "strongly coupled plasma" cannot be employed as a generic description for a wide variety of systems as can be done in the case of an ideal plasma.

We can illustrate this conclusion with some examples of strongly coupled plasmas. Consider a metallic plasma. Taking a metal of a singly-valent element, we assume that all the valence electrons contribute to the conductivity, and that the metallic ions form the metal lattice. Then the electron number density is $N_e = \rho/m$, where ρ is the density of the metal and m is the atomic mass. For example, in the case of copper, we have $N_e = 8.4 \times 10^{22}\,\mathrm{cm}^{-3}$ corresponding to room temperature, along with the parameters $\gamma = 1.5 \times 10^7$ and $\Gamma = 400$. Assuming this plasma to be classical, we find that it is a strongly coupled plasma. The other example of a nonideal plasma is an electrolyte, which is a solution with positive and negative ions. The high density of molecules in the solution makes it a strongly coupled plasma. These examples refer to stationary strongly coupled plasmas. The strong action of an intense, short pulse of energy on matter can generate a nonstationary strongly coupled plasma. Plasmas of this type can result from a strong explosion that compresses matter. Another example of this type is the plasma associated with laser fusion, where a target is irradiated by short laser pulses directed onto the target simultaneously from different directions. The energy transferred to the target causes extreme heating and compression, with a dense plasma created as a result.

These examples illustrate the above conclusion for a strongly coupled plasma: namely, that its properties are determined by interactions involving both neutral and charged particles. Different physical objects of this type do not have general properties determined only by the charged particles. Therefore, when one describes a system as a "strongly coupled plasma", the immediate implication is that it is not one of a general class with identical properties as in the case of an ideal plasma.

1.3.5
Quantum Plasmas

We can treat a dense, low-temperature plasma of metals as a degenerate Fermi gas. Because of the low temperature and high density, the system has quantum properties. Let us consider the limit when $T = 0$, and take into consideration the Pauli principle, according to which two electrons cannot be in the same state. With the positive charge distributed uniformly over the plasma volume, this plasma is a degenerate electron gas. At zero temperature the plasma electrons have distinct momenta p in the interval $0 \leq p \leq p_\mathrm{F}$. The Fermi momentum p_F is found from the relation

$$n = 2\int \frac{d\mathbf{p}d\mathbf{r}}{(2\pi\hbar)^3} ,$$

where n is the total number of plasma electrons, the factor 2 accounts for the two possible directions of the electron spin, and $d\mathbf{p}$ and $d\mathbf{r}$ are elements of the electron momentum and the plasma volume. Introducing the electron number density $N_\mathrm{e} = n/\int d\mathbf{r}$, we have $p_\mathrm{F} = (3\pi^2\hbar^3 N_\mathrm{e})^{1/3}$ for the Fermi momentum, and the Fermi energy is

$$\varepsilon_\mathrm{F} = \frac{p_\mathrm{F}^2}{2m_\mathrm{e}} = \frac{(3\pi^2 N_\mathrm{e})^{2/3}\hbar^2}{2m_\mathrm{e}} . \tag{1.93}$$

The parameters of a classical plasma satisfy the relation $\varepsilon_\mathrm{F} \ll T$ or $r_\mathrm{D}^2 N_\mathrm{e}^{1/3} \gg a_0$. We define a quantum plasma to be a charged particle system characterized by the small parameter

$$\eta = \frac{T}{\varepsilon_\mathrm{F}} , \tag{1.94}$$

exactly opposite to the condition for a classical plasma. The Fermi energy is a fundamental parameter of a degenerate electron gas, and it is the parameter that is used for the analysis of a quantum plasma. We introduce the parameter characterizing the ideal nature of a quantum plasma by analogy with the plasma parameter (1.3) as the ratio of the Coulomb interaction energy of electrons to the Fermi energy, or

$$\xi = \frac{e^2/r_\mathrm{W}}{\varepsilon_\mathrm{F}} = \frac{2^{5/3}}{3\pi a_0 N_\mathrm{e}^{1/3}} = \frac{0.337}{a_0 N_\mathrm{e}^{1/3}} , \tag{1.95}$$

where r_W is the Wigner–Seitz radius and a_0 is the Bohr radius. The ideal degenerate electron gas has a high density compared with the characteristic atomic density, or $N_e a_0^3 \gg 1$, that (1.95) is equivalent to $\xi \ll 1$. This means that the greater the electron number density, the more the properties of a degenerate electron gas determine the properties of a quantum plasma. In contrast, the role of the Coulomb interaction between charged particles of the plasma decreases with increase of the electron number density.

We can apply the model of a degenerate electron gas to describe the behavior of electrons in metals. Table 1.7 lists parameters of real metallic plasmas at room temperature. It can be seen that η is small at room temperature, meaning that the metallic plasma has the character of a quantum plasma. But the Coulomb interaction involving electrons and ions of metals is comparable to the exchange interaction potential of electrons, determined by the Pauli principle. Thus, a metallic plasma is a quantum plasma in which the potential of the Coulomb interaction of charged particles and the exchange interaction potential of the electrons have the same order of magnitude. As is seen, in all the cases of a metallic plasma according to the data in Table 1.7 the Coulomb interaction between the nearest electrons exceeds the exchange interaction between electrons, but these values are comparable.

Positive ions of real metals form a crystalline lattice at low temperatures. An important role in these crystals is played by the non-Coulomb interaction of free electrons with ions and bound electrons. Consider a simplified problem where electrons and ions of the metal participate only in the Coulomb interactions between them. The ions of this system form a crystalline lattice, and the energy per coupled pair of charged particles (one electron and ion) is

$$\varepsilon = \frac{3p_F^2}{10m_e} - \kappa e^2 N_e^{1/3} , \tag{1.96}$$

where the first term is the mean electron kinetic energy, the second term is the mean energy of the Coulomb interaction between charged particles, and κ depends on the lattice type. Here we take into account the redistribution of charged particles resulting from their interaction that leads to the attractive character of the mean interaction energy.

Accounting for $p_F \sim N_e^{1/3}$ and optimizing (1.96) for the specific plasma energy, we find that the optimal parameters for the plasma are

$$a_0 N_e^{1/3} = \frac{5\kappa\pi^{4/3}}{3^{5/3}} = 0.174\kappa , \quad \varepsilon_{min} = -\varepsilon_0 \kappa^2 , \tag{1.97}$$

Table 1.7 Parameters of metallic plasmas at room temperature.

Metal	Li	Na	Mg	Al	K	Cu	Ag	Cs	Au	Hg
N_e, 10^{22} cm^{-3}	4.6	2.5	8.6	18	1.3	8.4	5.9	0.85	5.9	8.5
η, 10^{-3}	5.5	8.2	3.6	2.2	13	3.7	4.6	17	4.6	3.6
ξ	1.8	2.2	1.4	1.1	2.7	1.4	1.6	3.1	1.6	1.4

where $\varepsilon_0 = 2.5/(3^{5/3}\pi^{4/3})(m_e e^4/\hbar^2) = 2.4\,\mathrm{eV}$. This manipulation shows that the system may have a stable configuration of bound ions and electrons (i.e., $\varepsilon_{\min} < 0$). The stable distribution of charged particles corresponds to $\xi = 1.9/\kappa$. The system so described is called a Wigner crystal. It can be seen that a Wigner crystal, like real metals, is characterized by an electron number density of the order of the typical atom number density a_0^{-3}. Note that the above results are based on simple models which do not account for details of interaction in metals, and hence the results have a qualitative character.

1.3.6
Ideal Electron–Gas and Ion–Gas Systems

The gaseous state condition for a weakly ionized gas relates not only to the interactions among charged particles or among neutral particles as separate groups but also to interactions between charged and neutral particles. The interaction between neutral atomic particles has a short-range character, whereas the interaction of a charged particle with neutral particles may be long range and is stronger than the interaction between neutral particles. Therefore, one can expect a violation of the condition for the gaseous state to occur in the interaction of one charged particle with surrounding particles in a dense gas. That is, in a system of atoms and a single charged particle, where the interaction of the atoms satisfies the gaseous state criterion, the interaction of the charged particle with the atoms does not have gaseous character, that is, the charged particle interacts with many atoms simultaneously. We now consider this phenomenon in detail.

We begin by examining the behavior of electrons in a dense gas. The gaseous character of the interaction between electrons and atoms implies the condition

$$\lambda = (N\sigma)^{-1} \gg \overline{r}\,, \tag{1.98}$$

where λ is the mean free path of an electron in the gas, σ is the cross section for electron–atom scattering, N is the number density of atoms, and $\overline{r}$ is the mean distance between atoms. We take $\overline{r}$ to be the Wigner–Seitz radius, so $\overline{r} = (4\pi N/3)^{-1/3}$. The electron–atom cross section is represented by $\sigma = 4\pi L^2$, where L is the scattering length for slow electrons scattered by atoms.

We can write the condition (1.98) as

$$N \ll N_{\mathrm{cr}}\,, \tag{1.99}$$

where $N_{\mathrm{cr}} = (4\pi L^3\sqrt{3})^{-1}$. Table 1.8 lists values of N_{cr} and the critical pressure $p_{cr} = N_{\mathrm{cr}}T$ for electron interaction at $T = 300\,\mathrm{K}$. It is seen that the gaseous state condition for interaction of electrons with atoms can be violated in a dense gas.

The other gaseous state conditions for the electron–atom interaction in a dense gas require that positions of neighboring atoms should not influence electron–atom scattering. These relationships give

$$L \gg \overline{r}\,, \quad p\overline{r}/\hbar \ll 1\,. \tag{1.100}$$

Table 1.8 Critical parameters for interaction of electrons and ions with a gas.

Gas, vapor	He	Ar	Kr	Xe	Cs
N_{cr}^e, 10^{21} cm^{-3}	200	90	7	1	0.03
p_{cr}^e, atm	8000	3000	300	50	1
N_{cr}^i, 10^{21} cm^{-3}	6.3	2.8	2.6	2.0	0.12

The first condition in (1.100) gives $N \ll 3\sqrt{3}N_{cr}$, so it is less restrictive than condition (1.99). The second condition in (1.100) has the form

$$N \ll N_1(T)\,, \tag{1.101}$$

where

$$N_1(T) = \frac{3\hbar^3}{4\pi(2m_e T)^{3/2}}\,.$$

This is independent of the identity of the gas. At a temperature $T = 300$ K, we have $N_1 = 2 \times 10^{22}$ cm^{-3}, corresponding to a gas pressure of 700 atm.

For a gas consisting of atoms plus ions arising from those atoms, we have the gaseous state condition $N\sigma^{3/2} \ll 1$, where σ is the cross section of the resonant charge exchange process, and $\overline{r}$ is the mean distance between particles. The charge exchange cross section is relatively large, and therefore the gaseous state condition for ions is violated at relatively small densities of the gas. Table 1.8 lists values of gas densities $N_{cr} = (\pi/2)^{1/2}\sigma^{-3/2}$ at which the criterion to have a gaseous state for interaction of ions and atoms is violated at a gas temperature of 1000 K.

We conclude that electron–atom or ion–atom interactions in a weakly ionized gas can be nonpairwise even when the interaction between neutral particles satisfies the gaseous state condition. Then charged particles interact simultaneously with several neutral particles, and the gaseous character of the interaction is lost.

1.3.7
Decrease of the Atomic Ionization Potential in Plasmas

A dense plasma alters the states of its constituent atoms, and thereby can change the atomic spectrum [71–73]. In particular, spectra of metallic plasmas differ significantly from spectra of isolated metal atoms. The greater the plasma density, the more drastic is the change in atomic parameters, including the ionization potential. Below we estimate the decrease of the atomic ionization potential as a function of the plasma density.

We can consider an atom as an isolated object starting from the principal quantum number n, for which the atomic size r_n is small compared to the average distance between ions as $N_i^{-1/3}$. Since the average size of an excited atom is $r_n \sim a_0 n^2$ (a_0 is the Bohr radius), the ionization potential for this quantum number and for charge Z of the Coulomb center is $J_n \sim Z^2 e^2/(a_0 n^2)$. From this we have for the decrease ΔJ in the atomic ionization potential in a plasma [74]

$$\Delta J = J_n \sim Z e^2 N_i^{1/3} . \tag{1.102}$$

In particular, according to calculations by the method of molecular dynamics, this value is

$$\Delta J = 3.2 Z e^2 N_i^{1/3} . \tag{1.103}$$

One can consider the ionization potential decrease for atoms in a plasma to be a result of the action of plasma microfields. Indeed, let us consider an excited electron located in the field of the Coulomb center with charge Z and an external electric field of strength E. Then the interaction potential U between an electron and the Coulomb center in the direction x of the electric field is (see Figure 1.6)

$$U = -\frac{Z e^2}{x} - e E x ,$$

where the origin of the frame of reference is taken as the Coulomb center. The maximum of this interaction potential U_{max} gives the decrease of the ionization potential:

$$\Delta J = -U_{max} = 2\sqrt{e E \cdot Z e^2} .$$

In particular, if we use a typical electric field strength in a plasma in accordance with (1.26), $E_0 = 2.6 Z e N_i^{2/3}$, we obtain for the decrease in the ionization potential $\Delta J = 3.2 Z e^2 N_i^{1/3}$ in accordance with (1.103).

Thus, the presence of microfields in a plasma may lead to the disappearance of excited atom states in a plasma. The disappearance of spectral lines was observed for the first time by Lanchos [75, 76] in 1930, and this effect means that the radiative lifetime for an observed atomic transition becomes equal to the time for the

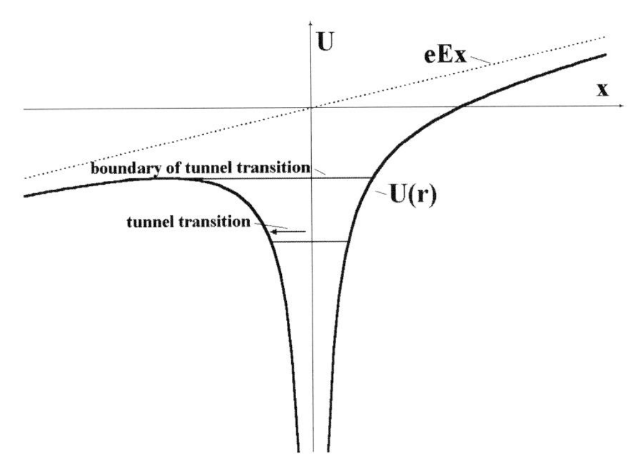

Figure 1.6 The potential energy in the plane of the electric field for an electron located in the field of a Coulomb center and electric field.

Table 1.9 Boundary values of the ion number density $(N_i)_n$ for disappearance of levels with principal quantum number n in a hydrogen-like plasma.

n	2	3	4	6	10
$(N_i)_n$, cm^{-3} for $Z = 1$	4.0×10^{20}	3.5×10^{19}	6.3×10^{18}	5.5×10^{17}	2.6×10^{16}
$(N_i)_n$, cm^{-3} for $Z = 4$	2.6×10^{22}	2.3×10^{21}	4.0×10^{20}	3.5×10^{19}	1.6×10^{18}

tunneling transition of the bound electron in the potential in Figure 1.6. We now consider a stronger action of the electric field when the decay time for an excited atomic level becomes equal to the orbital period for the excited electron. Then the boundary number density of ions $(N_i)_n$ for the disappearance of atomic levels with principal quantum number n follows from the relation

$$\frac{Z^2 e^2}{2 a_0 n^2} = \Delta J$$

and is given by

$$N_i = N_0 \frac{Z^3}{n^6}, \quad N_0 = 2.6 \times 10^{22}\,\mathrm{cm}^{-3}. \tag{1.104}$$

Table 1.9 contains certain values of the boundary ion number densities for a hydrogen-like plasma. Note that the number density of electrons N_e for a plasma with multicharged ions of charge Z is connected to the ion number density by the relation $N_e = Z N_i$. The values in Table 1.9 characterize the transition from an ideal plasma to a nonideal plasma or a plasma with strong coupling.

1.3.8 Spectrum of Atoms in a Plasma

The above analysis shows that one can divide states of electrons in the field of a Coulomb center in a plasma into three parts, bound electron states, free electron states, and localized electron states. Bound electron states correspond to discrete levels of an excited atom (ion) as it has in a vacuum. The energy ε_n of a state with principal quantum number n is given by

$$\varepsilon_n = -\frac{Z^2 e^2}{2 a_0 n^2}.$$

Since the statistical weight of states with principal quantum number n is $2n^2$, the density of states $\rho(\varepsilon)$ for discrete states is

$$\rho(\varepsilon) = \frac{dg}{d\varepsilon} = \frac{4 n^4 a_0}{Z^2 e^2} = \frac{Z^2 e^2}{a_0 \varepsilon^2}, \quad \varepsilon \le -\Delta J.$$

According to (1.103) at the boundary binding energy of a bound electron $\varepsilon_b = \Delta J$ the density of electron states is

$$\rho(-\varepsilon_b) \approx \frac{1}{10 a_0 e^2 N_i^{2/3}} .$$

For free electrons we use (1.53) for the statistical weight of free states with respect to bound ones. Using the Maxwell distribution function (1.49) for free electrons with the electron temperature T_e, we have for the density of free electron states

$$\rho_c(\varepsilon) = \frac{g_e g_i}{N_e} \cdot \frac{m_e^{3/2}}{\sqrt{2}\pi^2 \hbar^3} \cdot \varepsilon^{1/2} \exp\left(-\frac{\varepsilon}{T_e}\right) . \tag{1.105}$$

Evidently, this formula, as well as the Maxwell distribution (1.49) for free electrons, holds true if the electron energy exceeds remarkably the energy of interaction with plasma ions. Taking the quantity ΔJ given by (1.103) as the average potential energy for an electron, we define the boundary energy of free electrons as

$$\varepsilon_c = \Delta J = -\varepsilon_b .$$

Assuming $\varepsilon_c \ll T_e$, we obtain for the density of electron states in a hydrogen plasma ($Z = 1$)

$$\rho_c(\varepsilon) = \frac{g_e g_i}{N_e} \cdot \frac{m_e^{3/2} (\Delta J)^{1/2}}{\sqrt{2}\pi^2 \hbar^3} .$$

Taking $g_e = 2$, $g_i = 1$, we obtain

$$\rho(\varepsilon_c) \approx \frac{0.26}{N_i^{5/6} a_0^{5/2} \varepsilon_0} ,$$

where ε_0 is the atomic energy $\varepsilon_0 = e^2/a_0 = m_e e^4/\hbar^2$.

We now define an ideal plasma that is characterized by gas properties such that $\rho(\varepsilon_c) \gg \rho(-\varepsilon_b)$, that is,

$$\frac{\rho(\varepsilon_c)}{\rho(-\varepsilon_b)} \approx \frac{2.6}{N_i^{5/6} a_0^{5/2} \varepsilon_0} .$$

If we define an ideal plasma such that in an energy range of localized electrons the density of electron states decreases monotonically as the electron energy decreases, we obtain the criterion for an ideal plasma independently of the electron temperature

$$N_i a_0^3 \ll 300 , \tag{1.106}$$

that is, the ion density exceeds the atomic density ($N_0 = 6.76 \times 10^{24}\,\mathrm{cm}^{-3}$). This criterion shows that as long as the ion (electron) number density of a plasma is small compared with the atomic one N_0, the transferring range of energies between free and bound electron states, where electrons are bound simultaneously with many ions, is not important.

References

1 Langmuir, I. (1928) *Proc. Natl. Acad. Sci. USA*, **14**, 628.
2 Tonks, L. and Langmuir, I. (1929) *Phys. Rev.*, **33**, 195.
3 Suits, C.G. (Ed.) (1961) *The Collected Works of Irving Langmuir*, Pergamon Press, New York, pp. 111–120.
4 Le Chatelier, H.L. (1884) *C. r.*, **99**, 786.
5 Atkins, P.W. (1993) *The Elements of Physical Chemistry*, Oxford University Press, Oxford.
6 Fraunberger, F. (1985) *Illustrierte Geschichte der Elektrizität*, Aulis, Cologne.
7 Ross, S. (1991) *Nineteenth-Century Attitudes: Men and Science*, Kluwer, Dordrecht.
8 Carhardt, H.S. (1991) *Primary Batteries*, Cushing, Boston.
9 Crookes, W. (1878) *Philos. Trans.*, **170**, 135.
10 Goldstein, E. (1876) *Monat. Berl. Akad.*, **284**.
11 Goldstein, E. (1886) *Berl. Sitzungsber.*, **39**, 391.
12 Thomson, J.J. (1903) *Conduction of Electricity Through Gases*, Cambridge University Press, Cambridge.
13 Townsend, J.S. and Hurst, H.E. (1904) *Philos. Mag.*, **8**, 738.
14 Townsend, J.S. (1915) *Electricity in Gases*, Clarendon Press, Oxford.
15 Townsend, J.S. (1925) *Motion of Electrons in Gases*, Clarendon Press, Oxford.
16 Townsend, J.S. (1947) *Electrons in Gases*, Hutchinsons Scientific and Technical Publications, London.
17 Tonks, L. (1967) *Am. J. Phys.*, **35**, 857.
18 Brown, S.C. (1978) in: *Gaseous Electronics*, (eds M.N. Hirsh and H.J. Oskam), Academic Press, New York, pp. 1–18.
19 Debye, P. and Hückel, E. (1923) *Phys. Z.*, **24**, 185.
20 Langmuir, I. (1925) *Phys. Rev.*, **26**, 585.
21 Langmuir, I. and Tonks, L. (1927) *Phys. Rev.*, **34**, 376.
22 Tonks, L. and Langmuir, I. (1929) *Phys. Rev.*, **34**, 876.
23 Langmuir, I. (1929) *Phys. Rev.*, **33**, 954.
24 Holtsmark, J. (1919) *Ann. Phys.*, **58**, 577.
25 Davidson, R.C. (1974) *Theory of Nonneutral Plasmas*, Perseus Books, Reading, Massachusetts.
26 Davidson, R.C. (1990) *Physics of Nonneutral Plasmas*, Perseus Books, Reading, Massachusetts.
27 Child, C.D. (1911) *Phys. Rev.*, **32**, 492.
28 Langmuir, I. (1913) *Phys. Rev.*, **2**, 450.
29 Langmuir, I. (1914) *Phys. Z.*, **15**, 348.
30 Langmuir, I. (1914) *Phys. Z.*, **15**, 516.
31 Jaffe, G. (1920) *Ann. Phys.*, **63**, 145.
32 Penning, F.M. (1936) *Physica*, **3**, 563.
33 Kay, E.J. (1963) *J. Appl. Phys.*, **34**, 760.
34 Nezlin, M.V. (1993) *Physics of Intense Beams in Plasmas*, IOP Publishing, Bristol.
35 Bursian, V.P. and Pavlov, V.I. (1923) *Zh. Russ. Fiz.-Khim. Obshch.*, **55**, 71 (in Russian).
36 Smith, L.P. and Hartman, P.L. (1940) *J. Appl. Phys.*, **11**, 220.
37 Pierce, J.R. (1949) *Theory and Design of Electron Beams*, Van Nostrand, New York.
38 Pierce, J.R. (1944) *J. Appl. Phys.*, **15**, 721.
39 Frey, J. and Birdsall, C.K. (1966) *J. Appl. Phys.*, **37**, 2051.
40 Isihara, A. (1971) *Statistical Physics*, Academic Press, New York.
41 Landau, L.D. and Lifshitz, E.M. (1980) *Statistical Physics*, vol. 1, Pergamon Press, Oxford.
42 Birkhoff, G.D. (1931) *Proc. Natl. Acad. Sci. USA*, **17**, 656.
43 Neumann, J. (1932) *Proc. Natl. Acad. Sci. USA*, **18**, 70, 263.
44 Boltzmann, L. (1866) *Wien. Ber.*, **53**, 195.
45 Boltzmann, L. (1868) *Wien. Ber.*, **58**, 517.
46 Boltzmann, L. (1871) *Wien. Ber.*, **63**, 397, 679, 712.
47 Boltzmann, L. (1872) *Wien. Ber.*, **66**, 275.
48 Tolman, R.C. (1938) *The Principles of Statistical Mechanics*, Oxford University Press, London.
49 Feinman, R.P. (1972) *Statistical Mechanics. A Set of Lectures*, Perseus Books, Reading.
50 Maxwell, J.C.. (1860) *Philos. Mag.*, **19**, 19.
51 Maxwell, J.C. (1860) *Philos. Mag.*, **20**, 21.

52 Maxwell, J.C. (1871) *Theory of Heat*, Longmans, Green and Co, London.

53 Saha, M.N. (1921) *Proc. R. Soc. A*, **99**, 135.

54 Reif, F. (1965) *Statistical and Thermal Physics*, McGraw-Hill, Boston.

55 Planck, M. (1901) *Ann. Phys.*, **4**, 453.

56 Rayleigh, J. (1900) *Philos. Mag.*, **49**, 539.

57 Jeans, J.H. (1909) *Philos. Mag.*, **17**, 229.

58 Wien, W. (1896) *Wied. Ann. Phys. Chem.*, **58**, 662.

59 Neuman, W. (1987) *The Mechanism of the Thermoemitting Arc Cathode*, Academic-Verlag, Berlin.

60 Gale, W.F. and Totemeir, T.C. (eds) (2004) *Smithells Metals Reference Book*. 8th edn., Elsevier, Oxford.

61 Richardson, O.W. (1913) *Science*, **38**, 57.

62 Richardson, O.W. (1913) *Proc. R. Soc.*, **89**, 507.

63 Richardson, O.W. (1921) *Science*, **54**, 283.

64 Richardson, O.W. (1924) *Phys. Rev.*, **23**, 153.

65 Richardson, O.W. (1929) *Phys. Rev.*, **33**, 195.

66 Dushman, S. (1923) *Phys. Rev.* **21**, 623.

67 Treanor, C., Rich, J.W., and Rehm, R.G. (1968) *J. Chem. Phys.*, **48**, 1798.

68 Fisher, E.R. and Kummler, R.H. (1968) *J. Chem. Phys.*, **49**, 1075.

69 Fortov, V.E., Jakubov, I.I., and Khrapak, A.G. (2006) *Physics of Strongly Coupled Plasma*, Clarendon Press, Oxford.

70 Fortov, V.E. *et al.* (2005) *Phys. Rep.*, **421**, 1.

71 Lisitza, V.S. (1977) *Sov. Phys. Usp.*, **122**, 449.

72 Lisitza, V.S. (1987) *Sov. Phys. Usp.*, **153**, 379.

73 Lisitsa, V.S. (1994) *Atoms in Plasmas*, Springer, Berlin.

74 Unsöld, A. (1938) *Physik der Sternatmosphären*, Springer-Verlag, Berlin.

75 Lanczos, C. (1930) *Z. Phys.*, **62**, 518.

76 Lanczos, C. (1931) *Z. Phys.*, **68**, 204.

2
Elementary Processes in Excited and Ionized Gases

2.1
Elastic Collision of Atomic Particles

2.1.1
Elementary Act of Collisions of Particles in a Plasma

If we deal with an ionized gas consisting of atoms or molecules with an admixture of electrons and ions, the ionized gas is an ideal plasma, so each if its particles is free and can interact with surrounding particles in short time intervals. As a result of each act of strong interaction that is a collision of a test particle with surrounding particles, the state of the test particle may be changed, and our goal is to define the parameters of the elementary act of particle collision.

In considering the collision of a test particle A with particles B in a gas of these particles, where the initial state of particle A changes as a result of these collisions, we denote the initial state of particle A by subscript i and the final state by subscript f. Then the probability $P(t)$ that particle A remains in the initial state up to time t is governed by the equation

$$\frac{dP}{dt} = -\nu_{if} P \,, \qquad (2.1)$$

in compliance with the nature of this process. Here ν_{if} is the probability of this transition per unit time, and we assume that the transition takes place only between indicated states.

Let us use a coordinate frame where the test particle A is motionless. The transition probability per unit time ν_{if} is proportional to the incident flux j of particles B. One can introduce as a characteristic of the elementary act of particle collisions the cross section of this process, which is the ratio ν_{if}/j. It is important that this quantity is independent of the number density $[B]$ of the particles B. If all the particles B move with identical velocity $\mathbf{v}_B$, the flux of gas particles is given by $|\mathbf{v}_A - \mathbf{v}_B|[B]$, where $\mathbf{v}_A$ is the velocity of particle A. Thus, we have the following connection between the frequency of transition ν_{if} and the transition cross section σ_{if}:

$$\nu_{if} = [B]|\mathbf{v}_A - \mathbf{v}_B|\sigma_{if} \,. \qquad (2.2)$$

Fundamentals of Ionized Gases, First Edition. Boris M. Smirnov.

The cross section σ_{if} may depend on the relative velocity of the particles.

One can give another interpretation for the transition cross section σ_{if}. Let us place particles B on a plane with surface density n. Then the probability of the transition proceeding only between states i and f is $n\sigma_{if}$ ($n\sigma_{if} \ll 1$) after intersection of this plane by a particle A in the direction perpendicular to this plane.

When particles A and B can have arbitrary velocities, the transition probability per unit time may be represented in the form

$$\nu_{if} = [B]\langle|\mathbf{v}_A - \mathbf{v}_B|\sigma_{if}\rangle = [B]\langle k_{if}\rangle , \tag{2.3}$$

where the angle brackets signify an average over relative velocities of the particles, and the rate constant for the process

$$k_{if} = \langle|\mathbf{v}_A - \mathbf{v}_B|\sigma_{if}\rangle \tag{2.4}$$

is also a characteristic of the elementary act of collision. The rate constant for the process is useful when we are interested in the rate of transition averaged over the velocities of the particles.

It is useful to use the rate constants k_{if} for inelastic collisions in the balance equation for the number density N_i of particles A found in a given state i if transitions proceeds as a result of collisions with gas particles B. This equation has the form

$$\frac{dN_i}{dt} = [B]\sum_f k_{fi}N_f - [B]N_i\sum_f k_{if} . \tag{2.5}$$

The balance equation (2.5) may be extended by including other processes.

In elastic collisions of particles internal states of the colliding particles remain unchanged. Let us consider the classical case of particle collision that, in particular, takes place in thermal collisions of atoms in gases. Then the particle motion is described by Newton's equations

$$m_1\frac{d^2\mathbf{R}_1}{dt^2} = -\frac{\partial U}{\partial \mathbf{R}_1} , \quad m_2\frac{d^2\mathbf{R}_2}{dt^2} = -\frac{\partial U}{\partial \mathbf{R}_2} .$$

Here $\mathbf{R}_1$ and $\mathbf{R}_2$ are coordinates of colliding particles, m_1 and m_2 are their masses, and U is the interaction potential between the particles and depends on the relative distance between particles only, that is, $U = U(\mathbf{R}_1 - \mathbf{R}_2)$. Therefore, $-\partial U/\partial \mathbf{R}_i$ is the force acting on particle i due to the other particle, and $\partial U/\partial \mathbf{R}_1 = -\partial U/\partial \mathbf{R}_2$.

Introducing the coordinate of the center of mass $\mathbf{R}_c = (m_1\mathbf{R}_1 + m_2\mathbf{R}_2)/(m_1 + m_2)$ and the relative distance between colliding particles $\mathbf{R} = \mathbf{R}_1 - \mathbf{R}_2$, we rewrite the set of Newton's equations in terms of these variables:

$$(m_1 + m_2)\frac{d^2\mathbf{R}_c}{dt^2} = 0 , \quad \mu\frac{d^2\mathbf{R}}{dt^2} = -\frac{\partial U}{\partial \mathbf{R}} ,$$

where $\mu = m_1m_2/(m_1 + m_2)$ is the reduced mass of colliding particles. From this it follows that the center of mass travels with a constant velocity, and scattering is determined by the character of the relative motion of the particles in the center-of-mass frame [1].

This conclusion may be obtained also in terms of quantum mechanics [2]. Indeed, the Schrödinger equations for two colliding particles may be separated in the equation for the position of the center of mass and relative motion of colliding particles. Because the interaction potential of colliding particles depends on the relative distance $\mathbf{R} = \mathbf{R}_1 - \mathbf{R}_2$ between colliding particles, the center of mass of colliding particles moves with a constant velocity, and scattering of particles takes place in the center-of-mass frame of reference, where a particle with a reduced mass is scattered with respect to the force center.

Figure 2.1 gives the basic particle scattering parameters which are defined in the center-of-mass reference frame for a particle with a reduced mass with respect to the force center and include in the classical case the collision impact parameter ρ, the distance of closest approach r_0, and the scattering angle ϑ. If the interaction potential of particles $U(R)$ is isotropic, we have an additional relation between the collision parameters because of conservation of the particle momentum. Indeed, because the radial velocity of colliding particles is zero at the distance of closest approach r_0, the orbital momentum of colliding particles is $\mu v_\tau r_0$ at the distance of closest approach, where v_τ is the tangential velocity of particles at the distance of closest approach. On the other hand, the momentum of colliding particles is $\mu v \rho$, where v is the relative velocity of particles at large distances between particles. The energy conservation law gives the relation $\mu v_\tau^2/2 + U(r_0) = \mu v^2/2 \equiv \varepsilon$. Thus, conservation of the momentum and energy of colliding particles leads to the relationship [1]

$$1 - \frac{\rho^2}{r_0^2} = \frac{U(r_0)}{\varepsilon} . \tag{2.6}$$

Let us introduce the differential cross section of scattering as the number of scattering events per unit time and unit solid angle divided by the flux of incident particles. In the case of a central force field the elementary solid angle is $do = 2\pi d\cos\vartheta$, and particles are scattered into this element of solid angle from the

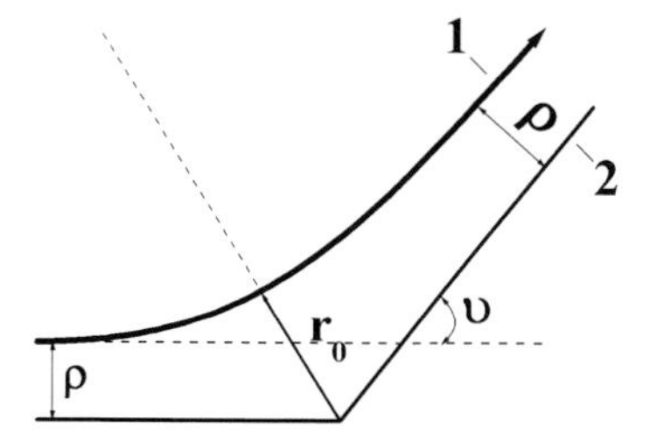

Figure 2.1 The trajectory of a particle in a central field in the center-of-mass frame of reference and the parameters of the particle collision: ρ is the collision impact parameter, ϑ is the scattering angle in the center-of-mass frame of reference, r_0 is the distance of closest approach, that is, the minimum distance between particles for a given impact parameter.

range of impact parameters between ρ and $\rho + d\rho$. The particle flux is Nv, where N is the number density of gas particles and v is the relative velocity of the collision. From this we find that the number of particles scattered per unit time into a given solid angle is $2\pi\rho d\rho N v$, so the differential cross section in the classical case is

$$d\sigma = 2\pi\rho d\rho \, . \tag{2.7}$$

Note that elastic scattering of particles determines transport parameters in gases and plasmas. If we consider transport processes from the standpoint of the trajectory of a test particle, transport parameters are determined by a significant change of the particle trajectory that corresponds to particle scattering into a large angle. Evidently, a typical scattering cross section for large angles is given by the relation when the interaction potential at the distance of closest approach is comparable to the kinetic energy of the colliding particles. Therefore, a typical cross section for scattering into large angles is given by

$$\sigma = \pi\rho_0^2 \, , \quad \text{where} \quad U(\rho_0) \sim \varepsilon \, . \tag{2.8}$$

In addition, we give one more definition of the differential cross section (2.7) on the basis of the thought experiment that is represented in Figure 2.2. In this case scattered particles pass through a gas layer of thickness l, where the number density of scattered particles is N and all the beam particles in the scattering angle $\vartheta \leq \vartheta_0$, where $\sin\vartheta_0 = R/L$, are collected by a detector. We take the velocity of incident particles v to be large compared with the thermal velocity of gas atoms, and for definiteness $R \ll L$, or $\vartheta_0 \ll 1$. If the intensity of the beam of incident particles is I_0 and the intensity of detected particles is I, the cross section of scattering into angles above ϑ_0 is given by

$$N l \sigma(\vartheta_0) = \frac{I_0 - I}{I_0} \, ,$$

and in this definition we assume $N l \sigma(\vartheta_0) \ll 1$. Correspondingly, the differential cross section is given by

$$d\sigma = \frac{1}{Nl}\frac{dI}{I_0} \, , \tag{2.9}$$

where the intensity variation dI is connected with a change of the angle ϑ_0 for detected particles, which may be achieved by variation of the distance L.

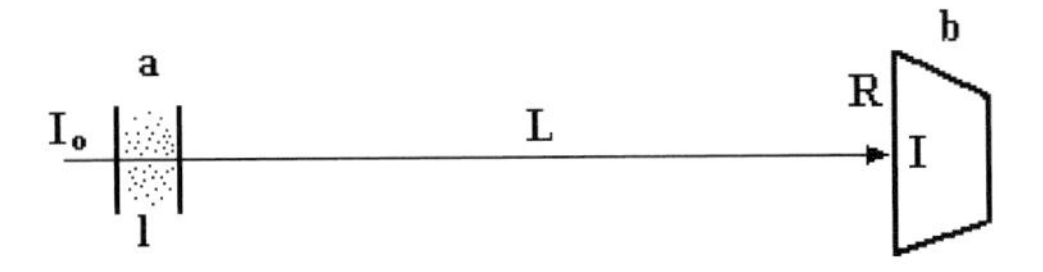

Figure 2.2 A thought experiment for definition of the differential cross section of particle scattering.

2.1.2
Model of Hard Spheres

A typical interaction potential of two atoms as a function of the distance between them is given in Figure 2.3. If the energy of colliding particles is large enough, $\varepsilon \gg D$, where D is the depth of the interaction well, and scattering is determined by the repulsive part of the interaction potential, one can approximate the interaction potential in this range by the potential with an infinite well that is given in Figure 2.3 and has the form

$$U(R) = 0\,, \quad r > R_0;\ U(R) = \infty\,, \quad r < R_0\,, \tag{2.10}$$

which corresponds to the hard-sphere model, and where R_0 is a radius of a hard sphere. The dependence of the collision impact parameter ρ on the distance of closest approach r_0 is given in Figure 2.4. The hard-sphere model for atom scattering is used widely in the study of the kinetics of neutral gases. The hard-sphere model has a long history and starts from the research of Maxwell [3–5], who on the basis of the model for gas molecules as colliding billiard balls found both the Maxwell distribution for an ensemble of free identical particles and the transport parameters for this ensemble.

The character of particle scattering within the framework of the hard-sphere model for colliding particles in the center-of-mass frame of reference consists of two straight lines (Figure 2.5 [6–8]), so their general point is located on the sphere surface, and the angles α of the trajectory lines with the line passing through this point and the sphere center are equal. These three lines are located in the same plane, and according to Figure 2.5 the connection between angles α and ϑ and between the collision impact parameter ρ and the scattering angle ϑ are given by

$$\vartheta = \pi - 2\alpha\,, \quad \rho = R_0 \sin\alpha\,.$$

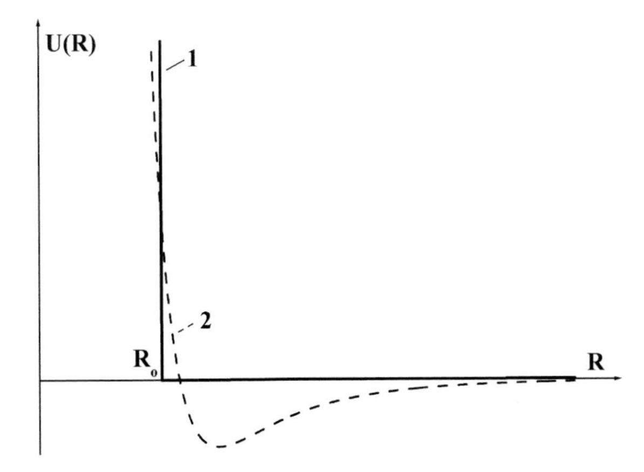

Figure 2.3 A typical interaction potential of two atoms (1) and the model potential (2) in the hard-sphere model.

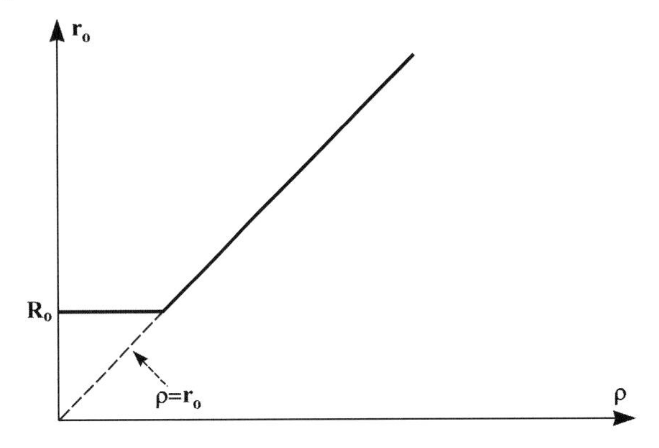

Figure 2.4 The dependence of the distance of closest approach on the collision impact parameter for the hard-sphere model.

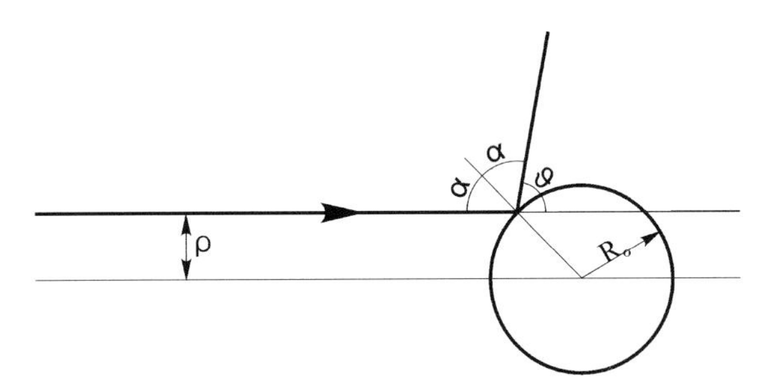

Figure 2.5 The character of scattering for the hard-sphere model: R_0 is the sphere radius, θ is the scattering angle, and the thick line is the particle trajectory.

For the differential cross section of particles scattering within the framework of the hard sphere model this gives

$$d\sigma = 2\pi\rho d\rho = \frac{R_0^2}{2} d\cos\theta \ ,$$

where the azimuthal symmetry of scattering is taken into account. From this we have for the diffusion σ^* and viscosity $\sigma^{(2)}$ cross sections of particle scattering on large angles

$$\sigma^* = \int (1 - \cos\vartheta) d\sigma = \pi R_0^2 \ , \quad \sigma^{(2)} = \int (1 - \cos^2\vartheta) d\sigma = \frac{2\pi}{3} R_0^2 \ . \quad (2.11)$$

As is seen, the average cross sections (2.11) are in agreement with the estimate (2.8) for the cross section of scattering into large angles.

The hard-sphere model corresponds to a sharply varied repulsive interaction potential according to the criterion

$$n = -\frac{d\ln U}{d\ln R} \gg 1 \ , \quad (2.12)$$

where the interaction potential $U(R)$ is a function of distance and is of the order of the energy of colliding particles. In particular, if the interaction potential is approximated by the dependence

$$U(R) = AR^{-n} , \tag{2.13}$$

criterion (2.12) has the form

$$n \gg 1 .$$

As a demonstration of this, Table 2.1 contains the parameters of the interaction potential for two inert gas atoms at distance R_0, where the interaction potential $U(R_0) = 0.3\,\mathrm{eV}$. These data follow from measurements [9] of small-angle scattering of inert gas atoms with kiloelectronvolt energies in inert gases. As is seen, criterion (2.12) holds true in this range of atom interaction.

If we apply the hard-sphere model for a certain interaction potential, we can estimate the hard-sphere radius R_0 from the relation

$$U(R_0) \sim \varepsilon ,$$

where ε is the energy of colliding particles. One can determine this value more accurately as a result of expansion of the cross section over a small parameter $1/n \ll 1$. Within the framework of this expansion, we have the dependence of the scattering angle ϑ for a given impact parameter ρ as a function of the distance of closest approach r_0 at a given value of the impact parameter ρ [10, 11]:

$$\vartheta = 2 \arcsin \sqrt{u} - 2 \ln 2 \frac{\sqrt{u(1-u)}}{1 + nu/2} u = \frac{U(r_0)}{\varepsilon} , \tag{2.14}$$

where the distance of closest approach r_0 depends weakly on the collision impact parameter ρ.

Formula (2.14) allows us to determine certain integral cross sections for particle scattering in a sharply varied repulsive interaction potential. In particular, for

Table 2.1 The parameter n for the dependence (2.13) of the interaction potential of two inert gas atoms and the distance R_0, where the repulsive interaction potential $U(R_0) = 0.3\,\mathrm{eV}$ [9]. The values of R_0 are given in parentheses and are expressed in angstroms.

Interacting atoms	He	Ne	Ar	Kr	Xe
He	5.9 (1.58)	5.6 (1.87)	5.2 (2.31)	5.5 (2.48)	5.2 (2.50)
Ne	–	7.6 (2.07)	6.6 (2.42)	7.6 (2.59)	6.8 (2.64)
Ar	–	–	6.1 (2.85)	6.9 (3.16)	5.9 (3.44)
Kr	–	–	–	7.7 (2.99)	7.1 (3.08)
Xe	–	–	–	–	6.4 (3.18)

widespread cross sections which determine transport coefficients of atomic particles in gases, we have, accounting for two terms of expansion of these cross sections over a small parameter, $1/n$ [12–14]

$$\sigma^* = \int (1 - \cos\vartheta) d\sigma = \pi R_*^2 , \quad \frac{U(R_*)}{\varepsilon} = 0.89 ;$$
$$\sigma^{(2)} = \int (1 - \cos^2\vartheta) d\sigma = \frac{2}{3}\pi R_2^2 , \quad \frac{U(R_2)}{\varepsilon} = 0.23 . \qquad (2.15)$$

These expressions conserve the form of the hard-sphere model and allow us to determine the sphere radius R_0.

Note that if the hard-sphere model holds true for the collision of a test atomic particle with gas atoms or molecules, one can introduce the mean free path of the test particle as $\lambda = 1/(N\sigma)$, where N is the number density of gas atoms and $\sigma = \pi R_0^2$ is the diffusion cross section of the test particle colliding with gas atoms. The concept of the mean free path was introduced by Clausius [15] and was used by Maxwell [3–5, 16] for determination of transport parameters of gases.

2.1.3
Collision Processes Involving Clusters

The hard-sphere model was applied above to the problem of collision of two gas atoms or molecules with a strongly varied repulsive interaction potential. In reality this model was designed in the first place for the analysis of many-particle systems. In particular, this model allows us to describe the properties of dense gases and condensed systems, allowing us to obtain the thermodynamic parameters of liquids [17, 18] and solids within the framework of this model. Since one of topics covered in this book is processes with clusters in an ionized gas, we continue with the hard-sphere model for clusters by changing the interaction potential (2.10) of atoms by adding of a narrow attraction potential at $R = R_0$. As a result, we obtain the liquid drop model that represents a system of interacting particles as a liquid drop consisting of small incompressible liquid drops. As the hard-sphere model, the liquid drop model for a system of bound particles has a universal character. In particular, it was introduced by Bohr [19] in nuclear physics and allows one to analyze various properties of atomic nuclei [20–22], including nuclear fission [23]. In considering a cluster consisting of many atoms, we reduce the interactions inside this system to electrostatic and exchange interaction between individual atoms or ions that chooses an optimal distance between the nearest atomic particles. We use the analogy of clusters with the macroscopic system of atoms where the competition between electrostatic and exchange (electron–electron) interactions chooses the optimal distance between atoms or ions [24, 25]. We then obtain the liquid drop model for a cluster that has a spherical shape and is cut off a macroscopic system. Assuming the number density of atoms in the cluster and bulk system to be identical and taking the number of cluster atoms to be n, we obtain from this for a cluster radius r_0

$$n = \frac{4\pi\rho r_n^3}{3m_a} = \left(\frac{r_0}{r_W}\right)^3 , \tag{2.16}$$

where m_a is the mass of an individual cluster atom and r_W is the Wigner–Seitz radius [24, 25], which is given by

$$r_W = \left(\frac{3m_a}{4\pi\rho}\right)^{1/3} . \tag{2.17}$$

The Wigner–Seitz radius is the fundamental cluster parameter, and Table 2.2 contains its values for some elements [26–28].

If we assume the size of the region where there is strong interaction between the colliding atom and cluster to be small in comparison with the cluster size, we can use the hard-sphere model for atom–cluster collision with the radius of the hard sphere equal to the cluster radius r_0. If an incident atom is reflected elastically from the cluster surface, the diffusion cross section of atom–cluster collision is

$$\sigma^* = \pi r_0^2 . \tag{2.18}$$

We have same expression for the atom–cluster diffusion cross section if an atom is captured by the cluster surface and then leaves this surface isotropically. Correspondingly, the rate constant for atom–cluster collision is

$$k = v_T \sigma^* = k_0 n^{2/3} , \quad k_0 = \sqrt{\frac{8T\pi}{m}} r_W^2 , \tag{2.19}$$

Table 2.2 Parameters of some metals and semiconductors: ρ is the liquid density at the melting point or at room temperature, and r_W is the Wigner–Seitz radius.

Element	ρ, g/cm^3	r_W, Å	Element	ρ, g/cm^3	r_W, Å	Element	ρ, g/cm^3	r_W, Å
Li	0.512	1.71	Zn	6.57	1.58	La	5.94	2.10
Be	1.69	1.28	Ga	6.08	1.66	Hf	12	1.81
Na	0.927	2.14	Rb	1.46	2.85	Ta	15	1.68
Mg	1.584	1.82	Sr	6.98	1.71	W	17.6	1.61
Al	2.375	1.65	Zr	5.8	1.84	Re	18.9	1.58
K	0.828	2.65	Mo	9.33	1.60	Os	20	1.56
Ca	1.378	2.26	Rh	10.7	1.56	Ir	19	1.59
Sc	2.80	1.85	Pd	10.4	1.60	Pt	19.8	1.57
Ti	4.11	1.66	Ag	9.32	1.66	Au	17.3	1.65
V	5.5	1.54	Cd	8.00	1.77	Hg	13.6	1.80
Cr	6.3	1.48	In	7.02	1.86	Tl	11.2	1.93
Fe	6.98	1.47	Sn	6.99	1.89	Pb	10.7	1.97
Co	7.75	1.44	Sb	6.53	1.95	Bi	10.0	2.02
Ni	7.81	1.44	Cs	1.843	3.06	U	17.3	1.77
Cu	8.02	1.47	Ba	3.34	2.54	Pu	16.7	1.70

where m is the atomic mass, $v_T = \sqrt{8T/\pi m}$ is the average thermal velocity of atoms, and n is the number of cluster atoms.

2.1.4
Cross Section of Capture

We now turn our attention to the other limiting case of atomic scattering, where the collision energy is small compared with the well depth D. The dependence of the collision impact parameter on the distance of closest approach, calculated on the basis of relation (2.6), is given in Figure 2.6. This curve is divided into two regions, whose boundary is separated by the arrow. Region 2 is not related to collisions of particles, and hence will not be considered. Effectively, the region of impact parameters is divided into two parts, and their boundary corresponds to the impact parameter ρ_c. At $\rho > \rho_c$, the values of the impact parameter and the corresponding distance of closest approach are comparable. If the collision impact parameter is smaller than ρ_c, the distance of closest approach r_0 is close to r_{min}, defined by $U(r_{min}) = 0$. Then collision capture occurs, and the bound system reduces its size until finally a short-range repulsion halts the contraction.

Since the capture is governed by the long-range part of the interaction potential, we can approximate the interaction potential by the dependence $U(R) = -C/R^n$. The impact parameter for capture ρ_c is determined as the minimum of the dependence $\rho(r_0)$ and, according to relation (2.6), the capture cross section is [1]

$$\sigma_c = \pi \rho_c^2 = \frac{\pi n}{n-2} \left(\frac{C(n-2)}{2\varepsilon} \right)^{2/n} . \tag{2.20}$$

The dependence of the cross section on parameters is similar to (2.8). In particular, in the case of polarization interaction of an ion and atom, $U(R) = -\alpha e^2/(2R^4)$ (α is the atomic polarizability), the polarization cross section for capture of an atom

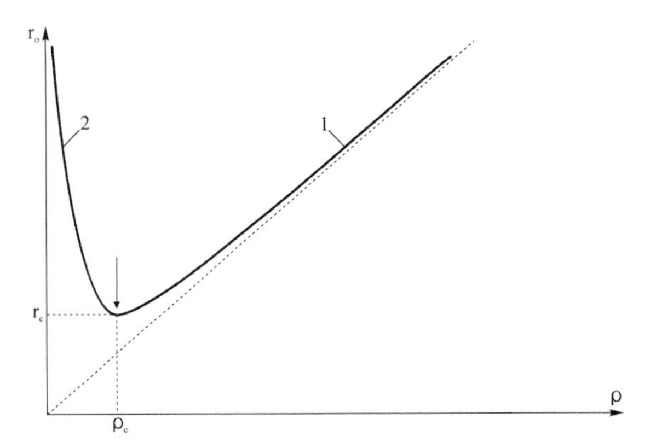

Figure 2.6 The dependence of the impact parameter on the distance of closest approach for an attractive interaction potential according to (2.6).

by an ion is

$$\sigma_c = 2\pi \left(\frac{\alpha e^2}{\mu v^2}\right)^{1/2} . \tag{2.21}$$

The final result shows that particle capture is determined by a long-range part of the interaction potential. Specifically, relation (2.6) gives, for $r_0 = r_c$, $-U(r_c)/\varepsilon = 2/(n-2)$, and because of $\varepsilon \ll D$, we have $|U(r_c)| \ll D$. That is, the capture is determined by the long-range part of the interaction. Since usually we have $n \gg 1$, in the region $r_0 > r_c$ the interaction potential is small compared with the kinetic energy of the particles. This means that the main contribution to the capture cross section is from impact parameters $\rho < \rho_c$ that correspond to particle capture.

Capture of particles is associated with a strong interaction. In fact, since

$$r_c \sim r_{\min}(D/\varepsilon)^{1/n} \gg r_{\min} ,$$

small distances between particles occur in the capture process. At these distances the attractive interaction potential greatly exceeds the kinetic energy of the particles. A strong interaction of particles in this region is also associated with a strong scattering of particles. Therefore, one can assume that the capture of particles leads to their isotropic scattering, and the diffusion scattering cross section (2.8) almost coincides with the capture cross section of (2.20). For instance, in the case of polarization interaction of particles, the diffusion cross section of particle scattering exceeds the capture cross section by 10%.

2.1.5
Total Cross Section of Scattering

The total cross section for elastic scattering of particles results from integrating over the differential cross sections for all solid angles, that is, $\sigma_t = \int d\sigma$. In classical case, the total cross section must be infinite. Classical particles interact and are scattered at any distances between them, and scattering takes place at any impact parameters. Therefore, the classical total cross section will tend to infinity if $\hbar \to 0$, where $\hbar$ is the Planck constant.

We can evaluate the total collision cross section by assuming that the colliding particles are moving along classical trajectories. The variation of the particle's momentum is given by the expression

$$\Delta \mathbf{p} = \int_{-\infty}^{\infty} \mathbf{F} dt , \tag{2.22}$$

where $\mathbf{F} = -\partial U/\partial \mathbf{R}$ is the force with which one particle acts upon the other, and U is the interaction potential between the particles. From (2.22) it follows that $\Delta p \sim U(\rho)/v$, where ρ is the impact parameter. According to the Heisenberg uncertainty principle, the value Δp can be determined up to an accuracy of $\hbar/\rho$. Therefore, the main contribution to the total scattering cross section is given by values of the impact parameter that satisfy the relation $\Delta p(\rho) \sim \hbar/\rho$. This leads to the estimate

$$\sigma_t \sim \rho_t^2\,, \quad \text{where} \quad \frac{\rho_t U(\rho_t)}{\hbar v} \sim 1 \tag{2.23}$$

for the total scattering cross section. In particular, if $U(R) = C/R^n$, the total cross section is

$$\sigma_t \sim \left(\frac{C}{\hbar v}\right)^{2/(n-1)} . \tag{2.24}$$

Since the scattering cross section is determined by quantum effects, it approaches infinity in the classical limit. This is demonstrated by (2.24), which indeed approaches infinity in the limit $\hbar \to 0$.

Particle motion will obey classical laws if the kinetic energy ε satisfies the condition

$$\varepsilon \gg \hbar/\tau\,, \tag{2.25}$$

where τ is a typical collision time. Since $\tau \sim \rho/v$ and $\varepsilon = \mu v^2/2$, from this it follows that $l = \mu\rho v/\hbar \gg 1$, where l is the collision angular momentum of the particles. If this criterion is fulfilled, the motion of the particles can be expected to follow classical trajectories. Furthermore, in this case the total cross section σ_t is greatly in excess of the cross section σ for large-angle scattering given by (2.8). In particular, for a monotonic interaction potential $U(R)$, (2.8) and (2.23) give $U(\rho_0)/U(\rho_t) \sim \mu\rho_t v/\hbar \gg 1$. It follows from the monotonic nature of $U(R)$ that $\rho_t \gg \rho_0$, so

$$\sigma_t \gg \sigma\,. \tag{2.26}$$

Note that if condition (2.25) is not satisfied and scattering has a quantum nature, then the large-angle scattering cross section and the total scattering cross section have the same order of magnitude. In particular, this is the situation for elastic scattering of electrons by atoms and molecules.

2.1.6 Gaseous State Criterion

The condition that defines the gaseous state of a system of particles can be formulated in terms of the collision cross section. A gas is a system of particles with weak interactions among them. This means that each particle follows a straight trajectory most of the time. Only occasionally does a particle interacts strongly enough with another particle to lead to large-angle scattering. This situation can be expressed by stating that the mean free path of a particle, $\lambda = 1/(N\sigma)$, is large compared with the interaction radius, $\sqrt{\sigma}$. Thus, the condition to be satisfied for a system to be in a gaseous state is

$$N\sigma^{3/2} \ll 1\,. \tag{2.27}$$

One can analyze this problem from another standpoint. We can express the gaseous state condition as

$$U(N^{-1/3}) \ll \mu v^2 .$$

This criterion signifies a weakly interacting system, since the interaction potential of a test particle with its neighbors is small compared with its mean kinetic energy. The notation $U(N^{-1/3})$ signifies that the potential is evaluated at the mean distance between particles, $N^{-1/3}$. On the basis of this expression and relation (2.8), we have $U(N^{-1/3}) \ll U(\rho_0)$. For a monotonic interaction potential this is equivalent to $N^{-1/3} \gg \rho_0$, and that inequality returns us to criterion (2.27).

The next step is to apply this criterion to a system of charged particles, that is, to a plasma. Because of the Coulomb interaction between charged particles, $|U(R)| = e^2/R$, criterion (2.8) gives

$$\sigma \sim e^4/T^2 \tag{2.28}$$

for a typical large-angle scattering cross section, where T is the average energy of the particles. Specifically, T is their temperature expressed in energy units. The condition (2.27) that an ensemble of particles be a true gas is transformed in the case of a plasma to the ideality plasma criterion (1.3):

$$\frac{Ne^6}{T^3} \ll 1 ,$$

where N is the number density of charged particles and T is their temperature.

2.1.7
Elastic Collisions of Electrons with Atoms

In analyzing elastic scattering of an electron by an atomic particle, we use the quantum character of this process. Assuming that the effective electron–atom potential is spherically symmetric, one can consider scattering of an electron by a motionless atom to be independent for different electron momenta l. This is the basis of the partial wave method [29], where the scattering parameters are the sum of these partial parameters of corresponding l. The characteristic of electron–atom scattering for a given l is the scattering phase δ_l, which describes the asymptotic wave function of the electron–atom system at large distances between the electron and the atom. The connection between the differential cross section $d\sigma = 2\pi|f(\vartheta)|^2 d\cos\vartheta$ of electron–atom scattering, the scattering amplitude $f(\vartheta)$, and the diffusion cross section σ^* is expressed through the scattering phases δ_l, as [2, 29]

$$f(\vartheta) = \frac{1}{2iq}\sum_{l=0}^{\infty}(2l+1)(e^{2i\delta_l} - 1)P_l(\cos\vartheta) ,$$

$$\sigma^* = \frac{4\pi}{q^2}\sum_{l=0}^{\infty}(l+1)\sin^2(\delta_l - \delta_{l+1}) , \tag{2.29}$$

where q is the electron wave vector, so the electron energy ε is expressed through the electron wave vector by the relation $\varepsilon = \hbar^2 q^2/2m_e$ (m_e is the electron mass). We account for the electron mass being small in comparison with the atom mass, so the reduced mass of colliding particles coincides practically with the electron mass. Note that the partial wave method is suitable for electron scattering by a structureless atom.

In the limit of small q scattering parameters (2.29) are expressed through the zero phase only, that is given by $\delta_0 = -Lq$ in this limit, where L is the scattering length. Correspondingly, the scattering parameters (2.29) are equal in this limit of small q:

$$f(\vartheta) = -L\,, \quad \sigma^*(0) = 4\pi L^2\,. \tag{2.30}$$

If we approximate the effective electron–atom interaction potential as a short-range one, the effective interaction potential $U(\mathbf{r})$ is expressed through the scattering length L by formula [30, 31]

$$U_{sh}(\mathbf{r}) = 2\pi L \frac{\hbar^2}{m_e} \delta(\mathbf{r})\,, \tag{2.31}$$

if the electron's wavelength exceeds the size of the atom's potential well. The scattering length coincides with the atom's effective radius [32], if the assumption is used that an electron cannot penetrate a region that is restricted by the effective radius. In addition, Table 2.3 contains values of the electron scattering lengths for scattering by inert gas atoms.

If the electron–atom scattering length is negative, the zero-scattering phase δ_0 becomes zero at low electron energy when other scattering phases δ_l are small. As a result, the electron–atom cross section (both diffusion and total) acquires a deep minimum at low electron energies that is known as the Ramsauer effect [35, 36]. As follows from Tables 2.3 and 2.4, the Ramsauer effect is realized in the case of electron scattering by argon, krypton, and xenon atoms. Figure 2.7 gives experimental values of the diffusion cross section of electron scattering by xenon atoms based on measurements [37–41], and Figure 2.8 represents the diffusion cross sections of electron scattering by inert gas atoms which result from the sum of measurements [34].

In the case where the electron–atom interaction potential is the sum of a short-range interaction potential (2.31) and the polarization interaction potential $U_l(r) = -\alpha e^2/2r^4$ as a long-range interaction potential, these interactions may be divided at low electron energies, and the scattering phases δ_l may be represented in the

Table 2.3 The electron scattering lengths for scattering by inert gas atoms [33, 34].

Atom	He	Ne	Ar	Kr	Xe
L, a_0	1.2	0.2	−1.6	−3.5	−6.5

Table 2.4 Electron scattering lengths for scattering by alkali metal atoms L_0 and L_1, and the parameters of the autodetaching state ^3P for the negative alkali metal ion with a spin of 1 [33].

	Li	Na	K	Rb	Cs
L_0/a_0	3.6	4.2	0.56	2.0	−2.2
L_1/a_0	−5.7	−5.9	−15	−17	−24
E_r, eV	0.06	0.08	0.02	0.03	0.011
Γ_r, eV	0.07	0.08	0.02	0.03	0.008

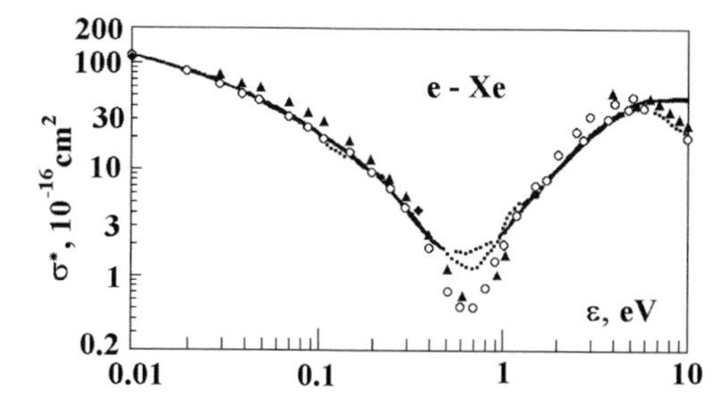

Figure 2.7 The diffusion cross section of electron scattering by a xenon atom as a function of the electron energy. Experimental data: solid curve – [37], open circles – [38], dotted curve – [39], black triangles – [40], black squares – [41]; the black rhombi give the cross section at zero energy at the minimum in accordance with (2.32) and the data in Table 2.3.

form of expansion over a small electron wave vector q [42, 43]. This expansion is based on a small parameter $q\alpha/La_0$ (a_0 is the Bohr radius), and accounting for the first term of this expansion leads to the following expression for the diffusion cross

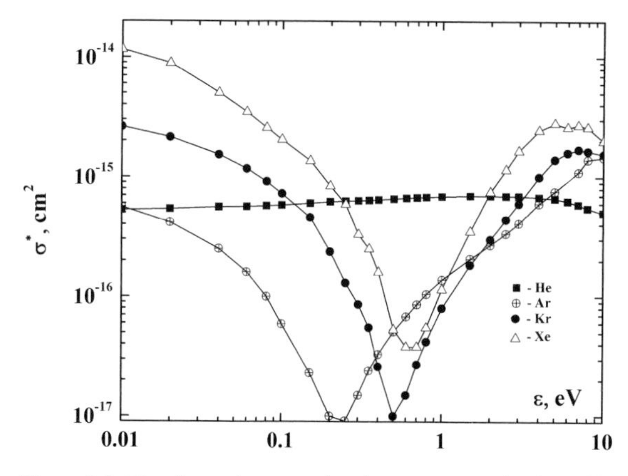

Figure 2.8 The dependence on the electron energy for the diffusion cross section of electron scattering by atoms of inert gases [34] according to measurements.

section of electron–atom scattering [44, 45]:

$$\sigma^* = 4\pi L^2 \left(1 - \frac{8}{5}x + \frac{2}{3}x^2\right) , \quad x = -\frac{\pi \alpha q}{2 L a_0} . \tag{2.32}$$

This cross section has a minimum if the wave vector is $q_{\min} = -12 L a_0/(5\pi\alpha)(x = 6/5)$ and the minimal cross section is 25 times less than that at zero electron energy. Although this analysis using short-range and polarization electron–atom interactions exhibits a deep minimum in the diffusion cross section of electron–atom scattering at a negative electron–atom scattering length, in reality the minimal cross section is lower than the cross section at zero electron energy by approximately two orders of magnitude.

Note that the above analysis relates to electron scattering by a structureless atom. In the case of electron scattering by atoms with incomplete electron shells, different channels of electron scattering are present in the scattering amplitude. As an example, we consider electron scattering by an alkali metal atom in the ground state, when the total electron–atom system may have spin $S = 0; 1$, and scattering proceeds independently for each channel, so at zero electron energy the diffusion cross section of electron–atom scattering is

$$\sigma^*(0) = \sigma_t(0) = \pi \left(L_0^2 + 3L_1^2\right) , \tag{2.33}$$

where L_0 and L_1 are the electron scattering lengths if the total spin of the electron–atom system is 0 and 1, respectively. These scattering lengths are given in Table 2.4. Note the importance of resonance ^3P in electron scattering by alkali metal atoms. The excitation energy of this autodetaching state E_r and its width Γ_r are given in Table 2.4.

2.1.8 Elastic Scattering of Charged Particles in a Plasma

Let us consider collision of two charged particles, for example, electron–electron or electron–ion scattering. We first illustrate the Coulomb interaction of particles, which leads to divergence of the diffusion cross section if particle collision proceeds in a vacuum. Taking the charge of both particles to be e and assuming the classical character of particle collision, we find the connection between the scattering angle ϑ and the collision impact parameter ρ in the case of scattering into small angles. Within the framework of standard model, this problem is reduced to motion of one particle with reduced mass μ in the Coulomb field e^2/R of the force center, where R is the distance of the particle from the center (or the distance between particles). Assuming in the zero approximation this particle moves along a straight trajectory, we have for a change of its momentum as a result of collision with a given impact parameter ρ [1]

$$\Delta p = \int_{-\infty}^{\infty} \frac{e^2 dt}{R^2} = \frac{2e^2}{\rho g} , \tag{2.34}$$

where g is the relative velocity of colliding particles. From this we have the connection between the scattering angle ϑ and the collision impact parameter ρ in the limit of small scattering angles,

$$\vartheta = \frac{\Delta p}{p} = \frac{2e^2}{\rho\mu g^2} ,$$

an for the differential cross section of particles interacting through the Coulomb interaction potential this gives [1]

$$d\sigma = 2\pi\rho d\rho = \frac{\pi e^4 d\Delta\varepsilon}{\varepsilon(\Delta\varepsilon)^2} . \tag{2.35}$$

This is the Rutherford formula [46] for scattering of two charged particles if their charges are equal to the electron charge. The criterion for validity of (2.35) for scattering into small angles has the form

$$\rho \gg \frac{e^2}{\varepsilon} , \tag{2.36}$$

and the energy change for a reduced particle in the center-of-mass frame of reference is

$$\Delta\varepsilon = \frac{\Delta p^2}{2m} = \frac{2e^4}{\rho^2\mu g^2} = \frac{e^4}{\rho^2\varepsilon} ,$$

where $\Delta\varepsilon$ is small compared with the particle energy ε. Note that in the limit of small scattering angles the differential cross section for collision of two charged particles is identical for both the same and opposite signs of the particle charge.

We now determine the diffusion cross section of collision of two charged particles in an ideal plasma, which is a measure of the exchange of energy and momentum between charged particles in a plasma and is very useful in expressing plasma properties. Assuming that small scattering angles give the main contribution to the diffusion cross section and using the Rutherford formula (2.35) for the differential cross section of scattering into small angles, we obtain for the diffusion cross section

$$\sigma^* = \int_0^\infty (1 - \cos\vartheta) \cdot 2\pi\rho d\rho = \int_0^\infty \vartheta^2 \pi\rho d\rho = \frac{4\pi e^4}{\mu^2 g^4} \int \frac{d\rho}{\rho} .$$

As is seen, the integral diverges in both the small and the large impact parameter limits. This means that in reality this integral must be cut short for small and large impact parameters, and the diffusion cross section of scattering of two charged particles can be represented in the form

$$\sigma^* = \frac{4\pi e^4}{\mu^2 g^4} \ln \frac{\rho_{max}}{\rho_{min}} ,$$

where the above integral is limited at small $\rho_{\min}$ and large $\rho_{\max}$ impact parameters. The lowest limit is determined by violation of the assumption of small scattering angles and is estimated as $\vartheta \sim 1$, or $\rho_{\min} \sim e^2/\varepsilon$. The upper limit is because of screening of the Coulomb field of interacting particles in a plasma, so the interaction potential of two charged particles in a plasma at a distance r from each other is $e^2 \exp(-r/r_D)/r$, where r_D is the Debye–Hückel radius. Hence, the upper limit is $\rho_{\max} \sim r_D$, and the diffusion cross section for the scattering of two charged particles in a plasma is [47]

$$\sigma^* = \frac{\pi e^4}{\varepsilon^2} \ln \Lambda \,, \quad \Lambda = \frac{r_D e^2}{\varepsilon} \,, \tag{2.37}$$

and the Coulomb logarithm $\ln \Lambda$ is determined within the accuracy up to a constant factor under logarithm.

Formula (2.37) leads to different diffusion cross sections for electron–electron and electron–ion scattering in a plasma. From this the diffusion cross section for scattering of a fast electron by a plasma electron is

$$\sigma^*_{ee} = \frac{4\pi e^4}{\varepsilon^2} \ln \Lambda \,, \tag{2.38}$$

where the energy of the fast electron is $\varepsilon = m_e v^2/2$, and v is the velocity of the fast electron. In the case of electron–ion scattering, if the electron velocity exceeds the ion velocity significantly, the diffusion cross section of electron–ion scattering is

$$\sigma^*_{ei} = \frac{\pi e^4}{\varepsilon^2} \ln \Lambda \,. \tag{2.39}$$

Let us estimate a typical value of the Coulomb logarithm for a plasma of a glow gas discharge, taking the number density of electrons and ions $N_e \sim 10^{12}\,\mathrm{cm}^{-3}$, a typical electron energy $\varepsilon \sim 1\,\mathrm{eV}$, and the temperature of atoms and ions $T = 400\,\mathrm{K}$. Under these conditions, the Debye–Hückel radius is $r_D \sim 10^{-4}\,\mathrm{cm}$, and the distance $\rho_{\min} \sim e^2/T_e$ for strong interaction of two charged particles involving an electron is $\rho_{\min} \sim 10\,\text{Å}$. For the Coulomb logarithm this gives $\ln \Lambda = 7$. As is seen, the value of the logarithm is large, which justifies the approximation used. Next, we use this value of the Coulomb logarithm in subsequent estimates.

2.1.9
Elastic Ion–Atom Collision and Resonant Charge Exchange

Two types of processes involving ions are of importance for ionized gases, elastic ion scattering of gas atoms and the resonant charge exchange process. The basic type of ion–atom interaction at large distances R between interacting particles has a polarization character and takes the form [2]

$$U(R) = -\frac{\alpha e^2}{2R^4} \,,$$

where α is the atom polarizability. Then the cross section of ion capture by an atom is given by (2.21) and has the form

$$\sigma_c = 2\pi\sqrt{\frac{\alpha e^2}{\mu g^2}}\,,$$

where g is the relative ion–atom velocity and μ is their reduced mass. The diffusion cross section of ion–atom scattering in the case of their polarization interaction exceeds this value by 10% and is [48–52]

$$\sigma^*(g) = 2.2\pi\sqrt{\frac{\alpha e^2}{\mu g^2}}\,. \tag{2.40}$$

Note that this process usually has classical character, which was used in derivation of (2.40). Indeed, the collision momentum l_c that corresponds to ion–atom capture is given by

$$l_c = \frac{\mu g \rho_c}{\hbar} = \frac{(2\alpha e^2 \mu^2 \varepsilon)^{1/4}}{\hbar}\,,$$

where $\varepsilon = \mu g^2/2$ is the energy of relative ion–atom motion. In particular, in the case of collision between a helium atom and a helium ion ($\alpha = 1.38 a_0^3$, the mass of the helium atom is $m_{He} = 7300 m_e$) at relative thermal energy ($\varepsilon = 0.026$ eV) this formula gives $l_c = 14$, which confirms the classical character of ion–atom capture.

The resonant charge exchange process, involving an atomic ion and a parent atom, is an ion collision process without energy transfer for the electron's degree of freedom. This process proceeds according to the scheme

$$A^+ + \widetilde{A} \rightarrow \widetilde{A^+} + A\,, \tag{2.41}$$

where the tilde marks a specific particle in this process. The resonant charge exchange process is important for transport of atomic ions in a parent gas because the cross section of resonant charge exchange exceeds the cross section of elastic ion–atom collisions at thermal collision energies.

The cross section of resonant charge exchange is large in slow ion–atom collisions because of the interference character of this process. Indeed, let us demonstrate this in the case of collision for a structureless ion or atom, where the valence electron is found in the s state (e.g., H^+–H or He^+–He). Then the eigenstate of the quasimolecule A_2^+, that is, the system consisting of ion A^+ and atom A at large distance from each other, may be composed of the states in Figure 2.9, where the ion belongs to the first or the second particle [53]. The wave functions ψ_1 and ψ_2 describe these states. Owing to the symmetry of this system, the eigenstate wave functions of the quasimolecule conserve or change their sign as a result of electron reflection with respect to the symmetry plane, that is, perpendicular to the axis joining nuclei and passing through its middle. The wave functions ψ_g and ψ_u of these states at large ion–atom distances are given by

$$\psi_g = \frac{1}{\sqrt{2}}(\psi_1 + \psi_2)\,, \quad \psi_u = \frac{1}{\sqrt{2}}(\psi_1 - \psi_2)\,,$$

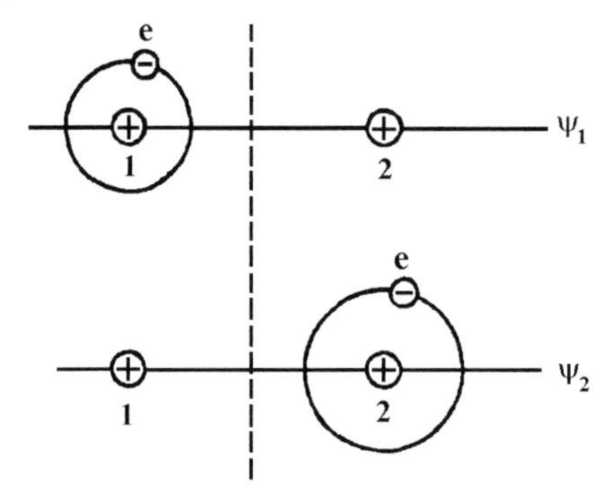

Figure 2.9 The dependence on the electron energy for the diffusion cross section of electron scattering by atoms of inert gases [34] according to measurements.

and satisfy the Schrödinger equations

$$\hat{H}\psi_{\rm g} = \varepsilon_{\rm g}\psi_{\rm g} \,, \quad \hat{H}\psi_{\rm u} = \varepsilon_{\rm u}\psi_{\rm u} \,,$$

where $\hat{H}$ is the Hamiltonian of electrons. One can construct, say, the wave function ψ_1, if a valence electron is located near the first atomic core, as a combination of the eigenstate wave functions as

$$\psi_1 = \frac{1}{\sqrt{2}}(\psi_{\rm g} + \psi_{\rm u}) \,.$$

We now assume the absence of transitions between states during evolution of the quasimolecule when the distance between the colliding ion and atom varies. Then the electron wave function in the course of variation of an ion–atom distance R is given by

$$\begin{aligned}\Psi(\mathbf{r}, R, t) = & \frac{1}{\sqrt{2}}\psi_{\rm g}(\mathbf{r}, R)\exp\left[-i\int_{-\infty}^{t}\frac{\varepsilon_{\rm g}(t')dt'}{\hbar}\right] \\ & + \frac{1}{\sqrt{2}}\psi_{\rm u}(\mathbf{r}, R)\exp\left[-i\int_{-\infty}^{t}\frac{\varepsilon_{\rm u}(t')}{\hbar}dt'\right] \,.\end{aligned}$$

Here $\mathbf{r}$ is the electron coordinate, and we account for the wave function of a stationary state is proportional to $\exp(-i\varepsilon t/\hbar)$, where ε is the state energy. From this formula we have for the probability of electron transition from one core to another one [54]

$$P(t) = |\langle\psi_2(t)|\Psi(t)\rangle|^2 = \sin^2\int_{-\infty}^{t}\frac{\Delta(R)}{2\hbar}dt \,,$$

where the exchange interaction potential is $\Delta(R) = \varepsilon_g(R) - \varepsilon_u(R)$. For the cross section σ_{res} of resonant charge exchange this gives

$$\sigma_{res} = \int_0^\infty 2\pi\rho d\rho \sin^2 \zeta(\rho) , \quad \zeta(\rho) = \int_{-\infty}^{\infty} \frac{\Delta(R)}{2\hbar} dt . \tag{2.42}$$

From this expression, the average probability of resonant charge exchange is 1/2 for small impact parameters ρ, so the cross section of this process may be represented in the form

$$\sigma_{res} = \frac{\pi}{2} R_0^2 , \quad \text{where} \quad \Delta(R_0) \sim \frac{1}{v} , \tag{2.43}$$

where v is the relative collision velocity. Because at large R we have $\Delta(R) \sim \exp(-\gamma R)$, and $\gamma = (2\hbar^2 J/m_e e^4)^{1/2}$ (J is the ionization potential), we obtain the following velocity dependence for the cross section of resonant charge exchange [55, 56]

$$\sigma_{res}(v) = \frac{\pi}{2\gamma^2} \ln^2 \left(\frac{C}{v} \right) ,$$

and this dependence may be represented in the form

$$\sigma_{res}(v) = \frac{\pi}{2} \left(R_0 + \frac{1}{\gamma} \ln \frac{v_0}{v} \right)^2 , \tag{2.44}$$

where $\pi R_0^2/2$ is the cross section of resonant charge exchange at the relative velocity v. In slow collisions $R_0\gamma \gg 1$, and Table 2.6 contains the values of this parameter along with the cross sections of resonant charge exchange at an ion energy of 1 eV in the laboratory frame of reference (an atom is motionless) [57–59]. These values were calculated on the basis of the asymptotic theory [60–63], which involves expansion of the cross section over a small parameter $1/(R_0\gamma)$. With the exception of atoms and ions of the first and second groups of the periodic table, the resonant charge exchange process is entangled with the processes of rotation of angular and spin momenta, and the cross sections in Table 2.5 are averaged over initial momentum directions. The accuracy of the averaged cross sections in Table 2.5 is several percent. If this cross section is approximated by the velocity dependence

$$\sigma_{res}(v) = \sigma_{res}(v_0) \left(\frac{v_0}{v} \right)^\kappa , \tag{2.45}$$

we have in this limiting case

$$\kappa = \frac{2}{\gamma R_0} \ll 1 .$$

Because of the approximation character of (2.45), the accuracy of this relation between parameters κ and $R_0\gamma$ is restricted. Table 2.6 contains values of this parameter for a collision energy of 1 eV.

Table 2.5 Parameters of the resonant charge exchange at a collision energy of 1 eV in the laboratory frame of reference. The parameter γ is given in atomic units, and the cross section of resonant charge exchange σ_{res} is expressed in 10^{-15} cm^2.

Element	γ	γR_0	κ	σ_{res}	Element	γ	γR_0	κ	σ_{res}
H	1.000	10	0.22	4.8	Mo	0.722	15	0.17	19
He	1.344	10	0.22	2.7	Pd	0.783	15	0.14	16
Li	0.630	14	0.16	22	Ag	0.746	15	0.16	17
Be	0.828	12	0.20	10	Cd	0.813	14	0.16	14
N	1.034	12	0.17	6.4	In	0.652	15	0.15	24
O	1.000	12	0.19	6.9	Sn	0.735	15	0.17	19
Ne	1.259	11	0.20	3.2	Xe	0.944	14	0.14	10
Na	0.615	15	0.17	26	Cs	0.535	17	0.15	44
Mg	0.750	14	0.18	15	Ba	0.619	16	0.15	30
Al	0.663	14	0.12	19	Ta	0.762	15	0.16	17
Ar	1.076	12	0.17	5.8	W	0.766	15	0.16	17
K	0.565	16	0.16	34	Re	0.761	15	0.16	17
Ca	0.670	14	0.18	15	Os	0.801	15	0.14	15
Cu	0.754	14	0.16	16	Ir	0.816	15	0.15	14
Zn	0.831	14	0.18	12	Pt	0.812	14	0.15	14
Ga	0.664	15	0.16	22	Au	0.823	15	0.16	14
Kr	1.014	13	0.16	7.5	Hg	0.876	14	0.15	12
Rb	0.554	16	0.15	38	Tl	0.670	15	0.16	22
Sr	0.647	15	0.16	25	Bi	0.732	16	0.14	22

Note that we are using the asymptotic theory of resonant charge exchange, which is guided by the classical character of nuclear motion [55, 64]. It is valid at large values of the collision momentum l_{res}:

$$l_{res} = \frac{\mu g R_0}{\hbar} \gg 1 \, .$$

In particular, in the case of resonant charge exchange involving helium ($He^+ + He$) at a thermal collision energy $\varepsilon = 0.026$ eV we have $l_{res} = 27$, that is, the classical criterion holds true.

It should be noted that the cross section of elastic ion–atom collision is less than the cross section of resonant charge exchange if the collision energy exceeds the thermal energy. Table 2.6 contains the ratio of the diffusion cross section (2.40) of elastic ion–atom collision to the cross section of resonant charge exchange at collision energy $\varepsilon = mg^2/2 = 0.1$ eV, where m is the mass of the atom or ion and g is the relative velocity of colliding particles. The statistical treatment of these data gives a value of 0.95 ± 0.14 for this ratio.

The cross section of resonant charge exchange depends weakly on the collision velocity, whereas the elastic cross section of ion–atom scattering increases with decreasing velocity. Hence, elastic ion–atom scattering becomes important at low

Table 2.6 The ratio between the diffusion cross section of elastic ion–atom scattering σ^* and the cross section of resonant charge exchange σ_{res} and x defined by (2.47) at a collision energy of 0.1 eV in the laboratory frame of reference.

Element	σ^*/σ_{res}	x, %	Element	σ^*/σ_{res}	x, %
H	0.79	3.2	Sr	1.1	5.9
He	0.78	3.1	Mo	0.97	4.8
Li	1.1	6.3	Pd	0.86	3.8
Be	1.1	5.9	Ag	0.93	4.4
N	0.81	3.3	Cd	1.1	6.4
O	0.64	2.1	Sn	0.74	2.8
Ne	0.90	4.2	Xe	0.99	5.0
Na	0.93	4.4	Cs	0.84	3.2
Mg	1.1	5.8	Ba	1.1	5.8
Al	0.83	3.5	Ta	1.1	5.8
Ar	1.1	6.0	W	1.2	7.6
K	0.96	4.8	Re	0.91	4.3
Ca	1.2	7.2	Os	0.95	4.6
Cu	0.75	2.9	Ir	0.95	4.6
Zn	1.1	5.8	Pt	0.88	4.0
Kr	1.0	5.4	Au	0.88	4.0
Rb	0.90	4.2	Hg	0.94	4.6

collision velocities. The influence of elastic ion–atom scattering on the resonant charge exchange cross section is twofold. On the one hand, curvature of the particle trajectory draws together the colliding ion and atom and increases the cross section of resonant charge exchange. On the other hand, at a low collision energy the resonant charge exchange process is determined by capture in ion–atom collisions, and the cross section of resonant charge exchange is half the capture cross section, that is, it is

$$\sigma_{res}(g) = \frac{1}{2}\sigma_{cap}(g) = \pi\sqrt{\frac{\alpha e^2}{2\varepsilon}} \ ,$$

where we assume the polarization character of ion–atom interaction. At larger collision velocities we have the following connection [1] between the collision impact parameter ρ and the distance of closest ion–atom approach r_{min} for the polarization interaction potential between them

$$\rho^2 = r_{min}^2\left(1 + \frac{\alpha e^2}{2r_{min}^4\varepsilon}\right) \ .$$

For the cross section of resonant charge exchange this gives

$$\sigma_{res} = \frac{\pi}{2}R_0^2 + \frac{\pi}{4}\frac{\alpha e^2}{R_0^2\varepsilon} \ , \qquad (2.46)$$

where $\pi R_0^2/2$ is the cross section of this process for straight trajectories of a colliding ion and atom. Combining these limiting cases, we conveniently represent the cross section of resonant charge exchange in the form [65]

$$\sigma_{\text{res}} = \begin{cases} \frac{\pi}{2} R_0^2 + \frac{\pi}{4} \frac{\alpha e^2}{R_0^2 \varepsilon}, & \varepsilon \geq \frac{\alpha e^2}{2 R_0^4} \\ \pi \sqrt{\frac{\alpha e^2}{2\varepsilon}} & \varepsilon \leq \frac{\alpha e^2}{2 R_0^4} \end{cases},$$

where R_0 corresponds to straight ion and atom trajectories. As is seen, in (2.46) the cross section depends on the parameter

$$x = \left| \frac{U(R_0)}{\varepsilon} \right| = \frac{\alpha e^2}{2 R_0^4 \varepsilon} = \left(\frac{\sigma_{\text{c}}}{4\sigma_{\text{res}}} \right)^2 , \tag{2.47}$$

where σ_{c} is the cross section of ion capture at the polarization ion–atom interaction. Values of x for a collision energy of 0.1 eV in the laboratory frame of reference are given in Table 2.6. As is seen, although the cross section of elastic ion–atom scattering is comparable to the cross section of resonant charge exchange at thermal energies, elastic scattering has a weak influence on the cross section of resonant charge exchange at thermal energies. Note that the accuracy of the cross sections for the resonant charge exchange process evaluated on the basis of the asymptotic theory for subthermal ion energies is better than 10%. As an example, in Figure 2.10 the cross sections of resonant charge exchange for krypton which were obtained on the basis of the asymptotic theory [57–59] and measurements [66–74] are compared.

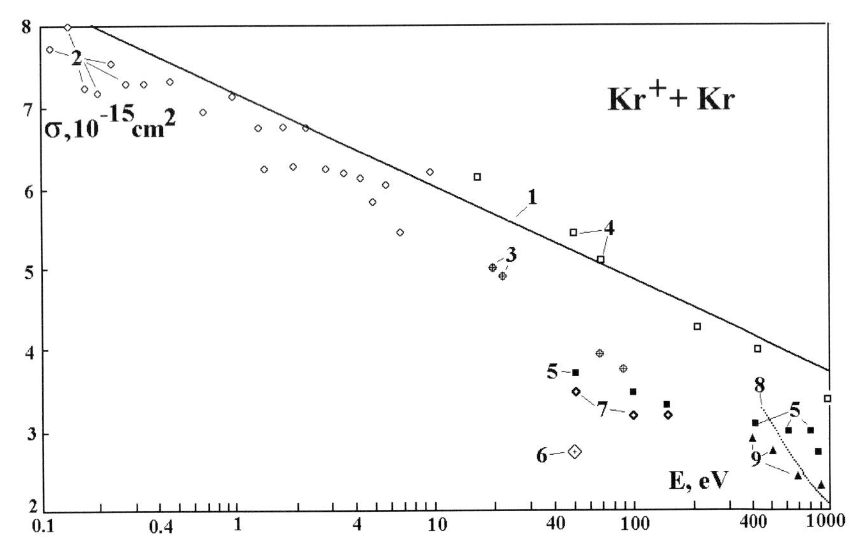

Figure 2.10 The cross sections of resonant charge exchange for krypton. 1 – asymptotic formula, experiment: 2 – [70], 3 – [71], 4 – [68], 5 – [69], 6 – [72], 7 – [66, 67], 8 – [73], 9 – [74].

2.2
Inelastic Processes Involving Electrons

2.2.1
Excitation and Quenching of Atoms by Electron Impact

The average electron energy in a gas is usually small compared with the excitation energy of atoms, and therefore excitation of atoms in a weakly ionized gas proceeds in the tail of the energy distribution function of electrons. Hence, we need the threshold cross section of atom excitation by electron impact for the analysis of the character of atom excitation in an ionized gas. In the other limiting case, when the electron energy exceeds significantly the excitation energy, the cross section is given by the Born approximation [2, 75–77]. The maximum cross sections corresponds to excitation of resonantly excited states when the excitation cross section is given by the Bethe formula [2, 75, 76, 78, 79]:

$$\sigma_{0*} = \frac{4\pi}{\Delta\varepsilon a_0}|(D_x)_{0*}|^2 \Phi\left(\frac{\varepsilon}{\Delta\varepsilon}\right) = \frac{2\pi e^4}{\Delta\varepsilon^2} f_{0*} \Phi\left(\frac{\varepsilon}{\Delta\varepsilon}\right) ,$$
$$\Phi(x) \to \frac{\ln C\sqrt{x}}{x} , \quad x \to \infty . \tag{2.48}$$

Here indices 0 and $*$ refer to the initial and final states of the excitation process, ε is the electron energy, $\Delta\varepsilon$ is the excitation energy, e is the electron charge, a_0 is the Bohr radius, $(D_x)_{0*}$ is the matrix element for the atom dipole moment projection between the transition states, f_{0*} is the oscillator strength for this transition, and C is a constant. Thus, in the case of excitation of resonantly excited states, the excitation cross section is expressed through parameters of radiative transitions for the atom. This analogy follows from an analogy between the interaction operator of a fast charged particle in an atom and the interaction operator of an electromagnetic wave and an atom [76, 79–82].

One can continue this formula on the basis of experimental data to electron energies ε which are comparable with the atom excitation energy $\Delta\varepsilon$ [83]. Then accounting for the threshold dependence of the excitation cross section [2, 84], we find for the cross section near the threshold for excitation of resonantly excited states [83, 85, 86]

$$\sigma_{0*}(\varepsilon) = \frac{2\pi e^4 f_{0*}}{\Delta\varepsilon^{5/2}} a\sqrt{\varepsilon - \Delta\varepsilon} , \tag{2.49}$$

and the numerical coefficient from experimental data is [83, 85]

$$a = 0.130 \pm 0.007 . \tag{2.50}$$

Formulas (2.48)–(2.50) can be used for determination of the excitation rate constants for atom excitation by electron impact in a gas discharge plasma [86, 87].

We determine below the cross section and rate constant for quenching of a resonantly excited atom by a slow electron. Atom quenching is an inverse process with respect to atom excitation by electron impact, and we find below the connection between the parameters of these processes, which proceed according to the scheme

$$e + A_0 \longleftrightarrow e + A_* \,, \tag{2.51}$$

where A_0 and A_* denote an atom in the ground and resonantly excited states. The connection between the parameters of direct and inverse processes is established on the basis of the principle of detailed balance for processes (2.51). Let us take one electron and one atom in a volume Ω, the atom can be found in states 0 and $*$, and transitions between these states result from collisions with the electron. Because this system is under equilibrium, there is a certain connection between the rate w_{0*} for transition $0 \to *$ and the rate w_{*0} for transition $* \to 0$. Introducing the interaction operator V which is responsible for these transitions, we have within the framework of the perturbation theory for the transition rates

$$w_{0*} = \frac{2\pi}{\hbar}|V_{0*}|^2 \frac{dg_*}{d\varepsilon}\,, \quad w_{*0} = \frac{2\pi}{\hbar}|V_{*0}|^2 \frac{dg_0}{d\varepsilon}\,.$$

Here $dg_0/d\varepsilon$ and $dg_*/d\varepsilon$ are the statistical weights per unit energy for the corresponding channels of the process. We use the definition of the cross sections of these processes

$$\sigma_{0*} = \frac{w_{0*}}{N v_0} = \Omega \frac{w_{0*}}{v_0}\,, \quad \sigma_{0*} = \frac{w_{*0}}{N v_*} = \Omega \frac{w_{*0}}{v_*}\,,$$

where $N = 1/\Omega$ is the number density of particles and v_0 and v_* are the electron velocities for the corresponding channels (for simplicity, we consider an atom to be motionless). For the matrix elements of the interaction operator, the time reversal operation gives $V_{0*} = V_{*0}^*$. This leads to the following relation between the cross sections of direct and inverse processes in electron–atom collisions [88, 89]:

$$\sigma_{0*} v_0 \frac{dg_0}{d\varepsilon} = \sigma_{*0} v_* \frac{dg_*}{d\varepsilon}\,. \tag{2.52}$$

The statistical weights of the corresponding channels of the processes (2.51) are

$$dg_0 = \Omega \frac{d\mathbf{p}_0}{(2\pi\hbar)^3} g_0\,, \quad dg_* = \Omega \frac{d\mathbf{p}_*}{(2\pi\hbar)^3} g_*\,,$$

where g_0 and g_* are the statistical weights for given atomic states. Then, finally formula (2.52) takes the form [88, 89]

$$\sigma_{\text{ex}} = \sigma_{\text{q}} \frac{v_*^2 g_*}{v_0^2 g_0}\,, \tag{2.53}$$

where $\sigma_{\text{ex}} = \sigma_{0*}$ is the excitation cross section and $\sigma_{\text{q}} = \sigma_{*0}$ is the quenching cross section. Near the threshold [84]

$$\sigma_{\text{ex}} = A\sqrt{\varepsilon - \Delta\varepsilon}\,, \quad \varepsilon - \Delta\varepsilon \ll \Delta\varepsilon\,, \tag{2.54}$$

where A is a constant, we have for the cross section of atom quenching resulting from collision with a slow electron of energy $\varepsilon = E - \Delta\varepsilon \ll \Delta\varepsilon$

$$\sigma_q = A \frac{g_0 \Delta\varepsilon}{g_* \sqrt{E - \Delta\varepsilon}} . \quad (2.55)$$

From this it follows that the rate constant for atom quenching by a slow electron (m_e is the electron mass)

$$k_q = v_f \sigma_q = A \frac{g_0 \Delta\varepsilon \sqrt{2}}{g_* \sqrt{m_e}} , \quad (2.56)$$

and the rate constant k_{ex} for atom excitation by electron impact is

$$k_{ex} = k_q \frac{g_*}{g_0} \sqrt{\frac{\varepsilon - \Delta\varepsilon}{\Delta\varepsilon}} . \quad (2.57)$$

It is of importance that the quenching rate constant k_q does not depend both on the electron energy and on the energy distribution function for slow electrons. It depends only on the parameters of the transition between atomic states, so the quenching rate constant is a convenient parameter characterizing also excitation of atoms by electron impact near the threshold. In particular, in the case of quenching of a resonantly excited state when this process is effective, the quenching rate constant within the framework of the perturbation theory using formula (2.49) gives approximately [92]

$$k_q = \text{const} \frac{g_0 f_{0*}}{g_* (\Delta\varepsilon)^{3/2}} = \frac{k_0}{(\Delta\varepsilon)^{7/2} \tau_{*0}} , \quad (2.58)$$

where f_{0*} is the oscillator strength for this transition, τ_{*0} is the radiative lifetime of the resonantly excited state, and λ is the wavelength of the emitted photon; const

Table 2.7 Parameter k_0 in formula (2.58) obtained from data in the references cited. This parameter is expressed in units of 10^{-5} cm^3/s if $\Delta\varepsilon$ is in electronvolts and τ_{*0} is in nanoseconds.

T_e, 10^3 K	6	8	10	12
$K(4^2P)$ [90]	–	4.1	4.2	3.9
$Rb(5^2P)$ [90]	5.5	5.7	4.4	3.8
$Cs(6^2P)$ [90]	3.4	3.1	2.8	3.4
$K(4^2P)$ [79]	5.2	4.8	5.0	5.4
$Rb(5^2P)$ [79]	4.6	4.8	5.0	5.0
$Cs(6^2P)$ [79]	4.3	4.4	4.5	4.8
$K(4^2P)$ [91]	3.4	3.7	4.1	4.2
$Rb(5^2P)$ [91]	3.7	3.7	4.0	4.0
$Cs(6^2P)$ [91]	3.9	4.3	4.6	4.9

and k_0 are numerical coefficients, and if the atom excitation energy $\Delta\varepsilon$ is expressed in electronvolts and the radiative lifetime is given in nanoseconds, the numerical coefficient is [8] $k_0 = (4.4 \pm 0.7) \times 10^{-5}\,\text{cm}^3/\text{s}$, if we use the data for the excitation cross sections of alkali metal atoms. As follows from the data in Table 2.7, the accuracy of (2.58) for the quenching rate constants of resonantly excited states is about 20%, and Table 2.8 contains the quenching rate constants of some resonantly excited atoms quenched by electron impact. The rate constant for quenching of atom metastable states by electron impact is lower than that for resonantly excited states because of a weaker coupling between these states during interaction with an electron. This is demonstrated by comparison of the data in Table 2.8 with those in Table 2.9, where these rate constants are given for metastable inert gas atoms.

Table 2.8 Parameters of resonantly excited atomic states of some atoms and the rate constant for quenching of these states by slow electron impact.

Atom, transition	$\Delta\varepsilon$, eV	λ, nm	f	τ_{*0}, ns	k_q, 10^{-8} cm^3/s
$H(2^1P \to 1^1S)$	10.20	121.6	0.416	1.60	0.79
$He(2^1P \to 1^1S)$	21.22	58.43	0.276	0.555	0.18
$He(2^1P \to 2^1S)$	0.602	2058	0.376	500	51
$He(2^3P \to 2^3S)$	1.144	1083	0.539	98	27
$Li(2^2P \to 2^2S)$	1.848	670.8	0.74	27	19
$Na(3^2P \to 3^2S)$	2.104	589	0.955	16.3	20
$K(4^2P_{1/2} \to 4^2S_{1/2})$	1.610	766.9	0.35	26	31
$K(4^2P_{3/2} \to 4^2S_{1/2})$	1.616	766.5	0.70	25	32
$Rb(5^2P_{1/2} \to 5^2S_{1/2})$	1.560	794.8	0.32	28	32
$Rb(5^2P_{3/2} \to 5^2S_{1/2})$	1.589	780.0	0.67	26	33
$Cs(6^2P_{1/2} \to 6^2S_{1/2})$	1.386	894.4	0.39	30	46
$Cs(6^2P_{3/2} \to 6^2S_{1/2})$	1.455	852.1	0.81	27	43

Table 2.9 The rate constant for quenching of metastable states of rare gas atoms by slow electron impact. The data are taken from [93].

Atom, transition	$\Delta\varepsilon$, eV	k_q, 10^{-10} cm^3/s
$He(2^3S \to 1^1S)$	19.82	31
$Ne(2^3P_2 \to 2^1S)$	16.62	2.0
$Ar(3^3P_2 \to 3^2S)$	11.55	4.0
$Kr(4^3P_2 \to 4^2S_0)$	9.915	3.4
$Xe(5^2P_2 \to 5^2S_0)$	8.315	19

2.2.2
Atom Ionization by Electron Impact

Processes in which ions and free electrons are either introduced into a plasma or removed from it are fundamental to the establishment of plasma properties. Below we analyze basic processes of this type. We consider first the ionization of an atom by electron impact:

$$e + A \to 2e + A^+ \, . \tag{2.59}$$

In this process the incident electron interacts with a valence electron, transfers to it part of its kinetic energy, and causes the detachment of the valence electron from the initially neutral atom. We can analyze this process in terms of a simple model developed by J.J. Thomson [94], in which it is assumed that electron collisions can be described on the basis of classical laws, and that the electrons do not interact with the atomic core in the course of the collision. Despite the fact that the atom is a quantum system, this model gives a correct qualitative description of the process because the cross sections for elastic collisions governed by the Coulomb interaction are identical in the classical and quantum cases.

In determining the ionization cross section, we ignore the interaction of incident and bound electrons with the charged core during energy transfer between these electrons, and this is a model assumption. Next, we assume the bound electron to be motionless at the beginning and consider scattering into small angles. Then the differential cross section of collision with exchange of energy $\Delta\varepsilon$ is given by the Rutherford formula (2.35), and the atom ionization occurs if the energy transferred to the valence electron exceeds the atomic ionization potential J. This leads to the following expression for the ionization cross section [94]:

$$\sigma_{\text{ion}} = \int_J^{\varepsilon} d\sigma = \frac{\pi e^4}{\varepsilon}\left(\frac{1}{J} - \frac{1}{\varepsilon}\right) . \tag{2.60}$$

This expression for the ionization cross section is called the Thomson formula. Although this result is for an atom with one valence electron, it can be generalized for atoms with several valence electrons.

Since the process was treated classically, the ionization cross section depends on classical parameters of the problem: m_e (the electron mass), e^2 (the interaction parameter), ε (the electron energy), and J (the ionization potential). The most general form of the cross section expressed through these parameters is

$$\sigma_{\text{ion}} = \frac{\pi e^4}{J^2} f\left(\frac{\varepsilon}{J}\right) , \tag{2.61}$$

where $f(x)$ is, within the framework of the approach we employ, a universal function which is identical for all atoms. In particular, for the Thomson model this function is

$$f(x) = \frac{1}{x} - \frac{1}{x^2} \, .$$

Figure 2.11 gives the values of the function $f(x)$ which follow from measuring cross sections of atom ionization by electron impact if an atom contains valence s electrons. The Thomson model gives a value that differs from the values obtained from measurements. In particular, at $x = 1.6$ the value of $f(x)$ according to the Thomson model is twice the measured value, and at $x = 100$ the value of $f(x)$ according to the Thomson model is half the measured value. In addition, the Thomson model gives the maximum $f(x)$ at $x = 2$, whereas according to measurements this maximum is observed at $x = 4$–5. Nevertheless, the Thomson model gives the correct behavior of the ionization cross section and is useful for certain estimations because of its simplicity. In particular, according to the Thomson model the ionization cross section is linear with respect to the energy excess $\varepsilon - J$ near the threshold and this differs slightly from the precise energy dependence [48], although practically the difference between the correct and model energy dependencies for the ionization cross section is not of importance.

There is another deficiency of the Thomson formula (2.60). In the course of its derivation we assume a bound electron to be motionless. If we do not make this assumption, then in the limit of high energies of an incident electron we obtain

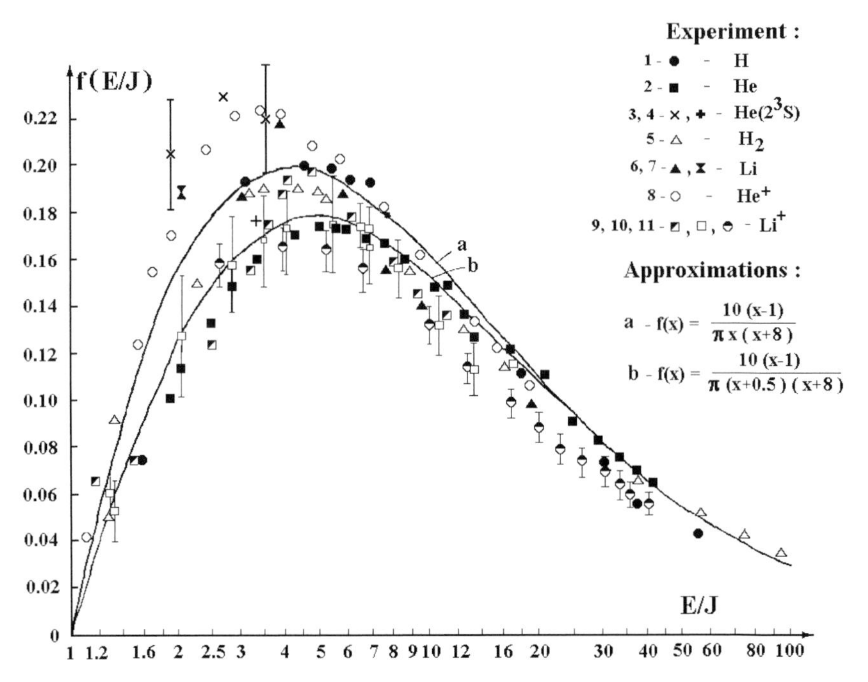

Figure 2.11 The reduced cross sections of ionization of atoms and ions by electron impact with valence s electrons versus the reduced electron energy. Experiment: 1 – [95]; 2 – [96]; 3 – [97]; 4 – [98]; 5 – [99]; 6 – [100]; 7 – [100]; 8 – [101]; 9 – [102]; 10 – [103]; 11 – [104]; 12 – [105]. Approximation $f(x) = 10(x-1)/[\pi x(x+8)]$ is given by the solid curve and approximation $f(x) = 10(x-1)/[\pi(x+0.5)(x+8)]$ is represented by the dashed curve.

instead of the Thomson formula (2.60) [8, 106, 107]

$$\sigma_{ion} = \frac{\pi e^4}{J\varepsilon}\left(1 + \frac{m_e \overline{v_e^2}}{3J}\right), \qquad (2.62)$$

where m_e is the electron mass, v_e is the velocity of a bound electron, and an overline means an average over velocities of a bound electron. In the limit $v_e = 0$ this formula is transformed into the Thomson formula (2.60). Since the average kinetic energy of a bound electron is compared with the atomic ionization potential, taking into account the velocity distribution for a bound electron leads to a remarkable change of the ionization cross section compared with the Thomson formula (2.60). In particular, if a bound electron is located mostly in the Coulomb field of an atomic core, we have $m_e \overline{v_e^2}/2 = J$, and (2.62) differs from the Thomson formula (2.60) by a factor of $5/3$. Accounting for this effect allows one to describe correctly the behavior of the ionization cross section at large electron energies.

Note that according to (2.48) the ionization cross section at high electron energies ε depends on this energy as $\ln \varepsilon/\varepsilon$ [78] according to the Born approximation, whereas classical theories, including the Thomson model, give $\sigma_{ion} \sim 1/\varepsilon$. The logarithmic dependence in the quantum approach is determined [108] by large collision impact parameters, whereas the ionization probability is small or is zero for classical considerations. One can consider this as a principal contradiction between the quantum and classical approaches. This also occurs for ionization of excited atoms when the classical description of an excited electron holds true. This contradiction was overcome by numerical analysis by Kingston [109–111], who proved that although the quantum (Born approximation) and classical approaches are described by different formulas, the values of the ionization cross sections for these cases are similar for excited atoms at high collision energies. This means that the different energy dependence for the classical and quantum ionization cross sections is not of importance.

The advantage of the Thomson model consists in its simplicity and the correct qualitative description of the ionization process. We now use the Thomson model for atom excitation with an electron transition in highly excited states. If the group of excited states is characterized by the principal quantum number n and the excitation energy of these states is $\Delta\varepsilon_n$, we assume the excitation energy $\Delta\varepsilon$ lies in the range

$$J - \frac{J_0}{(n-1/2)^2} \le \Delta\varepsilon \le J - \frac{J_0}{(n+1/2)^2},$$

where J_0 is the ionization potential of the hydrogen atom and J is the ionization potential of a given atom. For the excitation cross section in a highly excited state this gives

$$\sigma_n = \frac{2\pi e^4 \cdot J_0}{\varepsilon \cdot \Delta\varepsilon_n^3 n^3}. \qquad (2.63)$$

Let us also use the Thomson model to determine the average energy $\overline{\Delta\varepsilon}$ of secondary electrons in collisions of a fast electron with an atom. On the basis of the

Rutherford formula (2.35) we obtain

$$\overline{\Delta\varepsilon} = \frac{\int \Delta\varepsilon \, d\sigma}{\int d\sigma} = J \ln\left(\frac{\varepsilon}{J}\right) \tag{2.64}$$

in the limit that the energy of an incident electron is large compared with the atomic ionization potential $\varepsilon \gg J$.

2.2.3
Three Body Recombination of Electrons and Ions

If collision of two free atomic particles results in the formation of a bound state of these particles, it is necessary to transfer excess energy to other degrees of freedom. For example, for photorecombination of electrons and positive ions with the formation of atoms, it is necessary to transfer the excess energy to a photon. If a third atomic particle takes away energy, the process proceeds according to the scheme

$$A + B + C \to AB + C \, . \tag{2.65}$$

In this three body process, particles A and B form a bound system, and particle C carries away the energy released thereby. The balance equation for the number densities of particles in the collision process (2.65) has the form

$$\frac{d[AB]}{dt} = K[A][B][C] \, , \tag{2.66}$$

where $[X]$ is the number density of particles X and K is the rate constant for the three body process, with dimensionality centimeters to the sixth power per second. Equation (2.66) may be considered as the definition of the three body rate constant.

We now estimate the rate constant K for the three body process (2.65) following the Thomson theory [112], which allows us to determine the dependence of the rate constant on the interaction parameters of colliding atomic particles. We make the assumptions of the Thomson theory [112] that the binding energy of the end-product molecule AB is much larger than thermal energies, and the motion of the particles is governed by classical laws. The formation of a bound state of particles A and B occurs in the following way. As particles A and B approach each other, their energy increases as the potential energy of interaction is converted into kinetic energy. If a third particle C interacts strongly with particle A or particle B when these particles are close to each other, then the third particle may take excess energy from the initial kinetic energy of particles A or B. The bound state of particles A and B is thus formed as a result of a collision with the third particle.

This physical picture can be used as a basis for estimation of the rate constant for a three body process. Typical kinetic energies of the particles are of the order of the thermal energy T. We assume the mass of the third particle C to be comparable to the mass of either particle A or particle B. Since the energy exchange must exceed the initial kinetic energy of particles A and B, the interaction potential between these particles during collision with particle C must also be of the order of T. Therefore, let us define a critical radius b by the relation

$$U(b) \sim T \,, \tag{2.67}$$

where $U(R)$ is the interaction potential for particles A and B, and these particles may form a bound state if they are located inside the critical region in the course of collision with a third particle.

Now we can estimate the rate constant for the three body process. The rate of conversion of particle B into molecule AB is the product of two factors. One measures the probability for particle B to be located in the critical region near particle A as $[A]b^3$, and the other factor is the rate $[C]v\sigma$ of collisions with particles C. Here v is a typical relative collision velocity and σ is the cross section for the collision between particle C and either particle A or particle B, resulting in an energy exchange of the order of T. If the masses of the colliding particles are similar, this cross section is comparable to the cross section for elastic collision. This estimate for the rate of formation of molecule AB is thus

$$\frac{d[AB]}{dt} \sim [A][B]b^3[C]v\sigma \,.$$

Comparison of this expression with the definition of the rate constant for the three body process (2.66) gives the following estimate for the rate constant:

$$K \sim v b^3 \sigma \,. \tag{2.68}$$

Process (2.65) is a three body one if the number density of third particles C is small, and therefore the probability of collision of particles A and B with particle C in the region of strong interaction of particles A and B is small. This condition holds true if the critical radius is small compared with the mean free path of particles A and B in a gas of particles C, that is,

$$[C]\sigma b \ll 1 \,. \tag{2.69}$$

An example of the three body process is three body recombination of electrons and ions that proceeds according to the scheme

$$2e + A^+ \rightarrow e + A^* \,, \tag{2.70}$$

and is of importance for a dense plasma. Note that because the atomic ionization potential exceeds significantly the thermal energy of electrons, the recombination process includes many stages. But the recombination coefficient according to the Thomson theory is determined by the first slow stage, which corresponds to electron capture in an excited level with the binding energy of the order of the thermal energy of electrons T_e. The Thomson theory is applicable to this process since classical laws are valid for a highly excited atom.

The rate constant for process (2.70) is estimated according to (2.68), where the cross section of elastic scattering of two electrons with thermal energy is $\sigma \sim e^4/T^2$

and the critical radius for the Coulomb interaction of an electron and an ion is $b \sim e^2/T$. As a result, we have for the rate constant for process (2.70) [113]

$$K = \frac{\alpha}{N_e} = C \frac{e^{10}}{m_e^{1/2} T_e^{9/2}} . \tag{2.71}$$

In this expression, α is the recombination coefficient, defined as the rate constant for the decay of charged particles in the pairwise process, and $C \sim 1$ is a numerical coefficient. Formula (2.71) is valid if criterion (2.69) holds true; this now has the form $N_e e^6/T^3 \ll 1$ and coincides with the condition of plasma ideality (1.3).

The rate constant for three body electron–ion recombination (2.70) may be obtained in a simpler way on the basis of dimensional analysis. Indeed, the rate constant for process (2.70) must depend only on the interaction parameter e^2, the electron mass m_e, and the thermal energy T_e (the electron temperature). There is only one combination of these parameters that has the dimensions centimeters to the sixth power per second for the rate constant for the three body process, and it can be seen that this combination coincides with that given in (2.71). The value of the numerical coefficient C in (2.71) is of interest. According to the nature of the process, this value does not depend on the type of atom, because the nature of the process is such that this coefficient is determined by properties of highly excited states where the Coulomb interaction takes place between the atomic core and the electron. Additional analysis gives $C = 2 \times 10^{\pm 0.3}$ for this value.

2.2.4 Autoionizing and Autodetaching States in Collision Processes

Autoionizing and autodetaching states are of importance as intermediate states in collision processes. An autoionizing state is a bound state of an atom or positive ion whose energy is above the boundary of the continuous spectrum, and hence an electron can be released in the decay of such a state. For example, the autoionizing state $He(2s^2, {}^1S)$ is the state of the helium atom where both electrons are located in the excited 2s state, and as a result of such a decay one electron makes a transition to the ground state, and the other electron ionizes. The scheme of this process is

$$He(2s^2, {}^1S) \rightarrow He^+(1s, {}^2S) + e + 57.9\,eV . \tag{2.72}$$

An autodetaching state is identical to an autoionizing state, but occurs in a negative ion. The decay of such a state proceeds with the formation of a free electron and an atom or a molecule. An example of an autodetaching transition is

$$H^-(2s^2, {}^1S) \rightarrow H(1s, {}^2S) + e + 9.56\,eV . \tag{2.73}$$

Formation of autoionizing and autodetaching states determines the character of some collision processes involving electrons. These states give resonances in cross sections as a function of the electron energy in both elastic and inelastic scattering of electrons by atoms and molecules. They can be present as intermediate states in such processes. For instance, the cross section for vibrational excitation of molecules by electron impact increases by two to three orders of magnitude if this process proceeds through excitation of autodetaching states instead of by direct excitation. There are some processes that can proceed only through excitation of autoionizing or autodetaching states, such as dissociative attachment of electrons to molecules and dissociative recombination of electrons with positive molecular ions. Then the cross section of the process as a function of the electron energy exhibits resonances corresponding to autoionizing or autodetaching states. As a demonstration of this, Figure 2.12 gives a schematic display of states which shows the process of dissociative attachment of an electron to a molecule. The electronic states related to dissociative recombination of electrons and molecular ions have the same form as those for dissociative attachment of electrons to molecules and are represented in Figure 2.13. In both cases the first stage in the process, the electron capture by a molecular ion or molecule, has the resonant character. As

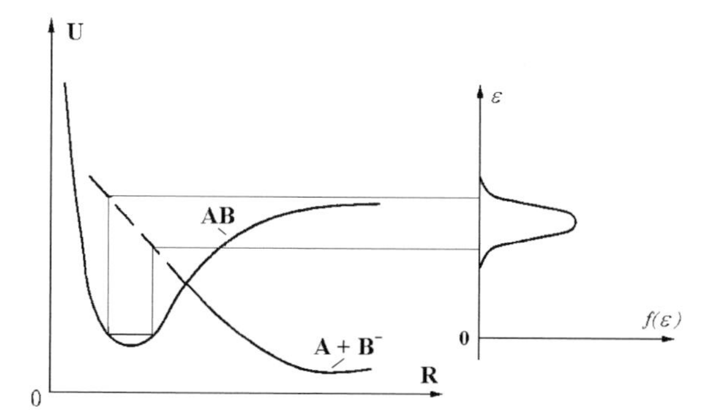

Figure 2.12 Electronic states of a molecule and its negative molecular ion which are responsible for electron attachment to a molecule. Variation of these states proceeds along a coordinate that determines this process (the distance between nuclei for a diatomic molecule). The probability of electron capture depending on the electron energy is indicated at the right.

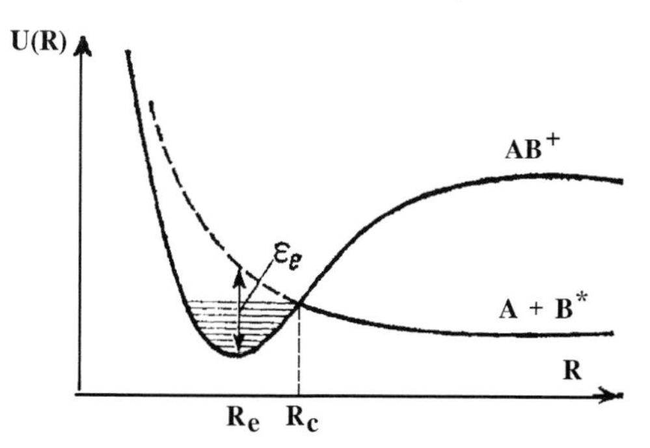

Figure 2.13 Electronic states of a molecule and its positive ion which determine the process of dissociative recombination; ϵ_e is the energy of the captured electron when the distance between nuclei is equal to the equilibrium distance R_e. The autoionizing state exists until the distance between nuclei does not exceed R_c.

a demonstration of this, Figure 2.14 gives the energy dependence for the cross section of dissociative attachment of an electron to a CO_2 molecule [114] and Figure 2.15 contains the lower electronic states of the CO_2 molecule and its negative molecular ion when the nuclei lie in one line. These electronic states are responsible for dissociative electron attachment to the CO_2 molecule [115].

Some collision processes of slow atomic particles are determined by transitions of the quasimolecule consisting of colliding atomic particles in autoionizing or autodetaching states. In this case an electronic state intersects the boundary of continuous spectra at some distance between the colliding particles. In the region of smaller distances, this state becomes the autoionizing or autodetaching state, and the system of colliding particles can decay with the release of an electron. In particular, the process of associative ionization, that is, the inverse process with respect to dissociative recombination, proceeds according to the scheme

$$A^* + B \rightarrow (AB)^{**} \rightarrow AB^+ + e\,, \tag{2.74}$$

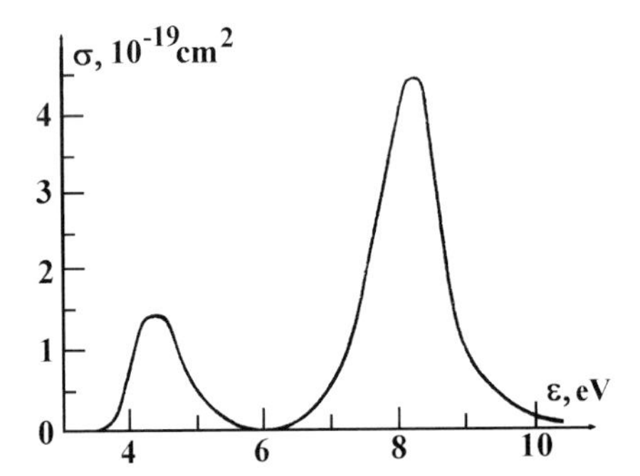

Figure 2.14 The cross section of dissociative attachment of an electron to a CO_2 molecule ($e + CO_2 \rightarrow O^- + CO$) as a function of the electron energy [114]. The maxima of the cross section at electron energies of 4.4 and 8.2 eV correspond to positions of the autodetachment levels of the negative ion CO_2^-.

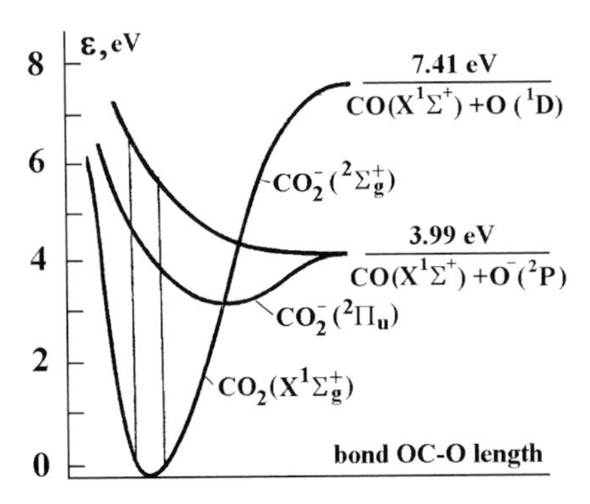

Figure 2.15 The electronic states of the CO_2 molecule and its negative ion which are responsible for electron attachment to this molecule [115].

where A^* is an excited atomic state and two stars denote an autodetaching state. The electronic states for this process are given in Figure 2.16, where the autoionizing state of a molecule corresponds to repulsion of colliding particles and has a form different from that in Figure 2.13, and the final state of the process relates to the electron release. At distances between nuclei $R < R_c$, the relevant state of AB^* is an autoionizing state, and the decay of this state leads to electron release. The other process of this group is destruction of a negative ion in collision with atoms

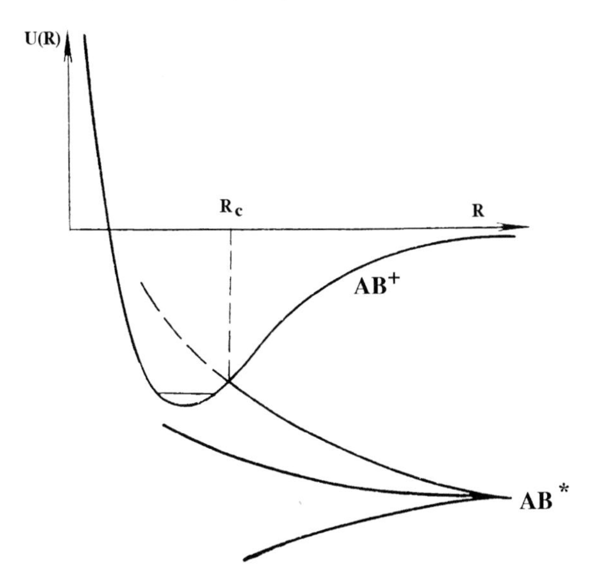

Figure 2.16 The character of molecular electronic states which determine the processes of dissociative recombination and associative ionization.

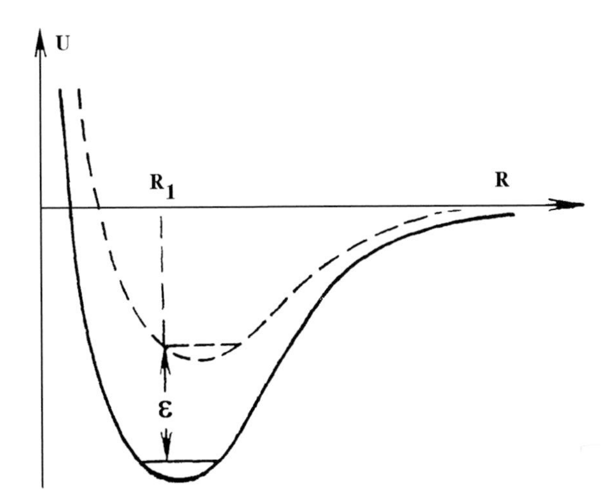

Figure 2.17 The character of molecular electronic states which determine the process of excitation of vibrational levels by electron impact through an autodetachment state.

or molecules in the associative detachment

$$A^- + B \to (AB)^{-**} \to AB^+ e \ . \qquad (2.75)$$

In the above processes, which proceed through formation of an autoionizing or autodetaching state, the initial and final states differ by bound and free electron states. But electron capture in an autodetaching state is possible, as is shown in Figure 2.17, where the system returns to the initial electron state and an electron remains free. This process may be responsible for excitation of vibrational molecular levels and has the resonant character, as other processes with formation of autoionizing or autodetaching states. In particular, this process is responsible for vibrational excitation of nitrogen molecules in a gas discharge plasma.

2.2.5
Dissociative Recombination

Dissociative recombination is a process in which a positive molecular ion is neutralized by recombination with a free electron, as a consequence of which the molecule dissociates into two parts. The scheme of this process is

$$e + AB^+ \to A^* + B \ . \qquad (2.76)$$

As an effective pairwise process, dissociative recombination is of importance for plasma properties. Figure 2.12 demonstrates the mechanism of dissociative recombination that proceeds through formation of autoionizing states of the molecule AB [116–118]. The excited state of molecule AB is an autoionizing state if the distance between the atoms is smaller than the distance of intersection R_c of this state with the boundary of the continuous spectrum. In the course of dissociative recombination, the electron colliding with the molecular ion is captured in a repulsive autoionizing state. The atoms move apart, and if the autoionizing state has not led to decay until the atoms reach the intersection distance R_c, the dissociative recombination process occurs. Dissociative recombination is a fairly complicated process. Note that there are many autoionizing states that can participate in this process, and vibrational excitation of the ions has a strong influence on the value of the recombination coefficient. Table 2.10 contains the values of the dissociative recombination coefficients at room temperature for molecular ions in the ground electronic and vibrational states. A rough dependence of this recombination coefficient α_{dis} on the electron temperature T_e at constant gas temperature has the form

$$\alpha_{\text{dis}} \sim T_e^{-0.5} \ . \qquad (2.77)$$

If complex or cluster ions participate in the dissociative recombination process, there is a strong interaction between the electron and the complex ion in a spatial region which determines the rate constant for this process. Then one can use a

Table 2.10 The rate constant for dissociative recombination α given at room temperature and expressed in units of 10^{-7} cm^3/s [11, 86].

Ion	α	Ion	α	Ion	α
Ne_2^+	1.8	H_3^+	2.3	NH_4^+	16
Ar_2^+	6.9	CO_2^+	3.6	CH_5^+	11
Kr_2^+	10	$H^+ \cdot CO$	1.1	$NH_4^+ \cdot NH_3$	28
Xe_2^+	20	N_4^+	20	$NH_4^+ \cdot (NH_3)_2$	27
N_2^+	2.2	O_4^+	14	$H_3O^+ \cdot H_2O$	24
O_2^+	2.0	H_3O^+	17	$H_3O^+ \cdot (H_2O)_2$	34
NO^+	3.5	$CO^+ \cdot CO$	14	$H_3O^+ \cdot (H_2O)_3$	44
CO^+	6.8	$CO^+ \cdot (CO)_2$	19	$H_3O^+ \cdot (H_2O)_4$	50

simple model [52] for the analysis of this process with introduction of the ion radius R_0, so penetration of an electron inside a sphere of a radius R_0 leads to the recombination process (2.76) because of the strong interaction between the electron and ion in this region. On the basis of (2.6), we have that the cross section of dissociative recombination or electron penetration in the region with distances r from the ion $r < R_0$ with positive charge e is [119]

$$\sigma_c = \pi R_0 \left(R_0 + \frac{e^2}{\varepsilon} \right) .$$

In the limit of low electron energies ε we obtain from this the following expression for the rate constant for this process averaged over the Maxwell energy distribution for electrons:

$$\alpha = \left\langle \sqrt{\frac{2\varepsilon}{m_e}} \sigma_c \right\rangle = \frac{2\sqrt{2\pi} R_0 e^2}{\sqrt{m_e T_e}} . \tag{2.78}$$

R_0 is several atomic units (a_0) for complex ions and is equal to the ion radius for cluster ions. Therefore, at room temperature the coefficient for dissociative recombination of an electron and a complex ion according to (2.77) is of the order of 10^{-6} cm^3/s. It is sufficiently large value if we compare it with the atomic rate constant ($k_0 = a_0^2 e^2/\hbar = 6.13 \times 10^{-9}$ cm^3/s) or with the coefficient for dissociative recombination for simple molecules (Table 2.10). Note that the above model, which describes the case of a strong interaction of the colliding electron and ion, gives an upper limit for the value of the dissociative recombination coefficient involving complex ions, and the radius of the sphere of strong electron–ion interaction R_0 is usually equal to several Bohr radii. Note that for large cluster ions, R_0 in (2.78) is the cluster radius.

2.2.6
Dielectronic Recombination

Dielectronic recombination of an electron and an ion proceeds through electron capture into an autoionizing state of the combined system, and the subsequent decay of the autoionizing state results from its radiative transition to a stable state. This process is of importance for recombination of electrons and multicharged ions because the radiative lifetime of the multicharged ion decreases strongly (proportional to Z^{-4}) with increase of its charge Z. The process of dielectronic recombination consists of the following stages:

$$e + A^{+Z} \to [A^{+(Z-1)}]^{**} , \quad [A^{+(Z-1)}]^{**} \to A^{+Z} + e ,$$
$$[A^{+(Z-1)}]^{**} \to A^{+(Z-1)} + \hbar\omega . \tag{2.79}$$

Denoting the rate constant for the first process as k, the radiative lifetime of the autoionizing state as τ, the width of the autoionizing level as Γ, and the energy of excitation of the autoionizing state $[A^{+(Z-1)}]^{**}$ above the ground state as E_a, we obtain the expression for the recombination coefficient for an electron and a multicharged ion for processes (2.79).

The balance equation for the number density N_{ai} of ions in a given autoionizing state is

$$\frac{dN_{ai}}{dt} = N_e N_Z k - N_{ai}\frac{\Gamma}{\hbar} - N_{ai}\frac{1}{\tau} ,$$

where N_e is the electron number density and N_Z is the number density of ions of charge Z. From this we find the number density of ions in the autoionizing state N_{ai} and the rate of recombination $J = N_{ai}/\tau$ to be

$$N_{ai} = \frac{N_e N_Z k}{\Gamma/\hbar + 1/\tau} , \quad J = \frac{N_{ai}}{\tau} = \frac{N_e N_Z k}{\Gamma\tau/\hbar + 1} .$$

The recombination coefficient α for processes (2.79) according to definition is given by

$$\alpha = \frac{J}{N_e N_Z} = \frac{k}{\Gamma\tau/\hbar + 1} .$$

Assuming there is thermodynamic equilibrium between a given autoionizing state and other ion states, we have for the number density of atoms in an autoionizing state according to the Saha formula (1.52)

$$\frac{N_Z N_e}{N_{ai}} = \frac{g_e g_Z}{g_{ai}} \left(\frac{m_e T_e}{2\pi\hbar^2}\right)^{3/2} \exp\left(\frac{E_a}{T_e}\right) ,$$

where g_e, g_Z, and g_{ai} are statistical weights for the participating atomic states, and T_e is the electron temperature. Thermodynamic equilibrium corresponds to $\tau \to \infty$ and for the rate constant k for electron capture in this autoionizing state gives

$$k = \frac{g_{ai}}{g_e g_Z} \left(\frac{2\pi\hbar^2}{m_e T_e}\right)^{3/2} \frac{\Gamma}{\hbar} \exp\left(-\frac{E_a}{T_e}\right) .$$

From this we obtain for the recombination coefficient for electrons and multi-charged ions for processes (2.79)

$$\alpha = \frac{g_{ai}}{g_e g_Z} \left(\frac{2\pi \hbar^2}{m_e T_e} \right)^{3/2} \frac{\Gamma/\hbar}{\Gamma\tau/\hbar + 1} \exp\left(-\frac{E_a}{T_e} \right) . \tag{2.80}$$

One can summarize the qualitative features of electron capture in an autoionizing state. The process corresponds to excitation of valence or internal electrons of the ion, accompanied by the capture of an incident electron in a bound state. The energy of the incident electron coincides with the excitation energy of the autoionizing state E_a. The above expression for the capture rate constant k is obtained from imposition of the condition of thermodynamic equilibrium among the atomic particles participating in the process. This requires the Maxwell distribution function for kinetic energies of the electrons, which leads to the following criterion for the validity of formula (2.80):

$$N_Z k \ll N_e k_{ee} , \tag{2.81}$$

where k_{ee} is the rate constant for the elastic electron–electron collisions that establish the Maxwell distribution function of electrons. If we have some other energy distribution function $f(\varepsilon)$ for electrons of energy ε, normalized by the relation $\int f(\varepsilon) d\mathbf{v} = 1$, generalization of (2.80) gives

$$\alpha = \frac{g_{ai}}{g_e g_Z} \left(\frac{2\pi \hbar}{m_e} \right)^3 f(E_a) \frac{\Gamma/\hbar}{\Gamma\tau/\hbar + 1} . \tag{2.82}$$

This expression also requires that criterion (2.81) be satisfied. This means that the electron energy distribution function is a Maxwell one in its dominant part. This makes it possible to introduce the electron temperature; but the tail of the distribution function, responsible for excitation of the autoionizing state, may be distorted.

2.2.7 Attachment of Electrons to Molecules

This process results in capture of an electron by a molecule in an autodetachment level and subsequent decay of a forming negative ion in an autodetachment state, usually by dissociation of this negative ion [120, 121]. Since this autodetachment state is characterized by a long lifetime, transfer of excess electron energy to atoms or molecules of a gas in collisions with them is impossible if the process of electron attachment occurs in a gas. In any case, the process of electron capture has a resonant character and results in transition between two electronic states. An example of the electron attachment process related to the CO_2 molecule is given in Figure 2.14 [114], and the corresponding electronic states in Figure 2.15 [115] lead to two resonances in this process.

Another example of this process is electron attachment to the oxygen molecule in the ground electronic state $O_2(^3\Sigma_g)$ that proceeds through formation of the autodetaching state $O_2^-(^2\Pi_u)$. Figure 2.18 shows the dependence on the electron energy

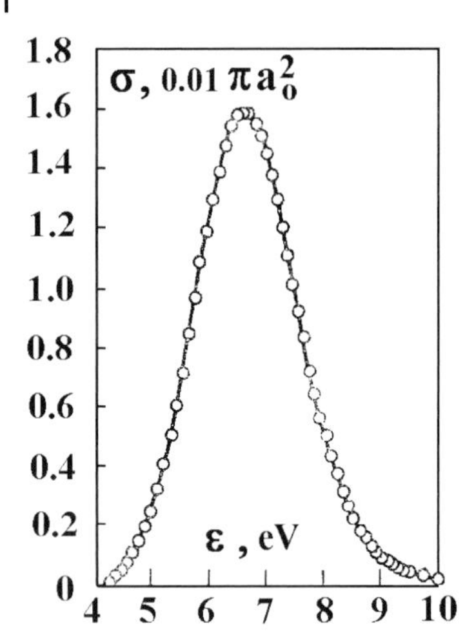

Figure 2.18 The cross section of electron attachment to the oxygen molecule as a function of the electron energy ϵ at room temperature [114].

for the cross section of electron attachment to the oxygen molecule [114, 121]. Note that the autodetaching state $O_2^-(^2\Pi_u)$ follows from the cross section dependence on the electron energy [122] for this process. One can see the resonant character of the electron attachment process and the parameters of this resonance are determined by the relative positions of the electronic states for the molecular and autodetaching states and by the coupling between these states. For vibrationally excited molecular states this resonance is broadened and is shifted, as is seen from Figure 2.19, where the left part of the resonance for electron attachment to the oxygen molecule is given at different temperatures [123]. As is seen, in this case the resonance increases and shifts to lower electron energies with increasing temperature. Note that the vibrational excitation energy for the oxygen molecule is 0.196 eV, so at a temperature of 1930 K approximately 69% of oxygen molecules are found in the ground vibrational state, and 21% of molecules are in the first excited vibrational state. Hence, from Figure 2.19 it follows that the resonance energy decreases with increasing vibrational state of the oxygen molecule, and the maximum cross section increases also.

Let us give one more example for the dependence of the maximum cross section in accordance with measurements [124] for electron attachment to HCl and DCl molecules in the temperature range 300–1200 K. If we denote the ratio of electron attachment cross sections for different vibrational states as $\eta_i = \sigma(v = i)/\sigma(v = 0)$, for the HCl molecule we obtain from these measurements that $\eta_1 = 38$ and $\eta_2 = 880$, whereas for the DCl molecule we have $\eta_1 = 32$ and $\eta_2 = 580$.

In reality, a resonance dependence of the cross section of electron attachment on the electron energy is typical for diatomic molecules with a small number of autodetaching states of a forming negative ion. For complex molecules with a large number of autodetaching states the dependence of the cross section on the electron energy becomes smooth, and the electron attachment process becomes possible at

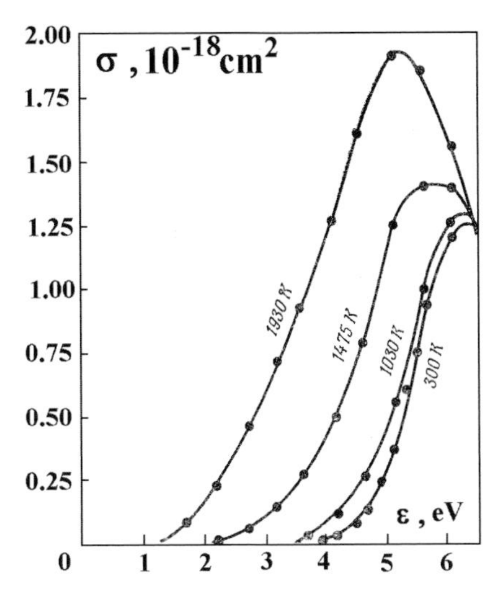

Figure 2.19 The cross section of electron attachment to the oxygen molecule as a function of the electron energy ϵ at different gas temperatures [123].

zero electron energy. We give in Table 2.11 the rate constants for electron attachment to halomethane molecules at room temperature. As is seen, the rate constant for this process is nonzero at low electron energies, although the position of the autodetachment level for electron capture depends on the type of molecule. The rate constants for electron attachment differ by an order of magnitude for different halomethane molecules which have identical structure.

The electron attachment process is of importance for electric breakdown of gases or for gas discharge. A small concentration of electronegative molecules may give an effective loss of electrons in a gas discharge plasma as a result of electron attach-

Table 2.11 The rate constants for electron attachment to halomethane molecules at room temperature [125, 126].

Molecule	Rate constant, cm^3/s	Molecule	Rate constant, cm^3/s
CH_3Cl	$< 2 \times 10^{-15}$	CF_2Cl_2	1.6×10^{-9}
CH_2Cl_2	5×10^{-12}	$CFCl_3$	1.9×10^{-7}
$CHCl_3$	4×10^{-9}	CF_3Br	1.4×10^{-8}
$CHFCl_2$	4×10^{-12}	CF_2Br_2	2.6×10^{-7}
CH_3Br	7×10^{-12}	$CFBr_3$	5×10^{-9}
CH_2Br_2	6×10^{-8}	CF_3I	1.9×10^{-7}
CH_3I	1×10^{-7}	CCl_4	3.5×10^{-7}
CF_4	$< 1 \times 10^{-16}$	CCl_3Br	6×10^{-8}
CF_3Cl	1×10^{-13}		

ment to molecules. This can prevent a gas from undergoing electric breakdown or can change the conditions of a stable gas discharge.

2.3
Elementary Processes Involving Ions and Atoms

2.3.1
Slow Inelastic Collisions of Heavy Atomic Particles

Collisions of heavy atomic particles, ions, atoms, and molecules, proceed in a plasma with small relative velocities compared with typical electron velocities. For example, the average thermal velocity of the helium ion at room temperature is $v_T = 1.2 \times 10^5\,\text{cm/s} = 5.7 \times 10^{-4} e^2/\hbar$, where $v_e = e^2/\hbar$ is the atomic velocity. Hence, we deal with slow collisions of heavy atomic particles in a plasma, where the velocity of the relative motion of colliding atomic particles is small compared with a typical atomic velocity v_e. This leads to a specific distribution of bound electrons in the course of collision when electrons respond to internal fields of the particles, and the electron distribution differs slightly from the distribution when the particles are motionless. Then one can analyze particle collisions within a framework of the electron terms of the quasimolecule consisting of colliding atomic particles with fixed nuclei, and the electron terms are the electron energies of the quasimolecule, where the distance between colliding particles is a parameter. In particular, in considering the resonant charge exchange process we compose this process as a result of interference between even and odd states of the quasimolecule.

Transition between two electron states is characterized by the ratio of the difference of their energies $\Delta\varepsilon$ to the quantum indefiniteness in energy $\hbar v/a$, where v is the collision velocity and a is a typical distance between nuclei associated with a significant change of the corresponding states. The above criterion is called the Massey parameter ξ [127]:

$$\xi = \frac{\Delta\varepsilon a}{\hbar v} . \tag{2.83}$$

If the Massey parameter is large, the probability of the corresponding transition is adiabatically small, that is, it behaves as $\exp(-a\xi)$, where $a \sim 1$ is a numerical coefficient. Therefore, transitions between electron states in slow atomic collisions can be a result of intersections or pseudointersections of corresponding electron states.

Consider as an example the charge exchange process

$$A^+ + B^- \to A^* + B . \tag{2.84}$$

In this case the Coulomb interaction takes place in the initial channel of the process, with only a weak interaction of neutral particles in the final channel. The intersection of electron states takes place at the distance between nuclei

$$R_c = \frac{e^2}{J - \text{EA}}, \tag{2.85}$$

where EA is the electron binding energy in negative ion B (the electron affinity of atom B) and J is the ionization potential of atom A. The transition transferring the electron from the field of one atomic particle to the other takes place near the intersection distance R_c.

Resonant collision processes are characterized by $\Delta\varepsilon = 0$ at infinite distances between colliding particles, and a small Massey parameter corresponds to these processes at finite distances between particles. Hence, resonant processes proceed effectively. One can consider a resonant process as an interference of the states between which the transition proceeds. Then the transition probability is approximately 1 if the phase shift is $\int \Delta\varepsilon(R)dt/\hbar \sim 1$. Thus, the cross section of a resonant process is of the order of R_0^2, where R_0 is the collision impact parameter for which the phase shift is of the order of unity, that is [44],

$$\sigma \sim \pi R_0^2 , \quad \text{where} \quad \int \frac{\Delta\varepsilon(R)dt}{\hbar} \sim 1 . \tag{2.86}$$

Let us consider as an example the excitation transfer process

$$A^* + A \to A + A^* .$$

This process can cause broadening of spectral lines. The interaction potential of the atoms in the ground and excited states is $U \sim \frac{D^2}{R^3}$, where D is the matrix element of the dipole moment operator between these states and R is the distance between the atoms. From (2.86), the cross section of this process ($\sigma \sim R_0^2$, where $(D^2/R_0^3)(R_0/v) \sim \hbar$) is

$$\sigma = \frac{C D^2}{\hbar v}, \tag{2.87}$$

where $C \sim 1$. Because $D \sim ea_0$, where $a_0 = \hbar^2/(m_e e^2)$ is the Bohr radius, and $v \ll e^2/\hbar$, from this it follows that the cross section of resonant excitation transfer is much larger than a typical atomic cross section, $\sigma \gg a_0^2$.

In the case of a nonresonant excitation transfer process

$$A^* + B \to A + B^*$$

the process is effective if the excitation energies for atoms A and B are nearby in accordance with the Massey criterion (2.83). As an example, we represent in Figure 2.20 the rate constant for the process of excitation transfer from metastable helium atoms to neon atoms in 2s states (for Pashen notations of levels) [128]. This process is the basis of operation for the He-Ne laser using a He-Ne mixture. In a gas discharge plasma, metastable helium levels are excited effectively, and transfer of their excitation to neon atoms creates an inverse population of levels that is necessary for laser operation.

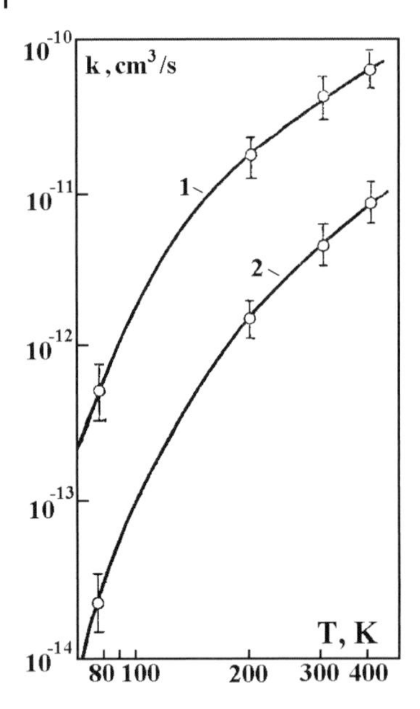

Figure 2.20 The rate constant for excitation transfer from a metastable helium atom to a neon atom depending on the neon temperature. 1 – $He(2^1S) + Ne \to He + Ne(2s)$, 2 – $He(2^3S) + Ne \to He + Ne(2s)$ [128].

2.3.2
Three Body Collision Processes Involving Ions

We now consider three body processes involving heavy atomic particles. In a dense plasma, the three body process of conversion of atomic ions into molecular ions is of importance. This process proceeds according to the scheme

$$A^+ + 2A \to A_2^+ + A \tag{2.88}$$

and determines the subsequent ion evolution in a plasma because the rate of dissociative recombination of electrons and molecular ions exceeds significantly the rate of recombination of electrons and atomic ions in the three body process. Hence, process (2.88) can lead to fast plasma decay. Process (2.88) occurs mostly because of the polarization interaction between the ion and the atom, so the interaction potential has the form $U(R) = -\alpha e^2/(2R^4)$, where α is the atom polarizability. The rate constant for this process can be constructed by dimensional analysis on the basis of the following parameters: the interaction parameter αe^2, the mass of the atom M, and the temperature or a typical thermal energy of the particles T. The only combination with the dimensionality of the rate constant for the three body process that can be constructed from these parameters is

$$K = \text{const}\frac{(\alpha e^2)^{5/4}}{M^{1/2}T^{3/4}}, \tag{2.89}$$

where the numerical factor const is of the order of 1. The same dependence of the rate constant for this process on the problem parameters follows from the Thomson theory (2.68) for three body processes. Table 2.12 contains the values of the rate constant for process (2.88) at room temperature and the values of the numerical factor in (2.89) that provide the experimental values of this rate constant.

In this consideration, we consider the process of conversion of atomic ions into molecular ones in three body collisions as a result of interaction of three colliding particles within the framework of one potential energy surface. But there are several electron states in the case of identical ions and atoms, and they include the even and odd states with respect to interaction of the ion with each atom. In other words, along with the ion–atom polarization interaction, the ion–atom exchange interaction may be of importance for this process. This is demonstrated in Figure 2.21, where the rate constant for the process

$$\mathrm{He}^+ + 2\mathrm{He} \to \mathrm{He}_2^+ + \mathrm{He}$$

is given at room temperature of atoms and a varied ion energy that depends on an external electric field [129]. Note that in the case of the ion–atom polarization interaction, the rate constant decreases with increasing ion energy, and the exper-

Table 2.12 The experimental rate constants for the three body process in (2.88) at room temperature and their comparison with (2.89) [27].

A	He	Ne	Ar	Kr	Xe
α, a_0^3	1.38	2.68	11.1	16.7	27.4
$K, 10^{-32}\,\mathrm{cm}^6/\mathrm{s}$	11	6	3	2.3	3.6
const	37	20	2.4	1.6	1.7

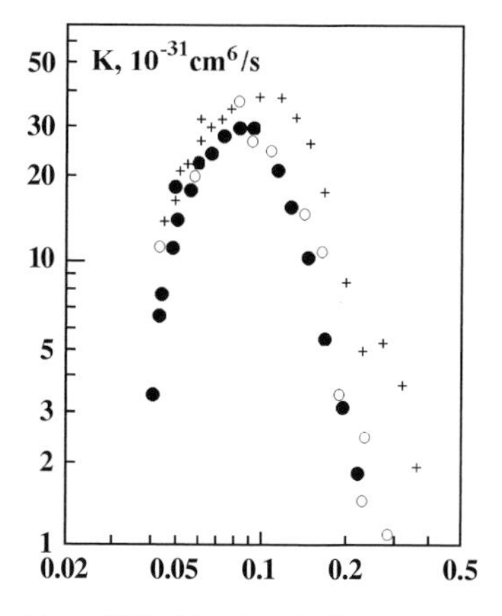

Figure 2.21 The three body rate constant for conversion of atomic helium into the molecular ion ($\mathrm{He}^+ + 2\mathrm{He} \to \mathrm{He}_2^+ + \mathrm{He}$) at room temperature as a function of the ion energy [129].

imental form of the rate constant for this process testifies to the role of the ion–atom exchange interaction.

Three body recombination of positive and negative ions follows the scheme

$$A^+ + B^- + C \to A^* + B + C\,. \tag{2.90}$$

The process gives rise initially to a bound state of the positive and negative ions A^+ and B^-, then a valence electron transfers from the field of atom B to the field of ion A^+, and the bound state decays into two atoms A^* and B. The second stage of the process occurs spontaneously during the approach of the negative and positive ions, so the three body recombination is determined by formation of the bound state of these ions.

We can estimate the rate constant for process (2.90) using the Coulomb interaction between the ions and a polarization interaction between each ion and atom. Assuming the atomic mass to be comparable to the mass of one of the ions, we use estimation (2.40) for the cross section of the atom–ion collision and the Thomson formula (2.68) for the rate constant for the three body recombination process. As a result, we have

$$K = \frac{\alpha}{N_\mathrm{i}} \sim \frac{e^6}{T^3}\left(\frac{\alpha e^2}{M}\right)^{1/2}\,. \tag{2.91}$$

Criterion (2.69) for the validity of the Thomson theory gives

$$\frac{[C]e^2(\alpha e^2)^{1/2}}{T^{3/2}} \ll 1\,. \tag{2.92}$$

Inserting the atom polarizability α, which is about several atomic units, into formula (2.92), we find that at room temperature the number density $[C]$ must be much less than $10^{20}\,\mathrm{cm}^{-3}$. Hence, the criterion for the three body character of recombination of positive and negative ions is fulfilled if the gas pressure is of the order of 1 atm.

2.3.3
Three Body Processes Involving Excited Atoms

Three body processes in a dense plasma may change the composition of this plasma. In particular, if excited atoms partake in this process, the formation of excited molecules may change both the radiative properties of this plasma and the subsequent rate of decay of excited states. As an example of this type, we consider the formation of excited (metastable and resonantly excited) molecules in inert gases. Atoms of these gases have a $n\mathrm{p}^6$ valence electron shell, and excited states have an $n\mathrm{p}^5(n+1)\mathrm{s}$ electron shell. The process of formation of excited molecules proceeds according to the scheme

$$A^* + 2A \to A_2^* + A\,. \tag{2.93}$$

Table 2.13 contains the rate constants for process (2.93) at room temperature.

An excimer molecule is an excited molecule A^*B such that the corresponding molecule AB in the ground state of atom A does not have a stable bound state, as usually occurs for an inert gas atom. Excimer molecules are used in excimer lasers, which operate on bound-free molecular transitions. Atoms of inert gases neon, argon, krypton, and xenon in the ground state have a filled electron shell (p^6) and cannot form any chemical bonds. The lowest excited states of these atoms have the electron shell configuration p^5s, so their valence electron is found in an s state, as are valence electrons of alkali metal atoms. Therefore, an alkali metal atom is a suitable model for an excited inert gas atom.

Excited states of inert gas atoms have ionization potentials that are similar to those of alkali metal atoms in the ground state. For example, the lowest resonantly excited states of the inert gas atoms $Ar(^3P_1)$ and $Kr(^3P_1)$ have ionization potentials of 4.14 and 3.97 eV, whereas the corresponding alkali metal atoms in the ground states $K(4^2S)$ and $Rb(5^2S)$ have ionization potentials of 4.34 and 4.18 eV. The analogy between excited atoms of inert gases and atoms of alkali metals in their ground states means that excited inert gas atoms form strong chemical bonds with halogen atoms. For example, the dissociation energy of the lowest state of the excimer molecule $KrF(^2B_{1/2})$ is 5.3 eV.

The radiative lifetime of excimer molecules is of the order of that for excited atoms of inert gases, specifically, in the range 10^{-7}–10^{-8} s. Therefore, generation of excimer molecules requires short pulses of energy, and electron beams or ultrahigh-frequency gas discharges are commonly used for this purpose. In these cases a pulse of electric energy is transformed into the UV radiation energy emitted by excimer molecules, and the efficiency of this transformation reaches tens of percents. Figure 2.22 contains radiative parameters of diatomic excimer molecules consisting of inert gas and halogen atoms.

The transformation of the initial electron energy to energy of radiation starts from the formation of excited atoms as a result of excitation by electron impact. In the following stage of the process, these excited atoms react with molecules containing a halogen or oxygen. This chemical reaction proceeds according to the so-called harpoon mechanism (see Figure 2.23). This refers to a mechanism in which an atom loses an electron during a collision, with that electron becoming attached to the other participant in the collision – a molecule that contains a halogen atom. The Coulomb attraction of the resulting oppositely charged ions causes a close ap-

Table 2.13 The rate constants for process (2.93) given in units of 10^{-33} cm^6/s at room temperature [27]. The accuracy classes are as follows: A – the accuracy is better than 10%, B – the accuracy is better than 20%, C – the accuracy is better than 30%, D – the accuracy is worse than 30%.

A^*	$Ne(^3P_2)$	$Ne(^3P_0)$	$Ne(^3P_1)$	$Ar(^3P_2)$	$Ar(^3P_0)$	$Ar(^3P_1)$	$Ar(^1P_1)$	$Kr(^3P_2)$
K	0.23(B)	0.5(C)	5.8(C)	10(A)	11(D)	12(D)	14(D)	36(C)
A^*	$Kr(^3P_0)$	$Kr(^3P_1)$	$Kr(^1P_1)$	$Xe(^3P_2)$	$Xe(^3P_0)$	$Xe(^3P_1)$	$Hg(^3P_0)$	$Hg(^3P_1)$
K	54(D)	30(C)	1.6(A)	55(D)	40(D)	70(C)	250(A)	160(C)

Excimer molecules.

	Ne	Ar	Kr	Xe
F	$B_{1/2}$ 2.00 *6.41* [108, 2.5] $C_{3/2}$ 1.99 *6.35*	$B_{1/2}$ [193, 4]	$B_{1/2}$ 2.51 *5.30* [248, 8] $C_{3/2}$ 2.44 *5.24* $D_{1/2}$ 2.47 *5.26*	$X_{1/2}$ 2.29 *0.15* $B_{1/2}$ 2.63 *5.30* [352, 16] $C_{3/2}$ 2.56 *5.03* [450, 100] $D_{1/2}$ 2.51 *5.46* [260, 11]
Cl		$B_{1/2}$ [175, 9]	$B_{1/2}$ [222, 19]	$X_{1/2}$ 3.2 *0.032* $B_{1/2}$ 3.22 *4.23* [308, 11] $C_{3/2}$ 3.14 *4.14* [330, 120] $D_{1/2}$ 3.18 *4.17* [236, 10]
Br		$B_{1/2}$ 2.81 *4.74*		$B_{1/2}$ 3.38 *4.30* [282, 15] $C_{3/2}$ 3.31 *3.95* [302, 120] $D_{1/2}$ 3.34 *3.98* [221, 9]
I				$B_{1/2}$ 3.62 *4.08* [254, 14] $C_{3/2}$ 3.57 *3.71* [292, 110] $D_{1/2}$ 3.59 *3.75* [203, 9]

Electronic state	Distance between nuclei at minimum, Å	Minimum of the potential energy, eV	Wavelength for the band center of the radiative transition, nm and radiative lifetime, ns
$B_{1/2}$	3.38	*4.30*	[282, 15]
$C_{3/2}$	3.31	*3.95*	[302, 120]
$D_{1/2}$	3.34	*3.98*	[221, 9]

Figure 2.22 Parameters of excimer molecules.

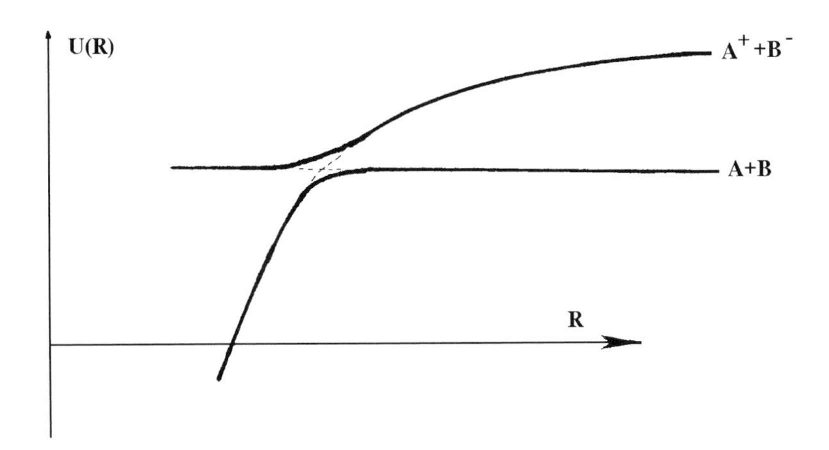

Figure 2.23 The behavior of the electronic states of the quasimolecule in the case of the harpoon reaction.

proach in the collision. As a result, the cross section for this type of collision is much larger than that for an ordinary gas-kinetic collision.

One can demonstrate this process by the example

$$\mathrm{Ar}(^3P_1) + \mathrm{Cl}_2 \rightarrow \mathrm{Ar}^*\mathrm{Cl} + \mathrm{Cl}\,. \tag{2.94}$$

We first compare the two electronic states $\mathrm{Ar}^* + \mathrm{Cl}_2$ and $\mathrm{Ar}^+ + \mathrm{Cl}_2^-$ at large distances between the colliding particles. The first state is of lower energy than the second at large distances by the amount $J - \mathrm{EA} = 1.7\,\mathrm{eV}$, where $J = 4.14\,\mathrm{eV}$ is the ionization potential of $Ar(^3P_1)$ and EA = 2.44 eV is the electron affinity for Cl_2. At large distances, a Coulomb interaction exists for the second state, and at the distance $R_c = e^2/(J - EA) = 16a_0$ (a_0 is the Bohr radius) these states intersect. In reality, a pseudointersection of the states occurs, so the states are separated by a gap at distance R_c (see Figure 2.22). At smaller distances between the particles the total system is found in the lowest state, and the Coulomb interaction occurs between the colliding particles, which decreases the distance of approach. If we assume that the rearrangement of chemical bonds occurs at a distance $R_0 < R_c$ between the colliding systems, then process (2.94) takes place. It follows from (2.85) that the maximal cross section of this process is $\pi R_c^2/2$. Since R_c is a large value, the rate constant for formation of excimer molecules under optimal conditions is greater than that corresponding to the gas-kinetic value. The rate constant is $7 \times 10^{-10}\,\mathrm{cm}^3/\mathrm{s}$ for process (2.94) at room temperature, and the rate constant for formation of other excimer molecules lies in the interval 10^{-10}–$10^{-9}\,\mathrm{cm}^3/\mathrm{s}$ at thermal energies.

Excimer molecules emit radiation in the UV range of the spectrum and are characterized by short lifetimes. The parameters of these transitions are given in Figure 2.22. Excimer molecules can be formed with atoms such as mercury, magnesium, and calcium instead of inert gas atoms, and oxygen atoms can replace the halogens. These properties of excimer molecules are used in excimer lasers which emit UV radiation [130–132].

2.3.4 Associative Ionization and the Penning Process

Associative ionization proceeds according to the scheme

$$A^* + B \to AB^+ + e\,. \tag{2.95}$$

This process is the inverse, in principle, to dissociative recombination (see Figure 2.16). But in reality these processes proceed through different vibrational states of a molecular ion AB^+ and different electron states of the molecule AB^*. It is favorable for the energetics of the process that the electron energy of the final state should be lower than that of the initial state. For this reason the dissociative recombination process proceeds through repulsive terms of molecule AB^*, whereas in the case of associative ionization it goes through attractive states of molecule AB^*. Hence, the cross section for associative ionization is of the order of gas-kinetic cross sections for those excited states for which this process is effective.

Thus, the atomic states which can partake in the associative ionization process are determined by the behavior of the electron states of the quasimolecule consisting of colliding atoms. As a demonstration of this, Figure 2.24 shows the electron states for two helium atoms, one of these is found in the ground state and the other one is in an excited state [133]. According to this figure, associative ionization may proceed with participation of the excited helium atom $He(3^1D)$, and in the first stage of the process of associative ionization the transition takes place in the repulsive term $He + He(2^1P)$ and then the molecular ion He_2^+ is formed.

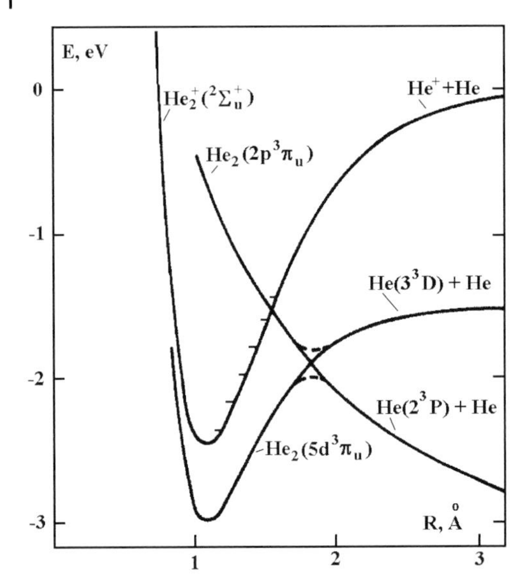

Figure 2.24 Electronic states of the quasimolecule composed of two helium atoms which are responsible for the process of associative ionization.

The cross section of associative ionization at room temperature is approximately 2×10^{-15} cm^2 [134–136], and for other excited states this cross section is less significant.

Another process of associative ionization includes two excited atoms, and their collision leads to formation of a molecular ion. Table 2.14 gives the parameters of this process with participation of two excited alkali metal atoms when the process proceeds according to the scheme

$$2M(^2P) \rightarrow M_2^+ + e - \Delta\varepsilon_i \,, \tag{2.96}$$

where $\Delta\varepsilon_i$ is the energy consumed in this process and k_{as} is the rate constant for this process. This process of associative ionization is of importance for a photoresonant plasma since as a result of this process atom excitations lead to ionization in an excited gas.

Table 2.14 Parameters of process (2.96) at 500 K [8].

A^*	$\Delta\varepsilon_i$, eV	k_{as}, cm^3/s
Na(3^2P)	<0	4×10^{-11}
K(4^2P)	0.1	9×10^{-13}
Rb(5^2P)	0.2	4×10^{-13}
Cs(6^2P)	0.33	7×10^{-13}

Excited atoms are present in a gas discharge plasma and may be of importance for its properties. One of the ionization channels in a gas discharge plasma is connected with the Penning process, which proceeds in a pairwise collision process according to the scheme [137, 138]

$$A + B^* \rightarrow e + A^+ + B\ , \tag{2.97}$$

where the ionization potential of atom A is lower than the excitation energy of atom B. Therefore, the state $A + B^*$ of the system of colliding atoms is above the boundary of the continuous spectrum, that is, it is an autoionizing state, and this state decays at moderate distances between colliding atoms when the width of this autoionizing state becomes enough for its decay during collision. Table 2.15 contains the cross sections of the Penning process involving metastable atoms of inert gases.

The Penning process (2.97) consists in decay of an autoionizing state of two interacting atoms, and the width of the state decreases with increasing separation between the atoms. Hence, there is an optimal distance between atoms that is comparable with the atomic size. The cross sections of the Penning process given in Figure 2.25 [139] and related to collisions of metastable helium atoms with atoms of heavy inert gases testify that the width of the autoionizing state for $He(2^1S)$ is larger than that for $He(2^3S)$. Next, the Penning process (2.97) has resonant character, which follows from the electron spectrum for this process, and the width of the electron spectrum is determined by the region of separations for the Penning process. Figure 2.26 [140] gives the electron spectrum for the Penning process involving xenon and mercury atoms. Each resonance corresponds to a certain channel of this process.

Note that metastable states include lower excited atomic states such that the radiative transition from these states to the ground atomic state is forbidden, and therefore these states are characterized by a large radiative lifetime and may accumulate in a plasma with a remarkable concentration. Hence, metastable atoms may be of importance for ionization processes in a plasma. In particular, a small concentration of heavy inert gases (approximately 10^{-3}) in helium decreases the break-

Table 2.15 The rate constant for the Penning process (2.97) at room temperature [27, 52, 92]. The rate constants are given in cubic centimeters per second.

Colliding atoms	$He(2^3S)$	$He(2^1S)$	$Ne(^3P_2)$	$Ar(^3P_2)$	$Kr(^3P_2)$	$Xe(^3P_2)$
Ne	4×10^{-12}	4×10^{-11}	–	–	–	–
Ar	8×10^{-11}	3×10^{-10}	1×10^{-10}	–	–	–
Kr	1×10^{-10}	5×10^{-10}	7×10^{-12}	5×10^{-12}	–	–
Xe	1.5×10^{-10}	6×10^{-10}	1×10^{-10}	2×10^{-10}	2×10^{-10}	–
H_2	3×10^{-11}	4×10^{-11}	5×10^{-11}	9×10^{-11}	3×10^{-11}	2×10^{-11}
N_2	9×10^{-11}	2×10^{-10}	8×10^{-11}	2×10^{-10}	4×10^{-12}	2×10^{-11}
O_2	3×10^{-10}	5×10^{-10}	2×10^{-10}	4×10^{-11}	6×10^{-11}	2×10^{-10}

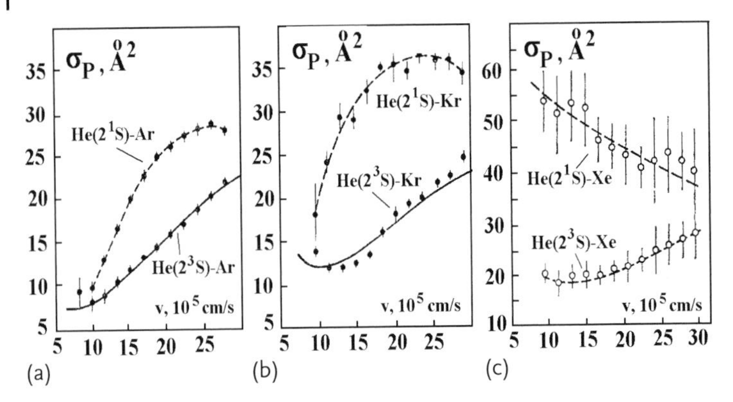

Figure 2.25 The cross section of the Penning processes involving metastable helium atoms and atoms of (a) argon; (b) krypton; and (c) xenon as a function of the relative velocity [139].

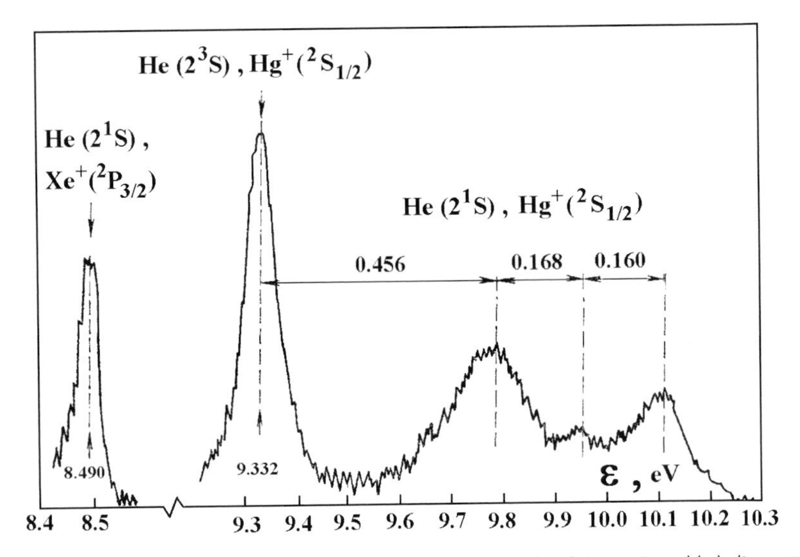

Figure 2.26 The electron spectrum in the Penning process involving metastable helium atoms colliding with xenon and mercury atoms [140].

down voltage of helium, and the Penning process (2.97) was discovered [137, 138] in this manner.

By its nature, the Penning process occurs in the decay of an autoionizing state $A^* B$. At large distances between nuclei the lifetime of this system is long because of the weak interaction between atomic particles. Therefore, this process occurs in close collisions, and its cross section is usually smaller than the gas-kinetic cross section.

2.3.5
Pairwise Recombination of Positive and Negative Ions

The process of charge transfer in the collision of negative and positive ions leads to mutual neutralization of ions and proceeds according to the scheme

$$A^+ + B^- \to A + B \,. \tag{2.98}$$

The process is determined by the behavior of electron states of this system, which are represented in Figure 2.23 for excited states, and the transition proceeds from the ion–ion electron state with Coulomb interaction to an electron state for interaction of two atoms. There are many excited atom states which are intersected by the ion–ion state, and a strong interaction takes place for such a state when which the Massey parameter (2.83) is of the order of 1. This parameter now has the form

$$\xi \sim \frac{R_c \Delta(R_c)}{\hbar v} \,, \tag{2.99}$$

where $\Delta(R_c)$ is the exchange interaction potential for this state at the distance R_c of intersection of this state with the Coulomb state and v is the collision velocity. Because the exchange interaction potential drops in an exponential way at large distances between atoms, the transition takes place in states which intersect the Coulomb state at not very large distances.

One can apply to this process the model (2.77) that the transition occurs if the distance of closest approach is below R_0. Then one can approximate the rate constant for this process by the following formula:

$$\alpha = \frac{2\sqrt{2\pi}\, R_0 e^2}{\sqrt{\mu T_i}} \,, \tag{2.100}$$

where μ is the reduced mass of colliding particles. Table 2.16 lists experimental values of the rate constants for process (2.98) at room temperature. These data relate to ions involving oxygen and nitrogen and are based on experiments [141–145]. From these data we find appropriate values of the parameter R_0 in (2.100), which are also represented in Table 2.16. The statistical treatment of these data gives the average value

$$R_0 = 40 a_0 \times 10^{\pm 0.3} \,.$$

Formula (2.100) may be generalized for collision of a multicharged ion with charge $|Z|e$ with a singly charged ion of opposite charge sign. The rate constant k for collisions of these ions where the distance of closest approach does not exceed R_0 is given by

$$k = \frac{2\sqrt{2\pi}|Z|e^2 R_0}{\sqrt{\mu T_i}} \,, \tag{2.101}$$

where μ is the reduced mass of colliding ions.

Table 2.16 Experimental values of the rate constants for mutual neutralization in pairwise collisions of positive and negative ions involving nitrogen and oxygen. The rate constants are given in units of 10^{-7} cm^3/s, and R_0 in (2.99) is expressed in a_0.

Process	Rate constant	R_0, a_0
$O^+ + O^-$	3 ± 1	34
$N^+ + O^-$	3 ± 1	32
$O_2^+ + O_2^-$	4 ± 1	74
$N_2^+ + O_2^-$	1.6 ± 0.5	28
$NO^+ + NO_2^-$	5 ± 2	44
$O_2^+ + NO_2^-$	4 ± 1	78
$NO^+ + NO_3^-$	0.4 ± 0.1	10
$O^+ + O_2^-$	2	28
$NO^+ + O^-$	5 ± 2	70

2.3.6
Processes Involving Formation of a Long-Lived Complex

A long-lived complex is an intermediate state formed as a result of the collision of two atoms. Subsequent evolution of this complex can lead to the formation of a bound state of these particles. For example, the formation of molecular ions in three body collisions can occur through the processes

$$A^+ + B \xrightarrow{k_c} (A^+ + B)^{**} \,, \quad (A^+ B)^{**} \xrightarrow{\tau} A^+ + B \,,$$
$$(A^+ B)^{**} + C \xrightarrow{k_q} AB^+ + C \,, \quad (A^+ B)^{**} + C \xrightarrow{k_d} A^+ + B + C \,. \qquad (2.102)$$

Here $(A^+ B)^{**}$ is an autodetaching state of the colliding particles, and the quantities written over the arrows are the rate constants k for the collision processes or the lifetime τ of the autodetaching state $(A^+ B)^{**}$.

Scheme (2.102) leads to the balance equations for the number densities of autodetaching states:

$$\frac{d}{dt}[(A^+ B)^{**}] = 0 = [A][B]k_c - \frac{[(A^+ B)^{**}]}{\tau} - [(A^+ B)^{**}][C](k_q + k_d) \,, \qquad (2.103)$$

where $[X]$ denotes the number density of particles X. The solution of this equation gives

$$[(A^+ B)^{**}] = \frac{[A][B]k_c}{1/\tau + [C](k_q + k_d)} \,.$$

Next, the balance equation for the third process (2.103) has the form

$$\frac{d[AB^+]}{dt} = k_q[(A^+ B)^{**}[C] = [A][B][C]\frac{k_c k_q \tau}{1 + [C](k_q + k_d)\tau} \,.$$

A comparison of this balance equation and the definition of the rate constant for three body processes shows that the third process (2.102) can be characterized as a three body process only if

$$[C](k_q + k_d) \ll 1 \,. \qquad (2.104)$$

Then the three body rate constant is

$$K = k_q k_d \tau \,. \qquad (2.105)$$

Table 2.17 Collision processes of electrons with atoms and molecules.

Process	Scheme	Comment number, section of [146]
Elastic collision of electrons with atoms and molecules	$e + A \to e + A$	[a], Section 2.1.7
Inelastic collision of electrons with atoms and molecules	$e + A \longleftrightarrow e + A^*$	Section 2.2.1
Ionization of atoms or molecules by electron impact	$e + A \to 2e + A^+$	Section 2.2.2
Transitions between rotational levels of molecules	$e + AB(J) \to e + AB(J')$	[b]
Transitions between vibrational levels of molecules	$e + AB(v) \to e + AB(v')$	[c]
Dissociative attachment of an electron to a molecule	$e + AB \to A + B^-$	Section 2.2.7
Dissociative recombination	$e + AB^+ \to A + B^*$	Section 2.2.5
Dissociation of a molecule by electron impact	$e + AB \to e + A + B$	[d]
Electron attachment to an atom in three body collisions	$e + A + B \to A + B^-$	Section 2.2.3
Electron–ion recombination	$2e + A^+ \to e + A$	
in three body collisions	$e + A^+ + B \to A^* + B$	Section 2.2.3
Dielectronic recombination	$e + A^{+Z} \to [A^{+(Z-1)}]^{**} \to A^{+(Z-1)} + \hbar\omega$	Section 2.2.6

a Typically, the cross section is of the order of the gas-kinetic cross section.

b J and J' are angular momenta of the molecule. The cross section is, typically, less by one or two orders of magnitude than the cross section for elastic collision of a slow electron and molecule. A selection rule determines the difference $J - J'$ depending on the molecular species and on the collision energy.

c Processes proceed via formation of an autodetaching state AB^-. The probability of the process depends on the relative position of the electron energy and the autodetaching state energy. The cross section for the process has a resonant energy dependence, and the maximum cross section is less than or of the order of the atomic cross section.

d The process proceeds via excitation of a molecule into a repulsive electron term. The dependence of the cross section on the electron energy is the same as for excitation of vibrational levels of the molecule by electron impact.

Table 2.18 Collision processes of atoms and molecules.

Process	Scheme	Comment number, section of [146]
Elastic collision of atoms and molecules	$A + B \to A + B$	[a], Section 2.1.9
Excitation of electron levels in collisions with atoms	$A + B \to A^* + B$	[b], Section 2.3.1
Ionization in collisions of atoms	$A + B \to A^+ + e + B$	[b], Section 2.3.1
Transitions between vibrational levels of molecules	$A + BC(v) \to A + BC(v')$	[c]
Transitions between rotational levels of molecules	$A + BC(J) \to A + BC(J')$	[d]
Quenching of excited electronic states in atomic collisions	$A^* + B(BC) \to A + B(BC)$	[b], Section 2.3.1
Associative ionization	$A^* + B \to AB^+ + e$	Section 2.3.5
Penning process	$A^* + B \to A + B^+ + e$	Section 2.3.5
Transfer of excitation	$A^* + B \to A + B^*$	Section 2.3.1
Spin exchange and transitions between hyperfine structure states	$A + B(\downarrow) \to A + B(\uparrow)$	[e]
Transitions between fine structure states	$A + B(j) \to A + B(j')$	[f]
Atom depolarization in collisions	$A + B(\downarrow) \to A + B(\uparrow)$	[g]
Formation of molecules in three body collisions	$A + B + C \to AB + C$	[h], Section 2.2.3
Chemical reactions	$A + BC \to AB + C$	[i]

a For thermal energies of collision, the cross section has the gas-kinetic magnitude (approximately 10^{-15} cm^2).

b Under adiabatic conditions, these processes have a small transition probability for collision energies of a few electronvolts, and the relative cross sections are smaller than typical atomic cross sections by several orders of magnitude.

c At thermal collision energies, the cross section for the transition is smaller than the gas-kinetic cross section by several orders of magnitude.

d At thermal collision energies, the cross section for the transition is of the order of a typical atomic cross section.

e Arrows indicate spin orientation of valence electrons. This is a resonant process. Consider, as an example, collision of two alkali metal atoms in the ground electronic state. Then, spin exchange corresponds to exchange of valence electrons, so the spin transition causes the transition between hyperfine structure levels of each atom.

f If the energy difference for fine structure levels is sufficiently small and the Massey parameter (2.83) is small for the transition, this process is resonant and is characterized by large cross sections.

g Arrows indicate directions of atomic angular momenta. This is a resonant process.

h For thermal energies, the rate constant for the process is of the order of 10^{-33}–10^{-32} cm^6/s.

i The cross section of the process is larger than the gas-kinetic process cross section.

For another limiting case,

$$[C](k_q + k_d) \gg 1\,, \tag{2.106}$$

the number density of particles C is so high that the rate constant for formation of the bound state A^+B does not depend on this number density, and the formation of the bound state A^+B can be considered to be two body in nature with rate constant

$$k = K[C] = \frac{k_c k_q}{k_q + k_d}\,. \tag{2.107}$$

Particles C affect the factor $k_q/(k_q + k_d)$ in (2.107), which is the probability that collision of autodetaching ion $(A^+B)^{**}$ and particle C leads to formation of the bound state AB^+.

Table 2.19 Collision processes involving ions.

Process	Scheme	Comment number, section of [146]
Elastic ion–atom collisions	$A^+ + B \to A^+ + B$	Section 2.1.9
Resonant charge exchange	$A^+ + A \to A + A^+$	Section 2.1.9
Nonresonant charge exchange	$A^+ + B \to A + B^+$	[a], [b]
Mutual neutralization of ions	$A^+ + B^- \to A + B$	Section 2.3.6
Decay of negative ions in atomic collisions	$A + B^- \to AB + e$ $A + B^- \to A + B + e$	Section 2.2.4
Ion–molecule reactions	$A^+ + BC \to AB^+ + C(AB + C^+)$	[c]
Conversion of atomic ions to molecular ions in three body collisions	$A^+ + B + C \to AB^+ + C$	Section 2.3.2
Ion–ion recombination in three body collisions	$A^+ + B^- + C \to A + B + C$	[d], Section 4.4.7

a The cross section of the process is larger than the gas-kinetic cross section.
b The cross section depends on the value of the Massey parameter.
c These reactions are akin to chemical reactions.
d At thermal collision energies, the rate constant of this process is of the order of 10^{-31} cm^6/s. Formation of an autodetaching state $(AB^+)^{**}$ in the course of the collision can result in this value being exceeded by several orders of magnitude.

In reality, both regimes can be fulfilled for formation of bound states through formation of a long-lived complex. For example, the lifetime of the long-lived complex consisting of electrons and large molecules is of the order of 10^{-6}–10^{-4} s, so criterion (2.106) is satisfied for number densities $[C] \gg 10^{15}$–10^{17} cm^{-3}. At small number densities of particles C, formation of the bound state proceeds in the three body process. Note that the Thomson formula for the three body process follows from (2.105) if the transition time for colliding particles is used as the lifetime of a long-lived complex.

2.3.7
Types of Elementary Processes

A summary of the elementary processes occurring in weakly ionized gases is given in Tables 2.17–2.19 [146]. The comments in the table footnotes contain details and special features of these processes.

2.4
Radiative Processes in Excited and Ionized Gases

2.4.1
Interaction of Radiation with Atomic Systems

Interaction of a plasma with radiation affects plasma properties, and in systems such as gas lasers or the plasma generated by laser radiation, these processes determine the plasma parameters. Electromagnetic fields cause transitions in atomic systems, and Table 2.20 gives a list of single-photon processes that are of interest for plasmas. The weakness of interaction is characterized by the fine structure constant $\alpha = e^2/(\hbar c) \approx 1/137$, which is the ratio of the average electron velocity in the

Table 2.20 Elementary interactions between atoms and radiation [146].

Process	Scheme of the process
Excitation as a result of photon absorption	$\hbar\omega + A \to A^*$
Spontaneous radiation of an excited atom	$A^* \to \hbar\omega + A$
Stimulated photon emission	$\hbar\omega + A^* \to 2\hbar\omega + A$
Atomic photoionization	$\hbar\omega + A \to A^+ + e$
Photodetachment of a negative ion	$\hbar\omega + A^- \to A + e$
Photodissociation of a molecule	$\hbar\omega + AB \to A + B$
Photorecombination of an electron and an ion	$e + A^+ \to A + \hbar\omega$
Radiative attachment of an electron to an atom	$e + A \to A^- + \hbar\omega$
Atomic photorecombination	$A + B \to AB + \hbar\omega$
Bremsstrahlung in electron–atom or electron–ion collisions	$e + A \to e + A + \hbar\omega$

hydrogen atom to the light velocity, and this testifies to the nonrelativistic character of the motion of valence electrons in atoms. Another small parameter related to radiative processes in plasmas is the ratio of the electric field of an electromagnetic wave to a typical atomic field. With the exception of specific cases, this ratio is small, and so radiative processes in atomic systems involving absorption or emission of photons usually proceed slowly on atomic timescales.

In external fields, emission of a photon is accompanied by an transition of an atom to a lower excited state or to the ground state. The lifetime τ of an excited atom with respect to this process is considerably longer than a typical atomic time. The reciprocal quantity $1/\tau$ (the rate of spontaneous radiation) with respect to a characteristic atomic frequency is measured by the cube of the fine structure constant, $[e^2/(\hbar c)]^3$, and hence is lower by at least six orders of magnitude than the frequency of the emitted photons. In particular, the radiative lifetime of the first resonantly excited state of the hydrogen atom, H(2p), is 2.4×10^{-9} s, whereas the characteristic atomic time is $\hbar^3/(me^4) = 2.4 \times 10^{-17}$ s. The weakness of typical electromagnetic fields in plasmas allows us to ignore multiphoton processes. In particular, we can ignore two-photon processes compared with single-photon processes. The lowest excited state of the atom from which it is possible to have a single-photon transition to the ground state is called a resonantly excited state. Radiative transitions involving resonantly excited states of atoms are the main subject of our consideration.

2.4.2 Spontaneous and Stimulated Emission

Radiative transitions between discrete states of an atom or molecule are summarized in a simple fashion in Figure 2.27. We designate by n_ω the number of photons in a given state. This value is increased by 1 as a result of a transition to the ground (lower) state and is decreased by 1 after absorption of a photon. Because the absorption rate is proportional to the number of photons present, we write the probability of photon absorption by one atom per unit time in the form

$$w(0, n_\omega \rightarrow *, n_\omega - 1) = A n_\omega \,, \tag{2.108}$$

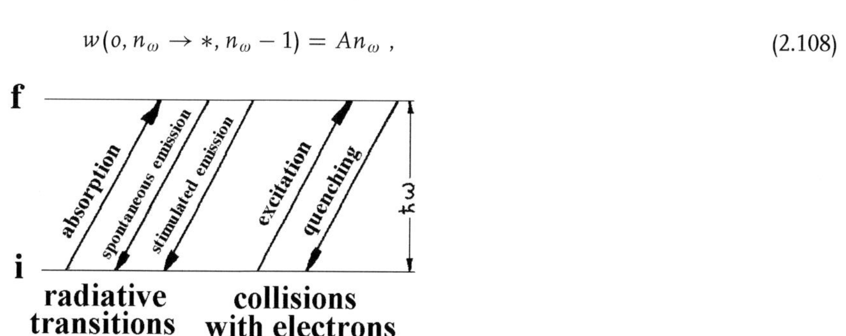

Figure 2.27 Collision and radiative transitions between two states.

where, in accord with Figure 2.27, we denote the lower state by subscript o and the upper state by subscript $*$. Equation (2.108) accounts for the fact that no transitions occur in the absence of photons ($n_\omega = 0$) and only single-photon transitions take place. The quantity A does not depend on the electromagnetic field strength and is determined only by properties of the atom.

The probability per unit time for an atomic transition with emission of a photon can be represented in the form

$$w(*, n_\omega \to o, n_\omega + 1) = \frac{1}{\tau_r} + B n_\omega \,. \tag{2.109}$$

Here $1/\tau_r$ is the reciprocal lifetime of the upper state with respect to spontaneous radiative transitions (those that proceed in the absence of the electromagnetic field) to the lower state and B refers to the radiation stimulated by the external electromagnetic field. Both values depend only on atomic properties. The quantities A and B in relations (2.108) and (2.109) are known as the Einstein coefficients [147].

Relationships among $1/\tau$, A, and B can be obtained by an analysis of the thermodynamic equilibrium existing between the atoms and photons. The relation between the number densities of atoms N_0 and N_* in the ground and excited states, respectively, are given by the Boltzmann law (1.42):

$$N_* = \frac{g_*}{g_0} N_0 \exp\left(-\frac{\hbar\omega}{T}\right) ,$$

where g_0 and g_* are the statistical weights of the ground and excited states, and the photon energy $\hbar\omega$ coincides with the energy difference between the two states. The mean number of photons in a given state is determined by the Planck distribution (1.56):

$$\overline{n}_\omega = \left[\exp\left(-\frac{\hbar\omega}{T}\right) - 1\right]^{-1} .$$

In thermodynamic equilibrium, the number of emissions per unit time must be equal to the number of absorptions per unit time. Applying this condition to a unit volume, we have

$$N_0 \, w(o, \overline{n}_\omega \to *, \overline{n}_\omega - 1) = N_* w(*, \overline{n}_\omega - 1 \to o, \overline{n}_\omega) \,.$$

According to (2.108) and (2.109), this relation takes the form

$$N_0 A \overline{n}_\omega = N_* \left(\frac{1}{\tau} + B\overline{n}_\omega\right) . \tag{2.110}$$

Using the above expressions for the connection between the equilibrium number densities of atoms and the equilibrium average number of photons in a given state, we obtain $A = g_*/(g_0\tau)$ and $B = 1/\tau$ for the Einstein coefficients. We then find the rates of the single-photon processes to be

$$w(o, n_\omega \to *, n_\omega - 1) = \frac{g_*}{g_0\tau_r} n_\omega \,, \; w(*, n_\omega \to o, n_\omega - 1) = \frac{1}{\tau_r} + \frac{n_\omega}{\tau_r} \,. \tag{2.111}$$

Note that thermodynamic equilibrium requires the presence of stimulated radiation, which is described by the last term and is of fundamental importance.

2.4.3
Radiative Transitions in Atoms

In considering radiation processes in a gas, we pay attention to two of them – dipole radiative transitions in atoms and photorecombination radiation involving electrons and ions of a plasma. We analyze interaction of atomic particles of an ionized gas with a weak electromagnetic field as occurs in gas discharges, so the intensity of this electromagnet field is small in comparison with the corresponding atomic value, and therefore typical times for radiative transitions are large compared with typical atomic times. One more small parameter affects this interaction when valence electrons partake in radiative transitions and their typical velocity is small compared with the speed of light c. A parameter of radiative transitions with a small value is the fine structure constant $\alpha = e^2/(\hbar c) = 1/137$, where e is the electron charge, $\hbar$ is the Planck constant, c is the light velocity, and $e^2/\hbar$ is a typical atomic velocity of valence electrons in their orbits. If we characterize a radiative transition between two atomic states by the radiative lifetime of the upper state with respect to the radiative transition τ_r, for the strongest transitions this lifetime is $\tau_\mathrm{r} \sim \tau_\mathrm{r}^o = \alpha^3 \tau_0$, where $\tau_0 = 2.42 \times 10^{-17}$ s is the atomic time and $\tau_\mathrm{r}^o = 6.2 \times 10^{-11}$ s.

In addition, we will consider the strongest interaction between the radiation field and an atomic system whose interaction operator has the form $\widehat{V} = -\mathbf{ED}$ [148], where $\mathbf{E}$ is the electric field strength of the radiation field, and $\mathbf{D}$ is the dipole moment operator of the atomic system. Therefore, the strongest radiative transitions are called dipole transitions and connect transition states with nonzero matrix elements of the dipole moment operator. We restrict ourselves below to only such radiative transitions. The upper excited state of a radiative transition is named a resonantly excited state if the dipole radiative transition in the ground state is possible from this state. For example, for the helium atom this state is $\mathrm{He}(2^1P)$ and for the resonantly excited argon atoms the states are $\mathrm{Ar}(3^3P_1)$ and $\mathrm{Ar}(3^1P_1)$. Table 2.21 lists parameters of radiative transitions involving the lowest excited states of helium and argon atoms. Note that the radiative lifetime for these transitions exceeds τ_r^o, because the interaction is shared between many excited states and the transition energy is less than the atomic one.

We also give the expression for the rate of dipole radiation [148, 150, 151], that is, the lifetime τ_r of a resonantly excited state with respect to spontaneous radiative transition to a lower state

$$\frac{1}{\tau_\mathrm{r}} = \frac{4\omega^3}{3\hbar c^3} |\langle o|\mathbf{D}|*\rangle|^2 g_0 = \frac{2\omega^2 e^2 g_0}{m_\mathrm{e} c^3 g_*} f_{o*} \,. \tag{2.112}$$

Here indices o and $*$ relate to lower and upper states of the radiative transition, $\mathbf{D}$ is the dipole momentum operator, ω is the transition frequency, g_0 and g_* are the statistical weights of lower and upper transition states, c is the light velocity, and the oscillator strength f_{o*} for transition between states o and $*$ is given by [152, 153]

$$f_{o*} = \frac{2m_\mathrm{e}\omega}{3\hbar e^2} |\langle o|\mathbf{D}|*\rangle|^2 g_* = \frac{2m_\mathrm{e}\omega}{3\hbar e^2} |\mathbf{D}|^2 g_* \,. \tag{2.113}$$

Table 2.21 Parameters of radiative transitions involving the lowest excited states of helium and argon atoms. $\Delta\varepsilon$ is the transition energy, λ is the wavelength of an emitted photon, and τ_r is the radiative lifetime with respect to this transition [149].

Radiative transition	$\Delta\varepsilon$, eV	λ, nm	τ_r, ns
$He(2^1P \to 1^1S)$	21.22	58.433	0.56
$He(2^3P \to 2^3S)$	1.144	1083	98
$He(2^1P \to 2^1S)$	0.602	2058	500
$Ar(3p^54s^3P_1 \to 1^1S)$	11.62	106.67	10
$Ar(3p^54s^1P_1 \to 1^1S)$	11.83	104.82	2

The radiative spectrum of an atom for excited states may have a complex form because of various forms of interactions inside the atom. As a demonstration of this, we represent in Figure 2.28 the radiative transitions of the argon atom, which include the upper states of transitions where an excited electron is found in 4s and 4p states. ΔT is the excitation energy of states which energy is counted from the lowest excited state, and τ_{rad} is the radiative lifetime for a given excited state that is summed over all transition states from this one, whereas $\tau_{ki} = 1/A_{ki}$ is

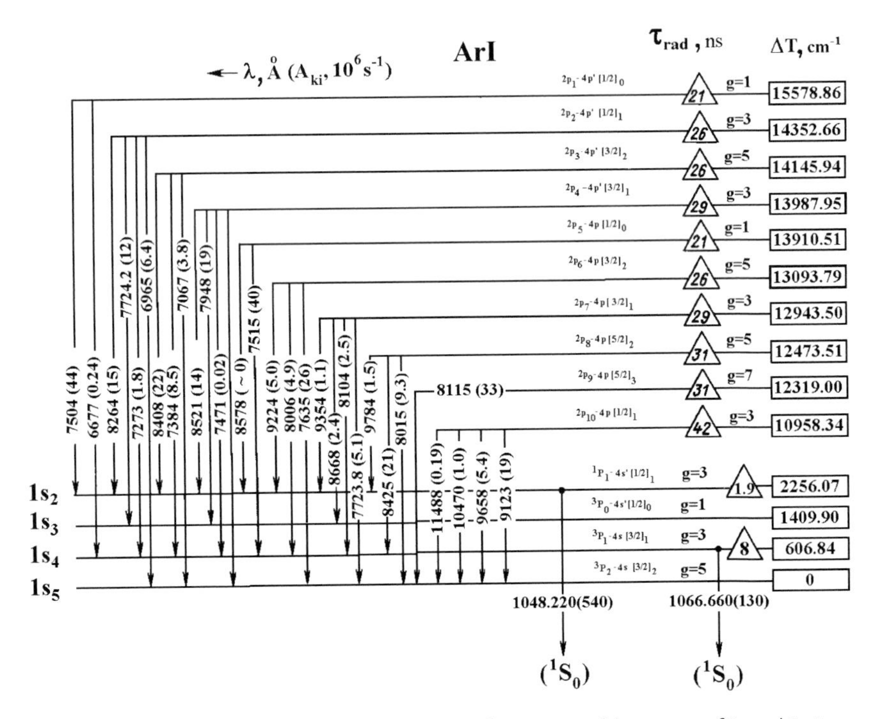

Figure 2.28 Radiative transitions for the argon atom from states of the groups of 2s and 2p in Pashen notation.

the radiative lifetime for a given transition with wavelength λ. Note that the excited states under consideration correspond to 4s and 4p valence electrons and a 3p hole, and the only quantum number of these states is the total momentum J that results from summation of momenta of the core (ion) and valence electrons. Along with the Paschen notation, which is $2s_5 \ldots 2s_2$ for states with a 4s valence electron and $2p_{10} \ldots 2p_1$ for states with a 4p valence electron, the notation is given for a certain scheme of the momentum summation. Namely, 4s and 4p correspond to the total momentum 3/2 for the total core momentum, and 4s′ and 4p′ relate to the case when the total core momentum is 1/2. The result of summation of the total core momentum with the orbital momentum of an excited electron is given in the square brackets, and the total atom momentum, which is a sum of this momentum and the spin of the valence electron, is given as a subscript. This scheme of momentum summation is based on some hierarchy of interactions that is not realized for a 4p valence electron and hence it is used to denote atomic levels.

2.4.4
Photoionization and Photorecombination Processes

The photoionization process proceeds according to the scheme

$$\hbar\omega + A \to e + A^+ \,, \tag{2.114}$$

and the opposite process, the process of photorecombination, is

$$e + A^+ \to \hbar\omega + A \,. \tag{2.115}$$

We first derive the connection between the cross sections of processes (2.114) and (2.115) on the basis of the principle of detailed balance, which has the form

$$g_{\text{rec}}\sigma_{\text{rec}} j_i = g_{\text{ion}}\sigma_{\text{ion}} j_p \,,$$

where σ_i and σ_r are the cross sections of photoionization and photorecombination, respectively, g_{ion} and g_{rec} are the statistical weights for these states, and j_e and j_p are the fluxes of electrons and photons in these channels. Taking the statistical weights as the number of states per electron or photon, we have

$$g_{\text{rec}} = g_e g_i \frac{4\pi k^2 dk}{(2\pi)^3} \,, \quad g_{\text{ion}} = g_a \cdot 2 \frac{4\pi q^2 dq}{(2\pi)^3} \,,$$

where g_e, g_i, and g_a are the statistical weights of an electron, an ion, and an atom with respect their electronic states, k and q are the wave vectors of a photon and an electron, the factor 2 accounts for two photon polarizations, and we consider the ion to be motionless. Because one photon is located in this volume Ω, we have for a photon flux $j_p = c/\Omega$ and an electron flux $j_e = v_e/\Omega$, where c is the light velocity and the electron velocity $v_e = \hbar q/m_e$. Using the conservation of energy for these transitions,

$$\hbar\omega = J + \frac{\hbar^2 q^2}{2m_e} \,,$$

where J is the atomic ionization potential, and the dispersion relation for the photon $\omega = kc$, we have

$$\sigma_{\text{rec}} = \frac{2g_a}{g_e g_i} \frac{k^2}{q^2} \sigma_{\text{ion}} \,. \tag{2.116}$$

We now give the expression for the photoionization cross section in the limiting cases. When atom A in processes (2.114) and (2.115) is a hydrogen atom in the ground state [152, 154, 155]

$$\sigma_{\text{ion}} = \frac{2^9 \pi^2}{3} \frac{e^2}{\hbar c} a_0^2 F(a_0 q) \,, \quad F(x) = \frac{\exp\left(-\frac{4}{x} \arctan x\right)}{(1 + x^2)^4 \left[1 - \exp\left(-\frac{2\pi}{x}\right)\right]} \,. \tag{2.117}$$

In the limit $x \to 0$ we have

$$F(0) = \frac{1}{\exp(4)} \left(\frac{\omega_0}{\omega}\right)^{8/3} \,,$$

where $\hbar\omega_0 = Ry = 13.6\,\text{eV}$ is the ionization potential for the hydrogen atom in the ground state. In the limit when the photoionization process proceeds near the threshold, this gives

$$\sigma_{\text{ion}} = \sigma_0 \left(\frac{\omega_0}{\omega}\right)^{8/3} \,, \tag{2.118}$$

where the photoionization cross section at the threshold σ_0 and the transition frequency ω_0 for the hydrogen atom are

$$\sigma_0 = \frac{2^9 \pi^2}{3 \exp(4)} \frac{e^2}{\hbar c} a_0^2 = 0.225 a_0^2 = 6.3 \times 10^{-18}\,\text{cm}^2 \,,$$
$$\omega_0 = \frac{m_e e^4}{\hbar^3} = 2.07 \times 10^{16}\,\text{s}^{-1} \,. \tag{2.119}$$

The principle of detailed balance (2.116) gives on the basis of (2.118) the following expression for the photorecombination cross section involving a slow electron in the ground state of the hydrogen atom:

$$\sigma_{\text{rec}} = \frac{2\omega^2}{c^2 q^2} \sigma_{\text{ion}} = \frac{2^9 \pi^2}{3 \exp(4)} \left(\frac{e^2}{\hbar c}\right)^3 \frac{\omega_0^{5/3}}{\omega^{2/3}(\omega - \omega_0)} a_0^2 = \sigma_1 \frac{\omega_0^{5/3}}{\omega^{2/3}(\omega - \omega_0)} \,, \tag{2.120}$$

where

$$\sigma_1 = \frac{2^8 \pi^2}{3 \exp(4)} \left(\frac{e^2}{\hbar c}\right)^3 a_0^2 = 0.225 a_0^2 = 1.7 \times 10^{-22}\,\text{cm}^2 \,. \tag{2.121}$$

In the case of a highly excited initial atomic state when a transferring electron is described by classical laws, the photoionization cross section is given by the Kramers formula [156]:

$$\sigma_{\text{ion}} = \frac{\sigma_K}{n^5} \left(\frac{\omega_0}{\omega}\right)^3 \,, \tag{2.122}$$

where n is the principal quantum number of a bound electron in the initial state, and an average is performed over other quantum numbers of this electron. The parameter σ_K in this formula is

$$\sigma_K = \frac{64\pi}{3\sqrt{3}} \frac{e^2}{\hbar c} a_0^2 = 7.9 \times 10^{-18}\,\text{cm}^2 . \tag{2.123}$$

Correspondingly, according to the principle of detailed balance (2.116), when the final atomic state is a highly excited state with principal quantum number n, for the photorecombination cross section, formula (2.122) gives

$$\sigma_{\text{rec}} = \frac{32\pi}{3\sqrt{3}} \left(\frac{e^2}{\hbar c}\right)^3 \frac{2\omega^2}{\omega(\omega - \omega_0/n^2)} \frac{a_0^2}{n^3} . \tag{2.124}$$

The Kramers formula corresponds to the limit when both transition states may be described classically [157]. Transitions between excited states conform to this limit, and this follows from general tables of parameters of radiative transitions of an electron that is under the action of the central field [158].

It is of interest to use the Kramers formula (2.122) for photoionization of the hydrogen atom in the ground state when the classical description of a transferring electron is not valid. In this case, according to this classical description, the threshold photoionization cross section is to σ_K of (2.123), which is 25% higher than the accurate value (2.119) of this quantity. This means that two limiting cases under consideration are able to give correct estimations of radiative parameters for real cases. Note that when the electron shell of an atom contains several valence electrons, general characteristics of the above dependencies are conserved [159].

In the same manner as photorecombination, one can consider the process of electron photoattachment to an atom that proceeds according to the scheme

$$e + A \to A^- + \hbar\omega . \tag{2.125}$$

In particular, Figure 2.29 shows the cross sections of electron photoattachment to halogen atoms [160], and we can use the principle of detailed balance (2.116) for these processes, where the statistical weights of the halogen atom and its negative ion are 6 and 1, respectively, so this formula gives the following connection between the cross section of photoattachment σ_{at} and the cross section of photodetachment σ_{det} of the negative ion:

$$\sigma_{\text{at}} = \frac{3(\text{EA} + \varepsilon)^2}{\varepsilon} \sigma_{\text{det}} , \tag{2.126}$$

where ε is the electron energy, EA is the atom electron affinity, and the photon energy is $\hbar\omega = \text{EA} + \varepsilon$ because of the energy conservation. Since for these halogen atoms $\text{EA} > 3\,\text{eV}$, the cross section of electron photoattachment σ_{at} is less by roughly by six orders of magnitude than the cross section of photodetachment σ_{det} of the negative ion in this range of electron energies.

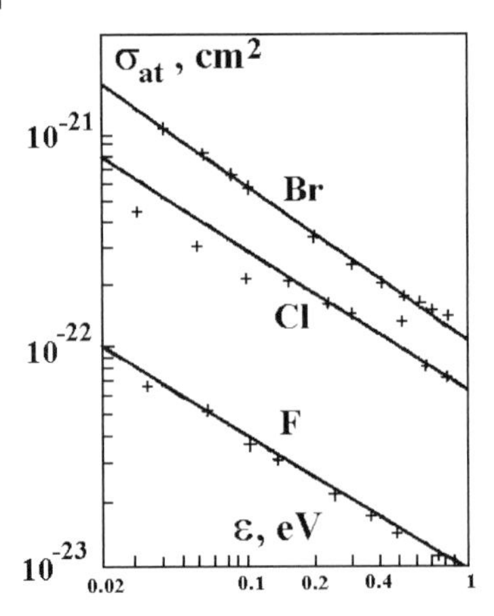

Figure 2.29 The cross section of photoattachment of electrons to halogen atoms [160].

2.4.5
Bremsstrahlung in Ionized Gases Involving Electrons

Along with bound–bound and bound–free radiative transitions which are accompanied by photon emission, free–free transitions are possible in collisions of two atomic particles, and below we consider bremsstrahlung resulting from electron scattering by ions and atoms. In these processes electrons are free both before and after collision. Our goal is to find the cross section of bremsstrahlung resulting from electron scattering by ions and atoms, and we start from the classical formula for radiation of an atomic particle. Then we use (2.112) for the radiative rate and take the initial i and final k states of this transition to be classical. The total intensity of this transition according to (2.112) is

$$I_{ik} = \frac{\hbar\omega}{\tau_r} = \frac{4\omega_{ik}^4}{3c^3}|\langle \mathbf{D}_{ik}\rangle|^2 = \frac{4}{3c^3}|\langle \ddot{\mathbf{D}}_{ik}\rangle|^2 ,$$

where $\hbar\omega$ is the transition energy, and in the classical limit we change the matrix element by the Furie component of the corresponding quantity. Note that this formula includes summation over final states that are close in energy and accounts for both states with $\omega > 0$ (emission) and states with $\omega < 0$ (absorption). Restricting ourselves to radiative processes with emission of photons, we conserve half of the total radiation intensity in this sum, which in the classical limit gives

$$I = \frac{2}{3c^3}|\ddot{\mathbf{D}}|^2 . \tag{2.127}$$

Since for a scattering electron $\mathbf{D} = e\mathbf{r}$, where $\mathbf{r}$ is the electron radius vector, the intensity of radiation due to scattering of an electron by an atomic particle is proportional to the acceleration squared in the particle field.

From (2.127) we have for the energy ΔE that an electron loses in one collision

$$\Delta E = \int_{-\infty}^{\infty} \frac{2e^2}{3c^3}\ddot{\mathbf{r}}^2 dt = \int_0^{\infty} S_\omega d\omega\,, \quad S_\omega = \frac{8\pi e^2\omega^4}{3c^3}|\mathbf{r}_\omega|^2\,,$$

where $S_\omega d\omega$ is the energy emitted during collision in the frequency range from ω to $\omega + d\omega$. Transferring to the rate of the radiative transition $S_\omega/\hbar\omega$, we obtain from this for the differential cross section of emission in electron–atomic particle collision in the classical case

$$\frac{d\sigma}{d\omega} = \int_0^{\infty} \frac{S_\omega}{\hbar\omega}\cdot d\sigma\,, \tag{2.128}$$

where $d\sigma$ is the differential cross section of electron scattering, and this formula holds true if the electron energy variation in this process is relatively small, that is,

$$\hbar\omega \ll \varepsilon\,,$$

where ε is the energy of a scattered electron.

We now consider bremsstrahlung as a result of electron scattering by atoms

$$e + A \to e + A + \hbar\omega\,. \tag{2.129}$$

On the basis of (2.128) and following [161, 162], we use the pulse approximation in which $\ddot{\mathbf{r}} = \Delta\mathbf{v}\delta(\mathbf{v})$, which gives

$$\ddot{\mathbf{r}}_\omega = \int_{-\infty}^{\infty} e^{i\omega t}\ddot{\mathbf{r}}\cdot\frac{dt}{2\pi} = \frac{\Delta\mathbf{v}}{2\pi}\,,$$

since $\omega\tau \ll 1$ for a typical collision time τ. Because the change of the electron velocity after elastic scattering by an atom is $\Delta v = 2v\sin(\vartheta/2)$, where ϑ is the scattering angle, we obtain for the radiated energy per unit frequency

$$S_\omega = \frac{8\pi e^2}{3c^3}|\ddot{\mathbf{r}}_\omega|^2 = \frac{4e^2v^2}{3\pi c^2}(1-\cos\vartheta)\,.$$

This gives for the differential cross section of bremsstrahlung in an electron–atom collision on the basis of (2.128)

$$\frac{d\sigma}{d\omega} = \int_0^{\infty} \frac{S_\omega}{\hbar\omega}\cdot d\sigma = \frac{4e^2}{3\pi\hbar\omega}\frac{v^2}{c^2}\sigma^*(v)\,, \tag{2.130}$$

where $\sigma^* = \int(1-\cos\vartheta)d\sigma$ is the diffusion cross section of electron–atom scattering. In deriving this formula we assumed that the electron scattering proceeds in a narrow range $\Delta r \sim a_0$ of electron distances from an atom. Together with the above criterion $\hbar\omega \ll \varepsilon$, this uses the condition $v \ll e^2/\hbar$, that is, the electron energy is small compared with typical atomic energies.

We now consider bremsstrahlung in an electron–ion collision that proceeds according to the scheme

$$e + A^+ \rightarrow e + A^+ + \hbar\omega \ . \tag{2.131}$$

Assuming that an electron moves along classical trajectories, we represent (2.128) in the form

$$\frac{d\sigma}{d\omega} = \int\limits_0^\infty \frac{S_\omega}{\hbar\omega} \cdot 2\pi\rho d\rho = \int\limits_0^\infty \frac{8\pi e^2\omega^4}{3c^3\hbar\omega} |\mathbf{r}_\omega|^2 2\pi\rho d\rho \ , \tag{2.132}$$

where ρ is the collision impact parameter. Considering scattering of a slow electron ($\varepsilon \ll m_e e^4/\hbar^2$) by an ion of charge e, we find below the dependence of the differential cross section of bremsstrahlung on the relevant parameters. The potential energy of the electron at a distance $r_{\min}$ of closest approach exceeds significantly its initial kinetic energy at given collision impact parameters ρ, and hence according to (2.6) we have the connection between these parameters $r_{\min} = \rho^2\varepsilon/e^2$, where ε is the kinetic energy of the electron far from the ion. Typical collision parameters for a given frequency ω of an emitted photon are estimated as $\omega \sim v/r_{\min}$, where v is the electron velocity at the distance of closest approach, that is, $v \sim \sqrt{e^2/m_e r_{\min}}$. Thus, we have

$$r_{\min} \sim \frac{e^{2/3}}{m_e^{1/3}\omega^{2/3}} \ , \quad \rho^2 \sim \frac{e^{8/3}}{m_e^{1/3}\omega^{2/3}\varepsilon} \ , \quad r_\omega \sim \frac{r_{\min}}{\omega} \sim \frac{e^{2/3}}{m_e^{1/3}\omega^{5/3}} \ .$$

From this consideration we have for the differential cross section of bremsstrahlung as a result of electron–ion scattering [45]

$$\frac{d\sigma_\omega}{d\omega} \sim \frac{\rho^2}{\hbar\omega}\frac{e^2\omega^4}{c^3}\frac{r_{\min}^2}{\omega^2} \sim \frac{e^4}{\hbar\omega}\frac{\omega^2 r_{\min}^3}{\varepsilon c^3} \sim \frac{e^6}{\hbar\omega m_e c^3\varepsilon} \ .$$

Note that this method of estimation of the cross section of electron–ion scattering does not allow us to determine the numerical factor in the expression for the cross section of bremsstrahlung, and this requires more cumbrous operations. Such an action is given, for example, in [163] and for this cross section gives

$$\frac{d\sigma}{d\omega} = \frac{8\pi e^6}{3\sqrt{3}m_e c^3\varepsilon\hbar\omega} \ . \tag{2.133}$$

This consideration holds true in the classical limit, so large collision momenta $l = m_e\rho v_0/\hbar \gg 1$ ($v_0 = \sqrt{2\varepsilon/m_e}$ is the initial electron velocity far from the ion) give the main contribution to the cross section. This leads to the criterion $\hbar\omega \ll m_e e^4/\hbar^2$. Another requirement corresponds to the criterion $\rho \gg r_{\min}$, which gives

$$\omega \gg \frac{m_e v_0^3}{e^2} \ .$$

2.4.6
Broadening of Spectral Lines

Radiation due to a certain radiative transition is monochromatic, that is, the width of frequencies of emitted photons is small compared with the frequency of this radiative transition. Nevertheless this narrow spectrum is of importance for reabsorption processes, and therefore we analyze now the mechanisms of broadening of spectral lines [45, 164]. Let us introduce the distribution function of emitted photons a_ω, so $a_\omega d\omega$ is the probability that the frequency of the emitted photon lies in a range between ω and $\omega + d\omega$. Because the probability is normalized to 1, the frequency distribution function of photons a_ω satisfies the relation

$$\int a_\omega d\omega = 1 , \tag{2.134}$$

and the distribution function of emitted photons is given by

$$a_\omega = \delta(\omega - \omega_0) , \tag{2.135}$$

where ω_0 is the frequency of an emitted photon. This means that the spectral line is narrow, that is, the line width $\Delta\omega$ is small compared with the frequency:

$$\Delta\omega \ll \omega_0 .$$

We first find the frequency distribution function a_ω due to a finite lifetime of transition states. Note that the amplitude of radiative transitions is expressed through the matrix element of the dipole moment operator between transition states, and we analyze the time dependence of the matrix element. Since the stationary wave function of state k contains a time factor $\exp(-i E_k t/\hbar)$, where E_k is the energy of this state, the matrix element for transition between stationary states i and k includes the time factor $\exp(-i\omega_0 t)$, where $\omega_0 = (E_i - E_k)/\hbar$. If we account for a finite lifetime τ_r of an upper state and represent the wave function of this state as $\exp(-i E_i t/\hbar - t/2\tau_r)$, the time dependence of the matrix element for transition between two states has the form $c(t) = (-i\omega_0 t - t/2\tau_r)$. Hence, the frequency dependence for a matrix element, the Furie component from the function $c(t)$, is given as $c_\omega \sim |i(\omega - \omega_0) + 1/2\tau|$, and therefore the frequency distribution function of emitted photons is

$$a_\omega = |c_\omega|^2 = \frac{1}{2\pi\tau} \cdot \frac{1}{(\omega - \omega_0)^2 + \nu^2} , \tag{2.136}$$

where $\nu = 1/(2\tau)$ is the spectral line width. Above we accounted for the normalization of the frequency distribution function according to condition (2.134). A frequency distribution function of this shape is named a Lorenz distribution function.

We now consider another mechanism of broadening of spectral lines due to motion of emitting atoms. Indeed, if an emitting particle moves with respect to a receiver with velocity v_x and emits a photon of frequency ω_0, according to the Doppler effect it is perceived by the receiver as having frequency

$$\omega = \omega_0 \left(1 + \frac{v_x}{c}\right) , \tag{2.137}$$

where c is the light velocity. Therefore, if radiating atoms are distributed over velocities, their radiation will be detected by a motionless receiver as being frequency distributed. This distribution is determined by the velocity distribution function $f(v_x)$ of radiating atoms.

If the distribution function over atom velocities is normalized to unity ($\int f(v_x) dv_x = 1$), the frequency distribution function follows from the relation

$$a_\omega d\omega = f(v_x) dv_x \, .$$

Let us consider a prevalent case with a Maxwell distribution over velocities, so that the distribution function has the form

$$f(v_x) = C \exp\left(-\frac{mv_x^2}{2T}\right) ,$$

where T is the atom temperature expressed in energy units, m is the radiating particle mass, and C is the normalization constant. From this we find for the frequency distribution function that is normalized according to condition (2.134)

$$a_\omega = \frac{1}{\omega_0}\left(\frac{mc^2}{2\pi T}\right)^{1/2} \cdot \exp\left[-\frac{mc^2(\omega-\omega_0)^2}{2T\omega_0^2}\right] . \tag{2.138}$$

Note that the ratio of a typical spectral line width due to Doppler broadening $\Delta\omega_D$ to the photon frequency ω_0, that is, the ratio of the atom thermal velocity to the light velocity, is relatively small:

$$\frac{\Delta\omega_D}{\omega_0} \sim \sqrt{\frac{T}{mc^2}} \, .$$

For example, for helium atoms at room temperature the right-hand side of this relation is 2.6×10^{-6}.

Interaction of a radiating atom with surrounding atomic particles may determine broadening of spectral lines in gases or plasmas. We divide this broadening into two limiting cases. In the first case, collision of individual particles of this plasma with a radiating atom proceeds fast, and these collisions are seldom, that is, at each moment interaction of a radiating atom is possible with one surrounding atomic particle only. This mechanism relates to a rare plasma. The other limiting case corresponds to interaction of a radiating atom with many surrounding atomic particles and relates to a dense plasma. The first case of broadening of spectral lines is described by the impact theory, and the second case relates to the quasistatic theory of broadening of spectral lines.

In the case of impact broadening of spectral lines, the phase of the transition matrix element changes during each interaction of a radiating atom with an incident atomic particle. Then the width of a spectral line is determined by collisions of a radiating atom with atomic particles of a plasma and is defined by the cross section of these collisions. Accurately, the frequency distribution function is given by the Lorenz formula (2.136), and in this formula we have [150] $1/\tau = N\langle v\sigma_t\rangle$, where N

is the number density of atomic particles, $\sigma_t = \int d\sigma$ is the total cross section of collision between a radiating atom and atomic particles, and averaging is done over the velocity distribution function of atomic particles.

Note that although perturbation of a radiating atom proceeds in both upper and lower transition states, we reduce the interaction to the upper state only because of a stronger interaction for this state. Next, the total cross section σ_t is based on the phase shift, which in turn is determined in the classical limit as $\int U(R)dt/\hbar$, so $U(R)$ is the interaction potential between colliding particles at a distance R between them. Hence, in the classical limit the total cross section σ_t of collision is estimated as

$$\sigma_t = \pi \rho_t^2 \,, \qquad \frac{U(\rho_t)}{\hbar} \sim \frac{v}{\rho_t} \,, \tag{2.139}$$

where v is the relative velocity of colliding particles and R_0 is the Weiskopf radius. The criterion for validity of this broadening mechanism corresponds to a small probability of perturbed particle location in a region of the order of a Weiskopf radius size that has the form

$$N\sigma_t^{3/2} \ll 1 \,. \tag{2.140}$$

In the case of the opposite criterion with respect to (2.140) the quasistatic theory of spectral line broadening is realized. This theory assumes perturbed particles to be motionless during times which are responsible for broadening. Assuming the interaction potential of a radiating atom with perturbed atomic particles is pairwise and isotropic, we find the frequency distribution function on the wing of a spectral line.

The wing of a spectral line is determined by the location of perturbed atomic particles at small distances from a radiating atom, where the probability of perturbed particles being located is small. For the frequency distribution function this gives

$$a_\omega d\omega = w(R)dR = N \cdot 4\pi R^2 dR \,,$$

where $w(R)dR$ is the probability of a perturbed particle being located at a distance from R to $R + dR$ from the radiating atom. This gives

$$a_\omega = 4\pi R^2 N\hbar \left(\frac{dU}{dR}\right)^{-1} \,, \tag{2.141}$$

where we assume a monotonic dependence $U(R)$. One can estimate the spectral line width for the quasistatic theory of broadening as

$$\Delta\omega \sim \frac{U(N^{-1/3})}{\hbar} \,. \tag{2.142}$$

We now write the criterion for validity of the quasistatic theory of broadening, which is based on the assumption that perturbed atomic particles are motionless during times which are responsible for broadening. We have for a typical time for a particle shift $\tau \sim N^{-1/3}/v$, because $N^{-1/3}$ is the average distance between perturbed particles. Since $\tau \sim (\Delta\omega)^{-1}$, on the basis of (2.142), we have for this criterion

$$\frac{U(N^{-1/3})N^{-1/3}}{\hbar v} \gg 1 .$$

Comparing this criterion with (2.139) for definition of the Weiskopf radius and assuming the monotonic dependence $U(R)$, we find from this the criterion for validity of the quasistatic theory in the form

$$\rho_t N^{1/3} \gg 1 . \qquad (2.143)$$

One can see this criterion is opposite of criterion (2.140). It is based on the assumption that the probability of two and more surrounding particles being located in a region where there is strong interaction with a radiating atom is small. A typical size of this region is $\rho_t \sim \sqrt{\sigma_t}$, and the criterion for the collision broadening of a spectral line for a typical frequency shift is (2.140), that is, the impact theory and the quasistatic theory of broadening of spectral lines relate to opposite cases of interaction of a radiating atom with surrounding atomic particles. Thus, we have various regimes of broadening of spectral lines in plasmas [164].

2.4.7
Cross Section and Absorption Coefficient for Resonant Photons

Let us consider the behavior of an atom that is located in an electromagnetic field of frequency ω. This field can cause atom transitions between lower o and excited $*$ states. Then the rates of the transitions between these states (the transition probabilities per unit time) according to (2.111) have the form

$$w(o, n_\omega \to *, n_\omega - 1) = A \cdot n_\omega ; \quad w(*, n_\omega \to o, n_\omega + 1) = \frac{1}{\tau_r} + B \cdot n_\omega ,$$

where τ_r is the radiative lifetime of an excited state $*$ with respect to spontaneous radiative transition to state o, n_ω is a number of photons with frequency ω, and A and B are the Einstein coefficients [147]. The arguments of the transition rates indicate that atom excitation is accompanied by absorption of one photon of a given frequency, and atom quenching leads to formation of one photon.

The expressions for the Einstein coefficients A and B may be obtained from the balance (2.110) of these radiative transitions under thermodynamic equilibrium:

$$A = \frac{g_*}{g_0 \tau_r} , \quad B = \frac{1}{\tau_r} . \qquad (2.144)$$

Let us derive the expression for the absorption cross section [45, 151, 165], which is the ratio of the transition rate $w(o, n_\omega \to *, n_\omega - 1)$ to the photon flux j_ω. The photon flux is $j_\omega = c dN_\omega$, where the number density of photons is $dN_\omega = 2n_\omega d\mathbf{k}/(2\pi)^3$, where $\mathbf{k}$ is the photon wave vector, $d\mathbf{k}/(2\pi)^3$ is the number of states in a range $d\mathbf{k}$ of the wave vector values, and the factor 2 accounts for the two independent polarization states. From this we have the photon flux on the basis of the dispersion relation $\omega = kc$ for photons: $j_\omega = \omega^2 d\omega/(\pi^2 c^2)$. Correspondingly we find for the absorption cross section σ_ω

$$\sigma_\omega = \frac{w(o, n_\omega \to *, n_\omega - 1)}{j_\omega} = \frac{An_\omega}{j_\omega} = \frac{\pi^2 c^2}{\omega^2} \cdot \frac{g_*}{g_o} \cdot \frac{a_\omega}{\tau_r} . \tag{2.145}$$

In the same manner one can derive the expression for the stimulated cross section σ'_ω as the ratio of the rate of photon formation under the action of the radiation field Bn_ω to the photon flux j_ω. We obtain

$$\sigma'_\omega = \frac{Bn_\omega}{j_\omega} = \frac{\pi^2 c^2}{\omega^2} \cdot \frac{a_\omega}{\tau_r} . \tag{2.146}$$

Let us introduce the absorption coefficient k_ω such that the intensity of radiation I_ω of frequency ω that propagates in a gas of atoms in direction x is given by

$$\frac{dI_\omega}{dx} = -k_\omega I_\omega . \tag{2.147}$$

One can see that $1/k_\omega$ is the mean free path of photons in a gas of atoms. According to the definition, the absorption coefficient k_ω is expressed through the absorption cross section σ_ω and the cross section of stimulated radiation σ'_ω as

$$k_\omega = N_0 \sigma_\omega - N_* \sigma'_\omega = N_0 \sigma_\omega \left(1 - \frac{N_*}{N_0} \frac{g_0}{g_*}\right) . \tag{2.148}$$

If the distribution of atom densities is determined by the Boltzmann formula (1.43), the expression for the absorption coefficient has the form

$$k_\omega = N_0 \sigma_\omega \left[1 - \exp\left(-\frac{\hbar\omega}{T}\right)\right] . \tag{2.149}$$

We now consider an important case of a transition between the ground state and an resonantly excited state when broadening of the spectral line is determined by collisions of these atoms with excitation transfer. The interaction operator that governs broadening of the spectral line in this case has the form

$$U(\mathbf{R}) = \frac{\mathbf{D}_1 \mathbf{D}_2 - 3(\mathbf{D}_1 \mathbf{n})(\mathbf{D}_2 \mathbf{n})}{R^3} .$$

Here $\mathbf{R}$ is the distance between colliding atoms, $\mathbf{n}$ is the unit vector directed along $\mathbf{R}$, and $\mathbf{D}_1$ and $\mathbf{D}_2$ are the dipole moment operators for the atoms indicated. Since the matrix element $d = |\langle o|\mathbf{D}|*\rangle|$ is nonzero, the total cross section for collision of these atoms that is responsible for broadening in this case, according to (2.139), is $\sigma_t \sim d^2/\hbar v$, where v is the collision velocity, and the rate constant $v\sigma_t$ that is responsible for broadening in the center of the spectral line is independent of the collision velocity. Table 2.22 gives values for some resonant transitions of atoms.

Table 2.22 Radiative parameters and the absorption coefficient for the spectral line center for resonant radiative transitions in atoms of the first and second groups of the periodic table [8, 146].

Element	Transition	λ, nm	τ, ns	g_*/g_0	$v\sigma_t$, 10^{-7} cm^3/s	k_0, 10^5 cm^{-1}
H	$1^2S \to 2^2P$	121.57	1.60	3	0.516	8.6
He	$1^1S \to 2^1P$	58.433	0.56	3	0.164	18
Li	$2^2S \to 3^2P$	670.8	27	3	5.1	1.6
Be	$2^1S \to 2^1P$	234.86	1.9	3	9.6	1.4
Na	$3^2S_{1/2} \to 3^2P_{1/2}$	589.59	16	1	3.1	1.1
Na	$3^2S_{1/2} \to 3^2P_{3/2}$	589.0	16	2	4.8	1.4
Mg	$3^1S \to 3^1P$	285.21	2.1	3	5.5	3.4
K	$4^2S_{1/2} \to 4^2P_{1/2}$	769.0	27	1	4.1	0.85
K	$4^2S_{1/2} \to 4^2P_{3/2}$	766.49	27	2	6.3	1.1
Ca	$4^1S \to 4^1P$	422.67	4.6	3	7.3	2.5
Cu	$4^2S_{1/2} \to 4^2P_{1/2}$	327.40	7.0	1	1.1	2.1
Cu	$4^2S_{1/2} \to 4^2P_{3/2}$	324.75	7.2	2	1.7	2.8
Zn	$4^1S \to 4^1P$	213.86	1.4	3	3.3	4.8
Rb	$5^2S_{1/2} \to 5^2P_{1/2}$	794.76	28	1	3.9	0.91
Rb	$5^2S_{1/2} \to 5^2P_{3/2}$	780.03	26	2	6.2	1.2
Sr	$5^1S \to 5^1P$	460.73	6.2	3	9.4	1.7
Ag	$5^2S_{1/2} \to 5^2P_{1/2}$	338.29	7.9	1	1.1	2.0
Ag	$5^2S_{1/2} \to 5^2P_{3/2}$	328.07	6.7	2	1.7	2.9
Cd	$5^1S \to 5^1P$	228.80	1.7	3	3.3	4.5
Cs	$6^2S_{1/2} \to 6^2P_{1/2}$	894.35	31	1	5.3	0.77
Cs	$6^2S_{1/2} \to 6^2P_{3/2}$	852.11	27	2	8.1	1.0
Ba	$6^1S \to 6^1P$	553.55	8.5	3	9.0	1.9
Au	$6^2S_{1/2} \to 6^2P_{1/2}$	267.60	6.0	1	0.49	1.9
Au	$6^2S_{1/2} \to 6^2P_{3/2}$	242.80	4.6	2	0.75	4.2
Hg	$6^1S \to 6^1P$	184.95	1.3	3	2.2	5.7

Note that in the case under consideration, broadening of a spectral line is determined by atomic collisions and corresponds to the criterion

$$N_0 \frac{d^2}{\hbar} \gg \omega \frac{v}{c} \,, \tag{2.150}$$

where v is a typical velocity of atoms. In addition, we assume $\hbar\omega \gg T$, which allows us to ignore stimulated radiation processes. Using (2.145) for the absorption cross section, we have for the absorption coefficient (2.149) in the spectral line center

$$k_0 = N_0 \sigma_\omega \sim N_0 \frac{c^2}{\omega^2} \frac{a_0}{\tau_r} \,.$$

Since the frequency distribution function at the central frequency is $a_0 \sim 1/\Delta\omega \sim \hbar/(N_0 d^2)$ and the radiative time according to (2.112) is estimated as $1/\tau_r \sim \omega^3 d^2/(\hbar c^3)$, we have the following estimate for the absorption coefficient at the spectral line center:

$$k_0 \sim \frac{\omega}{c} \sim \frac{1}{\lambda}, \tag{2.151}$$

where λ is the wavelength of radiation and the absorption coefficient k_0 for the spectral line center of resonant radiative transitions does not depend on the atom number density. It follows from this that the mean free path of resonant photons at the center of the resonant spectral line is proportional to the wavelength λ of resonant radiation, and Table 2.22 contains radiative parameters for some elements.

References

1. Landau, L.D. and Lifshitz, E.M. (1980) *Mechanics*, Pergamon Press, London.
2. Landau, L.D. and Lifshitz, E.M. (1980) *Quantum Mechanics*, Pergamon Press, Oxford.
3. Maxwell, J.C. (1860) *Philos. Mag.*, **19**, 19.
4. Maxwell, J.C. (1860) *Philos. Mag.*, **20**, 21.
5. Maxwell, J.C. (1871) *Theory of Heat*, Longmans, Green and Co., London.
6. Smirnov, B.M. (1982) *Introduction to Plasma Physics*, Nauka, Moscow, in Russian.
7. Smirnov, B.M. (2006) *Principles of Statistical Physics*, Wiley-VCH Verlag GmbH.
8. Smirnov, B.M. (2007) *Plasma Processes and Plasma Kinetics*, Wiley-VCH Verlag GmbH.
9. Leonas, V.B. (1972) *Sov. Phys. Usp.*, **107**, 29.
10. Smirnov, B.M. (1982) *Sov. Phys. Usp.*, **138**, 519.
11. Smirnov, B.M. (1992) *Sov. Phys. Usp.*, **162**(N12), 97.
12. Palkina, L.A., Smirnov, B.M., and Chibisov, M.I. (1969) *Zh. Tekh. Phys.*, **56**, 340.
13. Palkina, L.A. and Smirnov, B.M. (1974) *High Temp.*, **12**, 37.
14. Elezkii, A.V., Palkina, L.A., Smirnov, B.M. (1975) *Transport Phenomena in Weakly Ionized Plasma*, Atomizdat, Moscow, in Russian.
15. Clausisus, R. (1858) *Ann. Phys.*, **105**, 239.
16. Maxwell, J.C. (1873) *Nature*, **417**, 903.
17. Bird, G.A. (1976) *Molecular Gas Dynamics*, Clarendon Press, London.
18. Hansen, J.P. and McDonald, R. (1986) *Theory of Simple Liquids*, Academic Press, London.
19. Bohr, N. (1936) *Nature*, **137**, 344.
20. Bohr, N. and Kalckar, F. (1937) *K. Dan. Vidensk. Sels. Mat.-Fys. Medd.*, **14**, 1.
21. Gamov, G. (1937) *Structure of Atomic Nucleus and Nuclear Transformations*, Clarendon Press, Oxford.
22. Bohr, N. (1939) *Nature*, **143**, 330.
23. Bohr, N. and Wheeler, J.A. (1939) *Phys. Rev*, **56**, 426.
24. Wigner, E.P. and Seitz, F. (1934) *Phys. Rev.*, **46**, 509.
25. Wigner, E.P. (1934) *Phys. Rev.*, **46**, 1002.
26. Smirnov, B.M. (1999) *Clusters and Small Particles in Gases and Plasmas*, Springer, New York.
27. Smirnov, B.M. (2008) *Reference Data on Atomic Physics and Atomic Processes*, Springer, Heidelberg.
28. Smirnov, B.M. (2010) *Cluster Processes in Gases and Plasmas*, Wiley-VCH Verlag GmbH.
29. Faxen, Y. and Holtzmark, J. (1927) *Z. Phys.*, **45**, 307.
30. Fermi, E.E. (1934) *Nuovo Cim.*, **11**, 157.
31. Fermi, E.E. (1936) *Ric. Sci.*, **VII–II**, 13.
32. Bethe, H. (1949) *Phys. Rev.*, **76**, 38.

33 Lebedev, V.S. and Beigman, I.L. (1988) *Physics of Highly Excited Atoms and Ions*, Springer, Berlin.
34 Pack, J.L. *et al.* (1992) *J. Appl. Phys*, **71**, 5363.
35 Ramsauer, C. (1923) *Ann. Phys.*, **72**, 345.
36 Ramsauer, C. and Kollath, R. (1929) *Ann. Phys. (Leipzig)*, **3**, 536.
37 Frost, L.S. and Phelps, A.V. (1964) *Phys. Rev. A*, **136**, 1538.
38 Koizumi, T., Shirakawa, E., and Ogawa, E. (1986) *J. Phys. B*, **19**, 2334.
39 Register, D.F., Vuskovic, L., and Trajimar, S. (1986) *J. Phys. B*, **19**, 1685.
40 McEachran, R.P. and Stauffer, A.D. (1987) *J. Phys. B*, **20**, 3483.
41 Suzuki, M. *et al.* (1992) *J. Phys. D*, **25**, 50.
42 O'Malley, T.F., Spruch, L., and Rosenberg, T. (1961) *J. Math. Phys.*, **2**, 491.
43 O'Malley, T.F. (1963) *Phys. Rev.*, **130**, 1020.
44 Smirnov, B.M. (1968) *Atomic Collisions and Elementary Processes in Plasma*, Atomizdat, Moscow, in Russian.
45 Smirnov, B.M. (1972) *Physics of Weakly Ionized Gas* Nauka, Moscow, in Russian.
46 Rutherford, E. (1911) *Philos. Mag.*, **21**, 669.
47 Landau, L.D. (1937) *Zh. Exp. Theor. Fiz.*, **7**, 203.
48 Wannier, G.H. (1953) *Phys. Rev.*, **90**, 817.
49 Dalgarno, A., McDowell, M.R.C., and Williams, A. (1958) *Philos. Trans. A*, **250**, 411.
50 Mason, E.A., Schamp, H.W. (1958) *Ann. Phys.*, **4**, 233.
51 Kihara, T., Naylor, M.H., and Hirschfelder, J.O. (1960) *Phys. Fluids*, **3**, 715.
52 Smirnov, B.M. (1974) *Ions and Excited Atoms in Plasma*, Atomizdat, Moscow, in Russian.
53 Galitskii, V.M., Nikitin, E.E., and Smirnov, B.M. (1981) *Theory of Collisions of Atomic Particles*, Nauka, Moscow, in Russian.
54 Firsov, O.B. (1951) *Zh. Exp. Theor. Fiz.*, **21**, 1001.
55 Sena, L.A. (1946) *Zh. Exp. Theor. Fiz.*, **16**, 734.
56 Sena, L.A. (1948) *Collisions of Electrons and Ions with Atoms*, Gostekhizdat, Leningrad, in Russian.
57 Smirnov, B.M. (2000) *Phys. Scr.*, **61**, 595.
58 Smirnov, B.M. (2001) *JETP* **92**, 951.
59 Smirnov, B.M. (2001) *Phys. Usp.*, **44**, 221.
60 Smirnov, B.M. (1964) *Sov. Phys. JETP*, **19**, 692.
61 Smirnov, B.M. (1965) *Sov. Phys. JETP*, **20**, 345.
62 Smirnov, B.M. (1973) *Asymptotic Methods in Theory of Atomic Collisions*, Atomizdat, Moscow, in Russian.
63 Smirnov, B.M. (2003) *Physics of Atoms and Ions*, Springer, New York.
64 Sena, L.A. (1939) *Zh. Exp. Theor. Fiz.*, **9**, 1320, in Russian.
65 Smirnov, B.M. (1964) *Dokl. Akad. Nauk SSSR*, **157**, 325, in Russian.
66 Grosh, S.N. and Sheridan, W.F. (1957) *J. Chem. Phys.*, **26**, 480.
67 Grosh, S.N. and Sheridan, W.F. (1957) *Ind. J. Phys.*, **31**, 337.
68 Kushnir, R.M., Paluch, B.M., and Sena, L.A. (1959) *Izv. Acad. Nauk, Ser. Fiz.*, **23**, 1007, in Russian.
69 Gustafsson, E. and Lindholm, E. (1960) *Ark. Fys.*, **18**, 219.
70 Kaneko, Y., Kobayashi, N., and Kanomata, I. (1960) *J. Phys. Soc. Jpn.*, **27**, 992.
71 Galli, A., Giardani-Guidoni, A., and Volpi, G.G. (1962) *Nuovo Cim.*, **26**, 845.
72 Smith, D.L. and Kevan, L. (1971) *Am. Chem. Soc.*, **93**, 2113.
73 Kikiani, B.I., Saliya, Z.E., and Bagdasarova, I.G. (1975) *Zh. Tekh. Fiz.*, **45**, 586, in Russian.
74 Williams, J.F. (1968) *Can. J. Phys.*, **46**, 2339.
75 Mott, N.F. and Massey, H.S.W. (1965) *The Theory of Atomic Collisions*, Clarendon Press, Oxford.
76 Drukarev, G.F. (1987) *Collisions of Electrons with Atoms and Molecules*, Plenum Press, New York.
77 Nikitin, E.E., Smirnov, B.M. (1988) *Atomic and Molecular Collisions*, Nauka, Moscow, in Russian.
78 Bethe, H. (1930) *Ann. Phys.*, **4**, 325.
79 Vainstein, L.A., Sobelman, I.I., and Yukov, E.A. (1979) *Excitation of Atoms and Broadening of Spectral Lines*, Nauka, Moscow, in Russian.
80 Schevel'ko, V.P. and Vainstein, L.A. (1993) *Atomic Physics for Hot Plasmas*, Institute of Physics Publishing, Bristol.

81 Lisitsa, V.S. (1994) *Atoms in Plasmas*, Springer, Berlin.

82 Bureeva, L.A. and Lisitza, V.S. (1994) *Perturbed Atom*, Izd. Atom. Tekh., Moscow, in Russian.

83 Eletskii, A.V. and Smirnov, B.M. (1968) *Zh. Tekh. Fiz.*, **38**, 3.

84 Wigner, E.P. (1948) *Phys. Rev.*, **73**, 1002.

85 Eletskii, A.V. and Smirnov, B.M. (1983) *Sov. Phys. JETP*, **57**, 1639.

86 Eletskii, A.V. and Smirnov, B.M. (1986) *J. Sov. Laser Res.*, **7** 207.

87 Eletskii, A.V. and Sorokin, A.R. (1997) *Laser Phys.*, **7**, 431.

88 Gurevich, L.Je. (1940) *Grounds of Physical Kinetics*, GITTL, Leningrad, in Russian.

89 Smirnov, B.M. (1981) *Physics of Weakly Ionized Gases*, Mir, Moscow.

90 Antonov, E.E. and Korchevoi, Yu.P. (1977) *Ukr. Fiz. Zh.*, **22**, 1557.

91 Zapesochnyi, I.P., Postoi, E.N., and Aleksakhin, I.S. (1976) *Sov. Phys. JETP*, **41**, 865.

92 Smirnov, B.M. (1982) *Excited Atoms*, Energoatomizdat, Moscow, in Russian.

93 Kolokolov, N.B. (1984) in: *Chemistry of Plasma*, vol. 12, (ed. B.M. Smirnov), Energoatomizdat, Moscow, pp. 56–95, in Russian.

94 Thomson, J.J. (1912) *Philos. Mag*, **23**, 449.

95 Fite, W.L. and Brackman, R.T. (1958) *Phys. Rev.*, **112**, 1141.

96 Rapp, D. and Englander-Golden, P. (1965) *J. Chem. Phys.*, **43**, 1464.

97 Long, D.R. and Geballe, R. (1970) *Phys. Rev. A*, **1**, 260.

98 Vriens, L., Bonsen, T.F.M., Smith, J.A. (1968) *Physica*, **40**, 229.

99 Boyd, R.L.F. and Green, G.W. (1958) *Proc. Phys. Soc.*, **71**, 351.

100 McFahrland, R.H. and Kinney, J.D. (1965) *Phys. Rev. A*, **137**, 1058.

101 Dolder, K., Harrison, M.F.A., and Thonemann, P.C. (1961) *Proc. R. Soc.*, **264**, 367.

102 Lineberger, W.C., Hooper, J.P., and McDaniel, E.W. (1966) *Phys. Rev.*, **141**, 151.

103 Wareing, J.B. and Dolder, K.T. (1967) *Proc. R. Soc*, **91**, 887.

104 Zapesochnyi, I.P. and Aleksakhin, I.S. (1968) *JETP Lett.*, **55**, 76.

105 Rejoub, R., Lindsay, B.G., and Stebbings, R.F. (2002) *Phys. Rev. A*, **65**, 042713.

106 Thomas, L. (1927) *Proc. Camb. Philos. Soc*, **28**, 713.

107 Webster, D., Hansen, W., and Duveneck, F. (1933) *Phys. Rev.*, **43**, 833.

108 Seaton, M. (1962) in: *Atomic and Molecular Processes*, Academic Press, New York.

109 Kingston, A.E. (1964) *Phys. Rev. A*, **135**, 1537.

110 Kingston, A.E. (1966) *Proc. Phys. Soc.*, **87**, 193.

111 Kingston, A.E. (1968) *J. Phys. B*, **1**, 559.

112 Thomson, J.J. (1924) *Philos. Mag.*, **47**, 334.

113 Biberman, L.M., Vorob'ev, V.S., and Iakubov, I.T. (1987) *Kinetics of Nonequilibrium Low Temperature Plasma*, Consultants Bureau, New York.

114 Rapp, D. and Briglia, D.D. (1965) *J. Chem. Phys.*, **43**, 1480.

115 Clarke, I.D. and Wayne, R.P. (1969) *Chem. Phys. Lett.*, **3**, 73.

116 Guberman, S.L. (ed) (2003) *Dissociative Recombination of Molecular Ions with Electrons*, Springer, Berlin.

117 Florescu-Mitchell, A.I. and Mitchell, J.B.A. (2006) *Phys. Rep.* **430**, 277.

118 Larsson, M. and Orel, A.E. (2008) *Dissociative Recombination of Molecular Ions*, Cambridge University Press, Cambridge.

119 Smirnov, B.M. (1977) *Sov. Phys. JETP*, **45**, 731.

120 Smirnov, B.M. (1982) *Negative Ions*, McGraw-Hill, London.

121 Massey, H.S.W. (1976) *Negative Ions*, Cambridge University Press, Cambridge.

122 Van Brunt, R.L. and Kiefer, L.J. (1970) *Phys. Rev. A*, **2**, 1900.

123 Henderson, W.R., Fite, W.L., and Brackmann, R.T.. *Phys. Rev.* **183**, 157(1969)

124 Allan, M. and Wong, S.F. (1981) *J. Chem. Phys.*, **74**, 1678.

125 Burns, S.J., Matthews, J.M., and McFadden, D.L. (1996) *J. Chem. Phys.*, **100**, 19436.

126 Illenberger, E. and Smirnov, B.M. (1998) *Phys. Usp.*, **41**, 651.

127 Massey, H.S.W. (1949) *Rep. Prog. Phys.*, 12 248.

128 Jones, C.R., Niles, F.E., and Robertson, W.W. (1969) *J. Appl. Phys.*, **40**, 3967.

129 Ong, P.P. and Hasted, J.B. (1969) *J. Phys. B*, **2**, 91.

130 Baranov, V.Yu., Borisov, V.M., Stepanov, Yu.Yu. (1988) *Gas-Discharge Excimer Lasers with Rare Gas Halides*, Energoizdat, Moscow, in Russian.

131 Rhodes, Ch.K. (ed) (1998) *Excimer Lasers*, Springer, Berlin.

132 Basting, D. and Marowski, G. (eds) (2004) *Excimer Laser Technology*, Springer, Berlin.

133 Stevefelt, J. (1973) *Phys. Rev.*, **8**, 2507.

134 Teter, M.P., Niles, P.E., and Robertson, W.W. (1966) *J. Chem. Phys.*, **44**, 3018.

135 Wellenstein, H.F. and Robertson, W.W. (1972) *J. Chem. Phys.*, **56**, 1072.

136 Dubreuil, B. and Catherinot, A. (1980) *Phys. Rev. A*, **21**, 188.

137 Kruithof, A.A. and Penning, F.M. (1937) *Physica*, **4**, 430.

138 Kruithof, A.A. and Druyvestein, M.J. (1937) *Physica*, **4**, 450.

139 Woodard, M.R. *et al.* (1978) *J. Chem. Phys.*, **69**, 2978.

140 Hotop, H. and Niehaus, A. (1969) *Z. Phys.*, **228**, 68.

141 McGowan, S. (1967) *Can. J. Phys.*, **45**, 639.

142 Wilson, D.E. and Armstrong, D.A. (1970) *Can. J. Chem.*, **48**, 598.

143 Peterson, J.R. *et al.* (1971) *Phys. Rev. A*, **3**, 1651.

144 Barrow, P.D. and Sanche, L. (1972) *Phys. Rev. Lett.*, **28**, 333.

145 Smith, D., Church, M.J. (1976) *Int. J. Mass. Spectrom. Ion. Phys.*, **19**, 185.

146 Smirnov, B.M. (2001) *Physics of Ionized Gases*, John Wiley & Sons, Inc., New York.

147 Einstein, A. (1916) *Verh. Dtsch. Phys. Ges.*, **18**, 318.

148 Condon, E.U. and Shortley, G.H. (1967) *Theory of Atomic Spectra*, Cambridge University Press, Cambridge.

149 Radzig, A.A. and Smirnov, B.M. (1985) *Reference Data on Atoms, Molecules and Ions*, Springer-Verlag, Berlin.

150 Sobelman, I.I. (1979) *Radiative Spectra and Radiative Transitions*, Springer-Verlag, Berlin.

151 Krainov, V.P., Reiss, H., and Smirnov, B.M. (1997) *Radiative Transitions in Atomic Physics*, John Wiley & Sons, Inc., New York.

152 Bethe, H. and Salpeter, E. (1975) *Quantum Mechanics of One- and Two-Electron Atoms*, Springer-Verlag, Berlin.

153 Bethe, H.A. (1964) *Intermediate Quantum Mechanics*, Benjamin, New York.

154 Sommerfeld, A. and Schur, G. (1930) *Ann. Phys.*, **4**, 409.

155 Stobbe, M. (1930) *Ann. Phys.*, **7**, 661.

156 Kramers, H.A. (1923) *Philos. Mag.*, **46**, 836.

157 Bureeva, L.A. and Lisitza, V.S. (1997) *Perturbating Atom*, Izd. tom. Tekh., Moscow.

158 Bates, D.R. and Damgaard, A. (1949) *Philos. Trans.*, **242**, 101.

159 Starace, A.F. (1982) Theory of atomic photoionization, in: *Encyclopedia of Physics*, vol 31. Springer-Verlag, Berlin, pp. 1–121.

160 Frank, H. (1969) International Conference on Physics of Ionized Gases (ICPIG), Vienna, p. 4.

161 Son, E.E. (1978) *Kinetic Theory of Low Temperature Plasma*, Izdat. MFTI, Moscow, in Russian.

162 Son, E.E. (2006) *Lectures in Low Temperature Plasma Physics*, Massachusetts Institute of Technology, Plasma Science Fusion Center.

163 Landau, L.D. and Lifshitz, E.M. (1962) *Classical Theory of Fields*, Addison-Wesley, Reading.

164 Grim, H.R. (1997) *Principles of Plasma Spectroscopy*, Cambridge University Press, New York.

165 Hilborn, R.C. (1982) *Am. J. Phys.*, **50**, 982.

3
Physical Kinetics of Ionized Gases

3.1
Kinetics of Atomic Particles in Gases and Plasmas

3.1.1
The Boltzmann Kinetic Equation

An ensemble of identical atomic particles may be described in different manners. In hydrodynamic terms, we deal with average parameters of atoms of a gaseous system, in particular, with the drift velocity $\mathbf{w}(\mathbf{r})$ of gas atoms at a given coordinate $\mathbf{r}$. In kinetics, the distribution function is used at a given space point $f(\mathbf{v}, \mathbf{r})$, so $f d\mathbf{v}$ is the number density of atoms with velocities in a range between $\mathbf{v}$ and $\mathbf{v} + d\mathbf{v}$. In a general case, we define the distribution function $f(\mathbf{v}, J, \mathbf{r}, t)$ that relates to time t, and the parameter J characterizes all internal quantum numbers of atomic particles. The normalization condition for the distribution function has the form

$$N(\mathbf{r}, t) = \sum_J \int f(\mathbf{v}, J, \mathbf{r}, t) d\mathbf{v} , \tag{3.1}$$

where $N(\mathbf{r}, t)$ is the number density of atomic particles (atoms) at point $\mathbf{r}$ and time t. The distribution function gives detailed information about an ensemble of atomic particles and its evolution. In particular, the drift velocity of particles is given by

$$\mathbf{w}(\mathbf{r}, t) = \sum_J \int \mathbf{v} f(\mathbf{v}, J, \mathbf{r}, t) d\mathbf{v} . \tag{3.2}$$

The characteristic behavior of a test particle in a gas is such that most of the time it moves along straightforward trajectories, and sometimes it interacts with surrounding particles. This particle behavior in a gas is governed by the gaseous state criterion (2.27) $N\sigma^{3/2} \ll 1$, where σ is the cross section of pairwise collisions of gas particles. If this criterion holds true, the probability of collision of a test particle with any gas particle at a given time is of the order of $N\sigma^{3/2} \ll 1$, the probability of simultaneous collision of three gas particles is of the order of $(N\sigma^{3/2})^2 \ll 1$, and so on. Hence, most of the time the test particle does not interact with surrounding gas particles, and its state changes as a result of pairwise collisions with gas particles.

Fundamentals of Ionized Gases, First Edition. Boris M. Smirnov.

Using this character of particle evolution, one can represent the equation for the particle distribution function, the kinetic equation, in the form

$$\frac{df}{dt} = I_{\mathrm{col}}(f) , \tag{3.3}$$

where the left side of this equation describes evolution of a free test particle, and $I_{\mathrm{col}}(f)$ is the so-called collision integral, which takes into account the variation of the number of particles in a given state as a result of pairwise collisions.

The left-hand side of the kinetic equation that describes the motion of particles in external fields is

$$\frac{df}{dt} = \frac{f(\mathbf{v} + d\mathbf{v}, J, \mathbf{r} + d\mathbf{r}, t + dt) - f(\mathbf{v}, J, \mathbf{r}, t)}{dt} .$$

For a free test particle that does not collide with surrounding gas particles, we have the motion equation $d\mathbf{v}/dt = -\mathbf{F}/m$, where $\mathbf{F}$ is the force that acts on the test particle from external fields, m is the particle mass, and the particle velocity is $\mathbf{v} = d\mathbf{r}/dt$. Thus, we have

$$\frac{df}{dt} = \frac{\partial f}{\partial t} + \mathbf{v} \cdot \frac{\partial f}{\partial \mathbf{r}} + \frac{\mathbf{F}}{m} \cdot \frac{\partial f}{\partial \mathbf{v}} ,$$

and the kinetic equation (3.3) takes the form

$$\frac{\partial f}{\partial t} + \mathbf{v} \cdot \frac{\partial f}{\partial \mathbf{r}} + \frac{\mathbf{F}}{m} \cdot \frac{\partial f}{\partial \mathbf{v}} = I_{\mathrm{col}}(f) . \tag{3.4}$$

This is the Boltzmann kinetic equation [1] that describes the evolution of a particle system.

We note that the kinetics of an ensemble of identical particles based on the kinetic equation (3.4) uses other principles compared with classical hydrodynamics and thermodynamics of gases, where the average parameters of particles are taken for each spatial point and time. The concept of the distribution function for gas particles for gas dynamics was developed by Maxwell [2–4] and Boltzmann [5–9] and led to the kinetic equation (3.4) [1] that describes gas dynamics through evolution of the distribution function for gas atoms or molecules. Note that the kinetic description of the evolution of an ensemble of identical particles is suitable for gas and liquid systems [10–14].

3.1.2
Collision Integral for Gas Atoms

The collision integral contained in the kinetic equation characterizes the evolution of the system as a result of pairwise collisions of atomic particles. The collision integral is responsible for gas relaxation when this gas is moved out of equilibrium. The simplest model for the collision integral corresponds to the tau approximation, when the collision integral is approximated by the expression

$$I_{\mathrm{col}}(f) = -\frac{f - f_0}{\tau} , \tag{3.5}$$

where f_0 is the equilibrium distribution function. If the equilibrium state of a gas is disturbed, and the distribution function at the initial time is $f(0)$, the subsequent evolution of this gas within the framework of the tau approximation is described by the following kinetic equation:

$$\frac{df}{dt} = -\frac{f - f_0}{\tau} .$$

Its solution has the form

$$f = f_0 + [f(0) - f_0] \exp\left(-\frac{t}{\tau}\right) .$$

As is seen, the relaxation time τ is a parameter of the collision integral for the tau approximation. It can be estimated as $\tau \sim (N\sigma v)^{-1}$ according to the definition of the relaxation time, where v is a typical collision velocity and σ is a typical cross section of elastic scattering of atoms.

We now give the expression for the collision integral if it is determined by elastic collision of a test atom with gas atoms. Let us introduce the specific rate constant for elastic collision of two atoms $W(\mathbf{v}_1, \mathbf{v}_2 \to \mathbf{v}_1', \mathbf{v}_2')$, so $W d\mathbf{v}_1' d\mathbf{v}_2'$ is the probability per unit time and per unit volume for collision of two atoms with velocities $\mathbf{v}_1$ and $\mathbf{v}_2$ if their final velocities are in an interval from $\mathbf{v}_1'$ to $\mathbf{v}_1' + d\mathbf{v}_1'$ and from $\mathbf{v}_2'$ to $\mathbf{v}_2' + d\mathbf{v}_2'$, respectively. Then, by definition, the collision integral is

$$I_{\text{col}}(f) = \int (f_1' f_2' W' - f_1 f_2 W)\, d\mathbf{v}_1' d\mathbf{v}_2' d\mathbf{v}_2 .$$

where we use the notation $f_1 = f(\mathbf{v}_1)$, $f_2 = f(\mathbf{v}_2)$, $f_1' = f(\mathbf{v}_1')$, $f_2' = f(\mathbf{v}_2')$, $W = W(\mathbf{v}_1, \mathbf{v}_2 \to \mathbf{v}_1', \mathbf{v}_2')$, and $W' = W(\mathbf{v}_1', \mathbf{v}_2' \to \mathbf{v}_1, \mathbf{v}_2)$.

We now analyze the properties of the specific rate constant W for elastic collisions of atoms. The principle of detailed balance, which corresponds to time reversal $t \to -t$, gives $W = W'$. Indeed, on time reversal, the atom ensemble develops in the inverse direction with the same atom trajectories, which corresponds to change $\mathbf{v}_1 \leftrightarrow \mathbf{v}_1', \mathbf{v}_2 \leftrightarrow \mathbf{v}_2'$. This allows us to represent the collision integral in the form

$$I_{\text{col}}(f) = \int W\, (f_1' f_2' - f_1 f_2)\, d\mathbf{v}_1' d\mathbf{v}_2' d\mathbf{v}_2 . \tag{3.6}$$

The specific rate constant W has another property if atoms 1 and 2 are identical. Then replacing atoms gives

$$W(\mathbf{v}_1, \mathbf{v}_2 \to \mathbf{v}_1', \mathbf{v}_2') = W(\mathbf{v}_2, \mathbf{v}_1 \to \mathbf{v}_2', \mathbf{v}_1') . \tag{3.7}$$

Let us replace the specific rate constant in the expression (3.6) for the collision integral by the cross section of elastic collision of atoms. By definition, the elastic scattering cross section is the ratio of the number of scattering events per unit time to the flux of incident particles. Therefore, the differential cross section for elastic scattering of atoms is given by

$$d\sigma = \frac{f_1 f_2 W d\mathbf{v}_1 d\mathbf{v}_2 d\mathbf{v}_1' d\mathbf{v}_2'}{|\mathbf{v}_1 - \mathbf{v}_2| f_1 d\mathbf{v}_1 f_2 d\mathbf{v}_2} = \frac{W d\mathbf{v}_1' d\mathbf{v}_2'}{|\mathbf{v}_1 - \mathbf{v}_2|} .$$

Substitution of this expression into (3.6) gives for the collision integral

$$I_{col}(f) = \int (f_1' f_2' - f_1 f_2) |\mathbf{v}_1 - \mathbf{v}_2| d\sigma d\mathbf{v}_1' d\mathbf{v}_2' d\mathbf{v}_2 . \tag{3.8}$$

The nine integrations implied in (3.6) are replaced by five integrations in (3.8). This is a consequence of accounting for the conservation of momentum for the colliding particles (three integrations) and the conservation of their total energy (one more integration).

3.1.3
Equilibrium Distribution of Gas Atoms

The kinetic equation (3.4) allows us to analyze the equilibrium state of an atomic gas. As follows from (3.4), the equilibrium state of a gas $\partial f/\partial t = 0$ satisfies the equation $I_{col}(f) = 0$. As follows from (3.6) for the collision integral, this equation requires fulfillment of the following relation for any pair of colliding atoms:

$$f_1 f_2 = f_1' f_2' .$$

Let us rewrite this relation in the form

$$\ln f(\mathbf{v}_1) + \ln f(\mathbf{v}_2) = \ln f(\mathbf{v}_1') + \ln f(\mathbf{v}_2') .$$

From this it follows that $\ln f(\mathbf{v})$ is an additive function of the integrals of motion. Taking into account the conservation of total momentum and total energy of the atoms, we obtain the general form of the distribution function as

$$\ln f(\mathbf{v}) = C_1 + (\mathbf{C}_2 \cdot \mathbf{p}) + C_3 \varepsilon ,$$

where $\mathbf{p}$ and ε are the momentum and kinetic energy of an atom. This leads to the distribution function in the form

$$f(\mathbf{v}) = A \exp[-\alpha(\mathbf{v} - \mathbf{w})^2] .$$

This expression corresponds to the Maxwell distribution function, where A is the normalization constant, $\mathbf{w}$ is the average velocity of the distribution, and $\alpha = m/(2T)$, where m is the mass of an atom and T is the temperature of the gas. Thus, the equilibrium is established in a gaseous ensemble of identical particles as a result of collisions between them, and these collisions lead to the Maxwell distribution function for particles, which has the form

$$\varphi(\mathbf{v}) = A \exp\left[-\frac{m(\mathbf{v} - \mathbf{w})^2}{2T}\right] , \tag{3.9}$$

where A is the normalization constant in accordance with normalization condition (3.1) and $\mathbf{w}$ is the average velocity of particles.

Although we did not consider the dynamics for evolution of the ensemble of identical particles, it is clear that this ensemble tends to the equilibrium state with time. This now means that if the distribution function for particles differs from the equilibrium one at the beginning, it becomes the equilibrium one through a long time, and for the gaseous system in the absence of external fields the equilibrium distribution function is given by (3.9). Formally, one can find the contradiction between this statement and the principles of particle mechanics. Indeed, let the ensemble of particles develops from a nonequilibrium state to the equilibrium one. Let us reverse time, and then according to the principle of detailed balance the system would develop from the equilibrium state to a nonequilibrium state. This contradiction means there is an additional principle of statistical mechanics [15–17] that corresponds to a small shift of phases for collision events. From this we also conclude that the kinetics of an ensemble of identical particles includes additional assumptions which are valid for systems of many particles.

3.1.4 Collision Integral for Electrons in a Gas

The specifics of electron–atom collisions in a gas follow from the small ratio of the electron mass m_e to the mass M of an atom. Even if the electron momentum experiences a large change as a result of a collision with an atom, the electron energy varies little. Therefore, the velocity distribution of electrons is nearly symmetric with respect to directions of electron motion [18–21]. This allows one to represent the distribution function in the form of expansion over spherical harmonics [22–26], which results from the small ratio m_e/M and is the basis of strict kinetics for evolution of electrons in a gas [27]. Therefore, we have for the electron distribution function [12, 27]

$$f(\mathbf{v}) = f_0(v) + v_x f_1(v) , \tag{3.10}$$

where the x-axis is in the direction of the electric field $\mathbf{E}$.

The kinetic equation (3.4) now has the form

$$\frac{e\mathbf{E}}{m_e} \cdot \frac{\partial f}{\partial \mathbf{v}} = I_{col}(f) ,$$

where $\mathbf{E}$ is the electric field strength. Assuming the number density of electrons N_e to be small compared with the atom number density N_a, we find that the presence of electrons in a gas does not affect the Maxwell distribution function $\varphi(v_a)$ for the atoms, and the electron–atom collision integral has a linear dependence on the distribution function $f(\mathbf{v})$, so the electron–atom collision integral I_{ea} has the form

$$I_{ea}(f) = I_{ea}(f_0) + I_{ea}(v_x f_1) .$$

Let us integrate this equation over the angle θ between the directions of the electron velocity and the electric field strength while accounting for (3.10) and also repeat this operation after multiplication of this equation by $\cos\theta$. As a result, we obtain the above kinetic equation in the form

$$a\frac{v_x}{v}\frac{df_0}{dv} = I_{ea}(v_x f_1)\,, \quad \frac{a}{3v^2}\frac{d}{dv}\left(v^3 f_1\right) = I_{ea}(f_0)\,, \tag{3.11}$$

where $a = eE/m_e$.

First let us determine the collision integral from a nonsymmetric part $I_{ea}(v_x f_1)$ of the collision integral. Using (3.7) for the collision integral and accounting for conservation of the atom velocity $\mathbf{v}_a$ in an electron–atom collision, since the magnitude of $\mathbf{v}_a$ is small compared with that of the electron velocity $\mathbf{v}$, we have

$$I_{ea}(v_x f_1) = \int (\mathbf{v}' - \mathbf{v})_x v\, d\sigma\, f_1(v)\varphi(v_a)d\mathbf{v}_a\,.$$

The electron velocity after collision has the form $\mathbf{v}' = \mathbf{v}\cos\vartheta + \mathbf{k}v\sin\vartheta$, where ϑ is the scattering angle, and the unit vector $\mathbf{k}$ is directed perpendicular to the electron velocity $\mathbf{v}$. Accounting for random directions of vector $\mathbf{k}$, we obtain after integration

$$I_{ea}(v_x f_1) = -\nu v_x f_1(v)\,, \tag{3.12}$$

where $\nu = N_a v\sigma^*(v)$ is the rate of electron–atom collisions and $\sigma^*(v) = (1 - \cos\vartheta)d\sigma$ is the diffusion cross section of electron–atom scattering.

For determination of $I_{ea}(f_0)$ we take into account that the change of the electron energy in any single collision is small compared with the total electron energy. Therefore, the collision integral is a divergence of a flux in the electron energy space and is the right-hand side of the Fokker–Plank equation [28, 29], which takes into account a small variation of the electron energy in a single collision. Let us consider a group of processes with a small variation of a variable z in each individual event, that is, the system is diffusive in nature in the variable z. We define the probability $W(z_0, t_0; z, t)$ such that the value z occurs at time t if at time t_0 it was z_0. The normalization condition for this probability is

$$\int W(z_0, t_0; z, t)dz = 1\,.$$

Because of the continuous character of the evolution of the probability W, it satisfies the continuity equation

$$\frac{\partial W}{\partial t} + \frac{\partial j}{\partial z} = 0\,,$$

where the flux j can be represented in the form

$$j = AW - B\frac{\partial W}{\partial z}\,.$$

Here the first term is associated with the hydrodynamic flux, and the second one corresponds to the diffusion flux. By definition, the coefficients for these processes are

$$A(z,t) = \lim_{\tau\to 0} \frac{1}{\tau} \int (x-z)\, W(x,t;z,t+\tau) dx \;,$$
$$B(z,t) = \lim_{\tau\to 0} \frac{1}{2\tau} \int (x-z)^2\, W(x,t;z,t+\tau) dx \;. \tag{3.13}$$

The corresponding equation for the probability is called the Fokker–Planck equation [28, 29], and has the form

$$\frac{\partial W}{\partial t} = -\frac{\partial(AW)}{\partial z} + \frac{\partial^2(BW)}{\partial z^2} \;. \tag{3.14}$$

This equation can be generalized for the case when the normalization condition has the form

$$\int \rho(z)\, W(z_0,t_0;z,t) dz = 1 \;,$$

that is, the state density is $\rho(z)$. Then the quantity W in (3.13) must be replaced by ρW, and the Fokker–Planck equation then takes the form

$$\rho\frac{\partial W}{\partial t} = -\frac{\partial(\rho AW)}{\partial z} + \frac{\partial^2(\rho BW)}{\partial z^2} \;. \tag{3.15}$$

The right-hand side of this equation can be used as the collision integral of the spherical part of the electron distribution function because it describes the incremental changes of the electron energy. To accomplish this, we replace $W(z_0,t_0;z,t)$ in (3.15) with the distribution function f_0, and in place of z we substitute the electron energy ε. Then $\rho(\varepsilon)$ behaves as $\rho(\varepsilon) \sim \varepsilon^{1/2}$, and the collision integral takes the form

$$I_{\text{ea}}(f_0) = \frac{1}{\rho(\varepsilon)}\frac{\partial}{\partial\varepsilon}\left[-A\rho f_0 + \frac{\partial}{\partial\varepsilon}(B\rho f_0)\right] .$$

The connection between A and B is found from the condition that if the distribution function coincides with the Maxwell distribution function, the collision integral will be zero. This condition yields [30]

$$I_{\text{ea}}(f_0) = \frac{1}{\rho(\varepsilon)}\frac{\partial}{\partial\varepsilon}\left[\rho(\varepsilon)B(\varepsilon)\left(\frac{\partial f_0}{\partial\varepsilon} + \frac{f_0}{T}\right)\right] , \tag{3.16}$$

where T is the gas temperature. This collision integral is zero for the Maxwell distribution function for electrons when the electron temperature is equal to the gas temperature T.

In determining the quantity $B(\varepsilon)$ for the electron–atom collision integral, we account for its definition

$$B(\varepsilon) = \frac{1}{2}\left\langle \int (\varepsilon-\varepsilon')^2 N_{\text{a}} v\, d\sigma(\varepsilon\to\varepsilon')\right\rangle ,$$

where the angle brackets signify an average over atomic energies and $d\sigma$ is the electron–atom cross section corresponding to a given variation of the electron energy. We can make use of the constancy of the relative electron–atom velocity in the collision process, or $|(\mathbf{v} - \mathbf{v}_a)| = |\mathbf{v}' - \mathbf{v}_a|$, where $\mathbf{v}$ and $\mathbf{v}'$ are the electron velocities before and after the collision and $\mathbf{v}_a$ is the velocity of the atom, unvarying in a collision with an electron. From this it follows that $v^2 - (v')^2 = 2\mathbf{v}_a(\mathbf{v} - \mathbf{v}')$, which yields the result [30]

$$B(\varepsilon) = \frac{m_e^2}{2}\left\langle\frac{v_a^2}{3}\right\rangle \int (\mathbf{v} - \mathbf{v}')^2 N_a v\, d\sigma = T\frac{m_e}{M}\frac{m_e v^2}{2} N_a v \sigma^*(v) \,. \tag{3.17}$$

In the determination of this expression we use $\langle v_a^2/3\rangle = T/M$, where T is the gas temperature, M is the atom mass, and $|\mathbf{v} - \mathbf{v}'| = 2v\sin(\vartheta/2)$, where ϑ is the scattering angle and $\sigma^*(v) = \int(1 - \cos\vartheta)d\sigma$ is the diffusion cross section of electron–atom scattering. Thus, the collision integral from the spherical part of the electron distribution function has the form [12, 30]

$$I_{ea}(f_0) = \frac{m_e}{M}\frac{\partial}{v^2\partial v}\left[v^3\nu\left(\frac{T\partial f_0}{m_e v\partial v} + f_0\right)\right] , \tag{3.18}$$

where $\nu = N_a v\sigma^*(v)$ is the rate of electron–atom collisions.

3.1.5
Collision Integral for Fast Electrons in an Electron Gas

When a fast electron is moving in an ionized gas, its scattering by electrons and ions of this gas proceeds mostly into small angles because such scattering gives the main contribution to the diffusion cross section for collision of a fast electron and charged particles in a plasma. We determine below the spherical part of the electron distribution function $I_{ee}(f_0)$ for electron scattering by electrons and ions. In this way we account for the main part of the electron distribution function being almost spherically symmetric since because of the small parameter m_e/M (m_e is the electron mass, M is the atom mass) change of the electron momentum proceeds more effectively than change of the electron energy. Next, variation of the electron energy as a result of one collision with charged particles is relatively small, so the collision integral is given by (3.16), which has the form

$$I_{ee}(f_0) = \frac{\partial}{m_e v^2\partial v}\left[v B(\varepsilon)\left(\frac{\partial f_0}{m_e v\partial v} + \frac{f_0}{T_e}\right)\right] ,$$

where $B(\varepsilon)$ is determined by (3.13), and the Maxwell distribution function is used for the electron gas with electron temperature T_e, and the energy of a fast electron satisfies the criterion

$$\varepsilon \gg T_e \,.$$

Under this criterion, we can take the hydrodynamic flux into account in the Fokker–Planck equation (3.16), that is,

$$I_{ee}(f_0) = \frac{\partial}{m_e v^2\partial v}[v A(\varepsilon) f_0] \,.$$

This means that a fast electron loses energy only.

According to definition (3.13), we have for the rate of electron braking

$$A(\varepsilon) = \int (\varepsilon - \varepsilon') N_e v d\sigma = \int \frac{(\Delta p)^2}{2m_e} N_e v \cdot 2\pi \rho d\rho \,,$$

where $\Delta p = 2e^2/(\rho v)$ is the momentum which is transferred to a slow electron from the fast one for collision impact parameter ρ and N_e is the number density of slow electrons. From this we find for the rate of braking of a fast electron in a gas of slow electrons

$$A(\varepsilon) = \frac{4\pi e^4 N_e}{m_e v} \ln \Lambda \,, \tag{3.19}$$

where $\ln \Lambda$ is the Coulomb logarithm. Adding formally the diffusion term to the Fokker–Plank equation and requiring zero collision integral for the Maxwell distribution function, we obtain the collision integral for a fast electron in an electron gas:

$$I_{ee}(f_0) = \frac{4\pi e^4 N_e \ln \Lambda}{m_e v} \frac{\partial}{m_e v \partial v} \left(\frac{T_e \partial f_0}{m_e v \partial v} + f_0 \right) , \quad \varepsilon \gg T_e \,. \tag{3.20}$$

We now derive the equation for variation of the electron energy ε when a fast electron is moving in an electron gas. Then we use the kinetic equation in the form

$$\frac{\partial f_0}{\partial T} = I_{ee}(f_0) \,,$$

where the collision integral is given by (3.20) with $T_e = 0$. We take the distribution function for fast electrons as

$$f_0(u) = \frac{\delta(u - v)}{4\pi u^2} \,,$$

where the normalization condition for fast electrons is taken in the form

$$\int 4\pi u^2 f_0(u) du = 1 \,.$$

Let us multiply the kinetic equation for the distribution function for fast electrons by the electron energy ε and integrate over electron velocities. We obtain

$$\frac{d\overline{\varepsilon}}{dt} = -A(\overline{\varepsilon}) \equiv -\frac{4\pi e^4 N_e}{m_e v} \ln \Lambda \,, \tag{3.21}$$

where $\overline{\varepsilon}$ is the average electron energy, which exceeds significantly a thermal energy of slow electrons.

3.1.6
The Landau Collision Integral

Expression (3.20) gives the electron–electron collision integral in the limit when the energy of a test electron is high compared with the energy of most of the other electrons and is based on the assumption that the variation of the electron energy resulting from a single electron–electron collision is relatively small. This assumption is valid when the Coulomb logarithm is large, so small-angle scattering gives the main contribution to the electron–electron scattering cross section. This fact allows us to formulate the problem in another way by ignoring the assumption of high energy of a test electron. Because the electron momentum varies weakly in small-angle scattering, one can represent the collision integral by the three-dimensional Fokker–Planck equation. Such a form of the collision integral is named the Landau collision integral. We develop its form below [31].

We start from the general expression (3.6) for the collision integral and account for the symmetry of the rate W that follows from the principle of detailed balance for elastic collisions of identical particles and which is given by (3.7). Let us introduce the variation $\Delta\mathbf{v}$ of the electron velocity as a result of collision, that is, $\mathbf{v}_1' = \mathbf{v}_1 + \Delta\mathbf{v}$. Because of conservation of the total momentum in the collision we have $\mathbf{v}_1' = \mathbf{v}_1 + \Delta\mathbf{v}$. Using the symmetry of the rate W due to the principle of detailed balance, we represent it in the form

$$W(\mathbf{v}_1, \mathbf{v}_2 \to \mathbf{v}_1', \mathbf{v}_2') = W\left(\mathbf{v}_1 + \frac{\Delta\mathbf{v}}{2}, \mathbf{v}_2 - \frac{\Delta\mathbf{v}}{2}, \Delta\mathbf{v}\right),$$

and from the principle of detailed balance it follows that W is an even function of $\Delta\mathbf{v}$, that is,

$$W\left(\mathbf{v}_1 + \frac{\Delta\mathbf{v}}{2}, \mathbf{v}_2 - \frac{\Delta\mathbf{v}}{2}, \Delta\mathbf{v}\right) = W\left(\mathbf{v}_1 - \frac{\Delta\mathbf{v}}{2}, \mathbf{v}_2 + \frac{\Delta\mathbf{v}}{2}, -\Delta\mathbf{v}\right).$$

We use this symmetry in the expression for the electron–electron collision integral, which has the form

$$I_{ee}(f) = -\int \left[f(\mathbf{v}_1) f(\mathbf{v}_2) - f(\mathbf{v}_1 + \Delta\mathbf{v}) f(\mathbf{v}_2 - \Delta\mathbf{v})\right] \times W\left(\mathbf{v}_1 + \frac{\Delta\mathbf{v}}{2}, \mathbf{v}_2 - \frac{\Delta\mathbf{v}}{2}, \Delta\mathbf{v}\right), \tag{3.22}$$

and our goal is expansion of this expression over a small parameter $\Delta\mathbf{v}$.

The first term in the expansion of the collision integral over the small parameter $\Delta\mathbf{v}$ is

$$I_{ee}(f) = -\int \left[f(\mathbf{v}_2)\frac{\partial f(\mathbf{v}_1)}{\partial \mathbf{v}_1} - f(\mathbf{v}_1)\frac{\partial f(\mathbf{v}_2)}{\partial \mathbf{v}_2}\right] \Delta\mathbf{v}\, W\, d\Delta\mathbf{v}\, d\mathbf{v}_2 .$$

Since W is an even function of $\Delta\mathbf{v}$ and because of the symmetry of this function with respect to transformation $\mathbf{v}_1 \leftrightarrow \mathbf{v}_2$, this approximation gives zero.

In the second-order expansion over the small parameter $\Delta \mathbf{v}$ we have

$$I_{ee}(f) = -\int d\Delta\mathbf{v} d\mathbf{v}_2 W \left(\frac{1}{2}\Delta_\alpha \Delta_\beta \frac{\partial^2 f_1}{\partial v_{1\alpha} \partial v_{1\beta}} f_2 - \Delta_\alpha \Delta_\beta \frac{\partial f_1}{\partial v_{1\alpha}} \frac{\partial f_2}{\partial v_2 \beta} + \frac{1}{2}\Delta_\alpha \Delta_\beta f_1 \frac{\partial^2 f_2}{\partial v_{2\alpha} \partial v_{2\beta}} \right) - \int d\Delta\mathbf{v} d\mathbf{v}_2 \frac{1}{2}\Delta_\alpha \left(\frac{\partial W}{\partial v_{1\alpha}} - \frac{\partial W}{\partial v_{2\alpha}} \right) \Delta_\beta \left(\frac{\partial f_1}{\partial v_{1\beta}} f_2 - f_1 \frac{\partial f_2}{\partial v_{2\beta}} \right) ,$$

where we use the notation $f_1 \equiv f(\mathbf{v}_1)$, $f_2 \equiv f(\mathbf{v}_2)$, and $\Delta_\alpha \equiv \Delta \mathbf{v}_\alpha$; indices α and β denote components of the indicated vectors, and summation is made over two repeated indices. Integrating over parts, one can evaluate some terms of this expression. We have

$$\frac{1}{2}\int d\Delta\mathbf{v} d\mathbf{v}_2 W \Delta_\alpha \Delta_\beta \frac{\partial f_1}{\partial v_{1\alpha}} \frac{\partial f_2}{\partial v_{2\beta}} + \frac{1}{2}\int d\Delta\mathbf{v} d\mathbf{v}_2 \Delta_\alpha \Delta_\beta \frac{\partial W}{\partial v_{2\alpha}} \frac{\partial f_1}{\partial v_{1\beta}} f_2 = \frac{1}{2}\int d\Delta\mathbf{v} d\mathbf{v}_2 W \Delta_\alpha \Delta_\beta \frac{\partial f_1}{\partial v_{1\alpha}} \frac{\partial}{\partial v_{2\beta}} (W f_2) = 0 ,$$

$$\frac{1}{2}\int d\Delta\mathbf{v} d\mathbf{v}_2 W \Delta_\alpha \Delta_\beta f_1 \frac{\partial^2 f_2}{\partial v_{2\alpha} \partial v_{2\beta}} + \frac{1}{2}\int d\Delta\mathbf{v} d\mathbf{v}_2 W \Delta_\alpha \Delta_\beta \frac{\partial W}{\partial v_{2\alpha}} f_1 \frac{\partial f_2}{\partial v_{2\beta}} = \frac{\partial W}{\partial v_{2\alpha}} f_1 \frac{\partial f_2}{\partial v_{2\beta}} f_1 \frac{\partial}{\partial v_{2\alpha}} \left(W \frac{\partial f_2}{\partial v_{2\beta}} \right) = 0 ,$$

since the distribution function is zero at $v_{2\beta} \to \pm\infty$. After elimination of these terms we obtain

$$I_{ee}(f) = -\frac{1}{2}\int d\Delta\mathbf{v} d\mathbf{v}_2 \Delta_\alpha \Delta_\beta \times \left(W \frac{\partial^2 f_1}{\partial v_{1\alpha} \partial v_{1\beta}} f_2 - W \frac{\partial f_1}{\partial v_{1\alpha}} \frac{\partial f_2}{\partial v_{2\beta}} + \frac{\partial W}{\partial v_{1\alpha}} \frac{\partial f_1}{\partial v_{1\beta}} f_2 - \frac{\partial W}{\partial v_{2\alpha}} f_1 \frac{\partial f_2}{\partial v_{2\beta}} \right) = -\frac{\partial j_\beta}{\partial v_{1\beta}} ,$$

where the flux in electron-velocity space is

$$j_\beta = \int d\mathbf{v}_2 \left(f_1 \frac{\partial f_2}{\partial v_{2\beta}} - \frac{\partial f_1}{\partial v_{1\beta}} f_2 \right) D_{\alpha\beta} , \quad D_{\alpha\beta} = \frac{1}{2}\int \Delta_\alpha \Delta_\beta W d\Delta\mathbf{v} .$$

This symmetric form of the electron–electron collision integral is called the Landau collision integral after evaluation of the diffusion coefficient in a space of electron velocities. For evaluation of tensor $D_{\alpha\beta}$ we use (2.34) for variation of the electron velocity after electron–electron collision with scattering into a small angle:

$$\Delta_\alpha = \frac{2e^2 \rho_\alpha}{\rho^2 g m_e} ,$$

where ρ_α is a component of the collision impact parameter and $g = |\mathbf{v} - \mathbf{v}'|$ is the relative velocity of the colliding electrons. From this we obtain for the diffusion coefficient in electron-velocity space

$$D_{\alpha\beta} = \frac{1}{2}\int \Delta_\alpha \Delta_\beta W d\Delta\mathbf{v} = \frac{1}{2}\int \Delta_\alpha \Delta_\beta g d\sigma = \frac{4\pi e^4}{m_e^2 g}\int \frac{\rho_\alpha \rho_\beta}{\rho^3} d\rho .$$

Let us take the frame of reference where the relative velocity $\mathbf{g}$ is directed along the x-axis and the impact parameter is directed along the y-axis, that is, the plane of motion is the xy plane. Then only Δ_y is nonzero, so only tensor component D_{yy} is nonzero. For this component of the tensor we obtain

$$D_{yy} = \frac{2e^4}{m_e^2 g}\int \frac{1}{\rho^2} 2\pi\rho d\rho = \frac{4\pi e^4}{m_e^2 g}\ln \Lambda ,$$

where the integral over collision impact parameters is evaluated in a straightforward way, and $\ln \Lambda$ is the Coulomb logarithm.

Transferring to an arbitrary frame of reference, we take into account that tensor $D_{\alpha\beta}$ is symmetric with respect to its indices and can be constructed on the basis of symmetric tensors $\delta_{\alpha\beta}$ and $g_\alpha g_\beta$. Therefore, it has the form

$$D_{\alpha\beta} = \frac{2\pi e^4}{m_e^2 g^3}(g_\alpha g_\beta - g^2 \delta_{\alpha\beta}) \ln \Lambda .$$

As a result, the Landau collision integral, which accounts for collisions between electrons, has the form [31]

$$I_{ee}(f) = -\frac{\partial j_\beta}{\partial v_{1\beta}} , \quad j_\beta = \int d\mathbf{v}_2 \left(f_1 \frac{\partial f_2}{\partial v_{2\beta}} - \frac{\partial f_1}{\partial v_{1\beta}} f_2 \right) D_{\alpha\beta} ,$$

$$D_{\alpha\beta} = \frac{2\pi e^4}{m_e g^3}(g_\alpha g_\beta - g^2 \delta_{\alpha\beta}) \ln \Lambda . \tag{3.23}$$

The Landau collision integral is a version of the right-hand side of the Fokker–Planck equation.

Let us consider the case when a beam of fast electrons propagates through an electron gas, and braking of fast electrons results from collisions with slow electrons of the electron gas. Then the collision integral is given by (3.20), where the first term in parentheses is small and may be ignored, so the rate of variation of the average energy of fast electrons is given by (3.21). We now analyze the kinetics of the braking process using the Landau collision integral.

We divide electrons into two types, a beam of fast electrons with the distribution function $f(\mathbf{v}) = f_1(\mathbf{v})$ and an electron gas of slow electrons with the distribution function $f_2(v)$. Taking the x-axis as the beam direction, we find only two components of the diffusion tensors are nonzero:

$$D_{yy} = D_{zz} = \frac{2\pi e^4}{m_e^2 v}\ln \Lambda ,$$

where v is the velocity of a fast electron, which now coincides with the relative velocity of colliding electrons. Since the diffusion coefficient in the velocity space $D \equiv D_{yy} = D_{zz}$ is independent of the slow electron velocity, the first term in parentheses in (3.23) for electron flux is zero after integration by parts, because the distribution function is zero for infinite electron velocity. As a result, we obtain the kinetic equation for the distribution function $f(\mathbf{v})$ of fast electrons:

$$\frac{\partial f}{\partial t} = D N_e \left(\frac{\partial^2 f}{\partial^2 y^2} + \frac{\partial^2 f}{\partial^2 z^2} \right) , \quad D = \frac{2\pi e^4}{m_e^2 v} \ln \Lambda , \tag{3.24}$$

where $N_e = \int f_2 d\mathbf{v}_2$ is the number density of the electron gas.

Equation (3.24) describes the smearing of a beam of fast electrons moving through an electron gas. In the approach considered, the beam velocity does not change, but motion arises in the direction perpendicular to the beam. This equation is valid, until $v_y, v_z \ll v_x = v$. Solution of this equation gives the distribution function at this stage of beam evolution, but we are restricted now by average parameters. Indeed, for the diffusion equation we have

$$\overline{v_y^2} = \overline{v_z^2} = 2 D N_e t ,$$

where an average is made over velocities of fast electrons. In evaluating a change of the electron energy, we note an equal change of the velocity of colliding fast and slow electrons in the direction perpendicular to the motion. Hence, the energy $\Delta\varepsilon$ transferred from a fast electron to slow electrons in their collisions is

$$\Delta\varepsilon = \frac{m_e v_y^2}{2} + \frac{m_e v_z^2}{2} ,$$

and the average change of the fast electron energy for time t is

$$\overline{\Delta\varepsilon} = 2 m_e D N_e t .$$

From this we obtain the rate of variation for the average energy $\overline{\varepsilon}$ of a fast electron:

$$\frac{d\overline{\varepsilon}}{dt} = -\frac{\overline{\Delta\varepsilon}}{t} = -2 m_e D N_e = -\frac{4\pi e^4 N_e}{m_e v} \ln \Lambda ,$$

which coincides with (3.21).

3.2 Kinetics of Electrons in a Gas in an External Electric Field

3.2.1 Kinetics of Electrons in a Gas Resulting from Elastic Collisions with Atoms

The simplest method to generate a weakly ionized gas consists in applying an electric field to it. A gas discharge plasma is created in this way, and equilibrium in this system results from transition of the electric field energy to electrons and then this energy is transferred to atoms of the gas in electron–atom collisions. This leads to a self-maintaining state of this ionized gas. Analyzing the establishment of the electron velocity distribution function, we note that the cross section of elastic electron–atom collisions exceeds significantly the cross section of inelastic collisions. Hence, representing the electron distribution function as a sum (3.10) of the symmetric and antisymmetric harmonics, we obtain the collision integral (3.12) for the antisymmetric distribution function, and the set (3.11) of kinetic equations takes the form

$$a\frac{v_x d f_0}{v dv} = -\nu v_x f_1(v) , \quad \frac{a}{3v^2}\frac{d}{dv}\left(v^3 f_1\right) = I_{ea}(f_0) + I_{ee}(f_0) . \tag{3.25}$$

Here $\nu = N_a v\sigma^*$ is the rate of elastic electron–atom collisions, $a = eE/m_e$, and we include in the consideration both elastic electron–atom collisions and electron–electron collisions.

The first equation of the set (3.25) allows one to express the electron drift velocity through the symmetric part of the velocity distribution function as [27]

$$w_e = \int v_x^2 f_1 d\mathbf{v} = \frac{eE}{3m_e}\left\langle\frac{1}{v^2}\frac{d}{dv}\left(\frac{v^3}{\nu}\right)\right\rangle , \tag{3.26}$$

where an average is made over the spherical distribution function f_0 for electrons. This expression is independent of the character of establishment for the symmetric part of the electron distribution function f_0.

We now examine the limit of low electron number densities, when collisions between electrons are not essential in this process. Then, taking the electron–electron collision integral to be zero $I_{ee} = 0$ in the second equation of the set (3.25) and using (3.18) for the electron–atom collision integral, we obtain from the second equation of the set (3.25)

$$f_1 = \frac{3m_e}{aM}\nu\left(\frac{d f_0}{m_e v dv} + \frac{f_0}{T}\right) ,$$

where we account for $f_0 \to 0$ if $v \to \infty$. The solution of the set of equations for f_0 and f_1 yields

$$f_0(v) = A\exp\left(-\int_0^v \frac{m_e v' dv'}{T + Mu^2/3}\right) , \quad f_1(v) = \frac{m_e u}{T + Mu^2/3}\cdot f_0(v) , \tag{3.27}$$

where A is the normalization constant and $u = eE/(m_e\nu) = eE/[m_e N_a v\sigma^*(v)]$. In particular, if $\nu(v) = \text{const} = 1/\tau$, the electron drift velocity w_e and the average energy $\overline{\varepsilon}$ are given by

$$w_e = \frac{eE}{m_e\nu} = \frac{eE\tau}{m_e} , \quad \overline{\varepsilon} = \frac{3}{2}T + \frac{M}{2}w_e^2 \tag{3.28}$$

at any electric field strength. If $\sigma^*(v) = \text{const}$, in the limit of high electric field strengths $\overline{\varepsilon} \gg T$, the distribution functions (3.27) yield

$$w_e = 0.897 \left(\frac{m_e}{M}\right)^{1/4} \sqrt{\frac{eE\lambda}{m_e}}, \quad \overline{\varepsilon} = 0.427 \sqrt{\frac{M}{m_e}} eE\lambda = 0.530 M w_e^2, \quad (3.29)$$

where the mean free path for electrons is $\lambda = 1/(N_a \sigma^*)$.

As follows from (3.26), the electron drift velocity depends on the electric field strength E through the reduced electric field strength E/N_a, which is usually measured usually in Townsends (Td) [33] (1 Td $= 10^{-17}$ V cm^2). Table 3.1 contains the values of the electron drift velocity in some gases depending on the reduced electric field strength. In addition, Figures 3.1 and 3.2 give the measured drift velocity of electrons in helium and argon as a function of the reduced electric field strength.

Let us give the criterion for validity of this regime for electron drift in a gas. We ignored collisions between electrons compared with electron–atom collisions, which corresponds to assumption $I_{ea}(f_0) \gg I_{ee}(f_0)$ in the second equation of the set (3.25). This criterion may be represented in the form $B_{ea} \gg B_{ee}$, where the electron diffusion coefficient B_{ea} in the energy space for electron–atom collisions is given by (3.10), and this coefficient for electron–electron collisions in accordance with (3.19) and (3.20) is

$$B_{ee} = \frac{4\pi e^4 N_e T_e}{m_e v} \ln \Lambda .$$

Correspondingly, the criterion for this regime of electron drift takes the form

$$\frac{N_e}{N_a} \ll \frac{m_e}{M} \frac{T}{T_e} \frac{\varepsilon^2 \sigma^*_{ea}}{2\pi e^4 \ln \Lambda}, \quad (3.30)$$

Table 3.1 The experimental electron drift velocity w_e in gases expressed in units of 10^5 cm/s for the regime of a low electron number density [32].

E/N_a, Td	He	Ne	Ar	Kr	Xe	H_2	N_2
0.001	0.028	0.35	0.15	0.015	0.004	0.020	–
0.003	0.080	0.55	0.53	0.038	0.013	0.052	–
0.01	0.25	0.89	1.0	0.13	0.040	0.17	0.39
0.03	0.64	1.3	1.2	0.90	0.16	0.46	1.0
0.1	1.4	2.1	1.7	1.4	0.86	1.4	2.3
0.3	2.7	3.5	2.4	–	1.1	3.2	3.1
1	4.8	–	3.1	–	1.4	6.2	4.5
3	8.6	–	4.0	–	1.8	10	7.7
10	22	–	11	–	6.0	19	18
30	72	–	26	–	18	37	42
100	–	–	74	–	53	130	100

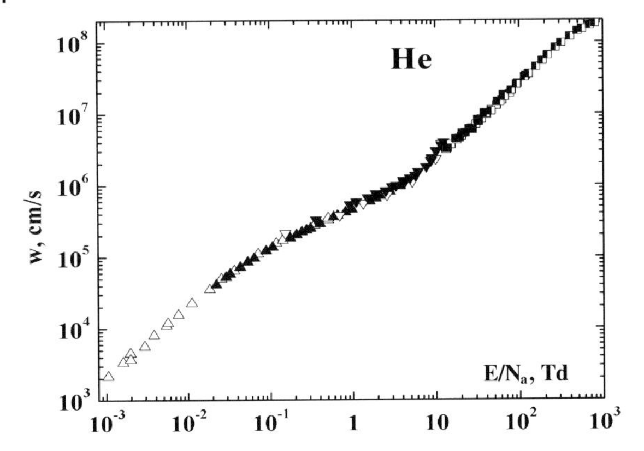

Figure 3.1 Electron drift velocity in helium depending on the reduced electric field strength in the regime of low electron number density according to experiments reported in [32].

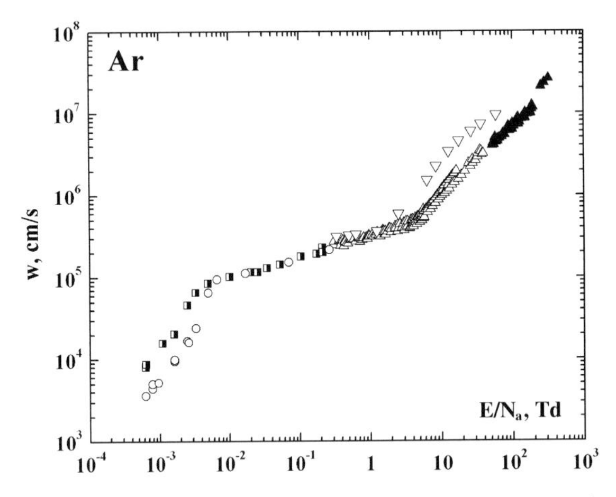

Figure 3.2 Electron drift velocity in argon depending on the reduced electric field strength in the regime of low electron number density according to experiments reported in [32].

where ε is the energy of a test electron, T is the gas temperature, and T_e is the electron temperature (or the effective electron temperature), which relates to the particle distribution function for electrons. In particular, if we consider a gas of helium ionized with $T = 400\,\mathrm{K}$, $T_e = 2\,\mathrm{eV}$, and $\ln \Lambda = 7$ and the electron energy corresponds to the excitation threshold of the helium atom $\varepsilon = 20\,\mathrm{eV}$, for which $\sigma^*_{ea} = 6\,\text{Å}^2$, we obtain the following criterion:

$$\frac{N_e}{N_a} \ll 3 \times 10^{-6} \, .$$

3.2.2
Evolution of Electrons in an Atomic Gas in an Electric Field

When an electron is moving in an atomic gas in an electric field, its energy varies weakly after a strong collision with gas atoms, and after $n \gg 1$ collisions this variation is proportional to $m_e\sqrt{n}/M$. But its drift velocity is established fast. Hence, one can define the electron drift velocity as a function of the current electron energy [34]. Its value follows from (3.26), where we take the distribution function to be $f_0(\varepsilon) = A\delta(\varepsilon - \varepsilon_0)$, where A is the normalization factor and ε_0 is the current electron energy. If the corresponding electron velocity is v, according to (3.26) we have

$$w_e(\varepsilon) = \frac{eE}{3m_e}\frac{1}{v^2}\frac{d}{dv}\left[\frac{v^3}{\nu(v)}\right], \tag{3.31}$$

where now $\varepsilon = m_e v^2/2$ is the current electron energy.

This consideration is valid when we consider evolution of an individual electron in a gas in an external electric field and ignore a chaotic change of its energy, which corresponds to the criterion

$$\varepsilon \gg \Delta\varepsilon \gg T\,. \tag{3.32}$$

Here ε is the current electron energy and $\Delta\varepsilon$ is the uncertainty in the electron energy or the width of its distribution function. Hence, here we consider the electron energy to be strictly determined in the course of its evolution. Let us determine the character of the variation of the electron energy ε if at the beginning it is relatively small. We multiply the kinetic equation (3.4) by the electron energy $\varepsilon = m_e v^2/2$ and integrate over electron velocities, which gives

$$\frac{d\overline{\varepsilon}}{dt} = eEw_e + \int \varepsilon\frac{I_{ea}}{N_e}d\mathbf{v}\,.$$

This equation has a simple physical sense: the first term on the right-hand side of this equation is the energy transferred from the field to the electron, and the second term accounts for collision of this electron with atoms. We take into account elastic electron–atom collisions only and use (3.18) for the collision integral. Because of criterion (3.32), one can replace the average electron energy $\overline{\varepsilon}$ by its current value ε and ignore the first term in (3.18) for the collision integral. As a result, we obtain

$$\frac{d\varepsilon}{dt} = eEw_e - 2\frac{m_e}{M}\varepsilon\nu\,. \tag{3.33}$$

This equation exhibits the character of variation of the electron energy when an electron travels in an atomic gas in an external electric field and has low energy at the beginning. We have that the electron energy increases in time, and in the course of its increase the relative energy portion given to gas atoms from an electron increases until the maximum electron energy ε_{max} is reached. This energy is given by

$$\varepsilon_{\max} = \frac{M}{2m_e} \frac{eEw_e(\varepsilon_{\max})}{\nu(\varepsilon_{\max})} .$$

Thus, within the framework of the scenario, when we ignore the chaotic character of variation of the electron energy that corresponds to the first term of the collision integral (3.18), the electron energy under the given conditions varies monotonically until it reaches its maximum value $\varepsilon_{\max}$. Accounting for the diffusion term of the collision integral (3.18) leads to broadening of the electron distribution function, but violation of criterion (3.32) proceeds with long time.

We demonstrate these results for the helium example when the diffusion cross section σ^* of electron–atom collisions is independent of the collision velocity. Then, introducing the mean free path for electrons $\lambda = 1/(N_a\sigma^*)$, we have

$$\varepsilon_{\max} = \sqrt{\frac{M}{6m_e}} eE\lambda = 0.96\overline{\varepsilon} ,$$

where the average electron energy $\overline{\varepsilon}$ in the limit of long times is given by (3.30). Next, one can represent the equilibrium electron energy in the form

$$\varepsilon_{\max} = ax ,$$

where $x = E/N_a$ is the reduced electric field strength and a is independent of the electric field strength. In particular, in the helium case, where $\sigma^* = 6\,\text{Å}^2$, we have $a = 0.58\,\text{eV/Td}$, and formula (3.29) gives the drift velocity of electrons in the limit of high electric field strengths. Next, we have from (3.26) in the limit of low electric field strengths

$$w_e = \frac{2eE\lambda}{3m_e}\left\langle\frac{1}{v}\right\rangle = \frac{2}{3}eE\lambda\sqrt{\frac{2}{\pi m_e T}} ,$$

where $\lambda = 1/N_a\sigma^*$ is the electron mean free path in the gas. Combining these limiting cases, we obtain for the electron drift velocity

$$w_e = \frac{0.53eE\lambda}{\sqrt{m_e T}\left[1 + 0.59\left(\frac{M}{m_e}\right)^{1/4}\sqrt{\frac{eE\lambda}{T}}\right]} .$$

In the helium case, this formula takes the form

$$w_e = \frac{v_0}{1 + C\sqrt{x}} ,$$

where $v_0 = 2.8 \times 10^6\,\text{cm/s}$, $C = 3.7$, and the reduced electric field strength $x = E/N_a$ is expressed in Townsends. This formula is compared with experimental data [32] for the electron drift velocity in helium in Figure 3.3.

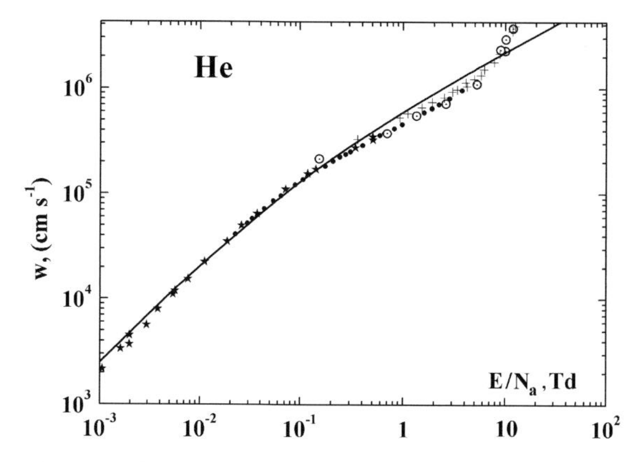

Figure 3.3 The electron drift velocity in helium in the regime of low electric field strength. Solid curve – theory, signs – experiment [32].

One can conclude there is general character of evolution of the electron energy when an electron travels in an atomic gas in an external electric field, and its energy exceeds the thermal atom energy T significantly. In the first stage of electron evolution, its energy increases monotonically up to ε_{max} and the equilibrium corresponds to this electron energy until criterion (3.32) holds true. In this stage of electron evolution its distribution function is described by a delta function. Subsequent collisions with gas atoms lead to broadening of the distribution function, which is given by (3.27).

3.2.3
Electrons in a Gas in the Regime of High Electron Number Density

The regime of electron drift in the limit of high electron number densities is defined by the inverse criterion with respect to criterion (3.30). In this regime $I_{ea}(f_0) \ll I_{ee}(f_0)$, and the second equation of the set (3.25) gives $I_{ee}(f_0) = 0$, which leads to the Maxwell distribution function (3.9). From this, for a motionless electron gas we have

$$f_0(v) = \varphi(v) = N_e \cdot \left(\frac{m_e}{2\pi T_e}\right)^{3/2} \cdot \exp\left(-\frac{m_e v^2}{2T_e}\right) ,$$

where this function is normalized by (3.1) and T_e is the electron temperature. Thus, the thermodynamic equilibrium takes place inside the electron subsystem in this limiting case. But since an external electric field violates such an equilibrium in the total ionized gas, the electron temperature T_e differs from the gas temperature T.

Using the Maxwell distribution function in the first equation of the set (3.25), we have

$$a \frac{m_e v}{T_e} f_0 = \nu v f_1 ,$$

which gives the following expression for the electron drift velocity:

$$w_e = \frac{eE}{3T_e}\left\langle\frac{v^2}{\nu}\right\rangle . \tag{3.34}$$

In Figures 3.4 and 3.5 we give the electron drift velocities in helium and argon calculated on the basis of this formula and using the experimental data for the electron–atom diffusion cross sections given in Figure 2.7.

The character of establishment of the electron temperature results from interaction between electrons and an electric field, which leads to energy transfer from

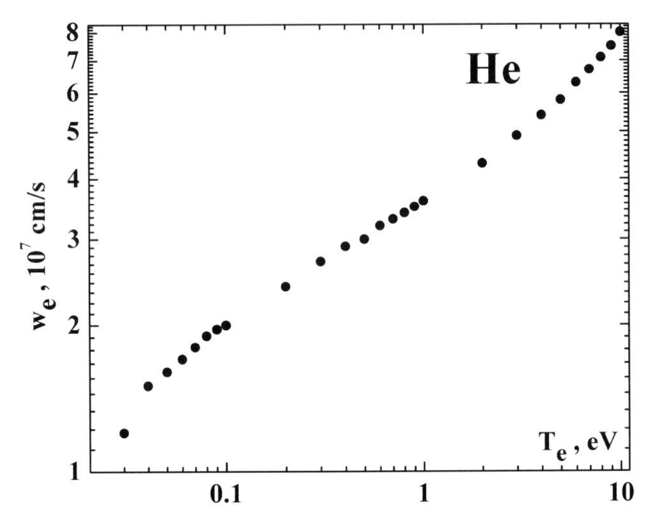

Figure 3.4 Electron drift velocity in helium as a function of the electron temperature according to (3.34) evaluated on the basis of the electron–atom elastic cross sections in Figure 2.8.

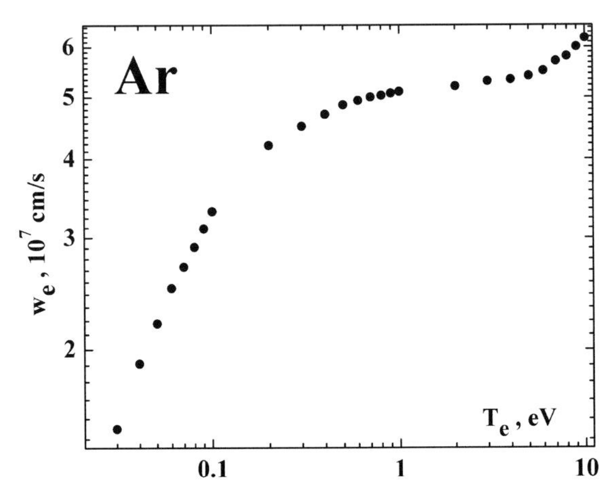

Figure 3.5 Electron drift velocity in argon as a function of the electron temperature according to (3.34) evaluated on the basis of the electron–atom elastic cross sections in Figure 2.8.

the electric field to electrons, and then from collisions between electrons and gas atoms, which leads to energy transfer from electrons to atoms. These processes establish the energy balance for electrons and determine the electron temperature T_e. To find this temperature, we use directly the stationary kinetic equation for electrons in a gas in an electric field. This equation has the form

$$\frac{e\mathbf{E}}{m_e}\frac{\partial f}{\partial \mathbf{v}} = I_{ea}(f) + I_{ee}(f_0) \,. \tag{3.35}$$

Because collisions between electrons are stronger in this regime than collisions between electrons and atoms, we have from this in the first approximation $I_{ee}(f_0) = 0$. This gives the Maxwell distribution function for electrons, where the electron temperature T_e is a parameter. The meaning of this is that the equilibrium of the energy distribution function is established by electron–electron collisions, whereas the drift velocity of the electrons is maintained by electron–atom collisions. To determine the electron temperature, we multiply this equation by $m_e v^2/2$ and integrate over electron velocities. We have $\int (m_e v^2/2) I_{ee}(f_0) d\mathbf{v} = 0$, because collisions between electrons do not change the total energy of the electron subsystem. Then we obtain the following balance equation:

$$eEw_e = \int \frac{m_e v^2}{2} I_{ea}(f_0) d\mathbf{v} = \frac{m_e^2}{M}\left(1 - \frac{T}{T_e}\right)\langle v^2 \nu \rangle \,,$$

where we use (3.18) for the collision integral of electrons and atoms. This is the balance equation for electrons, so its right-hand side is the energy per electron that is transferred from an external electric field, and the left-hand side is the specific energy transferred from an electron to gas atoms. From this we have for the difference between the electron and gas temperatures

$$T_e - T = \frac{Ma^2}{3}\frac{\langle v^2/\nu \rangle}{\langle v^2 \nu \rangle} \,, \tag{3.36}$$

where $a = eE/m_e$ and an average is made with the Maxwell distribution function for electrons.

Note that the regime of high electron number density holds true under the criterion opposite to (3.30), which has the form

$$\frac{N_e}{N_a} \gg \frac{m_e}{M}\frac{T}{T_e}\frac{\varepsilon^2 \sigma_{ea}^*}{2\pi e^4 \ln \Lambda} \,. \tag{3.37}$$

Under this condition one can ignore the term I_{ee} on the right-hand side of (3.35).

Let us consider the limiting cases for the dependence $\nu(v)$. If $\nu = \text{const}$, we have from (3.34) and (3.36)

$$w_e = \frac{eE}{m_e \nu} \,, \quad T_e - T = \frac{M}{3} w_e^2 \,.$$

If $\sigma^*(v) = \text{const}$, we introduce the mean free path of electrons as $\lambda = (N_a \sigma^*)^{-1}$. Then (3.34) and (3.36) yield

$$w_e = \frac{4eE\lambda}{\sqrt{2\pi T_e m_e}} \,, \quad T_e - T = \frac{3\pi}{32} M w_e^2 \,.$$

One can see that the drift velocity of electrons is small compared with a thermal energy, and the contribution of the drift velocity of electrons to the total kinetic energy of electrons is small: $m_e w_e^2/T_e \sim m_e/M$. Therefore, we used the Maxwell distribution function for electrons with zero drift velocity. One can rewrite the above formulas, expressing the electron drift velocity w_e and the average electron energy $\overline{\varepsilon}$ through the electric field strength. In the limit $T_e \gg T$, we have

$$w_e = \left(\frac{16}{3\pi\sqrt{3}}\right)^{1/2} \frac{\sqrt{eE\lambda}}{(m_e M)^{1/4}} = 0.990 \left(\frac{m_e}{M}\right)^{1/4} \sqrt{\frac{eE\lambda}{m_e}} ,$$
$$\overline{\varepsilon} = \frac{3}{2} T_e = \frac{\sqrt{3}}{4} \sqrt{\frac{M}{m_e}} eE\lambda = 0.433 \sqrt{\frac{M}{m_e}} eE\lambda . \tag{3.38}$$

These quantities have the same dependence on the problem parameters and are characterized by numerical factors which are close to those in the regime of low electron number densities according to (3.29).

Let us consider the case when the diffusion cross section of elastic electron–atom collision is independent of the velocity, as occurs in the helium case. Let us fix the electric field strength and let the electron density vary from small values to large ones. Then the drift velocity increases as a result of transition from the regime of low electric field strengths to the regime of high electric field strengths, but this value jumps by 10%. It should be noted that because of a small variation of the electron energy in elastic collisions with atoms, the remarkable difference between the electron energy and the thermal energy corresponds to low electric field strengths. Let us demonstrate this with the helium example when the diffusion cross section of elastic electron–atom collision varies weakly over a wide range of energies and is approximately $6\,\text{Å}^2$. Let us consider the regime of high electron number densities, so that (3.36) has the form

$$(T_e - T) T_e = \frac{M}{12 m_e} (eE\lambda)^2 ,$$

where $\lambda = 1(N_a\sigma^*)$ is the mean free path for electrons in helium. The solution of this equation for the electron temperature is

$$T_e = T\left[\frac{1}{2} + \sqrt{\frac{1}{4} + \left(\frac{E}{E_0}\right)^2}\right] , \quad E_0 = \sqrt{\frac{12 m_e}{M}} \frac{T}{e\lambda} . \tag{3.39}$$

E_0 characterizes the transition from low electric field strengths to high ones. In particular, for helium ($\sigma^* = 6\,\text{Å}^2$) at the gas temperature $T = 300\,\text{K}$ we have $E_0/N_a = 0.06\,\text{Td}$.

We also derive the equation for relaxation of the electron temperature T_e. We begin from a nonstationary kinetic equation (3.4). Multiplying this equation by the electron energy and integrating the result over electron velocities, we obtain for the average electron energy $\overline{\varepsilon} = 3T_e/2$

$$\frac{d\overline{\varepsilon}}{dt} = eEw_e - \frac{m_e^2}{M} \cdot \left(1 - \frac{T}{T_e}\right) \langle v^2 \nu \rangle .$$

Accounting for the electron drift velocity being established in a time of approximately $1/\nu$ and a time of approximately $M/(m_e \cdot \nu)$ being required for establishment of the electron temperature, we take the equilibrium value of the drift velocity in this equation. Assuming the difference $T_0 - T_e$ between the equilibrium and current electron temperatures to be relatively small, we have this equation in the form

$$\frac{d\overline{\varepsilon}}{dt} = \frac{3dT_e}{2dt} = -\frac{m_e^2}{M} \frac{T}{T_0^2} (T_e - T_0) \langle v^2 \nu \rangle .$$

Therefore, if we reduce this equation to the form

$$\frac{dT_e}{dt} = -\nu_\varepsilon (T_e - T_0) ,$$

the energy relaxation rate ν_ε is given by

$$\nu_\varepsilon = \frac{2m_e^2}{3M} \frac{T}{T_0^2} \langle v^2 \nu \rangle . \quad (3.40)$$

As is seen, at large electric field strengths this relaxation rate along with a small parameter m_e/M contains one more small parameter, T/T_e.

3.2.4
Conductivity of an Ionized Gas and a Plasma

The selective action of an external field on electrons of an ionized gas determines the behavior of the ionized gas in an external electric field. The field energy is transmitted first from the field to electrons and then it is transferred to gas atoms through their collisions with electrons. Because of a strong action of the electric field on electrons and a weak action on atoms, atoms have the Maxwell distribution function for energies, whereas the electron distribution function for energies may differ remarkably from the Maxwell one.

Electrons of a weakly ionized gas determine the electric properties of the gas, and the gas conductivity Σ is defined as the proportionality factor for the relationship between the electric current density **i** and the electric field strength **E** in Ohm's law:

$$\mathbf{i} = \Sigma \mathbf{E} .$$

The electric current is a sum of the electron and ion currents:

$$\mathbf{i} = -eN_e\mathbf{w}_e + eN_i\mathbf{w}_i ,$$

where N_e and N_i are the electron and ion number densities and $\mathbf{w}_e$ and $\mathbf{w}_i$ are the electron and ion drift velocities. Since $w_e \gg w_i$, the main contribution to the plasma conductivity is due to electrons because of the small electron mass. Hence, ignoring transport of ions, for the conductivity of an ionized gas we have

$$\Sigma = \frac{N_e e^2}{3m_e} \left\langle \frac{1}{v^2} \frac{d}{dv} \left(\frac{v^3}{\nu} \right) \right\rangle , \tag{3.41}$$

where ν is the rate of electron–atom collisions. If this rate $\nu = 1/\tau$ is independent of the collision velocity, this formula gives

$$\Sigma_0 = \frac{N_e e^2 \tau}{m_e} . \tag{3.42}$$

The same result corresponds to the tau approximation.

When electron collisions in an ionized gas occur with both atoms and ions, we represent the rate of collisions for electrons in the form

$$\nu = N_a v \sigma^*_{ea} + N_i v \sigma^*_{ei} ,$$

where N_a and N_i are the atom and ion number densities and σ^*_{ea} and σ^*_{ei} are the diffusion cross sections for electron scattering by atoms and ions. We are guided by a weakly ionized gas $N_a \gg N_i$, but because of a long-range electron–ion interaction, $\sigma^*_{ea} \ll \sigma^*_{ei}$ for thermal collisions. Therefore, collisions of electrons with ions may be of importance for the conductivity of an ionized gas in spite of a low degree of gas ionization. Using (2.39) for the diffusion cross section of electron–ion collisions, we represent the formula for the collision rate in the form

$$\nu = N_a \sigma^*_{ea} v + N_i \frac{4\pi e^4 \ln \Lambda}{m_e^2 v^3} ,$$

and in the case of a relatively high ionization rate, when the collision rate is determined by electron–ion collisions, according to (3.41) the conductivity of an ionized gas is given by the Spitzer formula [35]:

$$\Sigma = \frac{2^{5/2} T_e^{3/2}}{\pi^{3/2} m_e^{1/2} e^2 \ln \Lambda} . \tag{3.43}$$

In considering an intermediate case, for definiteness we assume the diffusion cross section of electron–atom collisions to be independent of the electron velocity. Then we have for the conductivity of an ionized gas

$$\Sigma = \Sigma_{ea} \Phi(\xi) , \quad \xi = \frac{N_i}{N_a} \frac{\pi e^4 \ln \Lambda}{T_e^2 \sigma^*_{ea}} , \quad \Phi(\xi) = \int_0^\infty \frac{e^{-x} x\,dx}{1 + \xi/x^2} , \tag{3.44}$$

where the conductivity Σ_{ea} is given by (3.41) and relates to the limiting case $N_a = 0$. It is convenient to represent the function $\Phi(\xi)$ in (3.44) in the form

$$\Phi(\xi) = \int_0^\infty \frac{e^{-x} x\,dx}{1 + \xi/x^2} = 1 - \xi \int_0^\infty \frac{e^{-x} x\,dx}{x^2 + \xi}$$
$$= 1 + \xi \left[\mathrm{chi}(\sqrt{\xi}) \cos(\sqrt{\xi}) + \mathrm{shi}(\sqrt{\xi}) \cos(\sqrt{\xi}) \right] , \tag{3.45}$$

where the specific functions chi(x) and shi(x) are given by [36]

$$\mathrm{chi}(x) = C + \ln x + \int_0^x \frac{\cosh t - 1}{t} dt\,, \quad \mathrm{shi}(x) = \int_0^x \frac{\sinh t}{t} dt\,,$$

and $C = 0.577$ is the Euler constant. The limiting expressions for this function are given by

$$\Phi(\xi) = 1 - \frac{\xi}{2} \ln \frac{1}{e^{2C}\xi}\,, \quad \xi \ll 1\,; \quad \Phi(\xi) = \frac{6}{\xi}\left(1 - \frac{20}{\xi}\right)\,, \quad \xi \gg 1\,.$$

Figure 3.6 represents the dependence $\Phi(\xi)$, and also its simple approximation $(1 + \xi/6)^{-1}$.

Let us consider the limiting cases for the electron drift velocity. At low values of ξ (a low degree of gas ionization) we have

$$\Sigma_\mathrm{e} = \Sigma_\mathrm{ea}\left(1 - \frac{\xi}{2} \ln \frac{1}{e^{2C}\xi}\right)\,.$$

At high values of ξ (a high degree of gas ionization), when scattering by ions dominates, the electron drift velocity is given by

$$\Sigma_\mathrm{e} = \frac{6}{\xi}\Sigma_\mathrm{ea}\left(1 - \frac{20}{\xi}\right) = \Sigma_\mathrm{ei}\left(1 - \frac{20}{\xi}\right)\,,$$

where w_ei is determined by (3.29).

Let us apply the above results for electron drift in helium. Being guided by the electron temperature $T_\mathrm{e} \sim 1\,\mathrm{eV}$, we take the diffusion cross section in scattering of an electron by a helium atom to be $\sigma_\mathrm{ea} = 6\,\text{Å}^2$. Correspondingly, ξ, which characterizes the ratio of ion and atom number densities, is

$$\xi = \frac{N_\mathrm{i}}{N_\mathrm{a}} \frac{\pi e^4 \ln \Lambda}{T_\mathrm{e}^2 \sigma_\mathrm{ea}^*} = \frac{N_\mathrm{i}}{N_\mathrm{a}} \frac{1100}{T_\mathrm{e}^2}\,,$$

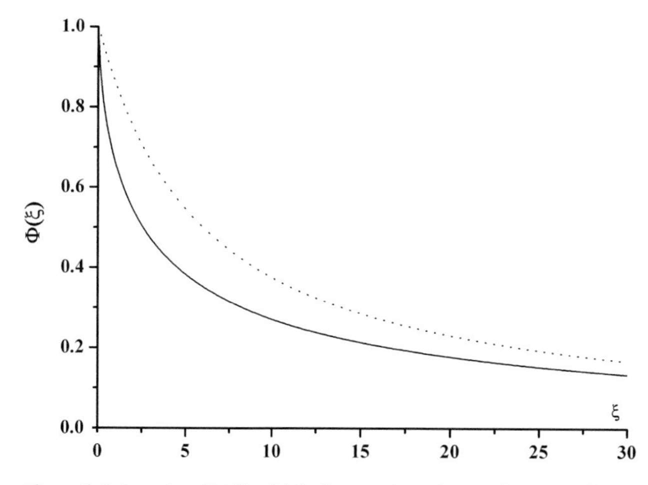

Figure 3.6 Function (3.45), which characterizes the conductivity of an ionized gas (3.44).

where the electron temperature T_e is in electronvolts and $\ln\Lambda = 10$. In particular, if $N_i/N_a = 10^{-4}$ and the electron temperature $T_e = 1$ eV, the electron drift velocity decreases by 6%, at $N_i/N_a = 10^{-5}$ the decrease of the electron drift velocity is about 2%, and in the case $N_i/N_a = 10^{-6}$, this drift velocity decrease is approximately 0.3%.

3.2.5
Electrons in a Gas in an Alternating Electromagnetic Field

As follows from the above consideration, there are two typical times for electron–atom collisions when an electron is moving in an atomic gas in an external field. The first one, $\tau = 1/\nu$ (ν is a typical rate of electron–atom collisions), characterizes an electron momentum variation, and the second time, approximately $M/(m_e\nu) \sim \tau M/m_e$, is a typical time for variation of the electron energy as a result of elastic collisions with atoms. We now consider electron motion in a gas in a harmonic electric field of strength $\mathbf{E}\cos\omega t$ under the condition $\omega\tau \gg m_e/M$, so the electron energy does not vary during the field period. This condition simplifies the problem [37–39]. The above condition for the field frequency corresponds to the following form of the distribution function similar to (3.10) [37]:

$$f(\mathbf{v}, t) = f_0(v) + v_x f_1 \exp(i\omega t) + v_x f_{-1} \exp(-i\omega t) ,$$

where the x-axis is directed along the field. Substituting this expansion into the kinetic equation and separating the corresponding harmonics by the standard method, we obtain the following set of equations instead of (3.11):

$$\frac{a}{2}\frac{d f_0}{d v} + (\nu + i\omega) v f_1 = 0 , \quad \frac{a}{2}\frac{d f_0}{d v} + (\nu - i\omega) v f_1 = 0 ,$$
$$\frac{a}{6v^2}\left[v^3 (f_1 + f_{-1})\right] = I_{ea}(f_0) . \tag{3.46}$$

From this we have for the electron drift velocity instead of (3.26)

$$w_e(t) = \int v_x^2 [f_1 \exp(i\omega t) + f_{-1} \exp(-i\omega t)] d\mathbf{v}$$
$$= \frac{eE}{3m_e}\left\langle \frac{1}{v^2}\frac{d}{dv}\left(v^3 \frac{\nu\cos\omega t + \omega\sin\omega t}{\omega^2 + \nu^2}\right)\right\rangle . \tag{3.47}$$

This expression corresponds to expansion over the small parameter $m_e/(M\omega\tau) \ll 1$, which allows us to ignore other terms of expansion over the spherical harmonics of the electron distribution function. Note that ω/ν may be found larger and smaller than one, and this ratio determines the phase shift for a drifting electron with respect to an external field. The above expressions are valid for criteria (3.30) and (3.37). We now write these formulas when criterion (3.37) holds true, which allows us to introduce the electron temperature T_e. Then, the electron drift velocity is

$$w_e(t) = \frac{eE}{3T_e}\left\langle v^2 \left(\frac{\nu\cos\omega t + \omega\sin\omega t}{\omega^2 + \nu^2}\right)\right\rangle , \tag{3.48}$$

and the difference between the electron and gas temperatures is [40]

$$T_e - T = \frac{Ma^2}{6\langle v^2 \nu \rangle} \left\langle \frac{\nu v^2}{\omega^2 + \nu^2} \right\rangle . \tag{3.49}$$

In the limit $\omega \ll \nu$, at $t = 0$ (3.48) coincides with (3.34), and (3.47) is transformed into (3.26) if we employ in (3.47) instead of the electric field strength E its effective value $E/\sqrt{2}$. In the other limiting case, $\omega \gg \nu$, the difference between the electron and gas temperatures is independent of the collision rate and is

$$T_e - T = \frac{Ma^2}{6\omega^2} . \tag{3.50}$$

3.2.6
Kinetics of Atom Excitation in Ionized Gases in an Electric Field

We now consider the character of atom excitation by electron impact in an ionized gas subjected to an external electric field. In analyzing atom excitation by electron impact in a plasma, we are restricted by a lower resonant excited atom state because in reality this process gives the main contribution to atom excitation. The rate of atom excitation by definition is given by

$$\frac{dN_*}{dt} = \int_{v_0}^{\infty} N_a k_{exc}(v) f_0(v) \cdot 4\pi v^2 dv , \tag{3.51}$$

where $v_0 = \sqrt{2\Delta\varepsilon/m_e}$ is the threshold electron velocity, where $\Delta\varepsilon$ is the atom excitation energy, $k_{exc}(v)$ is the rate constant for atom excitation resulting from collision with an electron of a velocity v, and the symmetric distribution function for electrons is normalized by the condition

$$\int_0^{\infty} f_0(v) 4\pi v^2 dv = N_e .$$

Formula (3.51) is valid if the electron distribution function above the excitation threshold is determined by elastic electron–atom collisions and the excitation process does not perturb it.

Let us find the rate of atom excitation in an ionized gas on the basis of (3.51) under the assumption that the excitation process does not change the electron distribution function above the excitation threshold. Then, the distribution function above the excitation threshold has the following form according to (3.27):

$$f_0(v) = f_0(v_0) \exp\left(-\frac{\varepsilon - \Delta\varepsilon}{T_{ef}}\right) , \quad T_{ef} = T + \frac{Mu^2}{3} ,$$

and we consider a typical case for a gas discharge plasma when excitation relates to the tail of the distribution function, that is,

$$\Delta\varepsilon \gg T_{ef} .$$

In evaluating the integral in (3.51), we take into account the threshold dependence for the rate constant for atom excitation by electron impact and express it through the rate constant for atom quenching k_q by a slow electron, where k_q is independent of the velocity of a slow electron. We have from the principle of detailed balance (2.57) near the excitation threshold

$$k_{ex} = k_q \frac{g_*}{g_0} \sqrt{\frac{\varepsilon - \Delta\varepsilon}{\Delta\varepsilon}} ,$$

where g_0 and g_* are the statistical weights of the ground and excited states of the atom. From this we have for the rate of atom excitation in an ionized gas for this limiting case

$$\frac{dN_*}{dt} = \left(\frac{2\pi T_{ef}}{m_e}\right)^{3/2} f_0(v_0)\nu_q \frac{g_*}{g_0} , \tag{3.52}$$

where $\nu_q = N_a k_q$ is the rate of atom quenching by a slow electron.

In reality, excitation of atoms in a plasma is a self-consistent problem, because atom excitation by electron impact in a plasma influences the velocity distribution function, and according to (3.51) this part of the electron distribution function, in turn, may determine the excitation rate. We now consider the opposite limiting case, when all the electrons whose energy exceeds the excitation threshold expend the energy on atom excitation. Then the distribution function for electrons decreases sharply above the excitation threshold, and we take here $f_0 = 0$. Note that although in reality the energy for this boundary condition exceeds the excitation threshold $\Delta\varepsilon$, we take it as

$$f_0(\Delta\varepsilon) = 0 .$$

We now analyze the nonstationary kinetic equation for the electron distribution function in the case of elastic electron–atom collisions. Then we have the following set of equations instead of (3.25) in the regime of a low number density of electrons:

$$\frac{\partial f_0}{\partial t} + \frac{a}{3v^2}\frac{\partial(v^3 f_1)}{\partial v} = I_{ea}(f_0) , \quad a\frac{\partial f_0}{\partial v} = -\nu v f_1 .$$

We ignore the time dependence of the antisymmetric part of the distribution function f_1 because establishment of the electron momentum proceeds fast, in contrast to the energy establishment. Reducing this set of equations to the equation for the symmetric part f_0 of the distribution function and using (3.18) for the electron–atom collision integral, we obtain

$$\frac{\partial f_0}{\partial t} + \frac{m_e}{M}\frac{\partial}{v^2 \partial v}\left\{v^3 \nu\left[\left(T + \frac{Ma^2}{3\nu^2}\right)\frac{\partial f_0}{m_e v \partial v} + f_0\right]\right\} = 0 . \tag{3.53}$$

For the rate of atom excitation under the given conditions this gives

$$\frac{dN_*}{dt} = -\frac{dN_e}{dt} = -\int 4\pi v^2 dv \frac{\partial f_0}{\partial t}$$
$$= -4\pi \frac{m_e}{M} v^3 \nu \left[\left(T + \frac{Ma^2}{3\nu^2}\right)\frac{\partial f_0}{\partial \varepsilon} + f_0\right]_{|\varepsilon=\Delta\varepsilon} .$$

Because the nonstationarity of the distribution function is determined by its tail and is relatively small, one can use in the first approximation the stationary expression for f_0 that is given by

$$\left(T + \frac{Ma^2}{3\nu^2}\right)\frac{df_0}{d\varepsilon} + f_0 = 0$$

with the boundary condition $f_0(\Delta\varepsilon) = 0$. Its solution has the form

$$f_0(\varepsilon) = \varphi_0(\varepsilon) - \varphi_0(\Delta\varepsilon) = \varphi_0(\varepsilon)\left[1 - \exp\left(\frac{\varepsilon - \Delta\varepsilon}{T_{\text{ef}}}\right)\right], \quad T_{\text{ef}} = T + \frac{Ma^2}{3\nu^2},$$

where the distribution function $\varphi_0(\varepsilon)$ is determined by (3.27) in ignoring the excitation process. As is seen, because of absorption of fast electrons, the electron energy distribution function differs from that in the absence of absorption because atom excitation near the boundary of electron absorption leads to a decrease of the electron energy by the atom excitation energy.

From the above formulas we obtain for the rate of atom excitation in this limiting case

$$\frac{dN_*}{dt} = \frac{m_{\text{e}}}{M} \cdot 4\pi v_0^3 \nu(v_0)\varphi_0(v_0) . \tag{3.54}$$

This expression does not contain the rate constant for excitation, which is assumed to be very large. Comparing (3.52) and (3.54), we have that the regime of (3.54) is realized if the following criterion holds true:

$$\frac{m_{\text{e}}}{M} \cdot \frac{g_0 \nu}{g_* \nu_q} \cdot \left(\frac{\Delta\varepsilon}{T_{\text{ef}}}\right)^{3/2} \ll 1 , \tag{3.55}$$

where $\nu_q = N_{\text{a}} k_{\text{q}}$ is the rate of quenching of excited atoms by a slow electron.

Let us consider the simple case when the elastic electron–atom collision rate $\nu(v)$ is independent of the electron velocity. Then the Maxwell distribution function is valid for electrons,

$$\varphi_0(\varepsilon) = N_{\text{e}}\left(\frac{m_{\text{e}}}{2\pi T_{\text{ef}}}\right)^{3/2} \exp\left(-\frac{\varepsilon}{T_{\text{ef}}}\right),$$

and for the rate of atom excitation (3.54) gives

$$\frac{dN_*}{dt} = \frac{4}{\sqrt{\pi}}\left(\frac{\Delta\varepsilon}{T_{\text{ef}}}\right)^{3/2} \exp\left(-\frac{\Delta\varepsilon}{T_{\text{ef}}}\right) N_{\text{e}} \frac{m_{\text{e}}}{M} \nu(v_0) ; \quad T_{\text{ef}} = T + \frac{Ma^2}{3\nu^2} .$$

This formula allows us to determine the excitation efficiency ξ, that is, the portion of input electric energy that is consumed in atom excitation. According to the definition, the excitation efficiency is given by

$$\xi = \frac{\Delta\varepsilon \frac{dN_*}{dt}}{eEw_{\text{e}} N_{\text{e}}},$$

since $\Delta\varepsilon \cdot dN_*/dt$ is the energy that is consumed in atom excitation by electrons, and $eEw_e N_e$ is the energy that is transmitted to electrons from the electric field. On the basis of the above formula for the excitation rate, we have for the excitation efficiency

$$\xi = \frac{4}{3\sqrt{\pi}} \left(\frac{\Delta\varepsilon}{T_{ef}}\right)^{3/2} \exp\left(-\frac{\Delta\varepsilon}{T_{ef}}\right) , \quad \Delta\varepsilon \gg T_{ef} . \tag{3.56}$$

Figure 3.7 shows the dependence of the efficiency of atom excitation ξ on the effective electron temperature under given conditions.

We represented above two limiting cases for atom excitation in an ionized gas. In reality we have several parameters whose combination leads to the corresponding limiting cases. These parameters are

$$\mu = \frac{m_e}{M} , \quad \gamma = \frac{T_{ef}}{\Delta\varepsilon} , \quad \delta = \frac{g_* \nu_q}{g_0 \nu} . \tag{3.57}$$

This means that we have additional limiting cases with respect to those given by (3.52) and (3.54). We now consider one more limiting case, when, on the one hand, the rate of excitation is expressed through the rate constant of excitation and, on the other hand, the distribution function varies significantly due to the excitation process near its threshold. As before, we assume the excitation threshold corresponds to the tail of the energy distribution function, that is, $\gamma < 1$. We account for the excitation process by inserting into the right-hand side of (3.53) the term $-\nu_{ex} f_0$, where ν_{ex} is the excitation rate, and this equation near the excitation threshold takes the form

$$\frac{\partial f_0}{\partial t} + \frac{2m_e}{M} \nu \Delta\varepsilon \frac{\partial}{\partial \varepsilon} \left(T_{ef} \frac{\partial f_0}{\partial \varepsilon} + f_0 \right) - \nu_{ex} f_0 = 0 ,$$

and below the excitation threshold this equation coincides with (3.53).

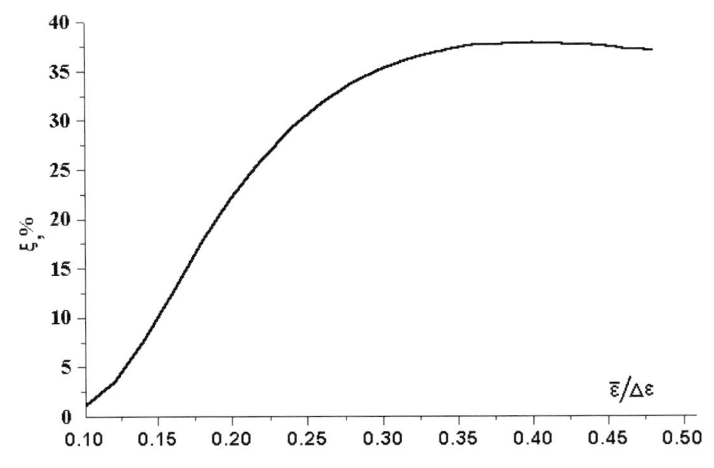

Figure 3.7 The efficiency of atom excitation in a plasma in a constant electric field as a function of the reduced average electron energy in the case when the rate of electron–atom elastic collisions is independent of the electron velocity.

We solve this equation in the case when the excitation process leads to a sharp variation of the distribution function with increasing electron energy. Therefore, we use the quasiclassical method taking the distribution function in the form

$$f_0(v) = \varphi_0(v)\exp(-S)\,,$$

where $\varphi_0(v)$ is the electron distribution function in the absence of the excitation process. The quasiclassical solution requires the validity of the criterion

$$\left(\frac{dS}{dv}\right)^2 \gg \frac{d^2S}{dv^2}\,.$$

In this approximation we obtain

$$\frac{dS}{dv} = \frac{\sqrt{3\nu_{ex}\nu}}{a}\,, \quad a = \frac{eE}{m_e}\,,$$

and near the threshold the electron distribution function is

$$f_0(v) = \varphi_0(v_0)\exp(-S) = \varphi_0(v_0)\exp\left(-\int\limits_{v_0}^{v}\frac{dv}{a}\sqrt{3\nu_{ex}\nu}\right)\,,$$

where v_0 is the threshold electron velocity. Transferring from the excitation rate to the quenching rate on the basis of (2.57), we have

$$S = \frac{2v_0}{5a}\sqrt{3\frac{g_*}{g_0}\nu_q\nu}\left(\frac{\varepsilon-\Delta\varepsilon}{\Delta\varepsilon}\right)^{5/4}\,.$$

We note that we assume the collision parameters, such as the rate of elastic electron–atom collisions, to be independent of the electron velocity near the excitation threshold. These formulas give the following expression for the rate of atom excitation by electrons if this process proceeds mostly near the excitation threshold [41, 42]:

$$\frac{dN_*}{dt} = \int 4\pi v^2 dv\, f_0(v_0)\exp(-S)\nu_{ex}(v)$$
$$= 4.30 a v_0^2\left(\frac{a}{v_0\nu_0}\right)^{1/5}\left(\frac{\nu_q g_*}{\nu_0 g_0}\right)^{2/5} f_0(v_0)\,, \qquad (3.58)$$

where the distribution function f_0 coincides with φ_0 in (3.54). It is convenient to express this rate of atom excitation through the parameters in (3.57), so (3.58) takes the form

$$\frac{dN_*}{dt} = 5.48 v_0^3\nu\varphi_0\mu^{3/5}\gamma^{3/5}\delta^{2/5}\,.$$

Comparing (3.58) with (3.52) and (3.54) for the rate of atom excitation in other limiting cases, we find that this case is realized under the condition

$$\left(\frac{\delta^{3/2}\gamma}{\mu}\right)^{2/5} \ll 1\,, \quad \left(\frac{\mu\gamma}{\delta^{3/2}}\right)^{3/5} \ll 1\,. \qquad (3.59)$$

One can see that this limiting case may be realized at low electric field strengths, which corresponds to $\gamma \ll 1$.

As a demonstration, we apply the above results to the helium case, where the quenching rate constant is $k_q = 3.1 \times 10^{-9}\,\mathrm{cm^3/s}$ [43], the rate constant for elastic electron scattering by a helium atom is $k_{ea} = 7.1 \times 10^{-8}\,\mathrm{cm^3/s}$ [44], and we have $\delta = 0.13$ in accordance with the definition. Next, $\mu = 1.4 \times 10^{-4}$ for the helium case, and γ in (3.57) is $4.3 \times 10^{-3} x^2$, where $x = E/N_a$ is the reduced electric field strength, which is measured in Townsends. In this case μ and δ are fixed, whereas γ varies with variation of the electric field strength. In this case the ratio of rates according to (3.54) and (3.52) is $8.4/x^3$, and the boundary of these mechanisms of atom excitation relates to the electric field strength of $x = 2\,\mathrm{Td}$. In the same manner, we find that (3.58) and (3.52) give the identical result at $x = 1.5\,\mathrm{Td}$, and the excitation rates according to (3.58) and (3.54) are equal at $x = 3.1\,\mathrm{Td}$. Thus, we conclude that the rate of excitation of the metastable $\mathrm{He}(2^3S)$ atom in an ionized gas of a low degree of ionization is determined by (3.52) at low electric field strengths and by (3.54) at high electric field strengths. Formula (3.58) may be used in a narrow intermediate range of electric field strengths, roughly between 1.5 and 3.1 Td.

We now consider atom excitation in an ionized gas in an electric field for a regime of electron evolution that corresponds to the opposite criterion with respect to (3.30) when the Maxwell distribution function for electrons holds true. The general character of atom excitation is the same as in the regime of a low density of electrons. In the limit when the excitation process does not influences the tail of the distribution function, the rate of atom excitation is given by (3.54), where $\varphi_0(v_0)$ is the Maxwell distribution function, and this formula has the form

$$\frac{dN_*}{dt} = \frac{4}{\sqrt{\pi}} \frac{m_e}{M} \left(\frac{T_e}{\Delta\varepsilon}\right)^{3/2} N_e \nu(v_0) \exp\left(-\frac{T_e}{\Delta\varepsilon}\right) . \qquad (3.60)$$

If the electron distribution function decreases sharply near the excitation threshold because of atom excitation, the rate of atom excitation by analogy with (3.53) is given by

$$\frac{dN_*}{dt} = -\int_{v_0}^{\infty} 4\pi v^2 dv \frac{\partial f}{\partial t}$$

$$= -\int_{v_0}^{\infty} 4\pi v^2 dv\, I_{ee}(f_0) = -\frac{4\pi v_0}{m_e} B_{ee}(v_0) \left(\frac{f_0}{T_e} + \frac{d f_0}{d\varepsilon}\right) ,$$

where the distribution function f_0 near the excitation threshold has the form

$$f_0(v) = N_e \left(\frac{m_e}{2\pi T_e}\right)^{3/2} \left[\exp\left(-\frac{\varepsilon}{T_e}\right) - \exp\left(-\frac{\Delta\varepsilon}{T_e}\right)\right] , \quad \varepsilon \le \Delta\varepsilon .$$

Using (3.20) for the electron–electron collision integral under assumption the excitation energy exceeds the electron temperature, that is, $\Delta\varepsilon \gg T_e$, we have for the rate of atom excitation that is determined by the rate of electron energy variation from zero energy up to the excitation threshold

$$\frac{dN_*}{dt} = N_e^2 k_{ee} \exp\left(-\frac{\Delta\varepsilon}{T_e}\right), \quad k_{ee} = 4\sqrt{2\pi} \cdot \frac{e^4 \Delta\varepsilon \ln \Lambda}{m_e^{1/2} T_e^{5/2}}, \tag{3.61}$$

where k_{ee} is the effective rate constant for elastic collisions of electrons due to the Coulomb interaction of electrons. Formula (3.61) is valid at high number densities of electrons according to the criterion

$$\frac{N_e}{N_a} \gg \frac{g_* k_q}{g_0 k_{ee}}, \tag{3.62}$$

where $g_e = 2$ is the statistical weight of an electron and g_* is the statistical weight of excited atoms. If criterion (3.62) hold true, fast establishment of the equilibrium occurs for the electron velocity distribution. This criterion is analogous to criterion (3.55) for the regime of low electron number density.

We give as a demonstration of these results the boundary degree of ionization of a plasma of inert gases for transition between (3.59) and (3.61) for the excitation rate of metastable states. This boundary degree of ionization is determined by

$$\left(\frac{N_e}{N_a}\right)_b = \frac{g_* k_q}{g_0 k_{ee}}$$

and is represented in Table 3.2 under typical conditions $T_e = 1\,\text{eV}$ and $\ln \Lambda = 10$ together with the rate constant k_{ee} according to (3.61).

In the analysis of the processes of atom excitation in a plasma above, we assumed that quenching of excited atoms proceeds through other processes rather than electron impact, as radiation of excited atoms and quenching of excited atoms on the walls. Let us consider now the quenching regime resulting from collision of excited atoms with electrons, that is, the equilibrium takes place according to the scheme

$$e + A^* \leftrightarrow e + A .$$

Then the equilibrium for atom excitation and quenching processes in collisions with electrons requires the equality of the rates:

$$\nu_{ex} f_0(v) v^2 dv = \nu_q \, f_0(v')(v')^2 dv' .$$

Table 3.2 The parameters of excitation of metastable inert gas atoms in accordance with (3.61) and (3.62).

Metastable atom	$\Delta\varepsilon$, eV	k_{ee}, 10^{-4} cm^3/s	$\left(\frac{N_e}{N_a}\right)_b$, 10^{-6}
He(2^3S)	19.82	5.8	5.4
Ne(2^3P_2)	16.62	2.9	0.69
Ar(3^3P_2)	11.55	2.0	2.0
Kr(4^3P_2)	9.915	1.7	2.0
Xe(5^2P_2)	8.315	1.4	13

Here v and v' are the electron velocities before and after atom excitation, and from the energy conservation we have $v^2 = 2\Delta\varepsilon/m_e + (v')^2$; $\nu_{ex} = N_0 k_{ex}$ and $\nu_q = N_* k_q$ are the rates of atom excitation and quenching by electron impact, so N_0 and N_* are the number densities of atoms in the ground and excited states, respectively, and k_{ex} and k_q are the rate constants for the corresponding processes, which are connected by the principle of detailed balance (2.57). Therefore, we have

$$\frac{N_0}{g_0} f_0(v) = \frac{N_*}{g_*} f_0\left(\sqrt{v^2 - v_0^2}\right), \quad v > v_0 = \sqrt{\frac{2\Delta\varepsilon}{m_e}}.$$

From this equality we obtain the following relation between the distribution functions for slow and fast electrons:

$$f_0(v) = \frac{f_0(v_0) f_0\left(\sqrt{v^2 - v_0^2}\right)}{f_0(0)}. \tag{3.63}$$

In the case of a Maxwell distribution function for slow electrons ($f_0(\varepsilon) \sim \exp(-\varepsilon/T_e)$), for fast electrons whose energy exceeds the threshold energy of atom excitation this formula gives

$$f_0(v) = f_0(v_0) \exp\left(\frac{\varepsilon - \Delta\varepsilon}{T_e}\right),$$

that is, it leads to a Maxwell distribution function for fast electrons. Thus, quenching collisions of electrons with excited atoms restore the Maxwell distribution function above the threshold of atom excitation.

The above cases of atom excitation by electrons in a plasma show that this process depends on the character of establishment of the electron distribution function near the threshold of excitation. The result depends both on the rate of restoring the electron distribution function in electron–electron or electron–atom collisions and on the character of quenching of excited atoms. Competition between these processes yields different ways of establishing the electron distribution function and different expressions for the effective rate of atom excitation in a gas and plasma. Thus, the excitation rates depend on the collision processes which establish the electron distribution function below and above the excitation threshold, and the equilibrium for excited atoms.

Ionization of atoms in a plasma proceeds similarly to atom excitation, but in contrast to excitation of the lowest electron level, excitation processes influence strongly the electron energy distribution function, and this leads to the dependence of the ionization rate on excitation processes. In the limit of low electron number density, the rate constant for atom ionization is given by

$$k_{ion} = \int_{v_0}^{\infty} v N_a \sigma_{ion}(v) f(v) d\mathbf{v},$$

where $v_0 = \sqrt{2J/m_e}$ is the threshold electron velocity for atom ionization, J is the atom ionization potential, and $\sigma_{ion}(v)$ is the ionization cross section. A spread quantity for the ionization process by electron impact in a gas in an external electric field is the first Townsend coefficient α, which is defined as

$$\alpha = \frac{v_0}{k_{\text{ion}}}$$

if we assume that the ionization process proceeds near the ionization threshold. The first Townsend coefficient is the reciprocal value of the electron mean free path with respect to ionization. Figures 3.8 and 3.9 show the measured values of the first Townsend coefficient for helium and argon [32].

The first Townsend coefficient is a convenient characteristic for description of the principal properties of gas discharge as the passage of an electric current through a gas under the action of an external electric field. Indeed, how to produce gas discharge and its basic forms were worked out in the nineteenth century, whereas the understanding of gas discharge as a self-maintaining phenomenon caused by

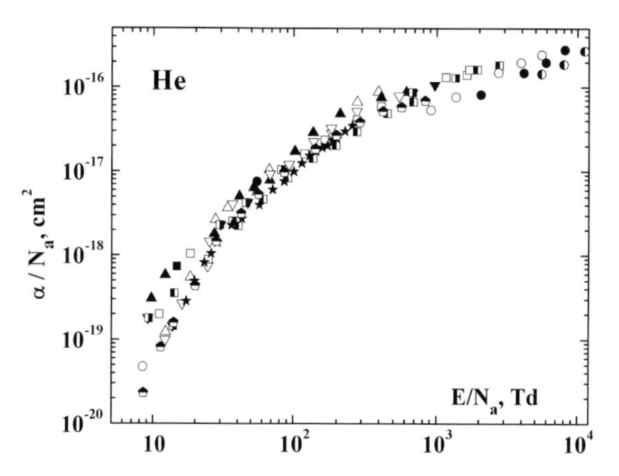

Figure 3.8 The first Townsend coefficient in helium [32].

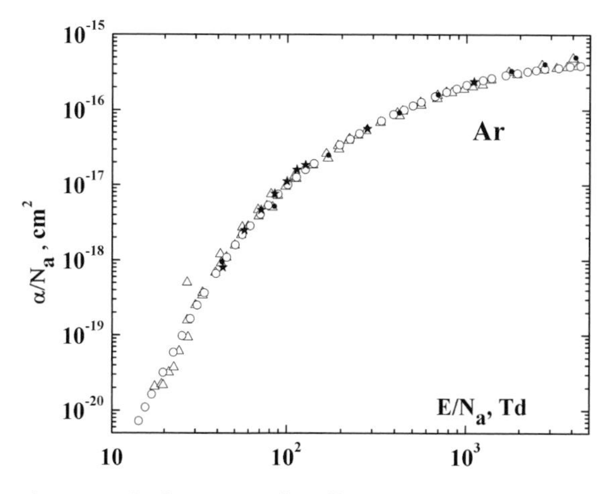

Figure 3.9 The first Townsend coefficient in argon [32].

passage of an electric current through a gas under the action of an electric field was achieved at the beginning of the twentieth century and was represented in the simplest form in books by Townsend [47–49]. The basis of self-sustaining gas discharge is the ionization balance resulting from formation and loss of charged particles (electrons and ions) and it is convenient to express this balance in terms of the first Townsend coefficient [47–49].

3.3 Radiation Transfer and Kinetics of Excitations in a Plasma

3.3.1 Equilibrium of Resonantly Excited Atoms in a Plasma

Excited atoms in a plasma result from inelastic collisions of atoms with electrons, and therefore the equilibrium number density N_* of atoms in a given excited state $*$ is determined by the Boltzmann formula (1.43) with the electron temperature T_e if only electron–atom collision processes are of importance. In the case of radiative transitions in accordance with the scheme in Figure 2.19, the balance equation for the number density of excited atoms has the form

$$\frac{dN_*}{dt} = N_e N_0 k_{ex} - N_e N_* k_q - \frac{N_*}{\tau} , \tag{3.64}$$

where k_{ex} and k_q are the rate constants for excitation and quenching of an atom by electron impact and τ is the lifetime of an excited atom resulting from its transport outside the plasma or photon emission. Because the cross sections of atom excitation and quenching of an excited atom state by electron impact are connected by the principle of detailed balance (2.53), the number density of excited atoms in the stationary case $dN_*/dt = 0$ is

$$N_* = \frac{N_0 N_e k_{ex}}{N_e k_q + 1/\tau} . \tag{3.65}$$

In the limit of a long lifetime $\tau \to \infty$ of the excited state, the solution of the balance equation (3.64) gives the Boltzmann distribution (1.43), and hence (3.65) may be represented in the form

$$N_* = \frac{N_*^B}{1 + 1/(N_e k_q \tau)} , \tag{3.66}$$

where N_*^B is the Boltzmann number density (1.43) of excited atoms, which corresponds to thermodynamic equilibrium between excited and nonexcited atomic states. Formula (3.66) has broader applicability than just application to resonantly excited states. It is necessary to take as τ the lifetime of an excited state with respect to its decay by any channel other than electron collisions; in particular, as a result of collisions with atoms or as a result of transport to walls. We assume here that processes involving excited atoms do not violate the Maxwell energy distribution of electrons with electron temperature T_e.

Formula (3.66) reflects the character of the equilibrium for resonantly excited atoms. Note that thermodynamic equilibrium is ascertained by comparison of the lifetime of excited atoms τ in a gas with the typical time $(N_e k_q)^{-1}$ for atomic quenching (not excitation!), and establishment of thermodynamic equilibrium for excited atoms requires the criterion

$$N_e k_q \tau \gg 1 . \tag{3.67}$$

3.3.2
Stepwise Ionization of Atoms

In analyzing the three body process (2.70) of electron–ion recombination, we consider this process as electron capture in a bound state with an ionization potential of the order of the thermal electron energy T_e, and subsequently the atom state varies as a result of successive collisions with electrons until transition to the ground state. We now consider the stepwise ionization of atoms that results from a series of reverse transitions until the atom is ionized under the action of collisions with electrons.

The mean electron energy in a gas discharge plasma is considerably lower than the atomic ionization potential

$$T_e \ll J , \tag{3.68}$$

and stepwise ionization of atoms by electron impact can take place only at a high number density of electrons, so there are no competing channels for transitions between excited states. The stepwise ionization process is the inverse process with respect to three body recombination of electrons and ions. In inverse processes, atoms undergo the same transformations but in opposite directions. As a result, on the basis of the principle of detailed balance, we obtain the connection between the rate constant for stepwise ionization k_{st} and the three body electron–ion recombination coefficient α according to (2.71). This gives the rate constant for stepwise ionization of atoms in a plasma in the limiting case when one can ignore atom radiation [50]:

$$k_{st} = \frac{\alpha}{N_e} \frac{g_e g_i}{g_a} \left(\frac{m_e T_e}{2\pi \hbar^2} \right)^{3/2} \exp\left(-\frac{J}{T_e} \right) = 2C \frac{g_i}{g_a} \frac{m_e e^{10}}{\hbar^3 T_e^3} \exp\left(-\frac{J}{T_e} \right) , \tag{3.69}$$

where $C \approx 2$ is a numerical factor, and is identical for various atoms.

Let us compare the rate constant for stepwise ionization (3.69) and the rate constant for direct ionization in a single collision, where we use the Thomson formula (2.60) for the ionization cross section, and because of criterion (3.68), the threshold cross section is used. The rate constant for direct ionization within the framework of the Thomson model is

$$k_{\text{ion}} = \int \frac{2\varepsilon^{1/2}}{\pi^{1/2} T_e^{3/2}} \exp\left(-\frac{J}{T_e}\right) \left(\frac{2\varepsilon}{m_e}\right)^{1/2} \sigma_{\text{ion}} d\varepsilon$$
$$= \frac{e^4}{J^2} \left(\frac{8\pi T_e}{m_e}\right)^{1/2} \exp\left(-\frac{J}{T_e}\right) , \tag{3.70}$$

where ε is the energy of the incident electron. Taking the atomic ionization potential J of the order of the atomic value $\varepsilon_0 = m_e e^4/\hbar^2$, we obtain from this

$$\frac{k_{\text{ion}}}{k_{\text{st}}} \sim \left(\frac{T_e}{\varepsilon_0}\right)^{7/2} \ll 1 , \tag{3.71}$$

because of criterion (3.68), that is, stepwise ionization is a more effective process in a low-temperature plasma than direct atom ionization by electron impact. Note that (3.69) for the rate constant for stepwise ionization holds true if the population of all the intermediate excited atom states corresponds to the Boltzmann formula (1.43). Usually this is not fulfilled, and (3.69) may be used as an estimate.

To analyze the character of stepwise ionization, we consider this process if the first stage of excitation of the atom ground state results in transition to a resonantly excited state. Then according to the definition of the ionization rate constant we have $N_0 k_{\text{st}} = k_{\text{st}}^* N_*$, where N_0 and N_* are the number densities of atoms in the ground and resonantly excited states, respectively, and k_{st} and k_{st}^* are the rate constants for stepwise atom ionization if it starts from the ground and resonantly excited atom states, respectively. We use (3.69) for the rate constant for stepwise ionization from the resonantly excited state and (3.66) for the number density of resonantly excited atoms. Then on the basis of the above definition of the ionization rate constant we have

$$k_{\text{st}} \approx 4 \frac{g_i}{g_a} \frac{m_e e^{10}}{\hbar^3 T_e^3} \frac{1}{1 + 1/(N_e k_q \tau)} \exp\left(-\frac{J}{T}\right) , \tag{3.72}$$

where g_i and g_a are the statistical weights for the ion and atom in the ground state, J is the ionization potential of an atom in the ground state, and τ is the effective lifetime of the first resonantly excited state. In the limit $N_e k_q \tau \gg 1$ (3.72) is converted into (3.69). This shows that additional channels for decay of excited states along with transitions in collisions with electrons lead to a decrease of the ionization rate constant.

3.3.3 Excited Atoms in a Helium Plasma

If excited atoms partake only in processes of excitation and quenching by electron impact, their number density is connected with the number density of atoms in the ground state by the Boltzmann formula (1.43). If the lifetime of excited atoms is determined also by other processes, the number density of excited atoms is less than that according to the Boltzmann formula (1.43), as it is given by (3.66), and the rate constant for stepwise ionization is lower than that given by (3.69). We analyze this in a helium plasma, being guided by the electron number density $N_e = 10^{11}-10^{14}\,\mathrm{cm}^{-3}$. The number density of electrons in glow and arc discharges may be included in this interval, and the indicated electron number density provides the regime of electron kinetics with dominant collisions between electrons that allows one to operate with the electron temperature T_e.

We restrict ourselves to four excited states of the helium atom, $\mathrm{He}(2^3S)$, $\mathrm{He}(2^1S)$, $\mathrm{He}(2^3P)$, and $\mathrm{He}(2^1P)$, and include them in the scheme of kinetics for excited states. The excitation energies with respect to the ground state energy are given in electronvolts in Figure 3.10. The parameters of radiative transitions are taken from Table 2.8 [51, 52]. The quenching rate constants resulting from electron impact are expressed in cubic centimeters per second and these values for resonantly excited states are given in Table 2.8, whereas the rate constant for quenching of the 2^3S state is taken from Table 2.9 and the measured rate constant [53] for the transition $e + \mathrm{He}(2^1S) \to e + \mathrm{He}(2^3S)$ is used. Next, considering the transition $e + \mathrm{He}(2^1P) \to e + \mathrm{He}(2^3P)$ proceeds from exchange between incident and bound electrons, we use the cross section of electron exchange [54] in evaluating the rate constant for this process. One can express the rate constants for inverse processes through these rate constants on the basis of the principle of detailed balance (2.57). Thus, in this way, ignoring higher excited states, we have all the information about the kinetics of these excited states, although with restricted accuracy. Solution of balance equations for the number density of atoms in these excited states allows us to have the populations of these states at different electron temperatures and electron number densities in the stationary case.

Because our goal is to obtain the qualitative character of kinetics of excited states in this case, we shall perform a rough analysis of the population of excited states in this case. For the resonantly excited state $\mathrm{He}(2^1P)$, (3.66) in the indicated range of

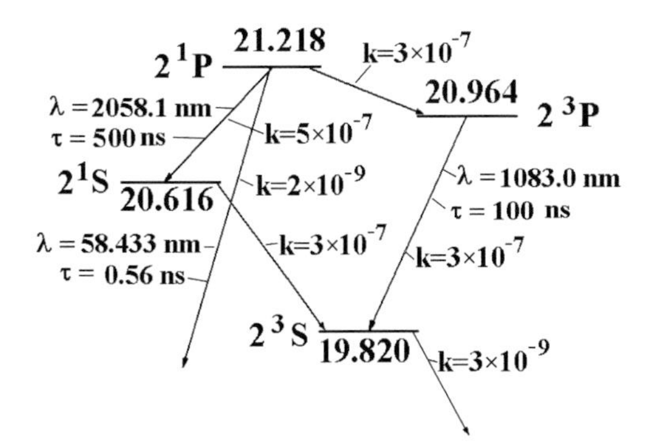

Figure 3.10 The lowest excited states of the helium atom. The energy of state excitation is expressed in electronvolts, and quantum numbers of states are given within the framework of the *LS* coupling scheme. The wavelengths of radiative transitions are indicated in nanometers and the radiative lifetimes of states are represented in nanoseconds.

N_e has the form

$$N(2^1P) = N_B(2^1P)\frac{N_e}{N_0},$$

where $N(2^1P)$ and $N_B(2^1P)$ are current and Boltzmann number densities of atoms in this state, and for an optically thick plasma we have $N_0 = 1 \times 10^{18}\,\text{cm}^{-3}$, whereas for a real optically thick plasma this value is several times lower. Thus, for the electron number densities we have

$$N(2^1P) \ll N_B(2^1P).$$

In the same manner we have

$$N(2^3P) = N_B(2^3P)\frac{N_e}{N_0},$$

and we have in this formula for optically thick plasma $N_0 = 4 \times 10^{13}\,\text{cm}^{-3}$. For simplicity, we take as in the previous case $N(2^3P) \ll N_B(2^3P)$.

In considering the balance for metastable state 2^3S involving electron collision processes, we have from the balance equation

$$\frac{N(2^1S)}{N(2^3S)} = \frac{N_B(2^1S)}{N_B(2^3S)}\frac{1}{1+\frac{k(2^1S\to 2^1P)}{k(2^1S\to 2^3S)}},$$

and since the rate constants $k(2^1S \to 2^1P)$ and $k(2^1S \to 2^3S)$ are comparable, the relative population of the 2^1S metastable state with respect to the 2^3S metastable state differs from that for the Boltzmann distribution, but not significantly.

For the lowest metastable state 2^3S, we obtain by ignoring the transition from the 2^1S state

$$N(2^3S) = N_B(2^3S)\frac{1}{1+\frac{k(2^3S\to 2^1S)+k(2^3S\to 2^3P)}{k(2^3S\to 1^1S)}}.$$

Being guided by electron temperatures in the range $T_e = 1$–$4\,\text{eV}$, we have that the rate constants for quenching are comparable with the excitation rate constants for transitions between excited states. Hence, from this formula we have

$$N(2^3S) \sim \frac{N_B(2^3S)}{100}.$$

Thus, we obtain from the above analysis that the population of atom excited states is lower than that under thermodynamic equilibrium. The reason is that emission of resonantly excited atoms leads to violation of the thermodynamic population of these levels, and this violation is transferred to metastable states through electron collision processes. Hence (3.69) for the rate constant for stepwise atom ionization by electron impact based on thermodynamic equilibrium for excited states does not work under real conditions where thermodynamic equilibrium is violated. We add to this that the Maxwell energy distribution of electrons may be violated in the tail of the distribution function because of inelastic processes involving electrons. This may enforce this effect.

3.3.4
Emission from a Flat Plasma Layer

Excited atoms can emit photons, and these photons can be absorbed by atoms in lower states of the radiative transitions, so the processes of radiation of excited atoms and absorption of atoms in lower states are determined by the character of plasma emission and propagation of these photons through a plasma. We study below the characteristics of interaction of resonant photons with a plasma, where resonant photons result from radiative transitions in atoms and their absorption by atoms causes an inverse transition between atom states. We consider firstly emission from a flat layer of a uniform plasma. For definiteness, we assume that the photons result from radiative transitions between the resonantly excited and ground states of atoms and take the statistical character of the elementary emission act, that is, the frequency distribution of photons is described by the photon distribution function a_ω, which is normalized according to (2.134) as $\int a_\omega d\omega = 1$. Absorption of photons of this frequency by plasma atoms is characterized by the plasma absorption coefficient k_ω, so $1/k_\omega$ is the mean free path for this photon in a plasma. One can introduce also the optical thickness of this plasma:

$$u_\omega = \int k_\omega dz \,, \tag{3.73}$$

where the photon propagates in the z direction.

Let us evaluate the photon flux from a plane plasma layer of thickness L for the geometry given in Figure 3.11. Here x is a direction perpendicular to the plasma surface, and θ is the angle between this direction and the direction of movement of a test photon. In the case of an optically thick plasma, $u_\omega = k_\omega L \ll 1$, all the photons reach the plasma surface, and the flux of photons j_ω from the plasma surface is given by

$$j_\omega d\omega = \int \frac{N_*(\mathbf{r}) d\mathbf{r}}{\tau_r} \cdot a_\omega d\omega \cdot \frac{\cos\theta}{4\pi r^2} \,.$$

We use here the layer uniformity along the plane direction, and therefore the photon flux is directed perpendicular to the plasma boundary, and $\mathbf{r}$ is a radiation point in the frame of reference where the origin lies on the plasma boundary. Next, $N_*(\mathbf{r})$ is the number density of radiating atoms at a given point, τ_r is the radiative lifetime

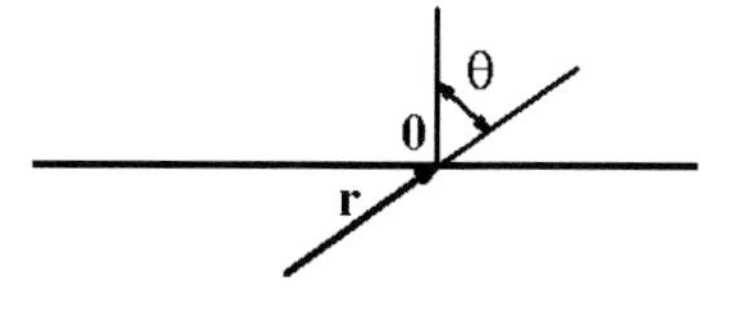

Figure 3.11 The geometry of photon emission. r is the distance that is traveled by the photon in a plasma, θ is the angle between the photon direction and the direction perpendicular to the plasma surface.

of this excited state with respect to a given transition, and so N_*/τ_r is the number of emission acts per unit volume and per unit time; $a_\omega d\omega$ is the probability that the frequency of an emitted photon lies in a given range. We assume also isotropic character of photon emission.

In the case of an arbitrary optical thickness of this plasma layer we introduce into this formula an additional factor, $\exp(-\int k_\omega dx)$, that an emitted photon reaches the plasma boundary, so the above formula is generalized to the form

$$j_\omega d\omega = \int \frac{N_*(\mathbf{r})d\mathbf{r}}{\tau} \cdot a_\omega d\omega \cdot \frac{\cos\theta}{4\pi r^2} \cdot \exp\left(-\int_0^r k_\omega dr'\right) . \tag{3.74}$$

We assume uniformity of the plasma layer in the longitudinal direction, that is, all the plasma parameters depend on a distance x from the plasma surface only. Taking the current optical thickness of a layer as $u = \int_0^x k_\omega dx'$ and the total optical thickness of the layer as $u_\omega = \int_0^L k_\omega dx$, that is, $du = k_\omega dx$, and using the cylindrical symmetry $d\mathbf{r} = 2\pi r^2 dr d\cos\theta$, we reduce the above formula for the photon flux to the form

$$j_\omega = \int_0^1 d\cos\theta \int_0^{u_\omega} du \frac{N_* a_\omega}{2\tau_r k_\omega} \exp\left(-\frac{u}{\cos\theta}\right) . \tag{3.75}$$

In considering the photon emission as a result of the atomic transition $* \to 0$, we use (2.149) if the number density of atoms in these states is subject to the Boltzmann law and use (2.145) for the absorption cross section. Then (3.75) reduces to the form

$$j_\omega = \frac{\omega^2}{2\pi^2 c^2} \int_0^1 d\cos\theta \int_0^{u_\omega} du \left[\exp\left(-\frac{u}{\cos\theta}\right) - 1\right]^{-1}$$
$$= j_\omega^{(0)} \int_0^1 \cos\theta \, d\cos\theta \left[1 - \exp\left(-\frac{u_\omega}{\cos\theta}\right)\right] , \tag{3.76}$$

where the photon flux $j_\omega^{(0)}$ under thermodynamic equilibrium according to (1.61) is given by

$$j_\omega^{(0)} = \frac{\omega^2}{4\pi^2 c^2} \left[\exp\left(\frac{\hbar\omega}{T}\right) - 1\right]^{-1} .$$

In the limiting cases of optically thick and optically thin plasmas, (3.76) gives

$$j_\omega = j_\omega^{(0)} , \quad u_\omega \gg 1 ; \quad j_\omega = 2 j_\omega^{(0)} u_\omega , \quad u_\omega \ll 1 .$$

3.3.5
Propagation of Resonant Radiation in a Dense Plasma

The above formulas allow one to analyze the character of radiation transfer in a plasma if reabsorption processes are of importance. For an optically dense plasma we have the equilibrium of absorption and emission processes at each plasma point, and the photon flux i_ω is isotropic inside the plasma. One can find this flux from the equilibrium between emitted and absorbed photons. Indeed, the rate of absorption acts per unit volume and in a frequency range from ω to $\omega + d\omega$ is $i_\omega k_\omega d\omega$, and the rate of photon emission per unit volume in this frequency range is $N_* a_\omega d\omega/\tau$. Hence, from the equality of these rates we have

$$i_\omega = \frac{a_\omega N_*}{k_\omega \tau_r} = 4 j_\omega^{(0)} .$$

This gives the connection between the photon flux inside an equilibrium plasma and outside it if the reabsorption processes create the equilibrium between atom excitation and the photon fluxes.

We now consider the process of propagation of resonant photons in a plasma from another standpoint, analyzing the behavior of a test emitted photon in a plasma [55–57]. Let us introduce the probability $P(r)$ that a resonant photon survives at a distance r from its formation, and this distance exceeds significantly the mean free path of the photon for the center of a spectral line $1/k_0$. Assuming statistical character for an emission frequency, we have this probability as the product of the probability $a_\omega d\omega$ for photon emission at a given frequency and the probability $\exp(-k_\omega r)$ of the photon surviving, that is,

$$P(r) = \int a_\omega d\omega \exp(-k_\omega r) . \tag{3.77}$$

Let us use this formula for the Lorentz (2.136) and Doppler (2.138) line shapes. In the case of the Lorenz shape of spectral lines we have, introducing a new variable $s = (\omega - \omega_0)/\nu$,

$$a_\omega d\omega = \frac{ds}{\pi(1+s^2)} , \quad k_\omega = \frac{k_0}{1+s^2} ,$$

where k_0 is the absorption coefficient for the spectral line center. Let us define $u = k_0 r$ as the optical thickness for the center line frequency; according to the problem condition, $u \gg 1$. Substituting this in (3.77), we obtain

$$P(r) = \frac{1}{\pi} \int_{-\infty}^{\infty} \frac{ds}{(1+s^2)} \exp\left(-\frac{u}{1+s^2}\right) .$$

For a dense plasma the main contribution to this integral follows from $s \gg 1$, and for the probability $P(r)$ for the photon to travel a given distance r we have for this limit

$$P(r) = \frac{1}{\sqrt{\pi u}} , \quad u \gg 1 . \tag{3.78}$$

For the Doppler shape of the spectral line we introduce a new variable

$$t = u \exp\left[-\frac{mc^2}{2T}\left(\frac{\omega-\omega_0}{\omega_0}\right)^2\right],$$

which gives the following expression for the probability of a resonant photon traveling a distance r:

$$P(r) = \frac{1}{\sqrt{\pi}u}\int_0^u e^{-t}dt\left(\ln\frac{u}{t}\right)^{-1} = \frac{1}{\sqrt{\pi}u\sqrt{\ln u + C}}, \quad u \gg 1, \tag{3.79}$$

where $C = 0.577$ is the Euler constant. Since the wings of the Doppler spectral line drop more sharply than those of the Lorentz spectral line, the probability of propagate large distances for the Lorentz spectral line is larger than that for the Doppler spectral line for the same optical thickness of the layer.

Reabsorption of a resonant photon results in its formation and subsequently in its absorption at another spatial point. We will characterize this process by the probability $G(\mathbf{r}', \mathbf{r})$, so $G(\mathbf{r}', \mathbf{r})$ is the probability for a photon formed at point $\mathbf{r}$ to be absorbed in a volume $d\mathbf{r}'$ near the point $\mathbf{r}'$. This probability satisfies the normalization condition

$$\int G(\mathbf{r}', \mathbf{r})d\mathbf{r}' = 1,$$

where the integral is taken over all space. We obtain the expression for this probability by analogy with that in (3.73) and (3.77), and it has the form

$$G(\mathbf{r}', \mathbf{r}) = \int \frac{a_\omega k_\omega d\omega}{4\pi|\mathbf{r}-\mathbf{r}'|^2}\exp\left(-\int k_\omega dx\right). \tag{3.80}$$

One can use the reabsorption process in the balance equation for excited atoms in a plasma where the reabsorption process is included along with the processes of atom excitation and quenching by electron impact and spontaneous remission by an excited atom. This equation has the form [58–60]

$$\frac{\partial N_*}{\partial t} = N_e N_0 k_{ex} - N_e N_* k_q - \frac{N_*}{\tau} + \frac{1}{\tau}\int N_*(\mathbf{r}')d\mathbf{r}'\int \frac{a_\omega k_\omega d\omega}{4\pi|\mathbf{r}-\mathbf{r}'|^2} \times \exp\left(-\int k_\omega dx\right) \tag{3.81}$$

and is named the Biberman–Holstein equation.

In particular, for the Lorenz profile of the spectral line (2.136) and for a uniform plasma this probability is

$$G(R) = \frac{k_0}{4\pi R^2}\int \frac{ds}{(1+s)^2}\exp\left(-\frac{k_0 R}{1+s^2}\right),$$

where we use the variable $s = 2(\omega - \omega_0)/\nu$ and k_0 is the absorption coefficient in the line center. In the limiting case of an optically thick plasma, $k_0 R \gg 1$, this formula gives

$$G(R) = \frac{1}{(4\pi k_0)^{1/2} R^{7/2}} . \tag{3.82}$$

Correspondingly, in the case of the Doppler profile of the spectral line (2.138), using a new variable $s = [(\omega - \omega_0)/\omega_0](mc^2/T)^{1/2}$, we have $a_\omega d\omega = \pi^{-1/2}\exp(-s^2)ds$ and $k_\omega = k_0 \exp(-s^2)$. Substituting this into (3.80) and introducing the variable $t = k_0 R \exp(-s^2)$, we obtain

$$G(R) = \frac{k_0}{4(\pi)^{3/2} R^2} \int_{-\infty}^{\infty} ds \exp(-s^2 - t) = \frac{1}{4(\pi)^{3/2} k_0 R^4} \int_0^{k_0 R} \frac{t e^{-t} dt}{\sqrt{\ln(k_0 R/t)}} .$$

Replacing the upper limit by infinity in the limiting case of an optically thick plasma, $k_0 R \gg 1$, and accounting for $\ln k_0 R \gg 1$, we obtain

$$G(R) = \frac{1}{4(\pi)^{3/2} k_0 R^4 \sqrt{\ln(k_0 R/t_0)}} ,$$

where t_0 is the solution of the equation $\int_0^\infty t e^{-t} dt \sqrt{\ln(t/t_0)} = 0$, that is, $t_0 = 1.52$. From this we have

$$G(R) = \frac{1}{4(\pi)^{3/2} k_0 R^4 \sqrt{\ln(k_0 R) - 0.42}} . \tag{3.83}$$

We find that propagation of resonant radiation in a gas or plasma has a specific character. In contrast to diffusion transfer of atomic particles in gaseous matter, radiation transfer proceeds at the wings of the spectral line for the radiative transition, and the greater the optical gas thickness, the greater the wings of the spectral line are responsible for radiation transfer. As an example of this character of radiation transfer, we consider emission of a cylindrical plasma region for the Lorenz profile (2.136) of the spectral line. In the limit of an optically thin plasma ($k_0 d \ll 1$, where d is the tube diameter), if all the photons formed leave the plasma region, we have for the flux j of photons [34]

$$j = \frac{N_*}{\tau_r}\frac{V}{S} = \frac{N_* d}{4\tau_r} ,$$

where $V = L \cdot \pi d^2/4$ and $S = L \cdot \pi d$ are, respectively, the volume and the surface area for a plasma region of length L. In the limit of an optically thick plasma, $k_0 d \gg 1$, we have for the photon flux

$$j = \frac{N_*}{\tau_r} \int \frac{P(r) \cos\theta \, d\mathbf{r}}{4\pi r^2} ,$$

where $P(r)$ is the probability that a resonant photon survives at a distance r from its formation, r is a distance of the point of photon formation from the surface point, and θ is the angle between the photon direction and the direction perpendicular to the plasma surface (Figure 3.4). Evaluating this integral, we obtain for the photon flux [34]

$$j = 0.274 \frac{N_* d^{1/2}}{k_0^{1/2} \tau_r} . \tag{3.84}$$

As is seen, the photon flux in an optically thick plasma is lower by $(k_0 d)^{1/2}$ than that in an optically thin plasma.

Let us introduce the effective radiative time τ_{ef} of the resonantly excited state as

$$\tau_{ef} = \tau_r \frac{J}{j} ,$$

where τ_r is the radiative lifetime of an isolated atom. The effective radiative lifetime depends both on the geometry of the volume occupied by the plasma and on the character of broadening of the spectral line. In particular, in the case when an optically thick uniform plasma is located inside a tube, and the Lorenz type (2.136) of spectral line broadening occurs, on the basis of the above result we have

$$\tau_{ef} = 0.91 \tau_r (k_0 d)^{1/2} . \tag{3.85}$$

3.3.6
Resonant Emission from a Nonuniform Plasma and Self-Reversal of Spectral Lines

Resonant radiation from a plasma containing resonantly excited atoms results from radiative transitions involving these exited states and ground states, and according to the above formulas the propagation of resonant photons through an optically thick plasma is not diffusive in nature. Namely, transport of photons over large distances compared with the mean free path of photons of the spectral line center is determined by photons with frequencies in the spectral line wings. The reason is that a photon emitted far from the center of a spectral line is more likely to propagate a long distances than is a photon emitted near the center of a spectral line, where repeated emissions and absorptions will occur with high probability. As a result, the principal contribution to long-distance propagation of resonant photons comes from the wings of the spectral line, where the mean free path of these photons is of the order of the dimensions of the gaseous system through which the photons propagate.

We now estimate on the basis of the above argumentation the flux of photons outside a gaseous system assuming that the photon transport process does not affect the density of excited atoms. We can take the mean free path of photons corresponding to the center of the spectral line to be small compared with the size L of the system, that is,

$$k_0 L \gg 1 ,$$

where the absorption coefficient k_0 for line-center photons is $k_0 = N_0 \sigma_{abs}(\omega_0) - N_* \sigma_{em}(\omega_0)$ according to (2.148), where ω_0 is the central photon frequency. Under these conditions, thermodynamic equilibrium is established between the atoms and the line-center photons whose free path length is small compared with the plasma size. Let i_ω be the flux of photons of frequency ω inside a plasma. Then the number of photons absorbed per unit volume per unit time in a frequency range from ω to $\omega + d\omega$ is given by $i_\omega k_\omega d\omega$, where k_ω is the absorption coefficient determined by (2.148). The reduced number of absorbed photons is equal to the corresponding number of emitted photons, which is given by $N_* a_\omega d\omega/\tau$. Then, on the basis of (2.145), (2.146), and (2.148), we obtain

$$i_\omega = \frac{a_\omega N_*}{k_\omega \tau_r} = \frac{\omega^2}{\pi^2 c^2}\left(\frac{N_0}{N_*}\frac{g_*}{g_0} - 1\right)^{-1} . \tag{3.86}$$

This photon flux is isotropic and can be detected at any point in the plasma medium that is separated from the system boundary by at least a photon mean free path. The photon flux outside a system with a flat surface is

$$j_\omega = \int_0^{\pi/2} i_\omega \cos\theta\, d(\cos\theta)\left(\int_{-\pi/2}^{\pi/2} d(\cos\theta)\right)^{-1} = \frac{i_\omega}{4} . \tag{3.87}$$

Here θ is the angle between the normal to the gas surface and the direction of photon propagation, and we have taken into account that the total photon flux outside the system is normal to the system surface. The flux of photons of frequency ω outside the gaseous system is

$$i_\omega = \frac{\omega^2}{\pi^2 c^2}\left(\frac{N_0}{N_*}\frac{g_*}{g_0} - 1\right)^{-1} , \quad k_\omega L \gg 1 . \tag{3.88}$$

If the plasma temperature is constant, we obtain (1.61) for the photon flux that corresponds to thermodynamic equilibrium, and which is given by

$$j_\omega = \frac{\omega^2}{4\pi^2 c^2}\left[\exp\left(\frac{\hbar\omega}{T}\right) - 1\right]^{-1} , \quad k_\omega L \gg 1 .$$

In considering radiation from a nonuniform plasma with a flat boundary and a uniform temperature along the boundary direction, we use (3.75) for the radiation flux from a flat layer of a plasma and with use of (2.148) for the absorption coefficient, the photon flux is reduced to the form

$$j_\omega = \frac{\omega^2}{2\pi^2 c^2}\int_0^1 d(\cos\theta)\int_0^{u_\omega} dx\, du \exp\left(-\frac{u}{\cos\theta}\right)\left(\frac{N_0}{N_*}\frac{g_*}{g_0} - 1\right)^{-1} . \tag{3.89}$$

Here the current optical thickness of a layer is defined as $u = \int_0^x k_\omega dx'$ and the total optical thickness of the layer is $u_\omega = \int_0^L k_\omega dx$, that is, $du = k_\omega dx$. This formula allows us to estimate the width of the spectral line of radiation that leaves the plasma. The boundaries of the spectral line can be estimated from the relation

$$u_\omega = \int_0^L k_\omega dx \sim k_\omega L \sim 1 . \tag{3.90}$$

This formula determines the width of the spectral line $\Delta\omega$ of the total radiation. This means that the radiation flux from a flat layer of a plasma j_ω is equal to the equilibrium radiation flux according to (1.61) for $|\omega - \omega_0| \leq \Delta\omega$ and is zero outside this range.

In the case of Lorentz broadening of the spectral line we have from (2.136) for the absorption coefficient at the line wing

$$k_\omega = k_0 \frac{\nu^2}{(\omega - \omega_0)^2} ,$$

and the width of the spectral line for the total radiation flux is estimated as

$$\Delta\omega \sim \nu\sqrt{k_0 L} , \quad k_0 L \gg 1 . \tag{3.91}$$

Correspondingly, for the Doppler profile (2.138) of the spectral line of the atom we obtain the width of the spectral line of total radiation:

$$\Delta\omega \sim \Delta\omega_D \sqrt{\ln(k_0 L)} , \quad k_0 L \gg 1 , \tag{3.92}$$

where $\Delta\omega_D$ is the width of the Doppler-broadened spectral line in the case of a small optical thickness of the plasma. Thus, resonant radiation from a plasma is characterized by a broadened spectral line compared with the spectral line from individual atoms, because the principal contribution to the emergent radiation arises largely from the wings of the spectra of individual atoms.

We now consider the profile of the spectral line for resonant radiation that leaves a nonuniform plasma under real conditions when the temperature at plasma boundaries is less than it is in the bulk of the plasma. The radiation emanating from a plasma for frequencies near the center of a spectral line originates in a plasma region near the boundaries. We use (1.61) for the radiative flux j_ω from the equilibrium plasma, and we substitute in this formula the temperature T of the layer that gives the main contribution to the radiation flux. This layer lies at distance x_ω from the boundary such that the optical thickness of this region is of

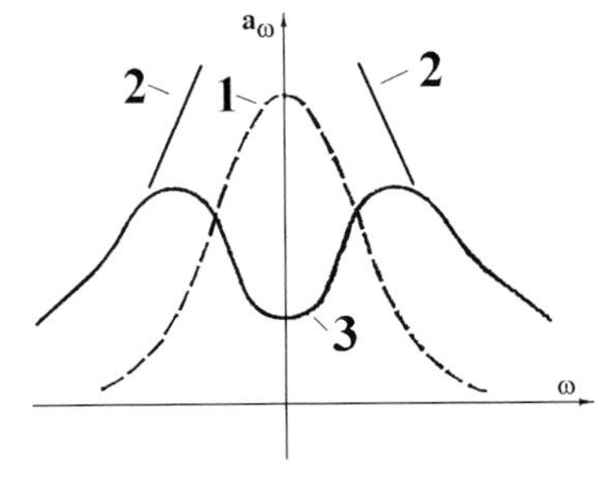

Figure 3.12 Self-reversal of spectral lines. 1 – profile of individual spectral line, 2 – spectral power of emitted radiation at constant temperature, 3 – spectral power of emitted radiation if the temperature drops to the boundary temperature.

the order of 1:

$$u(x_\omega) = \int_0^{x_\omega} k_\omega dx \sim k_\omega x_\omega \sim 1 \, .$$

This formula describes self-reversal of spectral lines with a dip at the line center as shown in Figure 3.12 for the case when the plasma temperature drops sharply near the plasma boundary. Evidently, the condition for a strong minimum of the radiation flux (3.92) for the spectral line center has the form

$$\frac{1}{k_0} \frac{\hbar\omega}{T} \frac{dT}{dx} \gg 1 \tag{3.93}$$

where $\hbar\omega > T$.

3.3.7
Radiation from the Solar Photosphere

Radiation transfer may be of importance for a plasma or excited gas of large size [61]. We analyzed above transfer of resonant radiation in the cases of uniform and strongly nonuniform radiating plasmas. We now consider a plasma with a slightly varied temperature, being guided by emission of the solar photosphere, which is created and governed by the processes

$$e + \mathrm{H} \longleftrightarrow \mathrm{H}^- + \hbar\omega \, . \tag{3.94}$$

Local thermodynamic equilibrium in the solar photosphere is supported by the processes

$$e + \mathrm{H} \leftrightarrow 2e + \mathrm{H}^+ \, ; \quad e + \mathrm{H}^- \leftrightarrow 2e + \mathrm{H} \, . \tag{3.95}$$

This equilibrium leads to Saha relations between the number densities of the corresponding particles, and for a quasineutral weakly ionized plasma $N_\mathrm{e} = N_\mathrm{p} \ll N_\mathrm{H}$ with a small concentration of negative ions ($N_- \ll N_\mathrm{e}$) we have

$$N_\mathrm{e} = N_\mathrm{p} = (N_0 N_\mathrm{H})^{1/2} \exp\left(-\frac{J}{2T}\right) , \quad N_- = \frac{N_\mathrm{H}^{3/2}}{4N_0^{1/2}} \exp\left(-\frac{\varepsilon_0}{T}\right) ,$$
$$N_0 = \left(\frac{m_\mathrm{e} T}{2\pi\hbar^2}\right)^{3/2} . \tag{3.96}$$

Here T is the temperature, N_H, N_e, N_p, and N_- are the number densities of hydrogen atoms, electrons, protons, and negative ions, respectively, $J = 13.605$ eV is the ionization potential of the hydrogen atom, and $\varepsilon_0 = J/2 - \mathrm{EA} = 6.048$ eV, where $\mathrm{EA} = 0.754$ eV is the electron affinity of the hydrogen atom. The radiation flux j_ω^E of the Sun at the position of the Earth is

$$j_\omega^\mathrm{E} = j_\omega \frac{R^2}{r^2} ,$$

where R is the solar radius, r is the distance between the Earth and the Sun, and j_ω is the radiation flux at the Sun's surface. As is seen, the spectral distribution for radiation near the Sun and the Earth is identical.

Let us represent the radiation flux emitted by the solar photosphere on the basis of (3.76) and (1.61) in the form

$$j_\omega = \frac{\omega^2}{2\pi^2 c^2} \int_0^1 d\cos\theta \int_0^\infty du_\omega \exp\left(-\frac{u_\omega}{\cos\theta}\right) F(u_\omega) ,$$

where $F(u_\omega) = \left[\exp(\hbar\omega/T) - 1\right]^{-1}$. Assuming the dependence $F(u_\omega)$ to be weak, we expand this function in a series

$$F(u_\omega) = F(u_0) + (u_\omega - u_0) F'(u_0) + \frac{1}{2}(u_\omega - u_0)^2 F''(u_0) .$$

u_0 is taken such that the second term is zero after integration. This yields $u_0 = 2/3$, and the radiation flux is

$$j_\omega = j_\omega^{(0)}(u_0) \left[1 - \frac{5F''(u_0)}{18F(u_0)}\right] , \tag{3.97}$$

where $j_\omega^{(0)} = \omega^2 (4\pi^2 c^2)^{-1} (\exp(\hbar\omega/T) - 1)^{-1}$ is the radiation flux of a blackbody at a temperature that corresponds to the point with optical thickness $u_\omega = 2/3$. The second term in the parentheses in (3.97) allows us to estimate the accuracy of the operation employed.

We now apply (3.97) to the solar atmosphere assuming the process

$$\hbar\omega + \mathrm{H}^- \to e + \mathrm{H} , \tag{3.98}$$

where H is a hydrogen atom. The dependence on the photon frequency for the cross section σ_{det} of the process (3.98) of electron photodetachment from the negative hydrogen ion is given in Figure 3.13, and we below use a simple approximation [65] for this cross section:

$$\sigma_{det}(\omega) = \sigma_{max} \cdot \frac{8\omega_0^{3/2}(\omega - \omega_0)^{3/2}}{\omega^3} , \tag{3.99}$$

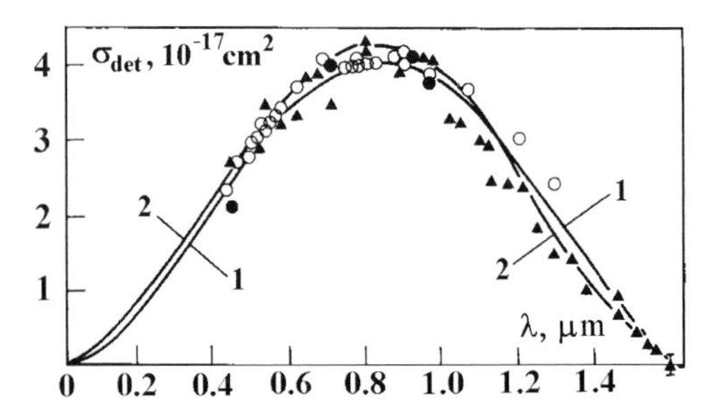

Figure 3.13 The cross section of photodetachment of the negative hydrogen ion. Experiment: open circles – [62], closed circles and triangles – [63], theory: 1 – [64], 2 – [65].

where the threshold frequency of photodetachment is $\hbar\omega = \text{EA} = 0.754\,\text{eV}$, and the maximum cross section is $\sigma_{\max} = 4 \times 10^{-17}\,\text{cm}^2$ and is attained at the frequency $\omega_{\max} = 2\omega_0$, which corresponds to the photon energy $\hbar\omega_{\max} = 2\text{EA} = 1.51\,\text{eV}$ or to the photon wavelength $\lambda = 0.8\,\mu\text{m}$.

Using parameters of the average quiet solar photosphere [66, 67], we have the effective temperature of radiation $T_{\text{ef}} = 5780\,\text{K}$ if we model the Sun's surface by a blackbody, and the free-fall acceleration on the solar boundary is $g = 2.74 \times 10^4\,\text{cm}^2/\text{s}$. This gives the length $l = 174\,\text{km}$ for the barometric distribution (1.48) of hydrogen atoms and $N_0 = 1.06 \times 10^{21}\,\text{cm}^{-3}$ in (3.96). According to the definition, the boundary of the photosphere, $z = 0$, is given as a layer with optical thickness $u(\lambda = 0.5\,\mu\text{m}) = 1$. At this wavelength we have the factor $\exp(-\hbar\omega_{\max}/T) = 0.008$ in (2.148). Ignoring this factor and taking according to the barometric formula (1.48) $N_{\text{H}}(z) \sim \exp(-z/l)$ and according to (3.96) $N_-(z) = N_-(0)\exp(-3z/2l)$, we have the optical density of the photosphere

$$u_\omega(z) = \frac{2l}{3} N_-(z)\sigma_{\text{det}}(\omega)\,, \tag{3.100}$$

where z is the photosphere altitude. From the condition $u(z = 0, \lambda = 0.5\,\mu\text{m}) = 1$ and accounting for $\sigma_{\text{det}}(\lambda = 0.5\,\mu\text{m}) = 0.8\sigma_{\max}$, we obtain for the number density of negative hydrogen ions at the photosphere boundary $N_-(0) = 2.7 \times 10^9\,\text{cm}^{-3}$. Then (3.96) gives for the number densities of hydrogen atoms and electrons at the photosphere boundary $N_{\text{H}}(0) = 1.6 \times 10^{17}\,\text{cm}^{-3}$ and $N_{\text{e}}(0) = 1.5 \times 10^{13}\,\text{cm}^{-3}$.

In accordance with (3.97), the condition $u_\omega = 2/3$ gives the altitude z_ω of the layer that is responsible for radiation at frequency ω. On the basis of (3.100) for the optical thickness we obtain $z_{\max} = 73\,\text{km}$ is responsible for radiation at $\lambda = 0.8\,\mu\text{m}$, which corresponds to the maximum photodetachment cross section. At this altitude we have $N_-(z_{\max}) = 1.4 \times 10^9\,\text{cm}^{-3}$, $N_{\text{H}}(z_{\max}) = 1.1 \times 10^{17}\,\text{cm}^{-3}$, and $N_{\text{e}}(z_{\max}) = 1.2 \times 10^{13}\,\text{cm}^{-3}$. Next, the photosphere altitude responsible for emission at $\lambda = 0.5\,\mu\text{m}$ is $z_\omega = 47\,\text{km}$, and at $\lambda = 0.4\,\mu\text{m}$ it is $z_\omega = 23\,\text{km}$. In these cases the photon flux at a given frequency is determined by (1.61). In addition to this, we have that radiation of the solar photosphere at frequencies of the order of the maximum frequency is determined by the photosphere layer of width $\Delta z \sim 1/k_\omega \sim 200\,\text{km}$. Hence, one can ignore the temperature variation in the radiating layer if

$$\frac{dT}{dz} \ll \frac{T^2}{\Delta z \hbar\omega} \sim 10\,\text{K/km}\,.$$

If the photosphere temperature varies slightly with the altitude, we use in (3.97) for the photon flux the temperature at altitude z_ω. Summing the radiative flux of the Sun's photosphere j_ω from radiation of layers with a different temperature, we use (3.97) as an expansion of j_ω over a small parameter, and below we find the validity of this expansion where the second term in parentheses in (3.97) is much less than 1. We have $N_-(z) \sim \exp(-3z/2l - \varepsilon_l/T)$, which gives

$$\frac{du}{dz} = -\frac{3u}{2l}\left(1 - \alpha\frac{dT}{dz}\right)\,, \quad \alpha = \frac{2l}{3}\frac{\varepsilon_0}{T^2}\,,$$

and under the above conditions $\alpha = 0.24\,\mathrm{km/K}$. Next, since $\hbar\omega \gg T$, we have

$$\frac{dF}{du} = -\frac{2l\hbar\omega}{3uT^2}\frac{dT}{dz}\left(1-\alpha\frac{dT}{dz}\right)^{-1} F\,,$$

$$\frac{d^2F}{du^2} = \left[\frac{2l\hbar\omega}{3uT^2}\frac{dT}{dz}\left(1-\alpha\frac{dT}{dz}\right)^{-1}\right]^2 F\,.$$

Correspondingly, we obtain the criterion

$$\frac{5}{18}\left[\frac{l\hbar\omega}{T^2}\frac{dT}{dz}\left(1-\alpha\frac{dT}{dz}\right)^{-1}\right]^2 \ll 1\,,$$

which allows us to ignore the second term in parentheses in (3.97) and allows us to represent the radiative flux in the form of blackbody radiation (1.61). The maximum of the left-hand side of this criterion corresponds to a large temperature gradient and is $5(\hbar\omega)^2/18\varepsilon_0^2$; this value is 0.04 for the photosphere parameters above. Thus, one can use the equilibrium relation (1.61) for the radiative flux j_ω at a given frequency for any linear dependence of the temperature on the altitude.

Taking into account a temperature variation, we have for the altitude of a layer z_ω that is responsible for emission at a given frequency

$$z_\omega = z_{\max} - \frac{2l\ln\left[\frac{\sigma_{\det}(\omega)}{\sigma_{\max}}\right]}{3\left(1-\alpha\frac{dT}{dz}\right)}\,, \qquad \alpha = \frac{2l}{3}\frac{\varepsilon_0}{T^2}\,. \tag{3.101}$$

Note that the measured value j_ω^{E} varies in time, and these formulas allow us to determine the current temperature of the photosphere boundary and its gradient on the basis of measurement of the spectrum j_ω^{E} for solar radiation at the position of Earth. Thus, if we assume the temperature of the solar atmosphere to be a smooth function of the height, we reduce the problem of the radiation from a plasma of variable temperature to a problem of a constant temperature for the radiating plasma. To solve this problem, it is necessary to use two parameters of the solar atmosphere: $N_{\mathrm{H}}(T_0)$ and dT/dz, where the temperature T_0 is close to the effective temperature of the radiation. Hence, two parameters of the solar atmosphere must be introduced into the problem for the determination of radiation fluxes from the photosphere and the spectrum of its radiation is similar to the blackbody spectrum.

3.3.8 Excitations in a Photoresonant Plasma

An examination of the propagation of resonant radiation in an excited gas shows a strong absorption near the center of the spectral line. This means that resonant radiation can be transformed into excitation of the gas. This is a property used as a diagnostic technique for the analysis of gases and flames. The optogalvanic method [68, 69] is based on measurement of a current through an excited or ionized gas as a function of the radiation wavelength. Absorption of radiation by certain atoms leads to the formation of electrons and ions that can be detected by changes in the electric current. Calibration of this method makes it possible to measure the content of some admixtures at very low concentrations.

A more widespread application based on absorption of resonant radiation concerns the generation of a photoresonant plasma. To accomplish this, the energy of the resonant radiation absorbed by a gas or vapor is transformed into the energy required to ionize the gas. There is a sequence of processes that determine the generation of a plasma by this means. Resonantly excited atoms are formed as a result of the absorption of resonant radiation. Collision of these excited atoms with each other leads to the formation of more highly excited atoms and subsequently to their ionization. Electrons released in the ionization process establish an equilibrium with the excited atoms that leads to plasma formation.

There are various applications of a photoresonant plasma. In particular, because some of the absorbed energy is transformed to energy of the plasma expansion at the end of the process, a photoresonant plasma is a convenient way to generate acoustic signals with adjustable parameters. Another type of application makes use of the high specific absorbed energy. Then a photoresonant plasma can be a source of multicharged ions. Other applications of photoresonant plasmas make use of both of these special properties of the plasma, as well as the possibility of transforming atoms into ions for ease of their detection.

In considering the formation of a photoresonant plasma, we will be guided by an alkali metal plasma, where the process of plasma formation proceeds according to the scheme

$$2A^* \rightarrow A_2^+ + e - \Delta\varepsilon_{\rm i} \,. \qquad (3.102)$$

Here collisions of two excited atoms may lead to atom ionization, and $\Delta\varepsilon_{\rm i}$ is the energy that is required for this process. Table 2.14 gives values of the rate constant $k_{\rm as}$ for this process for resonantly excited atoms of alkali metals at a temperature of 500 K.

We first consider the equilibrium of incident resonant radiation with the radiation flux inside the plasma. Evidently, the mean free path of incident radiation is small compared with a dimension of the region occupied by a plasma, and Table 2.22 shows that for a laboratory plasma of size $L \sim 1\,\rm cm$ this condition holds true even if the wavelength is far from the spectral line center. Below we ignore quenching of excited atoms by plasma electrons and find the number density of excited atoms N_* in this case. Let us analyze the balance of an incident radiative flux j_+ and an isotropic radiation flux $i(z)$ that is created by radiation of excited atoms, where z is the coordinate inside the gas region along the photon beam. For the stationary regime of propagation of resonant radiation through the plasma, we obtain the equality of incident and reflecting photon fluxes near the boundary, which gives $j_+(0) = i(0)/4$, that is, under equilibrium between incident resonant radiation and an absorbed gas we have $i(0) = 4j_+(0)$ at the plasma boundary.

Let us find the number density of excited atoms N_* from the balance equation between the rate of emission events N_*/τ per unit volume, where τ is the effective lifetime of the excited state and taking into account the reabsorption processes, and the rate of absorption events, which is $[j_+ + i(0)/2]k_\omega$ near the gas boundary, where k_ω is the absorption coefficient. We consider distances from the boundary exceeded the mean free path of photons, $1/k_\omega$, and emission of spontaneous radiation as the channel of loss of excited atoms. From this we obtain for the number density of excited atoms

$$N_* = 3k_\omega j_+ \tau_{\text{ef}} . \tag{3.103}$$

We assumed above the flux of incident radiation to be small, which allows us ignore stimulated radiation. In addition, collision processes involving excited atoms are assume to be weak compared with reabsorption processes.

Let us introduce the excitation temperature T_* on the basis of the Boltzmann formula (1.43) as

$$\frac{N_*}{N_0} = \frac{g_*}{g_0} \exp\left(-\frac{\hbar\omega}{T_*}\right) . \tag{3.104}$$

Taking into account the stimulated radiation in accordance with (2.148), we obtain the balance equation for excited atoms in the form

$$\frac{dN_*}{dt} = j_\omega k_0 - j_\omega k_0 \frac{N_*}{N_0}\frac{g_0}{g_*} - \frac{N_*}{\tau_{\text{ef}}} .$$

In the stationary case with using the excitation temperature T_* we have

$$j_\omega k_0 \left[1 - \exp\left(-\frac{\hbar\omega}{T_*}\right)\right] = \frac{N_*}{\tau_{\text{ef}}} .$$

Let us introduce the typical radiation flux

$$j_0 = \frac{N_0 g_*}{g_0 k_0 \tau_{\text{ef}}} . \tag{3.105}$$

Table 3.3 contains the values of the specific photon flux j_0/N_0 at $g_*/g_0 = 1$ and also of the specific typical intensity of radiation $I_0 = \hbar\omega j_0$. We have the following connection between the flux of resonant photons j_ω and the excitation temperature T_*:

$$j_\omega = \frac{j_0}{\exp\left(\frac{\hbar\omega}{T_*}\right) - 1} . \tag{3.106}$$

This connection between the flux of resonant photons in the spectral line center and the temperature of excitation may be represented in the form

$$T_* = \frac{\hbar\omega}{\ln\frac{1+\eta}{\eta}} ; \quad \eta = \frac{j_\omega}{j_0} = j_\omega \sigma_{\text{abs}} \tau_{\text{ef}} \frac{g_0}{g_*} , \tag{3.107}$$

and Figure 3.14 gives the dependence of the reduced excitation temperature T_* on the specific photon flux j_0/N_0 in accordance with (3.107). In particular, $T_* =$

Table 3.3 The typical specific photon flux of resonant radiation in vapors of alkali metals and the specific typical flux of the radiation intensity.

Vapor	j_0/N_0, 100 cm/s	I_0/N_0, 10^{-17} W cm
$\mathrm{Li}(2^2P)$	2.3	6.8
$\mathrm{Na}(3^2P_{1/2})$	5.6	19
$\mathrm{Na}(3^2P_{3/2})$	4.4	15
$\mathrm{K}(4^2P_{1/2})$	4.5	15
$\mathrm{K}(4^2P_{3/2})$	3.6	9.4
$\mathrm{Rb}(5^2P_{1/2})$	3.9	10
$\mathrm{Rb}(5^2P_{1/2})$	3.2	8.0
$\mathrm{Cs}(6^2P_{1/2})$	4.3	9.6
$\mathrm{Cs}(6^2P_{3/2})$	3.7	8.6

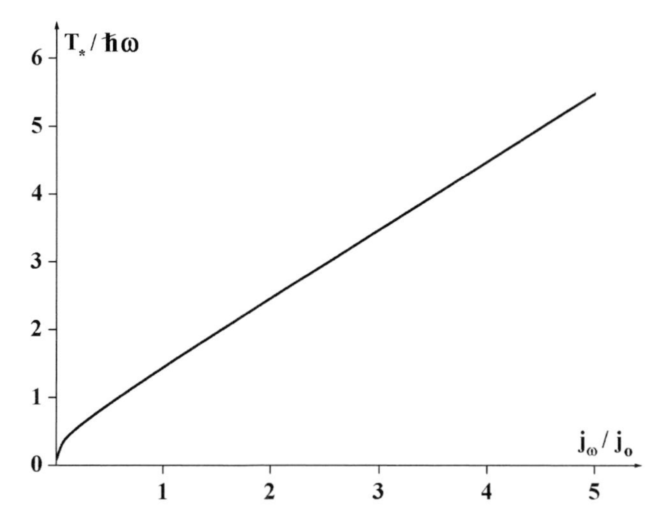

Figure 3.14 The relative excitation temperature T_* for for resonantly excited atoms versus the radiation flux according to (3.107).

$1.44\hbar\omega$ at $j_\omega = j_0$, that is, for alkali metal vapors the number density of resonantly excited atoms exceeds that of ground state atoms by 1.5 times at this photon flux.

We now consider the character of penetration of the flux of resonant radiation inside a plasma. Accounting for the processes of absorption and stimulated emission for resonant photons, we have the balance equation for the flux $j_\omega(z)$ for resonant photons propagating in a plasma:

$$\frac{dj_\omega}{dz} = -j_\omega\sigma_\omega N_0 + j_\omega\sigma'_\omega N_* = -j_\omega k_\omega ,$$

where σ_ω and σ'_ω are the cross sections of absorption and stimulated emission for propagating resonant photons in accordance with (2.145) and (2.146) and k_ω is the absorption coefficient of the plasma given by (2.149) in the case of thermodynamic equilibrium between atoms in the ground and resonantly excited states. From this we find that the photon flux is damped as it penetrates in a plasma and is given as $j_\omega(z) \sim \exp(-z/\lambda)$, where $\lambda = 1/k_\omega$. We ignore here the reabsorption processes that hold true for strong fluxes of incident radiation, and this gives $j_\omega \gg j_0$.

In the above regime of a photoresonant plasma, the equilibrium is supported between incident radiation and plasma emission at low intensities of incident radiation, which leads to a low number density of electrons. We now analyze another regime of evolution of a photoresonant plasma, where electron processes determine the energy balance in the plasma. In particular, in the regime of low intensity of incident radiation, the rate of associative ionization of excited atoms (process (3.102)) is small compared with the rate of photon departure, that is,

$$\frac{N_*}{\tau_{\mathrm{ef}}} \gg N_*^2 k_{\mathrm{as}} ,$$

where values of the rates of associative ionization k_{as} are given in Table 2.14 for an alkali metal plasma. Taking the Lorenz broadening mechanism of spectral lines, on the basis of (3.85) we obtain this criterion in the form

$$N_* \ll N_{\mathrm{ef}} = \frac{1}{\tau_{\mathrm{r}} k_{\mathrm{ion}} \sqrt{k_0 R}} , \tag{3.108}$$

where τ_{r} is the lifetime of an isolated atom. Taking typical parameters in this formula, $\tau \sim 10^{-8}\,\mathrm{s}$, $k_0 \sim 10^5\,\mathrm{cm}^{-1}$, $R \sim 1\,\mathrm{cm}$, and $k_{\mathrm{as}} \sim 10^{-12}\,\mathrm{cm}^3/\mathrm{s}$, we obtain criterion (3.108) as $N_* \ll 3 \times 10^{18}\,\mathrm{cm}^{-3}$. This criterion holds true for a weakly excited photoresonant plasma.

3.3.9
Kinetics of Electrons and Ionization Processes in Photoresonant Plasma

In considering the behavior of electrons in a photoresonant plasma, we note that the strongest processes involving electrons correspond to ground-state excitation and quenching of excited atoms according to the scheme

$$e + A^* \leftrightarrow e + A , \tag{3.109}$$

where A and A_* represent an atom in the ground and resonantly excited states. Then, for a Maxwell distribution of free electrons, we find from this equilibrium the electron temperature T_{e}

$$T_{\mathrm{e}} = \frac{\Delta\varepsilon}{\ln \frac{N_0 g_*}{N_* g_0}} ,$$

which coincides with the excitation temperature given by (3.104), $T_{\mathrm{e}} = T_*$. The electron number density is determined by ionization of excited atoms in collisions with electrons according to the scheme

$$e + A_* \leftrightarrow 2e + A^+ \,. \tag{3.110}$$

Evidently, this process is stepwise. As a result, electrons are formed both in collisions of two excited atoms according to process (3.102) and as a result of ionization of excited atoms by electron impact (process (3.110)). Correspondingly, the balance equation for the number density of electrons in the first stage of evolution of a photoresonant plasma takes the form

$$\frac{dN_e}{dt} = k_{as} N_*^2 + N_e N_* k_{ion} \,,$$

where N_e and N_* are the number densities of electrons and excited atoms, k_{ion} is the rate constant for ionization of an excited atom by electron impact, and k_{as} is the rate constant for process (3.104). The solution of this balance equation is

$$N_e = \frac{k_{as} N_*}{k_{ion}} \left[\exp(N_* k_{ion} t) - 1\right] \,. \tag{3.111}$$

As is seen, the process of associative ionization is of importance in increase of the electron number density in the first stage of evolution of a photoresonant plasma. Subsequently, increase of the electron number density is determined by ionization of atoms by electron impact.

To determine the electron temperature, we use the balance equation for the average energy of electrons per unit volume that results from processes (3.109):

$$\frac{d(N_e \, \varepsilon_e)}{dt} = \hbar\omega \, N_e (k_q N_* - k_{exc} N_0) \,,$$

where we assume a Maxwell distribution function for electrons, and $\varepsilon_e = 3T_e/2$ is the average electron energy, k_q is the rate constant for quenching of excited atoms by electron impact, and k_{exc} is the rate constant for atom excitation by electron impact. We assume the energy distribution function for electrons to be a Maxwell distribution function, whereas the electron temperature T_e varies in time. Taking the excitation temperature T_* in accordance with (3.104) and using the principle of detailed balance that connects rate constants k_q and k_{exc}, we reduce this balance equation to the form

$$\frac{dT_e}{dt} = \frac{2\hbar\omega}{3} k_q N_* \left[1 - \exp\left(\frac{\hbar\omega}{T_*} - \frac{\hbar\omega}{T_e}\right)\right] = -\frac{(T_* - T_e)}{\tau_T} \,,$$

where we assume the electron and excitation temperatures to be nearby ($|T_e - T_*| \ll T_*$), and a typical time for establishment of the electron temperature τ_T in a photoresonant plasma is

$$\tau_T = \frac{3T_*^2}{2(\hbar\omega)^2 k_q N_*} \,. \tag{3.112}$$

In the above regime of evolution of a photoresonant plasma, the energy of incident radiation is contained in outgoing radiation. We now consider another regime of evolution of a photoresonant plasma [70], where incident radiation energy is consumed on ionization of excited atoms in accordance with the following processes:

$$e + A^* \to e + A\,, \quad e + A^* \to 2e + A^+\,. \tag{3.113}$$

In this regime, electrons obtain energy by quenching of excited atoms and lose the energy as a result of the ionization process, which leads to the balance equation for the electron energy [70]:

$$\hbar\omega N_* N_e k_q = J_* N_* k_{ion} N_e\,,$$

where k_q is the rate constant for quenching of a resonantly excited atom by electron impact, k_{ion} is the rate constant for ionization of an excited atom in collision with an electron that is assumed to have stepwise character, and J_* is the ionization potential of a resonantly excited atom. This gives the equation for the electron temperature T_e^*:

$$k_{ion}\left(T_e^*\right) = \frac{\hbar\omega}{J_*} k_q\left(T_e^*\right)\,. \tag{3.114}$$

From this it follows that the electron temperature is independent of the number density of atoms, and its values are given in Table 3.4 for an alkali metal photoresonant plasma. Here (3.69) is used for the rate constant for stepwise ionization of excited atoms in collisions with electrons, and the values of the rate constants k_q for quenching the resonantly excited atom by electron impact are taken from Table 2.8. Note that because (3.69) gives overstated values of the rate constant for stepwise atom ionization, the values of the electron temperature in Table 3.4 are lower than the real values, but the simplicity of (3.69) allows us to analyze a general physical picture in this case.

Table 3.4 The electron temperature T_e^* according to (3.104), the equilibrium constant K_{ion} for ionization equilibrium at this temperature, and the intensity of incident radiation I_ω^* if the excitation temperature is equal to the electron temperature T_e^*.

Vapor	T_e^*, eV	ξ	K_{ion}, 10^{15} cm^{-3}	τ_*, 10^{-8} s	τ_T, 10^{-9} s	τ_{ion}, 10^{-8} s
Li(2^2P)	0.31	180	20	1.6	2.7	3.0
Na(3^2P	0.26	1400	13	1.1	2.6	2.1
K($4^2P_{1/2}$)	0.24	440	3.6	1.6	1.6	0.7
K($4^2P_{3/2}$)	0.24	440	7.5	1.5	1.6	2.6
Rb($5^2P_{1/2}$)	0.23	510	2.9	1.7	1.6	2.8
Rb($5^2P_{1/2}$)	0.22	630	5.8	1.6	1.6	2.7
Cs($6^2P_{1/2}$)	0.22	270	3.5	1.8	1.1	2.6
Cs($6^2P_{3/2}$)	0.21	480	6.0	1.6	1.1	2.4

Thus, there are two regimes for electron kinetics in a photoresonant plasma. At low radiation fluxes the average electron energy is determined by processes (3.109), and the electron temperature T_e coincides with the excitation temperature T_*. In the regime of high photon fluxes, the balance (3.114) for the electron temperature is determined by processes (3.113), and the electron temperature does not depend on the photon flux, whereas the excitation temperature increases with increasing photon flux according to (3.107). Evidently, in the second regime of electron kinetics, the loss of electron energy results from ionization of resonantly excited atoms rather than excitation of atoms in the ground state, and corresponds to the criterion

$$\xi = \frac{\exp\left(-\frac{\hbar\omega}{T_e}\right)}{\exp\left(-\frac{\hbar\omega}{T_*}\right)} \gg 1 . \tag{3.115}$$

The values of this parameter for photon flux $j_\omega = j_0$ are given in Table 3.4. As is seen, the second regime of electron evolution is realized for photon flux j_0 defined by (3.105).

Table 3.4 also contains the values of the equilibrium constant $K(T_e^*)$ at the maximum electron temperature T_e^* according to the relation $K_{ion} = N_e^2/N_*$, where N_e and N_* are the equilibrium values of the number densities of electrons and excited atoms at the ionization equilibrium. Note that the above evaluations relate to a weakly ionized plasma where the electron number densities are small compared with those of atoms, whereas remarkable ionization takes place when ionization equilibrium is attained. For example, in the sodium case at a typical number density of atoms $N = 1 \times 10^{16}\,\text{cm}^{-3}$ we have at $j_\omega = j_0$ that $N_* = 6 \times 10^{15}\,\text{cm}^{-3}$, and $N_e \approx 1 \times 10^{16}\,\text{cm}^{-3}$ at ionization equilibrium. This means that ionization processes destroy atoms, and then the character of interaction of resonant radiation with a plasma is changed.

Thus, we obtain strong action on a plasma in the regime of high radiation fluxes, and the sum of the interaction processes between photoresonant radiation and plasma leads to strong plasma ionization. Note that this interaction between photoresonant radiation and plasma proceeds in a narrow plasma region of length $1/k_\omega$ (k_ω is the absorption coefficient). Blooming of this region results from this interaction, and subsequently excitation of atoms proceeds in deeper plasma layers. Because of this character of interaction, the times for individual processes are of interest, and we determine them at $j_\omega = j_0$ assuming that the radiation frequency corresponds to the center of the spectral line. We have that a strong interaction proceeds in a region of length of approximately $1/k_0$, that is, of the order of 10^{-5} cm for alkali metal vapors. A typical time for equilibrium establishment between the number densities of atoms in the ground and resonantly excited state is

$$\tau_* = \frac{N_*}{j_\omega k_\omega} , \tag{3.116}$$

and the values of this time are given in Table 3.4 along with the values for the establishment of the electron temperature (according to (3.112)). We also give in

Table 3.4 typical times for establishment of ionization equilibrium τ_{ion} on the basis of (3.112),

$$\tau_{ion} = \tau_* + \frac{1}{k_{ion} N_*} \ln \left(\frac{k_{ion}}{k_{as} \sqrt{\frac{K}{N_*}}} \right) , \tag{3.117}$$

where we take into account that ionization proceeds after establishment of equilibrium for the resonantly excited state. As a result, resonant radiation propagates in a plasma in the form of an ionization wave.

References

1 Boltzmann, L. (1872) *Wien. Ber.*, **66**, 275.
2 Maxwell, J.C. (1860) *Philos. Mag.*, **19**, 19.
3 Maxwell, J.C. (1860) *Philos. Mag.*, **20**, 21.
4 Maxwell, J.C. (1867) *Philos. Trans. R. Soc.*, **157**, 49.
5 Boltzmann, L. (1866) *Wien. Ber.*, **53**, 195.
6 Boltzmann, L. (1868) *Wien. Ber.*, **58**, 517.
7 Boltzmann, L. (1871) *Wien. Ber.*, **63**, 397.
8 Boltzmann, L. (1871) *Wien. Ber.*, **63**, 679.
9 Boltzmann, L. (1871) *Wien. Ber.*, **63**, 712.
10 Batchelor, G.K. (1967) *An Introduction to Fluid Dynamics*, Cambridge University Press, Cambridge.
11 Currie, I.G. (1974) *Fundamental Mechanics of Fluids*, McGraw-Hill, London.
12 Lifshits, E.M. and Pitaevskii, L.P. (1981) *Physical Kinetics*, Pergamon Press, Oxford.
13 Huang, K. (1987) *Statistical Mechanics*, John Wiley & Sons, Inc., New York.
14 Acheson, D.J. (1990) *Elementary Fluid Dynamics*, Oxford University Press, Oxford.
15 Tolman, R.C. (1938) *The Principles of Statistical Mechanics*, Oxford University Press, London.
16 ter Haar, D. and Green, C.D. (1953) *Proc. Phys. Soc. A*, **66**, 153.
17 Smirnov, B.M. (2006) *Principles of Statistical Physics*, Wiley-VCH Verlag GmbH.
18 Lorentz, H.A. (1905) *Proc. Amst. Acad.*, 7, 438.
19 Lorentz, H.A. (1905) *Proc. Amst. Acad.*, 7, 585.
20 Lorentz, H.A. (1905) *Proc. Amst. Acad.*, 7, 684.
21 Enskog, D. (1912) *Ann. Phys.*, **38**, 731.
22 Davydov, B.I. (1935) *Phys. Z. Sowjetunion*, **8**, 59.
23 Morse, P.M., Allis, W.P., and Lamar, E.S. (1935) *Phys. Rev.*, **48**, 412.
24 Davydov, B.I. (1937) *Phys. Z. Sowjetunion*, **12**, 269.
25 Allis, W.P. and Allen, H.W. (1937) *Phys. Rev.*, **52**, 703.
26 Chapman, S. and Cowling, T.G. (1939) *The Mathematical Theory of Non-Uniform Gases*, Cambridge University Press, Cambridge.
27 Allis, W.P. (1956) *Handbuch der Physik*, vol. 21, (ed. S. Flugge), Springer, Berlin, p. 383.
28 Fokker, A.D. (1914) *Ann. Phys. (Leipzig)*, **43**, 810.
29 Planck, M. (1917) *Preuss. Akad. Wiss. Phys. Mat. Kl.*, **324**.
30 Gurevich, L.E. (1940) *Introduction to Physical Kinetics*, GIITL, Leningrad, in Russian.
31 Landau, L.D. (1937) *JETP*, **7**, 203.
32 Dutton, J. (1975) *J. Phys. Chem. Ref. Data*, **4**, 577.
33 Huxley, L.G.H. and Crompton, R.W. (1973) *The Diffusion and Drift of Electrons in Gases*, John Wiley & Sons, Inc., New York.
34 Smirnov, B.M. (1972) *Physics of a Weakly Ionized Gases*, Atomizdat, Moscow, in Russian.
35 Spitzer, L. (1962) *Physics of Fully Ionized Gases*, John Wiley & Sons, Inc., New York.

36 Gradshtein, I.S. and Ryzhik, I.M. (1962) *Tables of Integrals, Sums, Seria and Products*, GITTL, Moscow.

37 Margenau, H. (1946) *Phys. Rev.*, **69**, 508.

38 Ginzburg, V.L. (1960) *Propagation of Electromagnetic Waves in Plasma*, Fizmatgiz, Moscow, in Russian.

39 Ginzburg, V.L. and Gurevich, A.V. (1960) *Phys. Usp.*, **50**, 201.

40 Smirnov, B.M. (1981) *Physics of a Weakly Ionized Gases*, Mir, Moscow.

41 Smirnov, B.M. (2001) *Physics of Ionized Gases*, John Wiley & Sons, Inc., New York.

42 Smirnov, B.M. (2002) *Phys. Usp.*, **172**, 1411.

43 Smirnov, B.M. (2007) *Plasma Processes and Plasma Kinetics*, Wiley-VCH Verlag GmbH.

44 Pack, J.L. *et al.* (1992) *J. Appl. Phys.*, **71**, 5363.

45 Thompson, J.J. (1903) *Conduction of Electricity through Gases*, Cambridge University Press, Cambridge.

46 Townsend, J.S. and Hurst, H.E. (1904) *Philos. Mag.*, **8**, 738.

47 Townsend, J.S. (1915) *Electricity in Gases*, Clarendon Press, Oxford.

48 Townsend, J.S. (1925) *Motion of Electrons in Gases*, Clarendon Press, Oxford.

49 Townsend, J.S. (1947) *Electrons in Gases*, Hutchinsons Scientific and Technical Publications, London.

50 Smirnov, B.M. (1986) *High Temp.*, **24**, 239.

51 Wiese, W.L., Smith, M.W., and Glennon, B.M. (1966) *Atomic Transition Probabilities*, vol. 1, National Bureau of Standards, Washington.

52 Fuhr, J.R. and Wiese, W.L. (2003–2004) in: *Chemical Physics*, (ed. D.R. Lide), CRC, London, pp. 10104–10105.

53 Phelps, A.V. (1955) *Phys. Rev.*, **99**, 1307.

54 Schearer, L.D. (1968) *Phys. Rev.*, **171**, 81.

55 Biberman, L.M. and Veklenko, B.A. (1957) *Sov. Phys. JETP*, **4**, 440.

56 Veklenko, B.A. (1957) *Zh. Eksp. Teor. Fiz.*, **33**, 629.

57 Veklenko, B.A. (1957) *Zh. Eksp. Teor. Fiz.*, **33**, 817.

58 Biberman, L.M. (1947) *Zh. Eksp. Teor. Fiz.*, **17**, 416.

59 Holstein, T. (1947) *Phys. Rev.*, **72**, 1212.

60 Biberman, L.M. (1948) *Dokl. Akad. Nauk SSSR*, **49**, 659.

61 Molish, A.F. and Oehry, B.P. (1998) *Radiation Trapping in Atomic Vapours*, Oxford University Press, New York.

62 Smith, S.J. and Burch, D.S. (1959) *Phys. Rev.*, **116**, 1125.

63 Feldman, D. (1970) *Z. Naturforsch.*, **25**, 621.

64 Geltman, S. and Kraus, S. (1962) *Astropys. J.*, **136**, 935.

65 Smirnov, B.M. (1982) *Negative Ions*, McGraw-Hill, New York.

66 Allen, K.U. (1977) *Astrophysical Values*, Mir, Moscow.

67 Manska, J.T. (1992) *The Solar Transition Region*, Cambridge University Press, Cambridge.

68 Ochkin, V.N., Preobrazhensky, N.G., and Shaparev, N.Ya. (1998) *Optogalvanic Effect in Ionized Gas*, Gordon and Breach, London.

69 Ochkin, V.N. (2009) *Spectroscopy of Low Temperature Plasma*, Wiley-VCH Verlag GmbH.

70 Eletskii, A.V., Zaitsev, Yu.N., and Fomichev, S.V. (1988) *Zh. Eksp. Teor. Fiz.*, **94**, 98.

4
Transport Phenomena in Ionized Gases

4.1
Hydrodynamics of Ionized Gases

4.1.1
Macroscopic Gas Equations

The distribution function for atomic particles of a gaseous system gives detailed information about this system. It yields a variety of macroscopic parameters of the system by averaging the distribution function over particle velocities and internal quantum numbers. This operation leads to a transition from a kinetic description to a hydrodynamic description of the gaseous system and is accompanied by a decrease of the amount of information about the system. It is clear from a general consideration that the kinetic description is useful for nonequilibrium gaseous systems, whereas the hydrodynamic description is better for equilibrium systems where average gaseous state parameters characterize the system in total. In this way the macroscopic equations which follow from the kinetic equation are practically the equations of mass and heat transfer [1–3].

Below we obtain macroscopic equations for an ionized gas from the kinetic equation, starting from integration of the kinetic equation (3.4) over atom velocities. The right-hand side of the equation is the total variation of the density of particles per unit time due to collisions. Assuming that particles are not formed or destroyed inside the system volume, the right-hand side of the resulting equation is zero, and the kinetic equation becomes

$$\int \frac{\partial f}{\partial t} d\mathbf{v} + \int \mathbf{v} \cdot \frac{\partial f}{\partial \mathbf{r}} d\mathbf{v} + \frac{\mathbf{F}}{m} \cdot \int \frac{\partial f}{\partial \mathbf{v}} d\mathbf{v} = 0 .$$

We reverse the order of differentiation and integration in the first two terms, and introduce the definitions $\int f d\mathbf{v} = N$ and $\int \mathbf{v} f d\mathbf{v} = N\mathbf{w}$, where N is the particle number density and $\mathbf{w}$ is the mean velocity of the particles or the drift velocity. The third term is zero because the distribution function for infinite velocity is zero. Thus, we obtain

$$\frac{\partial N}{\partial t} + \operatorname{div}(N\mathbf{w}) = 0 . \tag{4.1}$$

Fundamentals of Ionized Gases, First Edition. Boris M. Smirnov.

This is a standard form for a continuity equation.

This equation corresponds to a one-component system where internal states of the particles are not distinguished. In fact, by definition, $\int I_{\text{col}}(f)d\mathbf{v}$ is the variation of the particle number density due to collision processes, so it is zero for the total number density summed over internal quantum numbers. If we discriminate internal states, the right-hand side of the continuity equation is akin to (4.1), so the continuity equation has the form of the balance equation:

$$\frac{\partial N_{\text{i}}}{\partial t} + \text{div}(N_{\text{i}}\mathbf{w}) = \sum_f k_{f\text{i}} N_j - \sum_f k_{if} N_j \,, \tag{4.2}$$

where indices i and f refer to internal states of particles and k_{if} is the rate constant for transition between states i and f.

Another macroscopic equation that is analogous to the equation of motion for particles follows from multiplication of the kinetic equation (4.1) by the particle momentum mv_α and integration over particle velocities, where v_α is a component of the particle velocity, and $\alpha = x, y$, or z. The right-hand side of the resulting equation is the variation of the total momentum of the particles. For a system of identical particles, collisions do not change the total momentum of the particles, so the right-hand side of the equation is zero and the macroscopic equation has the form

$$\int mv_\alpha \frac{\partial f}{\partial t} d\mathbf{v} + \int mv_\alpha v_\beta \frac{\partial f}{\partial x_\beta} d\mathbf{v} + F_\beta \int v_\alpha \frac{\partial f}{\partial v_\beta} d\mathbf{v} = 0 \,.$$

Here indices α and β both denote vector components ($\alpha, \beta = x, y, z$) with a summation over β, and x_β is a coordinate. If we change the order of the integration and summation in the first two terms and integrate the third term by parts, we obtain

$$\int v_\alpha \frac{\partial f}{\partial v_\beta} dv_\beta = v_\alpha f|_{-\infty}^{+\infty} - \int \delta_{\alpha\beta} dv_\beta = -N\delta_{\alpha\beta} \,,$$

where $\delta_{\alpha\beta}$ is the Kronecker symbol $\delta_{\alpha\beta} = 1$ if $\alpha = \beta$, and $\delta_{\alpha\beta} = 0$ if $\alpha \neq \beta$. Finally, we obtain

$$\frac{\partial}{\partial t}(mNv_\alpha) + \frac{\partial}{\partial x_\beta}(N\langle mv_\alpha v_\beta \rangle) - NF_\alpha = 0 \,,$$

where angle brackets denote averaging over the particle distribution function.

We define the pressure tensor as

$$P_{\alpha\beta} = \langle m(v_\alpha - w_\alpha)(v_\beta - w_\beta) \rangle \,. \tag{4.3}$$

Inserting this tensor into the above equation, we have

$$\frac{\partial}{\partial t}(mNv_\alpha) + \frac{\partial P_{\alpha\beta}}{\partial x_\beta} + \frac{\partial}{\partial x_\beta}(mNw_\alpha w_\beta) - NF_\alpha = 0 \,.$$

Subtracting from this the continuity equation in the form

$$m w_\alpha \left[\frac{\partial N}{\partial t} + \frac{\partial}{\partial x_\beta}(N w_\alpha) \right] = 0 \,,$$

we obtain

$$m \frac{\partial w_\alpha}{\partial t} + \frac{1}{N} \frac{\partial P_{\alpha\beta}}{\partial x_\beta} + m w_\beta \frac{\partial w_\alpha}{\partial x_\beta} - F_\alpha = 0 \,. \tag{4.4}$$

The form of this equation for the mean particle momentum depends on the representation of the pressure tensor $P_{\alpha\beta}$, which is determined by properties of the system. In the simplest case of an isotropic gas or liquid it can be expressed in the form

$$P_{\alpha\beta} = p \delta_{\alpha\beta} \,, \tag{4.5}$$

where $\delta_{\alpha\beta}$ is the Kronecker symbol and p is the pressure inside the gas or liquid. Insertion of this expression into (4.4) transforms the latter into the following equation, which in vector form is given by

$$\frac{\partial \mathbf{w}}{\partial t} + (\mathbf{w} \cdot \nabla)\mathbf{w} + \frac{\nabla p}{\rho} - \frac{\mathbf{F}}{m} = 0 \tag{4.6}$$

and is the Euler equation in the absence of external fields ($\mathbf{F} = 0$) [4–8]. Here $\mathbf{w}$ is the average velocity of the gas, $\rho = mN$ is the mass density, and $\mathbf{F}$ is the force acting on a single particle from external fields.

The macroscopic equations (4.1) and (4.4) relate to a one-component gas. One can generalize these equations for multicomponent gaseous systems. The right-hand side of the continuity equation is the variation of the number density of particles of a given species per unit volume and per unit time due to production or loss of this type of particles. In the absence of production of particles of one species as a result of decomposition of particles of other species, the continuity equation has the form of (4.2) regardless of processes involving other particles.

The right-hand side of (4.4) for a multicomponent system contains the variation per unit time of the momentum of particles of a given species as a result of collisions with particles of other types. If the mean velocities of the two types of particles are different, a momentum transfer will occur between them. The momentum transfer between two species is proportional to the difference between their mean velocities. Therefore, one can rewrite (4.4) as

$$\frac{\partial w_\alpha^{(q)}}{\partial t} + \frac{1}{m^{(q)} N^{(q)}} \frac{\partial P_{\alpha\beta}^{(q)}}{\partial x_\beta} + w_\beta^{(q)} \frac{\partial w_\alpha^{(q)}}{\partial x_\beta} - \frac{F_\alpha}{m_q} = \sum_s \frac{\left(w_\alpha^{(s)} - w_\alpha^{(q)} \right)}{\tau_{qs}} \,. \tag{4.7}$$

Superscripts q and s denote here the particle species and τ_{qs} is the characteristic time for the momentum transfer from species q to species s. Since the transfer does not change the total momentum of the system, the characteristic time for the momentum transfer satisfies the relation

$$\frac{m^{(q)} N^{(q)}}{\tau_{qs}} = \frac{m^{(s)} N^{(s)}}{\tau_{sq}} \,. \tag{4.8}$$

In particular, the Euler equation for a multicomponent system in vector form is given by

$$\frac{\partial \mathbf{w}_q}{\partial t} + (\mathbf{w}_q \cdot \nabla)\mathbf{w}_q + \frac{\nabla p}{\rho_q} - \frac{\mathbf{F}_q}{m_q} = \sum_s \frac{(\mathbf{w}_s - \mathbf{w}_q)}{\tau_{qs}} \,. \tag{4.9}$$

4.1.2
Equation of State for a Gas

The relation between the bulk parameters of a gas (pressure p, temperature T, and particle number density N) is given by the equation of state. Below we derive this equation for a homogeneous gas. For this purpose we employ a frame of reference where the gas (or a given volume of the gas) is at rest. The pressure is the force in this frame of reference that acts on a unit area of an imaginary surface in the system. To evaluate this force we take an element of this surface to be perpendicular to the x-axis, so the flux of particles with velocities in an interval from v_x to $v_x + dv_x$ is given by $dj_x = v_x f dv_x$, where f is the distribution function. Elastic reflection of a particle from this surface leads to inversion of v_x, that is, $v_x \to -v_x$ as a result of the reflection. Therefore, a reflected particle of mass m transfers to the area the momentum $2mv_x$. The force acting on this area is the momentum change per unit time. Hence, the gas pressure, that is, the force acting per unit area, is

$$p = \int_{v_x > 0} 2mv_x f dv_x = m \int v_x^2 f dv_x = mN \left\langle v_x^2 \right\rangle \,.$$

We account for the fact that the pressure on both sides of the area is the same.

In the above expression v_x is the velocity component in the frame of reference where the gas as a whole is at rest. Transforming back to the original axes, we have

$$p = mN \left\langle (v_x - w_x)^2 \right\rangle \,, \tag{4.10}$$

where w_x is the x component of the mean gas velocity. Because the distribution function is isotropic in the reference frame where the gas as a whole is at rest, the gas pressure is identical in all directions and is given by

$$p = mN \left\langle (v_x - w_x)^2 \right\rangle = mN \left\langle (v_y - w_y)^2 \right\rangle = mN \left\langle (v_z - w_z)^2 \right\rangle \,. \tag{4.11}$$

Let us use the definition of the gas temperature through the mean particle velocity in the frame of reference where the mean velocity is zero, that is,

$$\frac{3T}{2} = \frac{m}{2} \left\langle (\mathbf{v} - \mathbf{w})^2 \right\rangle \,.$$

Then (4.10) leads to the following relationship between pressure and temperature

$$p = NT\,. \tag{4.12}$$

Equation (4.12) is the equation of state for an ideal gas.

4.1.3
The Navier–Stokes Equation

Note that above in deriving the generalized Euler equation (4.6) for a one-component system we ignored collisions between atoms, and collisions between atoms or molecules of different components are taken into account in (4.7)–(4.9) through typical times τ_{qs}. We now include collisions between atoms in the balance equation for the average atom momentum for a one-component system. Such collisions lead to the frictional force between neighboring gas layers which are moving with different velocities. This effect is expressed through the gas viscosity coefficient η, which is taken into account as arising from the above-mentioned frictional force. According to the definition of the viscosity coefficient η, if a gas (or liquid) is moving along the x-axis and varies in the z direction, the frictional force f per unit area is given by

$$f = -\eta \frac{\partial w_x}{\partial z}\,, \tag{4.13}$$

where w_x is the average gas velocity.

We include the gas viscosity in the expression for the pressure tensor, which takes the form

$$P_{\alpha\beta} = p\delta_{\alpha\beta} + P'_{\alpha\beta}\,,$$

where the latter term accounts for the gas viscosity. On the basis of the viscosity definition and transferring to an arbitrary frame of reference, one can represent the pressure tensor $P'_{\alpha\beta}$, which is symmetric, in the form

$$P'_{\alpha\beta} = -\eta \left[\frac{\partial w_\alpha}{\partial x_\beta} + \frac{\partial w_\beta}{\partial x_\alpha} + a\delta_{\alpha\beta} \frac{\partial w_\alpha}{\partial x_\beta} \right].$$

The factor a may be found as follows. The forces of viscous friction in a gas occur because neighboring gas layers move with different velocities. If a gas could be decelerated as a whole, this friction mechanism would have no effect and the force due to the gas viscosity would vanish. Hence, the trace of the pressure tensor is zero. This yields $a = -2/3$, and the viscosity term in the pressure tensor can be written in the form

$$P'_{\alpha\beta} = -\eta \left[\frac{\partial w_\alpha}{\partial x_\beta} + \frac{\partial w_\beta}{\partial x_\alpha} - \frac{2}{3}\delta_{\alpha\beta} \frac{\partial w_\alpha}{\partial x_\beta} \right]. \tag{4.14}$$

With the viscosity part of the pressure tensor taken into account, (4.4) takes the form of the Navier–Stokes equation [6–10]:

$$\frac{\partial \mathbf{w}}{\partial t} + (\mathbf{w}\cdot\nabla)\mathbf{w} = -\frac{\nabla p}{\rho} + \nu\Delta\mathbf{w} + \frac{\nu}{3}\nabla\mathbf{w} + \frac{\mathbf{F}}{m}\,, \tag{4.15}$$

where $\nu = \eta/\rho$ is the kinematic viscosity coefficient. This equation describes evolution of a moving gas by accounting for the gas viscosity and the action of external fields.

4.1.4
Macroscopic Equation for Ion Motion in a Gas

We now derive the motion of ions in a gas subjected to an external electric field **E** by accounting for collisions of ions with gas atoms. The collision integral is given by (3.8) in this case, and if the number density of ions is small compared with the number density of atoms, elastic collisions with atoms govern the character of the motion of ions in a gas. We first use the tau approximation (3.5) for the collision integral. Multiplying the kinetic equation (3.4) by the ion momentum $m\mathbf{v}$ and integrating it over ion velocities, we obtain the balance equation in the stationary case in the form

$$e\mathbf{E} = \frac{m\mathbf{w}_i}{\tau} , \qquad (4.16)$$

where $\mathbf{w}_i$ is the average ion velocity and τ is the characteristic time between successive collisions of the ion with atoms. This time can be estimated as $\tau \sim (N_a v \sigma)^{-1}$, where N_a is the number density of atoms, v is a typical relative velocity of an ion–atom collision, and σ is a typical cross section for collisions with scattering into large angles. The left-hand side of this equation is the force on the ion from the electric field, and the right-hand side is the frictional force arising from collisions of the ion with the gas atoms. We shall determine the frictional force below without resorting to the tau approximation.

If we multiply the kinetic equation (3.4) by the ion momentum $m\mathbf{v}$ and integrate it over ion velocities $d\mathbf{v}$, we obtain the following balance equation by accounting for expression (3.8) for the collision integral:

$$e\mathbf{E}N_i = \int m\left(\mathbf{v}_i' - \mathbf{v}_i\right) g d\sigma f_1 f_2 d\mathbf{v}_i d\mathbf{v}_a . \qquad (4.17)$$

The quantities $\mathbf{v}_i$ and $\mathbf{v}_a$ are the initial velocities of the ion and atom, respectively, of masses m and m_a, N_i and N_a are the number densities of ions and atoms, g is the relative velocity of ion–atom collision that is conserved in the collision, and f_i and f_a are the distribution functions for ions and atoms. From the principle of detailed balance, which ensures invariance under time reversal in the evolution of the system, it follows that

$$\int \mathbf{v}_i f_i' f_a' d\sigma d\mathbf{v}_i d\mathbf{v}_a = \int \mathbf{v}_i' f_i f_a d\sigma d\mathbf{v}_i d\mathbf{v}_a .$$

The invariance under time reversal ($t \to -t$) of $d\sigma d\mathbf{v}_i d\mathbf{v}_a$ must be examined. Let us express the ion velocity $\mathbf{v}_i$ in terms of the relative ion–atom velocity $\mathbf{g}$ and the center-of-mass velocity $\mathbf{V}$ by the relation $\mathbf{v}_i = \mathbf{g} + m_a\mathbf{V}/(m + m_a)$. We find that $m(\mathbf{v}_i - \mathbf{v}_i') = \mu(\mathbf{g} - \mathbf{g}')$. The relative velocity after collision has the form $\mathbf{g}' = \mathbf{g}\cos\vartheta + \mathbf{k}g\sin\vartheta$, where ϑ is the scattering angle and $\mathbf{k}$ is a unit vector directed perpendicular to the vector $\mathbf{g}$. Because of the random distribution of $\mathbf{k}$, the integration over scattering angles gives $\int(\mathbf{g}-\mathbf{g}')d\sigma = \mathbf{g}\sigma^*(g)$, where $\sigma^*(g) = \int(1-\cos\vartheta)d\sigma$ is the diffusion cross section of ion–atom scattering. Hence, (4.17) takes the form

$$e\mathbf{E}N_{\rm i} = \int \mu \mathbf{g} g \sigma^*(g) f_{\rm i} f_{\rm a} d\mathbf{v}_{\rm i} d\mathbf{v}_{\rm a} \,. \tag{4.18}$$

In the case of polarization interaction between the ion and the atom, the diffusion cross section is similar in magnitude to the cross section of polarization capture (2.40), and is inversely proportional to the relative velocity g of the collision. Since

$$\int \mathbf{g} f_{\rm i} f_{\rm a} d\mathbf{v}_{\rm i} d\mathbf{v}_{\rm a} = (\mathbf{w}_{\rm i} - \mathbf{w}_{\rm a}) N_{\rm i} N_{\rm a} = \mathbf{w}_{\rm i} N_{\rm i} N_{\rm a} \,,$$

where $\mathbf{w}_{\rm i}$ is the average ion velocity and $\mathbf{w}_{\rm a} = 0$ is the average atom velocity, (4.18) leads to

$$\mathbf{w}_{\rm i} = \frac{e\mathbf{E}}{\mu N_{\rm a} g \sigma^*(g)} \,, \tag{4.19}$$

where the diffusion cross section $\sigma^*(g)$ is given by (2.40). Note that this formula is valid at any electric field strength, including very high field strengths, when the ion distribution function is very different from the Maxwell distribution dependence on the gas temperature. The integral equation (4.18) can be a basis for the analysis of ion behavior in a gas in an external electric field.

4.1.5 The Chapman–Enskog Approximation for Ion Mobility in Gas

On the basis of (4.19) one can construct a method for numerical evaluation of the ion mobility in a weak electric field, that is, the Chapman–Enskog approximation [11, 12]. This corresponds to low electric field strengths according to the criterion

$$eE\lambda \ll T \,, \tag{4.20}$$

where λ is the mean free path of ions in a gas. Then the velocity distribution function for ions is similar to the Maxwell one $\varphi(v)$ and can be represented in the form of the following expansion:

$$f(\mathbf{v}) = \varphi(v)\left[1 + v_x \psi(v)\right] \,,$$

where the x-axis is directed along the electric field, and the function $\psi(v)$ may be found from the integral equation (4.19).

In constructing the Chapman–Enskog approximation [11, 12] for evaluation of the ion mobility, we expand the function $\psi(v)$ over the Sonin polynomials $S_n^k(u^2)$, where $u = v/\sqrt{2T/M}$ is the reduced ion velocity (m is the ion mass, T is the gas temperature). The optimal index value for the ion mobility is $n = 3/2$, and in the first approximation we restrict ourselves to the Sonin polynomial with $k = 0$, which is equal to 1. Hence, in the first Chapman–Enskog approximation the quantity ψ is independent of the ion velocity and may be found from (4.19), which has the following form in this approximation:

$$eEN_{\rm i} = \psi \int \mu g_x g \sigma^*(g) v_x \varphi(v)\varphi(v_a) d\mathbf{v} d\mathbf{v}_{\rm a}$$
$$= \frac{\psi}{3}\mu \int g^2 \sigma^*(g)\varphi(v)\varphi(v_a) d\mathbf{v} d\mathbf{v}_{\rm a} \,.$$

From the definition of the ion drift velocity $w_{\rm i}$ we have

$$w_{\rm i} = \int \mathbf{v} f(\mathbf{v}) d\mathbf{v} = \frac{\psi}{3}\langle v^2 \rangle \,.$$

This leads to the following connection between the ion drift velocity in a weak electric field $K_{\rm I}$ and the cross section of ion–atom collision in the first Chapman–Enskog approximation:

$$K_{\rm I} = \frac{3e\sqrt{\pi}}{8N_{\rm a}\overline{\sigma}\sqrt{2T\mu}} \,, \quad \overline{\sigma}(T) = \frac{1}{2}\int_0^\infty \sigma^*(x) e^{-x} x^2 dx \,, \quad x = \frac{\mu g^2}{2T} \,. \tag{4.21}$$

4.1.6
Excitation of an Ionized Gas in an Electric Field

A plasma is a convenient means to accomplish energy transfer from an external electric field to a gas. It is used in plasma generators, gas lasers, and other devices for transformation of electric energy to other forms of energy. We can find the specific power transformed by a gas discharge plasma in the low electric current regime when interaction of charged particles is nonessential, and electric currents in a gas do not change its properties significantly. One electron moving in a gas in an external electric field of strength E transfers to gas atoms the power $eEw_{\rm e}$, where $w_{\rm e}$ is the electron drift velocity. Therefore, the power transformed per unit volume of a plasma is

$$P = N_{\rm e} eEw_{\rm e} = N_{\rm e} N_{\rm a} w_{\rm e} \,, \tag{4.22}$$

where $N_{\rm e}$ is the electron number density and $N_{\rm a}$ is the number density of atoms or molecules of the gas, and the electron drift velocity $w_{\rm e}$ depends on the ratio $eE/N_{\rm a}$.

For an understanding of the practical physical quantities associated with the energy transformation process, Table 4.1 gives the parameters in (4.22) for mean electron energies of 1 and 3 eV, and the rate of the temperature change in this table is determined by formula

$$\frac{dT}{dt} = N_{\rm e}\frac{eE}{N_{\rm a}}\frac{w_{\rm e}}{c_p} \tag{4.23}$$

Table 4.1 Parameters of the excitation of a weakly ionized gas in an external electric field. Here N_a is the number density of gas atoms or molecules, $1\,\text{Td} = 1 \times 10^{-17}\,\text{V cm}^2$, η is the energy fraction as a percentage that goes into elastic (elas), vibrational (vib), rotational (rot), and electronic (elec) degrees of freedom; the specific released energy $P/(N_e N_a)$ is expressed in units of $10^{-29}\,\text{W cm}^3$, and the specific temperature variation $(1/N_e)(dT/dt)$ is given in units of $10^{-7}\,\text{K cm}^3/\text{s}$.

Gas	ε, eV	E/N, Td	w_e, 10^6 cm/s	η_{elec}	η_{vib}	η_{rot}	η_{elec}	$\frac{P}{N_e N}$	$(1/N_e)(dT/dt)$
He	1	1.2	0.53	100	–	–	–	0.1	0.3
	3	3.6	0.94	100	–	–	–	0.54	1.6
Ar	1	0.2	0.2	100	–	–	–	0.006	0.02
	3	1.4	0.35	100	–	–	–	0.08	0.23
H_2	1	15	2.4	16	79	5	–	5.8	12
	3	60	7	6	93	1	–	67	140
N_2	1	4	0.93	8	83	9	–	0.	1.2
	3	105	11	0.2	53	0.5	46	180	380
CO	1	17	3.1	0.7	98	1	–	8.4	17
	3	130	13	0.1	54	0.1	46	270	560
CO_2	1	2.4	7	0.1	85	15	–	2.7	4.3
	3	60	13	0.1	97	0.1	3	120	200

in the first stage of gas heating, where c_p is the heat capacity per molecule at constant pressure. Table 4.1 also contains the pathways of energy consumption; η is the energy fraction transformed to the corresponding degree of freedom and expressed as a percentage. In the first stage of this process in molecular gases, the energy is transformed primarily to vibrational excitation. This is a property employed in gas discharge of molecular lasers that generate radiation by vibrational–rotational transitions of molecules.

To provide high efficiency for this method of energy input, we compare it with heating of a gas through the walls of an enclosure containing this gas. Assuming that heat transfer occurs by thermal conductivity, we have the estimate

$$q \sim \kappa \frac{\Delta T}{l}$$

for the heat flux, where κ is the thermal conductivity coefficient, ΔT is the difference between temperatures for the walls of the enclosure, and l is a dimension of the enclosure. From this it follows that the specific power introduced by heating the walls of the enclosure is

$$P \sim \kappa \frac{\Delta T}{l^2} \, .$$

For a numerical estimate we take $\Delta T \sim 1000\,\text{K}$, $\kappa \sim 10^{-3}\,\text{W/(cm K)}$ (a value corresponding to the thermal conductivity coefficient of air at $T \sim 1000\,\text{K}$), and $l \sim 1\,\text{cm}$. The result is $P \sim 1\,\text{W/cm}^3$. Such a specific power is reached at atmospheric pressure and a number density of electrons of $N_e \sim 10^9\,\text{cm}^{-3}$. Because the electron number density can be higher by several orders of magnitude, the electrical method for energy input into a gas through electrons is more effective than gas heating.

4.2 Transport Phenomena in Neutral Gases

4.2.1 Transport of Particles in Gases

Let us consider a weakly nonuniform gas when a thermodynamic parameter that characterizes its state varies slowly in space. A flux occurs in this gas that tends to equalize this thermodynamic parameter over the spatial region. Depending on a certain slowly varied parameter, this flux determines a certain transport phenomenon in a gas. The number density of atoms or molecules of each species and the temperature and the mean velocity of atoms or molecules are parameters that may vary. If some of these parameters vary in this region, appropriate fluxes arise to equalize these parameters over the total volume of a gas or plasma. The flux is small if a variation of the appropriate parameter is small on distances of the order of the mean free path of atoms or molecules. Then a stationary state of the system with fluxes exists, and such states are conserved during times much longer than typical times between particle collisions. This corresponds to the criterion

$$\lambda \ll L , \tag{4.24}$$

where λ is the mean free path of particles in a collision and L is a typical size of the gaseous system under consideration or the distance over which a parameter varies noticeably. If this criterion is fulfilled, the system is in a stationary state to a first approximation, and transport of particles, heat, or momentum occurs in the second approximation in terms of an expansion over a small parameter according to criterion (4.24). Various types of such transport will be considered below.

The coefficients of proportionality between fluxes and corresponding gradients are called kinetic coefficients or transport coefficients. In particular, the diffusion coefficient D is introduced as the proportionality factor for the relationship between the particle flux $\mathbf{j}$ and the gradient of concentration c of a given species, or

$$\mathbf{j} = -DN\nabla c , \tag{4.25}$$

where N is the total particle number density. If the concentration of a given species k is low ($c_k \ll 1$), that is, this species is an admixture to a gas, the flux of particles of a given species can be written as

$$\mathbf{j} = D_k \nabla N_k , \tag{4.26}$$

where N_k is the number density of particles of a given species.

The thermal conductivity coefficient κ is defined as the proportionality factor for the relationship between the heat flux $\mathbf{q}$ and the temperature gradient:

$$\mathbf{q} = -\kappa \nabla T . \tag{4.27}$$

The viscosity coefficient η is the proportionality factor for the relationship between the frictional force acting on a unit area of a moving gas and the gradient of the mean gas velocity in the direction perpendicular to the surface of a gas element. If the mean gas velocity **w** is parallel to the x-axis and varies in the z direction (see Figure 4.1), the frictional force is proportional to $\partial w_x/\partial z$ and acts on an xy surface in the gas. Thus, the force f per unit area is

$$f = -\eta \frac{\partial w_x}{\partial z} \tag{4.28}$$

and this definition of the viscosity coefficient for gases is also valid for liquids.

Assume that transport processes in a gas are determined by elastic collisions of gas atoms and obtain the dependence of kinetic coefficients on the cross section of elastic scattering of atoms and other parameters of a gaseous system, starting from the diffusion coefficient. The diffusive flux is the difference between fluxes in opposite directions. Each of these, in order of magnitude, is $N_k v$, where N_k is the number density of atoms of a given species and v is the thermal velocity of atoms. Thus, the net flux behaves as $j \sim \Delta N_k v$, where ΔN_k is the difference between the number densities of atoms which determine oppositely directed fluxes of atoms. Atoms that reach a given point without collisions have distances from it of the order of the mean free path $\lambda \sim (N\sigma)^{-1}$, where σ is a typical cross section for elastic collisions of atoms and N is the total number density of atoms. Hence $\Delta N_k \sim \lambda \nabla N_k$, and the diffusive flux behaves as $j \sim \lambda v \nabla N_k$. Comparing this estimation with the definition of the diffusion coefficient (4.25), we obtain for the diffusion coefficient

$$D \sim v\lambda \sim \frac{\sqrt{T}}{N\sigma\sqrt{m}} \,. \tag{4.29}$$

Here T is the gas temperature and m is the mass of atoms of a given species, assumed to be of the same order of magnitude as the masses of other particles composing the gas. In this analysis we do not need to account for the sign of the

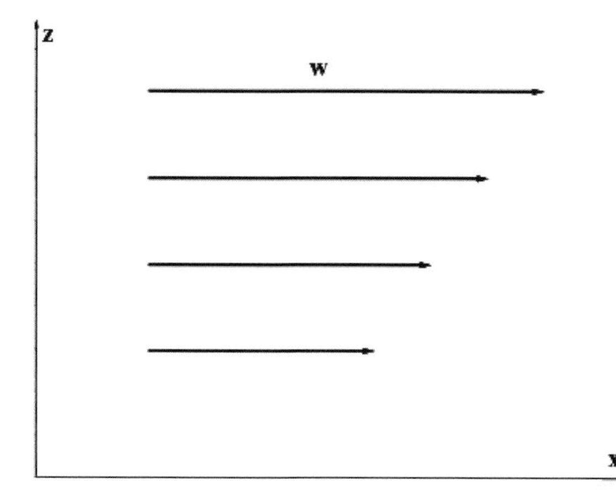

Figure 4.1 The geometry of fluxes in the definition of the viscosity coefficient.

flux, because it is simply opposite to the number density gradient and tends to equalize the atom number densities at neighboring points. The same can be said about the signs of the fluxes and gradients for the other transport phenomena.

4.2.2 Diffusive Motion of Particles in Gases

Let us consider the characteristics of the diffusion process [13, 14]. Ignoring the processes of formation and destruction of atoms of type k in a spatial region, we have the continuity equation (4.1) in the form

$$\frac{\partial N_k}{\partial t} + \operatorname{div} \mathbf{j} = 0 ,$$

where the flux of atoms of a given species is given by (4.26). For simplicity, we consider the case $N_k \ll N$, where N is the total number density of atoms, and omit below subscript k. Then, the diffusive motion of atoms in a gas is described by

$$\frac{\partial N}{\partial t} = D \Delta N . \tag{4.30}$$

It follows from this that a typical time for atom transport over a distance of the order of L is $\tau_L \sim L^2/D$. Using the estimate (4.29) for the diffusion coefficient, we find that $\tau_L \sim \tau_0 (L/\lambda)^2$, where $\tau_0 \sim \lambda/v$ is a typical time between successive collisions. Condition (4.24) leads to the condition $\tau_L \gg \tau_0$, which is the identifying feature allowing us to consider this process as time-independent.

To study the diffusion of test particles, we introduce the probability $W(\mathbf{r}, t)$ that a test particle is at point $\mathbf{r}$ at time t. If we assume this particle is located at the spatial origin at time zero, the probability will be spherically symmetric. The normalization condition is

$$\int_0^\infty W(r,t) \cdot 4\pi r^2 dr = 1 . \tag{4.31}$$

The probability W satisfies (4.30), which in the spherically symmetric case takes the form

$$\frac{\partial W}{\partial t} = \frac{D}{r} \frac{\partial^2}{\partial r^2} (r W) .$$

To find mean values for the diffusion parameters, we multiply this equation by $4\pi r^4 dr$ and integrate the result over all r. The left-hand side of the equation yields

$$\int_0^\infty 4\pi r^4 dr \frac{\partial W}{\partial t} = \frac{d}{dt} \int_0^\infty r^2 W \cdot 4\pi r^2 dr = \frac{d}{dt} \overline{r^2} ,$$

where $\overline{r^2}$ is the mean square of the distance from the origin. Integrating twice by parts and using the normalization condition (4.30), we transform the right-hand side of the equation into

$$D \int_0^\infty 4\pi r^4 dr \frac{1}{r} \frac{\partial^2}{\partial r^2}(rW) = -3D \int_0^\infty 4\pi r^2 dr \frac{\partial}{\partial r}(rW)$$
$$= 6D \int_0^\infty W \cdot 4\pi r^2 dr = 6D \, .$$

The resulting equation is $d\overline{r^2} = 6D dt$. Since at zero time the particle is located at the origin, the solution of this equation has the form

$$\overline{r^2} = 6Dt \, . \tag{4.32}$$

Because the motion in different directions is independent and has a random character, it follows from this that

$$\overline{x^2} = \overline{y^2} = \overline{z^2} = 2Dt \, . \tag{4.33}$$

The solution of (4.30) can be obtained from the normal distribution (1.77), which is appropriate for this process. Diffusion consists of random displacements of a particle, and the result of many collisions of this particle with its neighbors fits the general concept of the normal distribution. In the spherically symmetric case we have

$$W(\mathbf{r}, t) = w(x, t) w(y, t) w(z, t) \, ,$$

and substituting $\Delta = \langle x^2 \rangle = 2Dt$ into (1.77), we obtain

$$w(x, t) = (4\pi Dt)^{-1/2} \exp\left(-\frac{x^2}{4Dt}\right)$$

for each w function. This yields

$$W(r, t) = (4\pi Dt)^{-3/2} \exp\left(-\frac{r^2}{4Dt}\right) . \tag{4.34}$$

4.2.3
The Einstein Relation

If a particle is subjected to an external field while traveling in a vacuum, it is uniformly accelerated. If this particle travels in a gas, collisions with gas particles create a frictional force, and the mean velocity of the particle in the gas is established both by the external field and by the collisions with other particles. The proportionality coefficient for the relationship between the mean velocity $\mathbf{w}$ of a particle and the force $\mathbf{F}$ acting on the particle from an external field is the particle mobility. Thus, the definition of the mobility b of a particle is

$$\mathbf{w} = b\mathbf{F} . \tag{4.35}$$

We now assume that the test particles in a gas are in thermodynamic equilibrium with the gas subjected to the external field. According to the Boltzmann formula (1.43), the distribution for the number density of the test particles is $N = N_0 \exp(-U/T)$, where U is the potential due to the external field and T is the temperature of the gas. The diffusive flux of the test particles according to (4.26) is $\mathbf{j} = -D\nabla N = D\mathbf{F}N/T$, where $\mathbf{F} = -\nabla U$ is the force acting on the test particle. Because thermodynamic equilibrium exists, the diffusive flux is compensated for by the hydrodynamic particle flux, $\mathbf{j} = \mathbf{w}N = b\mathbf{F}N$. Equating these fluxes, we find that the kinetic coefficients are related as

$$b = \frac{D}{T} . \tag{4.36}$$

This expression is known as the Einstein relation in accordance with Einstein's study of Brownian motion [13, 15, 16]. In reality, this relation was obtained earlier by Nernst [17] and Townsend and Bailey [18, 19], but Einstein used this relation for the analysis of Brownian motion (see [20, 21]). The Einstein relation is valid for small fields that do not disturb the thermodynamic equilibrium between the test and gas particles. On the basis of the Einstein relation (4.36) and the estimation (4.29) for the diffusion coefficient of a test particle in a gas, one can estimate the particle mobility as

$$b \sim \frac{1}{N\sigma\sqrt{mT}} .$$

Note that the mobility of a charged particle is introduced as the proportionality coefficient for the relationship between the drift velocity of a test charged particle $\mathbf{w}$ and the electric field strength $\mathbf{E}$ as

$$\mathbf{w} = K\mathbf{E} . \tag{4.37}$$

Correspondingly, the Einstein relation (4.36) between the mobility K and the diffusion coefficient D for a charged particle has the form

$$K = \frac{eD}{T} . \tag{4.38}$$

4.2.4 Heat Transport

Heat transport can be treated in a manner analogous to that employed for particle transport. The heat flux is defined as

$$\mathbf{q} = \int \mathbf{v} \cdot \frac{mv^2}{2} f d\mathbf{v} , \tag{4.39}$$

where f is the velocity distribution function for the particles, and the relation between the heat flux and the temperature gradient is given by (4.27). For estimation of the thermal conductivity coefficient, we use the same procedure as in the case of the diffusion coefficient. Take the heat flux through a given point as the difference between the fluxes in opposite directions, and express the difference of the heat fluxes in terms of the difference between temperatures. From (4.39), the heat flux can be estimated as $q \sim N v \Delta T$ because the energies of particles reaching this point from opposite sides are different. Because only particles located at a distance of about λ reach the given point without collisions, we have $\Delta T \sim \lambda \nabla T$. Substituting this into the equation for the heat flux and comparing the result with (4.27), we find that our estimate for the thermal conductivity coefficient is

$$\kappa \sim N v \lambda \sim \frac{v}{\sigma} \sim \frac{\sqrt{T}}{\sigma \sqrt{m}} . \tag{4.40}$$

The thermal conductivity coefficient is independent of the particle number density. An increase in the particle number density leads to an increase in the number of particles that transfer heat, but this then causes a decrease in the distance for this transport. These two effects mutually cancel.

Our next step is to derive the heat transport equation for a gas where thermal conductivity supplies a mechanism for heat transport. We denote by $\overline{\varepsilon}$ the mean energy of a gas particle, and for simplicity consider a single-component gas. Assuming the absence of sources and sinks for heat, we obtain the heat balance equation by analogy with the continuity equation (4.1) for the number density of particles, namely,

$$\frac{\partial}{\partial t}(\overline{\varepsilon} N) + \operatorname{div} \mathbf{q} = 0 .$$

We take the gas to be contained in a fixed volume. With $\partial \overline{\varepsilon} / \partial T = c_V$ as the heat capacity per gas particle, the above equation takes the form

$$\frac{\partial T}{\partial t} + \mathbf{w} \cdot \nabla T = \frac{\kappa}{c_V N} \Delta T , \tag{4.41}$$

where $\mathbf{w}$ is the mean velocity of the particles, we use the continuity equation (4.1) for $\partial N / \partial t$, and employ (4.27) for the heat flux. For a motionless gas this equation is analogous to the diffusion equation (4.30), and its solution can be obtained by analogy with (4.34).

4.2.5
Thermal Conductivity Due to Internal Degrees of Freedom

An additional channel of heat transport arises from the energy transport associated with internal degrees of freedom. Excited atoms or molecules that move through a region with a relatively low temperature can transfer their excitation energy to the gas, and successive transfers of this nature amount to the transport of energy. The inverse process can also take place, in which ground-state atoms or molecules pass through a region with a relatively high temperature, are excited in this region, and then transport this excitation energy to other regions in the gas. This mechanism of heat transport is significant if a typical distance over which excited and nonexcited atoms reach equilibrium is small compared with the length of this system.

The heat flux can be represented as a sum of two terms:

$$\mathbf{q} = -\kappa_{tr} \nabla T - \kappa_i \nabla T ,$$

where κ_{tr} is the thermal conductivity coefficient due to the transport of translational energy, whereas the second term is due to transport of energy in the internal degrees of freedom. The thermal conductivity coefficient is then just the sum of these two terms, or

$$\kappa = \kappa_{tr} + \kappa_i . \tag{4.42}$$

We shall now analyze the second term.

The internal state of the gas particle is denoted by subscript i. Because of the presence of a temperature gradient, the number density of particles in this state is not constant in space, and the diffusion flux is given by

$$\mathbf{j}_i = -D_i \nabla N_i = -D_i \frac{\partial N_i}{\partial T} \nabla T .$$

This leads to the heat flux

$$\mathbf{q} = \sum_i \varepsilon_i \mathbf{j}_i = -\sum_i \varepsilon_i D_i \frac{\partial N_i}{\partial T} \nabla T ,$$

where ε_i is the excitation energy of the ith state. If the diffusion coefficient is the same for excited and nonexcited particles, the thermal conductivity coefficient due to internal degrees of freedom is

$$\kappa_i = \sum_i \varepsilon_i D_i \frac{\partial N_i}{\partial T} = D \frac{\partial}{\partial T} \sum_i \varepsilon_i N_i = D \frac{\partial}{\partial T} (\bar{\varepsilon} N) = D c_V , \tag{4.43}$$

where $\bar{\varepsilon} = \sum_i \varepsilon_i N_i / N$ is the mean excitation energy of the particles, $N = \sum_i N_i$ is the total number density, and $c_V = \partial \bar{\varepsilon} / \partial T$ is the heat capacity per particle. Using estimates (4.29) and (4.40) for the diffusion coefficient and the thermal conductivity coefficient, one can conclude that $\kappa_i \sim \kappa_{tr}$ if the excitation energy of internal states is of the order of a thermal energy of the particles.

Another example of this mechanism for thermal conductivity occurs when a dissociated gas undergoes recombination upon entering a cold region. We examine the example of a gas that consists of atoms coexisting in thermodynamic equilibrium with a small admixture of diatomic molecules representing the bound system of these atoms. The dissociation and recombination proceed in accordance with the simple scheme

$$A + A \longleftrightarrow A_2 .$$

The number densities of atoms N_a and molecules N_m are connected by the Saha formula (1.52), so $N_a^2/N_m = F(T)\exp(-D/T)$, where D is the dissociation energy and $F(T)$ has a weak temperature dependence compared with the exponential dependence. Because $N_a \gg N_m$ and $D \gg T$, we have $\partial N_m/\partial T = (D/T^2)N_m$, and from (4.43) it follows that

$$\kappa_i = \left(\frac{D}{T}\right)^2 D_m N_m \,, \tag{4.44}$$

where D_m is the diffusion coefficient for molecules in an atomic gas. Comparing this expression with the thermal conductivity coefficient (4.40) due to translational heat transport, we have

$$\frac{\kappa_i}{\kappa_t} \sim \left(\frac{D}{T}\right)^2 \frac{N_m}{N_a} \,. \tag{4.45}$$

In the regime we have been treating, we have the $N_a \gg N_m$ and $D \gg T$. Therefore, the ratio (4.45) can be about unity for comparatively low concentrations of molecules in the gas. Note that these results are valid if the dissociation equilibrium is established fast compared with molecular transport for the distances under consideration.

4.2.6
Thermal Capacity of Molecules

Equation (4.41) for heat transfer allows us to obtain expressions for thermal capacities. This equation refers to a gas contained within a fixed volume. If the process proceeds in a small part of a large volume of gas, it is necessary to replace the thermal capacity at constant volume c_V in (4.41) by the thermal capacity at constant pressure c_p. It is convenient to rewrite (4.41) in the form

$$\frac{\partial T}{\partial t} + \mathbf{w} \cdot \nabla T = \chi \Delta T \,, \tag{4.46}$$

where $\chi = \kappa/(c_V N)$ is the thermal diffusivity coefficient.

To find the connection between the thermal capacities c_V and c_p, let the temperature of a element of gas containing n molecules vary by dT. Then the energy variation of this gas element is $dE = nc_p dT$. This energy change can also be stated as $dE = nc_V dT + pdV$, where p is the gas pressure and V is the volume. The equation of state (4.12), $p = Tn/V$, allows us to rewrite the second energy change relation as $dE = nc_V dT + ndT$. Comparing the two expressions for dE yields

$$c_p = c_V + 1 \,. \tag{4.47}$$

The thermal capacity c_V per molecule can be written as the sum of terms due to different degrees of freedom, that is,

$$c_V = c_{tr} + c_{rot} + c_{vib} \,, \tag{4.48}$$

where the thermal capacities c_{tr}, c_{rot}, and c_{vib} correspond to translational, rotational, and vibrational degrees of freedom. Because the mean kinetic energy of particles in thermal equilibrium is $3T/2$, we have $c_{tr} = 3/2$. The classical energy of the rotational degree of freedom is $T/2$ and coincides with the translational energy. Hence, for a diatomic molecule with two rotational degrees of freedom we have $c_{rot} = 1$ in the classical limit $B \ll T$ (B is the rotational constant). Polyatomic molecules have three rotational degrees of freedom instead of two as for a diatomic molecule. Hence, the rotational thermal capacity of polyatomic molecules is 3/2 if the thermal energy is much greater than the rotational excitation energy. This condition is satisfied at room temperature for rotational degrees of freedom of most molecules, but can be violated for the vibrational degrees of freedom. Therefore, we determine the vibrational thermal capacity by any available relation between the excitation energy for the first vibrational level and the thermal energy of the molecule.

Using the Planck formula (1.56) for the number of excitations for the harmonic oscillator, we have the following expression for the average oscillator excitation energy above the minimum of the potential energy

$$\varepsilon_{vib} = \frac{\hbar\omega}{2} + \frac{\hbar\omega}{\exp(\hbar\omega/T) - 1} ,$$

where ω is the frequency of the oscillator. We find that the vibrational thermal capacity corresponding to a given vibration is

$$c_{vib} = \frac{\partial \epsilon_{vib}}{\partial T} = \left(\frac{\hbar\omega}{T}\right)^2 \frac{\exp(-\hbar\omega/T)}{[1 - \exp(-\hbar\omega/T)]^2} . \tag{4.49}$$

This expression yields $c_{vib} = 1$ in the classical limit $\hbar\omega \ll T$. For the opposite limit, $\hbar\omega \gg T$, the vibrational thermal capacity is exponentially small.

The above analysis yields

$$c_V = \frac{3}{2} + \frac{n_{rot}}{2} + c_{vib} \tag{4.50}$$

for the thermal capacity of a molecule, where n_{rot} is the number of rotational degrees of freedom of the molecules, and the vibrational thermal capacity is a sum of individual vibrations, for which (4.49) can be used. Note that the analysis uses the assumption of thermodynamic equilibrium between different degrees of freedom.

4.2.7 Momentum Transport and Gas Viscosity

Transport of momentum takes place in a moving gas such that the mean velocity of the gas particles varies in the direction perpendicular to the mean velocity. Particle transport then leads to an exchange of particle momenta between gas elements with different average velocities. This creates a frictional force that slows those gas particles with higher velocity and accelerates those having lower velocities. We can estimate the value of the viscosity coefficient by analogy with the procedures employed for diffusion and thermal conductivity coefficients. The force acting per unit area as a result of the momentum transport is $F \sim Nvm\Delta w_x$, where Nv is the particle flux and $m\Delta w_x$ is the difference in the mean momentum carried by particles moving in opposite directions at a given point. Since particles reaching this point without colliding are located from it at distances of the order of the mean free path λ, we have $m\Delta w \sim m\lambda \partial w_x/\partial z$. Hence, the force acting per unit area is $F \sim Nvm\lambda \partial w_x/\partial z$. Comparing this with (4.28), and using $(T/m)^{1/2}$ instead of v and $(N\sigma)^{-1}$ instead of λ, we obtain the estimate

$$\eta \sim \frac{\sqrt{mT}}{\sigma} \tag{4.51}$$

for the viscosity coefficient η. The viscosity coefficient is found to be independent of the particle number density. As was true for the thermal conductivity coefficient, this independence comes from the compensation of opposite effects occurring with the momentum transport. The number of momentum carriers is proportional to the number density of atoms, and a typical transport distance is inversely proportional to it. The effects offset each other.

To determine the force acting on a spherical particle of radius R moving in a gas with velocity w, we assume that the particle radius is large compared with the mean free path, $R \gg \lambda$, and that the velocity w is not very high, so the resistive force arises from viscosity effects. The total resistive force is proportional to the particle area, so (4.28) gives $F \sim \eta R w$. A more precise determination of the numerical coefficient gives the Stokes formula for the frictional force that acts from the gas to the spherical particles due to collisions of gas molecules [22, 23]:

$$F = 6\pi\eta R w . \tag{4.52}$$

4.2.8 The Chapman–Enskog Approximation for Kinetic Coefficients of Gases

The kinetic coefficients of gases are determined by elastic scattering in collisions of gas atoms or molecules, since the elastic cross section exceeds significantly the cross section for inelastic collisions. The Chapman–Enskog approximation for kinetic coefficients accounts [24–26] for the simple connection between the kinetic coefficients and the cross sections of elastic collisions of atoms or molecules and is an expansion over a small numerical parameter. Therefore, we are restricted by the first Chapman–Enskog approximation only. The expression for the ion mobility in the first Chapman–Enskog approximation was derived earlier and is given by (4.21). From this, on the basis of the Einstein relation (4.37) we obtain the diffusion coefficient D of a test particle in a gas in the first Chapman–Enskog approximation as [11, 12]

$$D = \frac{3\sqrt{\pi T}}{8N\sqrt{2\mu}\overline{\sigma}} , \quad \overline{\sigma} \equiv \Omega^{(1,1)}(T) = \frac{1}{2}\int_0^\infty \exp(-t)t^2\sigma^*(t)dt , \quad t = \frac{\mu g^2}{2T} , \tag{4.53}$$

and this expression holds true for both a charged and a neutral test particle if the mobility of a neutral atomic particle is taken as the limit $e \to 0$. Here T is the gas temperature in energy units, as usual, μ is the reduced mass of a test particle and a gas atom or molecule, $\sigma^*(g)$ is the diffusion cross section for collision of a test particle and a gas atom at relative velocity g, and brackets mean an averaging over the Maxwell distribution function for gas atoms and test particles.

By analogy with (4.53) for the diffusion coefficient of a test atomic particle in a gas, we have the following expression for the thermal conductivity coefficient [11, 12]:

$$\kappa = \frac{25\sqrt{\pi T}}{32\overline{\sigma_2}\sqrt{m}} . \tag{4.54}$$

Here m is the mass of a gas atom or molecule, and the average cross section $\overline{\sigma_2}$ for collision of gas atoms is given by

$$\overline{\sigma_2} \equiv \Omega^{(2,2)}(T) = \int_0^\infty t^2 \exp(-t)\sigma^{(2)}(t)dt , \quad t = \frac{\mu g^2}{2T} ,$$

$$\sigma^{(2)}(t) = \int (1 - \cos^2\vartheta)d\sigma . \tag{4.55}$$

The viscosity coefficient η in the first Chapman–Enskog approximation is given by [11, 12]

$$\eta = \frac{5\sqrt{\pi T m}}{24\overline{\sigma_2}} , \tag{4.56}$$

where the average cross section $\overline{\sigma_2}$ for collision of gas atoms is given by (4.55). As is seen, we have the following relation between the coefficients of thermal conductivity and viscosity in the first Chapman–Enskog approximation:

$$\kappa = \frac{15}{4m}\eta . \tag{4.57}$$

It should be noted that elastic collisions of atoms or molecules at thermal energies are described more or less by the hard-sphere model, where the interaction potential of colliding particles is given by (2.10). Then the average cross sections of collisions for the first Chapman–Enskog approximation are

$$\overline{\sigma} = \sigma_0 ; \quad \overline{\sigma_2} = \frac{2}{3}\sigma_0 , \tag{4.58}$$

where $\sigma_0 = \pi R_0^2$ and R_0 is the hard-sphere radius. Correspondingly, (4.53), (4.54), and (4.56) for the kinetic coefficients in the first Chapman–Enskog approximation are simplified and take the form

$$D = \frac{3\sqrt{\pi T}}{8\sqrt{2\mu}N\sigma_0} ; \quad \kappa = \frac{75\sqrt{\pi T}}{64\sigma_0\sqrt{m}} ; \quad \eta = \frac{5\sqrt{\pi T m}}{16\sigma_0} . \tag{4.59}$$

For the classical character of movement of atomic particles, the elastic cross sections are expressed through the interaction potential between colliding particles. Correspondingly, kinetic coefficients of gases are expressed through the interaction potential of particles. In particular, if the interaction potential $U(R)$ varies sharply with a change of distance R between colliding particles, the diffusion coefficient is expressed through the interaction potential between a test particle and gas atom by [27–30]

$$D = \frac{3\sqrt{T}}{8\sqrt{2\pi\mu} N R_0^2} , \quad \frac{U(R_0)}{T} = 2.25 . \qquad (4.60)$$

This formula assumes repulsion between colliding particles when the connection between the diffusion cross section and a sharply varied interaction potential of particles is given by (2.15). In the same manner, one can express the thermal conductivity κ and viscosity η coefficients of a gas through the repulsion interaction potential of atomic particles [27–30]:

$$\kappa = \frac{75\sqrt{T}}{64\sqrt{\pi m} R_2^2} , \quad \eta = \frac{5\sqrt{Tm}}{16\sqrt{\pi} R_2^2} , \quad \frac{U(R_2)}{T} = 0.83 . \qquad (4.61)$$

Note that the classical character of collision of atomic particles requires that large collision momenta l give the main contribution to the collision cross sections, that is, $l = \mu\rho v/\hbar \gg 1$, where μ is the reduced mass of colliding particles, ρ is a typical collision impact parameter, and v is a typical collision velocity. Let us make the estimation for collision of two nitrogen molecules at temperature $T = 300$ K when the average velocity of relative motion of molecules is $v_T = \sqrt{8T/(\pi\mu)} = 6.7 \times 10^4$ cm/s. Taking a typical collision impact parameter as $\rho = \sqrt{\sigma_g/\pi}$, where the gas-kinetic cross section for collision of two nitrogen molecules in thermal collisions is $\sigma_g = 39\,\text{Å}^2$ [34], we find $\rho = 3.5\,\text{Å}$. From this we have for a typical collision momentum $l = \mu\rho v/\hbar = 31$, which justifies application of the classical theory in this case.

Below we represent the values of kinetic coefficients for simple gases. Table 4.2 contains the self-diffusion coefficients at the normal number density of atoms or molecules, $N = 2.687 \times 10^{19}\,\text{cm}^{-3}$, which corresponds to normal gas conditions ($T = 273$ K, $p = 1$ atm). It is convenient to approximate the temperature depen-

Table 4.2 The diffusion coefficients D of atoms or molecules in the parent gas expressed in square centimeters per second and reduced to the normal number density [31].

Gas	D (cm^2/s)	Gas	D (cm^2/s)	Gas	D (cm^2/s)
He	1.6	H_2	1.3	H_2O	0.28
Ne	0.45	N_2	0.18	CO_2	0.096
Ar	0.16	O_2	0.18	NH_3	0.25
Kr	0.084	CO	0.18	CH_4	0.20
Xe	0.048				

dence of the self-diffusion coefficient as

$$D = D_0 \left(\frac{T}{300} \right)^{\gamma} , \qquad (4.62)$$

where the gas temperature T is given in Kelvin. It should be noted that within the framework of the hard-sphere model according to (4.60) $\gamma = 1.5$, since at a constant pressure (4.11) gives that the number density of atoms is inversely proportional to the gas temperature. Comparison of this value with that following from the results of measurements [31, 32], given in Table 4.3, shows the validity of the hard-sphere model for the self-diffusion coefficient where the numerical coefficient of (4.62) is $\gamma = 1.5$.

Table 4.3 contains the parameters in (4.62) for the diffusion coefficients of inert gases. Note the symmetry of the diffusion coefficient, so the diffusion coefficient for atomic particles A in a gas consisting of atomic particles B is equal to the diffusion coefficient of atomic particles B in a gas consisting of atomic particles A. The diffusion coefficients for atoms of the first group of the periodic table and for mercury atoms in inert gases are given in Table 4.4.

Table 4.3 The parameter D_0 in (4.62), which is given in square centimeters per second, is reduced to the normal number density and relates to the gas temperature $T = 300$ K. The exponent γ in (4.62) is given in parentheses. Data from [31, 32] are used.

Atom, gas	He	Ne	Ar	Kr	Xe
He	1.66 (1.73)	1.08 (1.72)	0.76 (1.71)	0.68 (1.67)	0.56 (1.66)
Ne		0.52 (1.69)	0.32 (1.72)	0.25 (1.74)	0.23 (1.67)
Ar			0.195 (1.70)	0.15 (1.70)	0.115 (1.74)
Kr				0.099 (1.78)	0.080 (1.79)
Xe					0.059 (1.76)

Table 4.4 Diffusion coefficients for metal atoms in inert gases at temperature $T = 300$ K, which are reduced to the normal number density and are expressed in square centimeters per second [33].

Atom, gas	He	Ne	Ar	Kr	Xe
Li	0.54	0.29	0.28	0.24	0.20
Na	0.50	0.31	0.18	0.14	0.12
K	0.54	0.24	0.22	0.11	0.087
Cu	0.68	0.25	0.11	0.077	0.059
Rb	0.39	0.21	0.14	0.11	0.088
Cs	0.34	0.22	0.13	0.080	0.057
Hg	0.46	0.21	0.11	0.072	0.056

The gas-kinetic cross section as the collision parameter of atoms or molecules at room temperature is defined on the basis of (4.59) for the diffusion coefficient within the framework of the hard-sphere model, and the measured values of the diffusion coefficient are used in this formula. Correspondingly, the connection between the gas-kinetic cross section σ_g and the diffusion coefficient D of atomic

Table 4.5 Gas-kinetic cross sections σ_g expressed in units of 10^{-15} cm^2 [34].

Colliding particles	He	Ne	Ar	Kr	Xe	H_2	N_2	O_2	CO	CO_2
He	1.3	1.5	2.1	2.4	2.7	1.7	2.3	2.2	2.2	2.7
Ne		1.8	2.6	3.0	3.3	2.0	2.4	2.5	2.7	3.1
Ar			3.7	4.2	5.0	2.7	3.8	3.9	4.0	4.4
Kr				4.8	5.7	3.2	4.4	4.3	4.4	4.9
Xe					6.8	3.7	5.0	5.2	5.0	6.1
H_2						2.0	2.8	2.7	2.9	3.4
N_2							3.9	3.8	3.7	4.9
O_2								3.8	3.7	4.4
CO									4.0	4.9
CO_2										7.5

Table 4.6 The thermal conductivity coefficient κ of gases at atmospheric pressure in units of 10^{-4} W/(cm K) [35].

T, K	100	200	300	400	600	800	1000
H_2	6.7	13.1	18.3	22.6	30.5	37.8	44.8
He	7.2	11.5	15.1	18.4	25.0	30.4	35.4
CH_4	–	2.17	3.41	4.88	8.22	–	–
NH_3	–	1.53	2.47	6.70	6.70	–	–
H_2O	–	–	–	2.63	4.59	7.03	9.74
Ne	2.23	3.67	4.89	6.01	7.97	9.71	11.3
CO	0.84	1.72	2.49	3.16	4.40	5.54	6.61
N_2	0.96	1.83	2.59	3.27	4.46	5.48	6.47
Air	0.95	1.83	2.62	3.28	4.69	5.73	6.67
O_2	0.92	1.83	2.66	3.30	4.73	5.89	7.10
Ar	0.66	1.26	1.77	2.22	3.07	3.74	4.36
CO_2	–	0.94	1.66	2.43	4.07	5.51	6.82
Kr	–	0.65	1.00	1.26	1.75	2.21	2.62
Xe	–	0.39	0.58	0.74	1.05	1.35	1.64

Table 4.7 The viscosity coefficient of gases in units of 10^{-5} g/(cm s) [31, 36].

T, K	100	200	300	400	600	800	1000
H_2	4.21	6.81	8.96	10.8	14.2	17.3	20.1
He	9.77	15.4	19.6	23.8	31.4	38.2	44.5
CH_4	–	7.75	11.1	14.1	19.3	–	–
H_2O	–	–	–	13.2	21.4	29.5	37.6
Ne	14.8	24.1	31.8	38.8	50.6	60.8	70.2
CO	–	12.7	17.7	21.8	28.6	34.3	39.2
N_2	6.88	12.9	17.8	22.0	29.1	34.9	40.0
Air	7.11	13.2	18.5	23.0	30.6	37.0	42.4
O_2	7.64	14.8	20.7	25.8	34.4	41.5	47.7
Ar	8.30	16.0	22.7	28.9	38.9	47.4	55.1
CO_2	–	9.4	14.9	19.4	27.3	33.8	39.5
Kr	–	–	25.6	33.1	45.7	54.7	64.6
Xe	–	–	23.3	30.8	43.6	54.7	64.6

particles in a gas is given by

$$\sigma_g = \frac{0.47}{ND}\sqrt{\frac{T}{\mu}}\,, \tag{4.63}$$

where N is the number density of gas atoms or molecules, T is the room gas temperature, and μ is the reduced mass of a test particle and a gas particle. The values of the gas-kinetic cross sections σ_g on the basis of (4.63) and measured values of the diffusion coefficients are given in Table 4.5.

The thermal conductivity coefficients are given in Table 4.6 for inert gases and simple molecular gases at a pressure of 1 atm, and the viscosity coefficients at atmospheric pressure are represented in Table 4.7 according to the corresponding measurements. Note also the simple relation (4.57) within the framework of the first Chapman–Enskog approximation between the thermal conductivity coefficient and the viscosity coefficient.

4.3
Transport of Electrons in Gases

4.3.1
Diffusion and Mobility of Electrons in Gases in Electric Field

Comparing the drift of electrons in gases in an external electric field with that of neutral atomic particles in gases, we note that in contrast to neutral particles with a Maxwell distribution function for velocities, the distribution function for electrons can differ remarkably from the Maxwell one. On the other hand, the analysis of the electron behavior is simplified because of a small mass compared with the mass of other atomic particles in a gas. This allows one to represent the integral of electron–atom collisions in the analytical form, as (3.12) and (3.18) for the collision integral in the case of elastic electron–atom collisions. Formula (3.26) gives the electron drift velocity in a gas in external fields with elastic collisions between electrons and atoms. We now derive the expression for the transverse diffusion coefficient of electrons in a gas under these conditions.

By definition, the diffusion coefficient of electrons D_e connects the electron number density gradient ∇N_e and the electron flux $\mathbf{j}_e$ by the relation

$$\mathbf{j}_e = -D_e \nabla N_e \,.$$

On the other hand, the kinetic equation for the electron distribution function has the following form if the electron number density varies in space:

$$v_x \nabla f = I_{ea}(f) \,,$$

where I_{ea} is the integral of electron–atom collisions. We take the electron distribution function in the standard form (3.10)

$$f = f_0(v) + v_x f_1(v) \,,$$

where the x-axis is directed along the gradient of the electron number density. Because the distribution function is normalized to the electron number density $f \sim N_e$, we have $\nabla f = f \nabla N_e / N_e$, and the kinetic equation for the electron distribution function takes the form

$$\frac{v_x f_0 \nabla N_e}{N_e} = -\nu v_x f_1$$

if we use (3.12) for the collision integral taking into account elastic electron–atom collisions. From this we obtain for the nonsymmetric part of the electron distribution function

$$f_1 = -\frac{f_0 \nabla N_e}{\nu N_e} \,.$$

This leads to the following expression for the electron flux:

$$\mathbf{j}_e = \int \mathbf{v} f d\mathbf{v} = \int v_x^2 f_1 d\mathbf{v} = -\frac{\nabla N_e \int v_x^2 f_0 d\mathbf{v}}{\nu N_e} = -\nabla N_e \left\langle \frac{v_x^2}{\nu} \right\rangle \,,$$

where angle brackets mean averaging over the electron distribution function. Comparing this formula with the definition of the diffusion coefficient according to the equation $\mathbf{j}_e = -D_e \nabla N_e$, we find for the diffusion coefficient of electrons in a gas

$$D_e = \left\langle \frac{v^2}{3\nu} \right\rangle \,.$$

Note that if electrons are located in an external electric field, along with a nonsymmetric part due to the electron number density gradient, there is another nonsymmetric part of the distribution function due to an electric field. In deriving (4.64) we separate these parts, which is valid if they correspond to different directions. Hence, if an electric field influences the distribution function for electrons, the expression (4.64) for the electron diffusion coefficient holds true for the transverse diffusion coefficient only.

In the absence of an electric field, if the velocity distribution function for electrons is a Maxwell one, (3.26) gives

$$w_e = \frac{eE}{3T}\left\langle \frac{v^2}{\nu} \right\rangle = \frac{eED_e}{T} , \qquad (4.64)$$

which corresponds to the Einstein relation (4.38). In addition, the diffusion coefficient of electrons is identical in different directions. This holds true also for the regime of a high number density of electrons when we have the Einstein relation in the following form instead of (4.38)

$$w_e = \frac{eED_e}{T_e} , \qquad (4.65)$$

where T_e is the electron temperature. Table 4.8 gives the reduced zero-field mobilities of electrons in inert gases at room temperature, $K_e N_a$ (K_e is the electron mobility, N_a is the number density of gas atoms), and the values of the reduced diffusion coefficients of electrons, $D_e N_a$.

An external electric field influences the character of electron drift in a gas and the value of the diffusion coefficient. If $\nu = \text{const}$, the diffusion coefficient is proportional to the average electron energy, and the diffusion coefficients in the transverse and longitudinal directions are the same. In other cases only the diffusive motion in the transverse direction with respect to the electric field is separated from electron drift, and the dependence on the electric field strength is connected with the dependence of the diffusion cross section of elastic electron–atom collision on the electron energy. We give in Figures 4.2 and 4.3 the coefficients of transverse diffusion of electrons in helium and argon on the basis of experimental data [37]. As is seen, in the argon case with the Ramsauer effect for elastic electron–atom scattering, the dependence of the diffusion coefficient on the electric field strength has a nonmonotonic character.

Let us consider the case when the diffusion cross section of elastic electron–atom collision is independent of the electron energy, which corresponds to the helium

Table 4.8 Electron drift parameters in inert gas atoms at zero field and room temperature [37].

Gas	He	Ne	Ar	Kr	Xe
$D_e N_a$, 10^{21} (cm s)$^{-1}$	7.2	72	30	1.6	0.43
$K_e N_a$, 10^{23} (cm s V)$^{-1}$	2.9	29	12	0.62	0.17

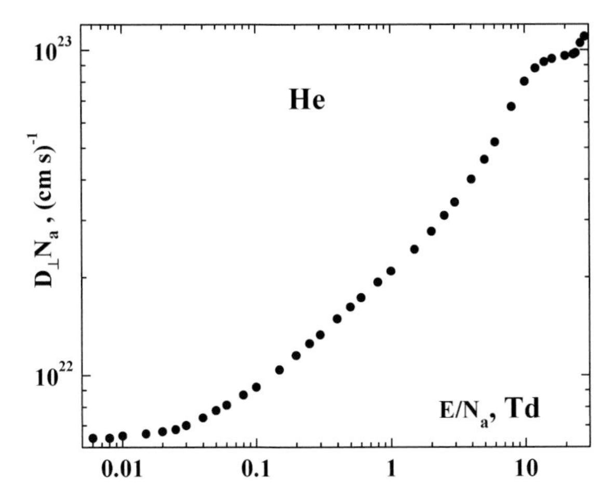

Figure 4.2 The reduced coefficient of transverse diffusion for electrons moving in helium in an external electric field. The values refer to the regime of a low number density of electrons and are obtained on the basis of experimental data collected in [37].

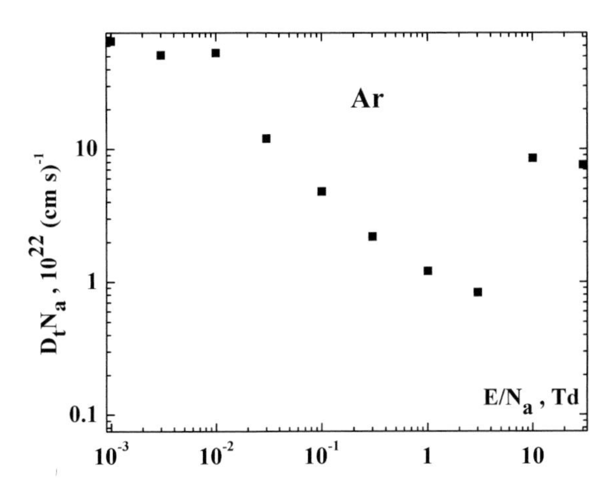

Figure 4.3 The reduced coefficient of transverse diffusion for electrons moving in argon in an external electric field. The values refer to the regime of a low number density of electrons and were obtained on the basis of experimental data reported in [37].

case where $\sigma^* \approx 6\,\text{Å}^2$. We have on the basis of (3.27) for the isotropic part of the electron distribution function

$$f_0(\varepsilon) = \frac{N_e}{\pi \Gamma(3/4)} \left(\frac{m_e}{2\varepsilon_0}\right)^{3/2} \exp\left(-\frac{\varepsilon^2}{\varepsilon_0^2}\right) , \tag{4.66}$$

where this distribution function is normalized to the number density of electrons and

$$\varepsilon_0 = eE\lambda\sqrt{\frac{m_a}{3m_e}} ,$$

where E is the electric field strength, m_e and m_a are the electron and atom masses, and $\lambda = (N_a\sigma^*)^{-1}$ is the mean free path of electrons in a gas with number density of atoms N_a. Substituting this distribution function into (4.64), we obtain for the electron diffusion coefficient in this case

$$D_\perp = 0.292 \left(\frac{m_a}{m_e}\right)^{1/4} \lambda \sqrt{\frac{eE\lambda}{m_e}} . \tag{4.67}$$

In the limit of high electric field strength E the diffusion coefficient of electrons is proportional to $\sqrt{E}$. Taking $\sigma^* = 6\,\text{Å}^2$ in the helium case, we obtain

$$D_\perp = D_0\sqrt{x} ,$$

where $D_0 = 1500\,\text{cm}^2/\text{s}$, the diffusion coefficient, is reduced to the normal number density of atoms $N_a = 2.69 \times 10^{19}\,\text{cm}^{-3}$, and the reduced electric field strength $x = E/N_a$ is expressed in Townsend. Note that the case of high electric field strengths at room temperature of helium, if the average electron energy exceeds remarkably the thermal energy of atoms, corresponds to the criterion $x \gg 0.05\,\text{Td}$.

4.3.2
Diffusion of Electrons in a Gas in a Magnetic Field

We now determine the transverse diffusion coefficient for electrons in a strong magnetic field if the directions of the electric and magnetic fields coincide. This corresponds to the condition $\omega_H \gg \nu$, where $\omega_H = eH/(m_e c)$ is the electron Larmor frequency. The projection of the electron trajectory on a plane perpendicular to the field consists of circles whose centers and radii change after each collision. The diffusion coefficient, by its definition, is $D_\perp = \langle x^2\rangle/t$, where $\langle x^2\rangle$ is the square of the displacement for time t in the direction x perpendicular to the field. We have $x - x_0 = r_H \cos\omega_H t$, where x_0 is the x-coordinate of the center of the electron's rotational motion and $r_H = v_\rho/\omega_H$ is the Larmor radius, where v_ρ is the electron velocity in the direction perpendicular to the field, and ω_H is the Larmor frequency for an electron. From this it follows that

$$\langle x^2\rangle = n\left\langle (x - x_0)^2\right\rangle = \frac{n v_\rho^2}{2\omega_H^2} ,$$

where n is the number of collisions. Since $t = n/\nu$, where ν is the frequency of electron–atom collisions, we obtain

$$D_\perp = \left\langle \frac{v_\rho^2 \nu}{2\omega_H^2} \right\rangle = \left\langle \frac{v^2 \nu}{3\omega_H^2} \right\rangle , \qquad \omega_H \gg \nu ,$$

where angle brackets mean averaging over electron velocities. Combining this result with (4.64), we find that the transverse diffusion coefficient for electrons in a gas, moving perpendicular to electric and magnetic fields, is

$$D_\perp = \frac{1}{3}\left\langle \frac{v^2 \nu}{\omega_H^2 + \nu^2} \right\rangle . \tag{4.68}$$

4.3.3 Thermal Conductivity of Electrons in an Ionized Gas

Because of their small mass, electrons are effective carriers of heat, and their transport can give a contribution to the thermal conductivity of a weakly ionized gas. Since the thermal conductivity of a gas is determined by different carriers, the thermal conductivity of an ionized gas is the sum of the thermal conductivity due to atoms and electrons independently. Below we evaluate the thermal conductivity coefficient due to electrons in an ionized gas with a low degree of ionization, where the number density of electrons N_e is small compared with that of atoms N_a, that is, $N_e \ll N_a$. We consider the regime of a high electron number density for electron drift in an external electric field, and the criterion to be fulfilled is the inverse of criterion (3.30). In this regime the Maxwell distribution function for electron velocities $\varphi(v)$ is realized, and the electron temperature T_e is the parameter of this distribution.

The gradient of the electron temperature creates a heat flux $\mathbf{q}_e$, and the electron thermal conductivity κ_e is defined by the equation

$$\mathbf{q}_e = -\kappa_e \nabla T_e \, . \tag{4.69}$$

As usual, we assume the electron temperature gradient to be relatively small, so criterion (4.24) for the mean free path of electrons in a gas λ is fulfilled, and L in this criterion is a typical distance over which the electron temperature T_e varies remarkably.

To determine the electron thermal conductivity, we represent the velocity distribution function for electrons in a standard way similar to (3.10):

$$f(\mathbf{v}) = \varphi(v) + (\mathbf{v}\nabla \ln T_e) f_1(v) \, .$$

Assuming that variation of the electron momentum is determined by electron–atom collisions, we obtain the kinetic equation in the form

$$\mathbf{v}\nabla f = I_{ea}(f) \, .$$

We assume that the spatial dependence of the electron distribution function is due to the gradient of the electron temperature, and the equation of state for the electron gas is given by (4.11) for the electron pressure $p_e = N_e T_e$, which is constant in space. On the basis of (3.12) for the collision integral for the nonsymmetric part of the electron distribution function, we reduce the kinetic equation to the form

$$\varphi(v)\left(\frac{m_e v^2}{2T_e} - \frac{5}{2}\right)\mathbf{v}\nabla T_e = -\nu f_1 \, .$$

For the nonsymmetric part of the distribution function this gives

$$f_1(v) = -\frac{\varphi(v)}{\nu}\left(\frac{m_e v^2}{2T_e} - \frac{5}{2}\right) .$$

We now determine the heat flux due to electrons:

$$\mathbf{q}_e = \int \frac{m_e v^2}{2} v_x f(\mathbf{v}) d\mathbf{v} = \int \frac{m_e v^2}{2} v_x^2 \nabla \ln T_e f_1(v) d\mathbf{v} .$$

Accounting for the connection between the symmetric and nonsymmetric parts of the distribution function and (4.69) for the heat flux due to electrons, we obtain for the thermal conductivity coefficient of electrons

$$\kappa_e = N_e \left\langle \frac{v^2}{3\nu} \frac{m_e v^2}{2T_e} \left(\frac{m_e v^2}{2T_e} - \frac{5}{2} \right) \right\rangle , \tag{4.70}$$

where angle brackets mean averaging over the Maxwell electron distribution function.

Let us consider the limiting cases. We take the dependence $\nu \sim v^n$ for the rate of electron–atom collisions, that is, $\nu(v) = \nu_0 z^{n/2}$, where $z = m_e v^2/(2T_e)$. Then (4.70) gives

$$\kappa_e = \frac{4}{3\sqrt{\pi}} \cdot \frac{T_e N_e}{\nu_0 m_e} \left(1 - \frac{n}{2}\right) \Gamma\left(\frac{7-n}{2}\right) . \tag{4.71}$$

In particular, if $\nu = \text{const}$, this formula gives

$$\kappa_e = \frac{5 T_e N_e}{2\nu_0 m_e} .$$

If $n = 1$, that is, $\nu = v/\lambda$(λ is the mean free path), we have from (4.71)

$$\kappa_e = \frac{2}{3\sqrt{\pi}} N_e \lambda \sqrt{\frac{2T_e}{m_e}} .$$

To determine the contribution of the electron thermal conductivity to the total thermal conductivity coefficient, it is necessary to connect the gradients of the electron T_e and atomic T temperatures. Let us consider the case when the difference between the electron and gas temperatures is determined by an external electric field, and the connection between the electron and gas temperatures is given by (3.36). If $\nu \sim v^n$, this formula gives

$$\nabla T_e = \frac{\nabla T}{1 + n - \frac{nT}{T_e}} .$$

In particular, in the case $T_e \gg T$, the total thermal conductivity coefficient is

$$\kappa = \kappa_a + \kappa_e \frac{\nabla T_e}{\nabla T} = \kappa_a + \frac{\kappa_e}{1+n} , \tag{4.72}$$

where κ_a is the thermal conductivity coefficient of the atomic gas. If we use (4.40) as an estimation for the gas thermal conductivity coefficient, replacing the electron parameters in this formula by the atom parameters, we obtain that the electron thermal conductivity can contribute to the total value at low electron number densities $N_e < N_a$ because of the small electron mass and the high electron temperature.

4.3.4 Thermal Diffusion of Electrons

Along with directed transport processes, such as transport of particles due to a gradient of the particle number density, transport of momentum due to a gradient of the drift velocity, heat transport due to a temperature gradient, and cross-fluxes are possible. Below we consider the simplest of them, namely, the flux of electrons due to a gradient of the electron temperature. This flux is characterized by the thermodiffusion coefficient D_T, which is defined as

$$\mathbf{j} = -D_T \nabla \ln T_e \,. \tag{4.73}$$

To determine the thermodiffusion coefficient D_T, we analyze the kinetic equation (3.4) for electrons and take into account electron–atom collisions; this has the form

$$\mathbf{v} \cdot \nabla f = I_{ea}(f) \,. \tag{4.74}$$

Taking the electron velocity distribution function in a standard form similar to (3.10), we represent it as

$$f(\mathbf{v}) = \varphi(v) + v_x f_1(v) \,,$$

where $\varphi(v)$ is the Maxwell distribution function for the electrons and the temperature gradient is along the x-axis. Using (3.12) for the collision integral from the nonsymmetric part of the distribution function for electrons, we obtain

$$v_x \frac{\partial \varphi}{\partial x} = -\nu v_x f_1 \,,$$

where ν is the rate of elastic electron–atom collisions.

The electron flux created by the nonsymmetric part of the distribution function is

$$j_x = \int v_x f_x d\mathbf{v} = \int v_x^2 f_1 d\mathbf{v} = -\frac{1}{3}\int \frac{v^2}{\nu}\frac{\partial f_0}{\partial x} d\mathbf{v} = -\frac{d}{dx}\left[N_e \left\langle \frac{v^2}{3\nu} \right\rangle\right] ,$$

where the angle brackets mean averaging over electron velocities with a Maxwell distribution function for electrons. We obtain the expression for the electron flux if a parameter of the electron distribution varies in space. Taking into account that this dependence is contained in the electron temperature, we obtain for the electron flux

$$j_x = -\nabla T_e \frac{d}{dT_e}\left[N_e \left\langle \frac{v^2}{3\nu} \right\rangle\right] .$$

Comparing this formula with the definition (4.73) of the thermodiffusion coefficient, we obtain

$$D_T = T_e \frac{d}{dT_e}\left[N_e \left\langle \frac{v^2}{3\nu} \right\rangle\right] = T_e \frac{d}{dT_e}(N_e D) \,, \tag{4.75}$$

where D is the electron diffusion coefficient in a gas according to (4.64). Since the electron pressure $p_e = N_e T_e$ is constant in space, this expression may be written as

$$D_T = N_e T_e^2 \frac{d}{d T_e} \left(\frac{D}{T_e} \right) . \tag{4.76}$$

In particular, if the rate of electron–atom collisions does not vary in space, $\nu =$ const, this equation gives $D_T = 0$. In the case of the dependence $\nu \sim v^n$ we have

$$D_T = -n N_e D . \tag{4.77}$$

This means that the direction of the electron flux with respect to the temperature gradient depends on the sign of n.

4.3.5
Cross-Fluxes in Electron Thermal Conductivity

A characteristic property of the electron thermal conductivity is such that cross-fluxes can be essential in this case. In the case the electron thermal conductivity of a weakly ionized gas in an external electric field, when a temperature gradients exist, we have the following expressions for fluxes

$$\mathbf{j} = N_e K \mathbf{E} - D_T N_e \nabla \ln T_e , \quad \mathbf{q} = -\kappa_e \nabla T_e + \alpha e \mathbf{E} . \tag{4.78}$$

We first consider the case of electron transport when displacement of electrons as a whole cannot violate the plasma quasineutrality that corresponds to plasma regions far from electrodes and walls. Then the mobility K in (4.78) is the electron mobility, and one can ignore the ion mobility including the ambipolar diffusion. The expression for the electron thermodiffusion coefficient is given by (4.76), and (4.70) gives the thermal conductivity coefficient. Below we determine the coefficient α in (4.78) by the standard method by means of expansion of the electron distribution function over the spherical harmonics, similar to (3.10). Then the first equation of the set that is analogous to (3.25) yields $f_1 = eE f_0/(\nu T_e)$, and the coefficient α is

$$\alpha = \frac{m_e N_e}{6 T_e} \cdot \left\langle \frac{v^4}{\nu} \right\rangle = \frac{4 T_e N_e}{3\sqrt{\pi} m_e \nu_0} \Gamma \left(\frac{7}{2} - \frac{n}{2} \right) , \tag{4.79}$$

where we take $\nu = \nu_0 \left(v/\sqrt{2T_e/m_e} \right)^n$. For $n = 0$ this gives

$$\alpha = \frac{5 T_e N_e}{2 m_e \nu} ,$$

and for $n = 1$, when $\nu = v/\lambda$, this formula yields

$$\alpha = \sqrt{\frac{2T_e}{m_e}} \cdot \frac{\lambda}{3\sqrt{\pi}} = \frac{2\lambda N_e}{3 v_T} ,$$

where $v_T = [8T_e/(\pi m_e)]^{1/2}$ is the mean electron velocity.

Equation (4.78) together with the corresponding expressions for the kinetic coefficients allow us to determine the electron heat flux under different conditions in the plasma. Let us determine the effective thermal conductivity coefficient in the direction perpendicular to an external electric field **E**. If the plasma is placed into a metallic enclosure, the transverse electric field is absent, $E = 0$, and the first equation (4.78) coincides with (4.69). If the walls are dielectric ones, we have $\mathbf{j} = 0$, which corresponds to the regime of ambipolar diffusion when electrons travel together with ions. In the electron scale of values this gives $\mathbf{j} = 0$, that is, an electric field of strength $\mathbf{E} = D_T \nabla \ln T_e/(N_e K)$ arises. Let us represent the heat flux in the form

$$\mathbf{q} = -C\kappa_e \nabla T_e ,$$

where the coefficient C is

$$C = 1 - \frac{e\alpha D_T}{\kappa_e T_e N_e K} .$$

Let us take the velocity dependence for the rate of electron–atom collisions as usual as $\nu \sim v^n$, that is, $\nu(v) = \nu_0 z^{n/2}$, where $z = m_e v^2/(2T_e)$. Using the Einstein relation $D = KT_e/e$, we get

$$C = 1 + \frac{\alpha n}{\kappa_e} .$$

As a result, we obtain

$$C = \frac{n+2}{2-n} . \tag{4.80}$$

As is seen, the effective thermal conductivity coefficient for electrons in the two cases considered of metallic and dielectric walls depends on n. For $n = 0$, this value is identical in both instances. For $n = 1$, it is greater by a factor of 3 in the second case than in the first.

4.3.6 Townsend Energy Coefficient

The Townsend energy coefficient for electrons is introduced as

$$\eta = \frac{eED_\perp}{Tw} = \frac{eD_\perp}{TK_e} , \tag{4.81}$$

where $D_\perp$ is the transverse diffusion coefficient for electrons in a gas, K_e is their mobility, and T is the gas temperature. For a Maxwell distribution function for electrons with the gas temperature, $\eta = 1$ according to the Einstein relation (4.38). In the regime of high electron number densities with a Maxwell distribution function for electron velocities, which is characterized by the electron temperature T_e, the Townsend energy coefficient of electrons is

$$\eta = \frac{T_e}{T} \, .$$

Experimentally, the Townsend energy coefficient may be determined [38, 39] by a simultaneous measurement of the transverse diffusion coefficient of electrons and their mobility in an electric field when a pulsed electron beam of a certain width is injected in a drift chamber with a certain electric field strength. Then the electron drift velocity follows from measurement of time the beam takes to pass through a drift chamber, and the coefficient of transverse diffusion results from broadening of this beam during the drift time.

Under these conditions, the effective electron temperature T_{ef}, which is named the characteristic energy [37, 38], in the regime of a low electron number density on the basis of the Townsend energy coefficient is

$$T_{ef} = \eta T = \frac{e D_\perp}{K_e} \, . \tag{4.82}$$

Using (3.26) for the electron drift velocity and (4.64) for the diffusion coefficient, we obtain for the Townsend energy coefficient

$$\eta = \left\langle \frac{m_e v^2}{\nu} \right\rangle \left(T \left\langle \frac{1}{v^2} \frac{d}{dv} \frac{v^3}{\nu} \right\rangle \right)^{-1} \, . \tag{4.83}$$

If the rate of elastic electron–atom collisions ν is independent of the electron velocity, this equation gives

$$\eta = \left\langle \frac{m_e v^2}{3T} \right\rangle \, . \tag{4.84}$$

In this case the Townsend energy coefficient is the ratio of the mean electron energy to the mean energy of the atom. Such a relation holds both for the case $\nu = \text{const}$ and for a Maxwell distribution function for electrons.

In the regime of a low number density of electrons, when the electron velocity distribution function is given by (3.27), the expression (4.83) for the Townsend energy coefficient takes the following form in the limit of low electron number density

$$\eta = \frac{\langle v^2/\nu \rangle}{\langle v^2/\nu (1 + M u^2/3T) \rangle} \, , \tag{4.85}$$

where $u = eE/(m_e \nu)$. In particular, if the electric field is strong and the electron–atom cross section does not depend on the electron velocity $\nu = v/\lambda$, where the mean free path of electrons λ is independent of the collision velocity, we have

$$\eta = 1.14 \left\langle \frac{m_e v^2}{3T} \right\rangle \, .$$

This demonstrates the extent to which the Townsend energy coefficient is the ratio between the average electron and atom energies. Thus, the Townsend energy coefficient is a convenient parameter characterizing the mean electron energy with respect to that of the atom. Figures 4.4 and 4.5 give the dependence of the characteristic electron temperature on the electric field strength for electrons moving in helium and argon, respectively, in accordance with experimental data [37].

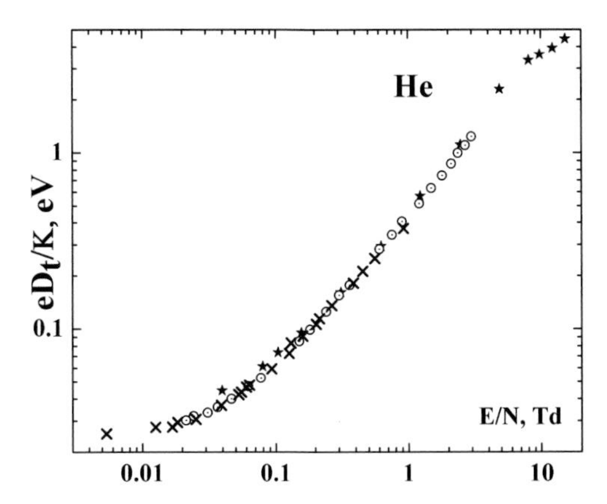

Figure 4.4 The characteristic temperature of electrons drifting in helium under the action of an electric field according to experimental data reported in [37].

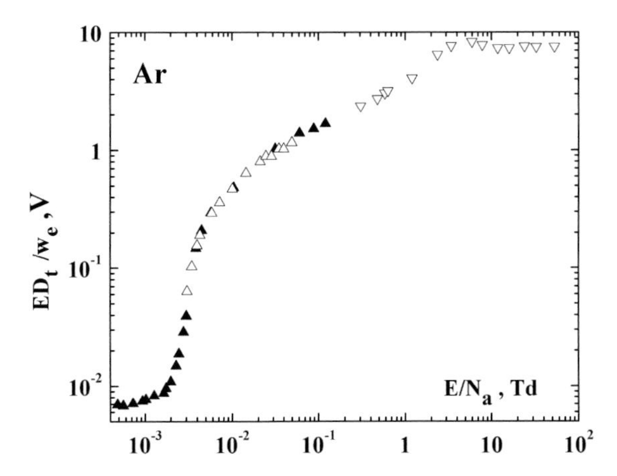

Figure 4.5 The characteristic temperature of electrons drifting in argon under the action of an electric field according to experimental data reported in [37].

4.4 Transport of Atomic ions and Clusters in Plasma

4.4.1 Zero-Field Mobility of Ions in Gases

The ion mobility $K_{\rm i}$ is defined as the ratio of the ion drift velocity w to the electric field strength E according to (4.37)

$$K_{\rm i} = \frac{w_{\rm i}}{E} ,$$

and is independent of the electric field strength at low fields according to criterion (4.20):

$$e E \lambda \ll T .$$

At low field the connection between the ion mobility and the cross section of ion–atom scattering in the first approximation of the Chapman–Enskog method is given by (4.21), and according to the Einstein relation (4.38) the diffusion coefficient of ions in gases is given by

$$D_{\rm i} = \frac{3\sqrt{\pi T}}{8 N_{\rm a} \bar{\sigma} \sqrt{2\mu}} , \tag{4.86}$$

where the average cross section of ion–atom scattering is given in (4.21). The diffusion coefficient does not depend on the direction until the distribution function for ions is isotropic.

We have two types of ion–atom scattering depending on the ion and atom species. In the first case, the types of ion and atom are different, and the scattering cross section in (4.21) and (4.86) corresponds to elastic ion–atom scattering. In the second case, atomic ions move in a parent atomic gas, and the resonant charge exchange process that proceeds according to the scheme

$$A + \widetilde{A^+} \rightarrow A^+ + \widetilde{A} \tag{4.87}$$

determines the ion mobility. Indeed, if the cross section of the resonant charge exchange process exceeds significantly the cross section of elastic ion–atom collision, the effective ion scattering proceeds as a result of the Sena effect [40, 41], as shown in Figure 4.6. Ion scattering results in charge transfer from one atomic core to the other one, and hence the charge exchange cross section characterizes the ion mobility.

In the case of ion drift in a foreign gas in a weak electric field we use (2.40) for the ion–atom diffusion cross section of elastic scattering assuming polarization character of ion–atom interaction. Then the diffusion cross section is inversely proportional to the collision velocity, and the ion mobility follows from (4.19). We obtain the Dalgarno formula [42] for the ion mobility K in the form

$$K = \frac{36}{\sqrt{\alpha\mu}} . \tag{4.88}$$

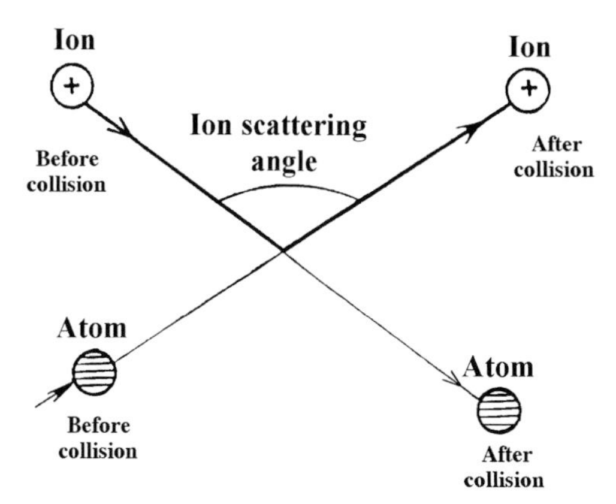

Figure 4.6 The Sena effect [40] in ion–atom scattering. As a result of the resonant charge exchange process, an incident atom becomes an ion, which leads to ion scattering, although incident particles move along straight trajectories.

Here the mobility expressed in square centimeters per volt per second is reduced to the normal number density of atoms $N_a = 2.689 \times 10^{19}\,\text{cm}^{-3}$, the atom polarizability α is measured in atomic units (a_0^3), and the ion–atom reduced mass μ is given in atomic mass units (1.66×10^{-24} g). The accuracy of (4.88) may be found from comparison of it with experimental data [21, 43–45] for the ion mobility in gases, whose accuracy is up to 2%. Table 4.9 contains the experimental mobilities of alkali metal ions in atomic and molecular gases [44, 46–50]. The ratios of experimental mobilities for alkali metal ions in simple gases to those according to (4.88) are given in Table 4.10 and characterize the accuracy of the Dalgarno formula (4.88), which is estimated as 20%.

It is convenient to improve the Dalgarno formula (4.88) on the basis of these experimental data. The statistical treatment of experimental data allows one to reduce the Dalgarno formula to the form [34]

$$K = \frac{41 \pm 4}{\sqrt{\alpha\mu}}\,, \tag{4.89}$$

Table 4.9 Experimental mobilities of positive alkali metal ions in gases. The mobilities are given in square centimeters per volt per second and are reduced to normal conditions ($N_a = 2.689 \times 10^{19}\,\text{cm}^{-3}$).

Ion, gas	He	Ne	Ar	Kr	Xe	H_2	N_2	O_2
Li^+	22.9	10.6	4.6	3.7	2.8	12.4	4.15	3.96
Na^+	22.8	8.2	3.1	2.2	1.7	12.2	2.85	1.63
K^+	21.5	7.4	2.7	1.83	1.35	13.1	2.53	2.73
Rb^+	20	6.5	2.3	1.45	1.0	13.0	2.28	2.40
Cs^+	18.3	6.0	2.1	1.3	0.91	12.9	2.2	2.27

Table 4.10 The ratio of experimental mobilities of the ions given in Table 4.9 to those according to the Dalgarno formula (4.88) [43].

Ion, gas	He	Ne	Ar	Kr	Xe	H_2	N_2
Li^+	1.27	1.17	1.06	1.07	1.06	1.02	0.89
Na^+	1.38	1.24	1.07	1.06	1.09	1.13	0.97
K^+	1.35	1.28	1.08	1.10	1.07	1.14	0.98
Rb^+	1.29	1.26	1.08	1.07	1.0	1.14	0.99
Cs^+	1.19	1.18	1.07	1.08	1.07	1.15	0.99

where the same notation is used as in (4.88), and the accuracy of 10% reflects the coincidence of its results with experimental data. For instance, the ratio of experimental mobilities of alkali metal ions in helium to those according to (4.89) is 1.13 ± 0.06. One can conveniently use (4.89) for estimation of the ion mobility in a foreign gas. In addition, Table 4.11 contains statistically averaged experimental values K_{exp} of the mobilities for molecular ions of inert gases in parent gases, and theoretical data for these mobilities K_{pol} according to the Dalgarno formula (4.88).

On the basis of (4.89) for the ion mobility and the Einstein relation (4.38) we obtain the following expression for the diffusion coefficient of ions D_i in foreign gases if an electric field is absent

$$D_i = \frac{1.0 \pm 0.1}{\sqrt{\alpha\mu}}, \tag{4.90}$$

where the diffusion coefficient is expressed in square centimeters per second and is reduced to the normal number density of atoms, and the other notations are the

Table 4.11 The zero-field mobilities of molecular ions of inert gases in a parent gas [51].

Ion	T, K	K_{exp}, $cm^2/(V\,s)$	K_{pol}, $cm^2/(V\,s)$	$D_i N_a$, $10^{18}\,cm^{-1}\,s^{-1}$
He_2^+	295	16.7 ± 0.1	21 ± 2	13 ± 1
He_2^+	77	16.4 ± 0.3	21 ± 2	3.3 ± 0.3
He_3^+	77	18.1 ± 0.1	20 ± 2	3.7 ± 0.4
He_4^+	77	17.9 ± 0.1	19 ± 2	3.6 ± 0.4
Ne_2^+	295	6.5 ± 0.6	6.8 ± 0.7	5.0 ± 0.5
Ne_2^+	77	5.4	6.8 ± 0.7	1.1 ± 0.1
Ne_3^+	77	5.4 ± 0.1	6.4 ± 0.6	1.1 ± 0.1
Ar_2^+	295	1.86 ± 0.04	2.4 ± 0.2	1.4 ± 0.1
Ar_2^+	77	1.8 ± 0.1	2.4 ± 0.2	0.36 ± 0.04
Ar_3^+	77	1.7 ± 0.1	2.2 ± 0.2	0.34 ± 0.03
Kr_2^+	295	1.1 ± 0.1	1.3 ± 0.1	0.86 ± 0.09
Xe_2^+	295	0.75 ± 0.05	0.59 ± 0.06	0.59 ± 0.06

same as in (4.88). Table 4.11 contains the diffusion coefficients of molecular ions of inert gases in parent gases.

We now consider the mobility of atomic ions in a parent atomic gas taking into account that the cross section of the resonant charge exchange process exceeds significantly the cross section of elastic ion–atom scattering. Then straight trajectories of colliding ions and atoms give the main contribution to the diffusion cross section, so the scattering angle is $\vartheta = \pi$. Correspondingly, the diffusion cross section is [52]

$$\sigma^* = \int (1 - \cos\vartheta) d\sigma = 2\sigma_{\text{res}} , \tag{4.91}$$

where σ_{res} is the cross section of the resonant charge exchange process. Then, assuming the resonant charge exchange cross section σ_{res} to be independent of the collision velocity, we obtain from (4.21) for the ion mobility in a parent gas in the first Chapman–Enskog approximation [43, 53, 54]

$$K_{\text{I}} = \frac{3e\sqrt{\pi}}{16 N_{\text{a}} \sigma_{\text{res}} \sqrt{TM}} ,$$

where M is the mass of the ion or atom, and we assume here the cross section σ_{res} of resonant charge exchange to be independent of the collision velocity. The second Chapman–Enskog approximation gives a correction [55], $\Delta = 1/40$. Accounting for a weak velocity dependence of the cross section of resonant charge exchange, we obtain in the second Chapman–Enskog approximation [43, 54]

$$K_{\text{II}} = \frac{0.341 e}{N_{\text{a}} \sqrt{T m_{\text{a}}} \sigma_{\text{res}}(2.1 v_{\text{T}})} , \tag{4.92}$$

where the argument of the cross section indicates the collision velocity for which the cross section is taken, and $v_{\text{T}} = \sqrt{8T/(\pi m_{\text{a}})}$ is the average atom velocity. This formula may be represented in a convenient form [43, 54]

$$K = \frac{1340}{\sqrt{T m_{\text{a}}} \sigma_{\text{res}}(2.1 v_{\text{T}})} , \tag{4.93}$$

which is related to the normal number density of gas atoms $N_{\text{a}} = 2.69 \times 10^{19}\,\text{cm}^{-3}$; the mobility is expressed in square centimeters per volt per second, the gas temperature is given in Kelvin, the atom mass m_{a} is expressed in atomic mass units $(1.66 \times 10^{-24}\,\text{g})$, and the cross section is given in units of $10^{-15}\,\text{cm}^2$.

In analyzing the mobility of an atomic ion in a parent gas in a weak electric field, we ignore elastic ion–atom scattering. We now take it into account assuming its contribution to be small and is determined by the ion–atom polarization interaction [43, 56]. In this case the cross section of resonant charge exchange is given by (2.46), and the zero-field mobility of an atomic gas in a parent gas is given by [43]

$$K = \frac{K_0}{1 + x + x^2} , \quad x = \frac{\sqrt{\alpha e^2/T}}{\sigma_{\text{res}}(\sqrt{7.5T/M})} , \tag{4.94}$$

where K_0 is the mobility of an atomic ion in a parent gas in the absence of elastic scattering, x is given by (2.47), and we account for a weak velocity dependence of the cross section of resonant charge exchange [56]. We also take into account in this formula the limiting case of low collision velocities. Table 4.12 gives the contribution of elastic scattering at room temperature, and elastic scattering leads to a mobility decrease.

As follows from the data in Table 4.12, because of elastic ion–atom scattering the ion mobility in parent gases at room temperature decreases by approximately 10%. Table 4.13 gives the experimental values of the ion mobilities for atomic ions of inert gases in a parent gas at an indicated temperature. These values are averaged over various measurements, and the indicated accuracy corresponds to statistical averaging of experimental data. The zero-field mobilities of atomic ions of inert gases in a parent gas at some temperatures evaluated on the basis of (4.93) and (4.94) are given in Table 4.13 in parentheses.

As a demonstration of these results, we show in Figure 4.7 the temperature dependence for the mobility of atomic helium ions in helium which includes (4.94) for the mobility due to resonant charge exchange and accounting for elastic ion–atom scattering as a correction, and experimental results [57–62].

Table 4.12 Relative variation $\Delta K/K$ of the zero-field mobility of ions A^+ in a parent gas A due to elastic scattering at room temperature for inert gases and alkali metal vapors.

A	$\Delta K/K$, %	A	$\Delta K/K$, %
He	5.8	Li	12
Ne	7.8	Na	8.2
Ar	11	K	8.9
Kr	6.0	Rb	7.7
Xe	9.2	Cs	7.5

Table 4.13 Experimental zero-field mobilities of atomic ions of inert gases in a parent gas at indicated temperatures. Theoretical values according to (4.94) are given in parentheses. The mobilities are given in square centimeters per volt per second and are reduced to the normal number density of atoms.

T, K	≈ 300	195	77
He	10.4 ± 1.0 (11)	12 ± 1 (13)	16 ± 1 (17)
Ne	4.1 ± 0.1 (4.0)	4.3 ± 0.3 (4.6)	5.3 ± 0.5 (6.0)
Ar	1.5 ± 0.1 (1.6)	1.9 ± 0.3 (1.8)	2.1 ± 0.1 (2.3)
Kr	0.85 ± 0.1 (0.81)	(0.92)	(1.1)
Xe	0.53 ± 0.01 (0.52)	(0.60)	(0.78)

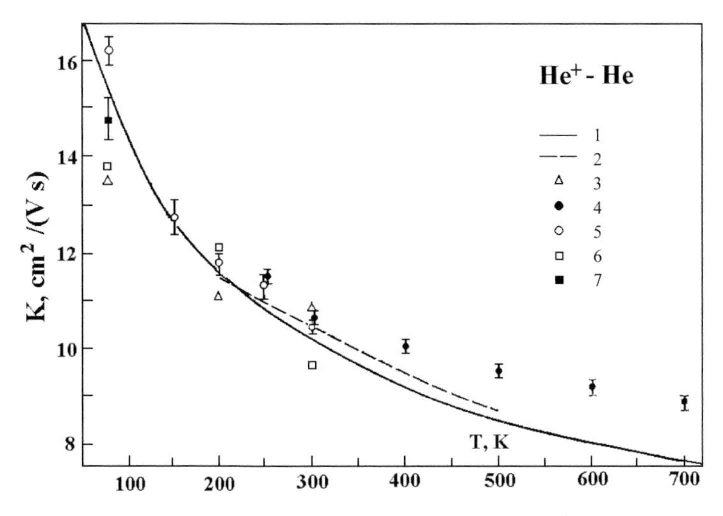

Figure 4.7 The temperature dependence for the mobility of He^+ ions in helium at low electric field strengths. Theory: 1 – formula (4.94), 2 – [57], experiment: 3 – [58], 4 – [59], 5 – [60], 6 – [61], 7 – [62].

4.4.2
Mobility of Ions at High and Intermediate Field Strengths

At high electric fields in accordance with the criterion

$$eE\lambda \gg T$$

the ion drift velocity is much greater than the ion thermal velocity, and in the case of drift of atomic ions in a parent gas, because of the absence of elastic scattering, ion velocities in the field direction are much greater than in other directions. Moving in a strong electric field, the ion accelerates under the action of the field, stops as a result of the charge exchange, and this process is repeated. In considering this limiting case and following [40, 63], we find the ion drift velocity under the assumption that the charge exchange cross section σ_{res} does not depend on the collision velocity. Ions are accelerated in the direction of the electric field, and the velocity in the field direction is large in comparison with that in transverse directions. We introduce the distribution function $f(v_x)$ for ions for velocities v_x in the field direction, which is analogous to the probability $P(t)$ of absence of charge exchange with time t after the previous exchange event. This probability satisfies the equation

$$\frac{dP}{dt} = -\nu P ,$$

where $\nu = N v_x \sigma_{res}$ is the rate of the resonant charge exchange process. The solution of this equation is

$$P(t) = \exp\left(-\int_0^t \nu dt'\right) .$$

Accounting for the equation of motion for the ion

$$M\frac{dv_x}{dt} = eE$$

allows us to connect the ion velocity with the time after the last collision by the expression $v_x = eEt/M$, so $P(t)$ is the velocity distribution function for ions whose mass is M. Assuming the cross section of the resonant charge transfer process σ_{res} to be independent of the collision velocity, we find the the velocity distribution function for ions

$$f(v_x) = C\exp\left(-\frac{Mv_x^2}{2eE\lambda}\right), \quad v_x > 0, \tag{4.95}$$

where C is a normalization factor, and the mean free path of ions in a parent gas is $\lambda = 1/(N_a\sigma_{res})$. The ion drift velocity w_i and the mean ion energy $\overline{\varepsilon}$ are

$$w_i = \overline{v_x} = \sqrt{\frac{2eE\lambda}{\pi M}}, \quad \overline{\varepsilon} = \frac{M\overline{v_x^2}}{2} = \frac{eE\lambda}{2}. \tag{4.96}$$

The ion drift velocity exceeds significantly the thermal velocity of atoms.

The above operation allows one to take into account the energy dependence of the cross section of resonant charge exchange. Indeed, approximating the velocity dependence for the charge exchange cross section by the dependence given in (2.45), we have

$$\sigma_{res}(v) = \sigma_{res}(v_0)\left(\frac{v_0}{v}\right)^\kappa, \quad \kappa = \frac{2}{\gamma R_0},$$

where the argument gives the collision velocity, and $\kappa \ll 1$. Repeating the above operations in deriving the distribution function for ions, we obtain for the distribution function in this case instead of (4.95)

$$f(v_x) = C\exp\left[-\frac{Mv_x^2 N_a\sigma_{res}(v_x)}{(2-\kappa)eE}\right], \quad v_x > 0.$$

From this we obtain for the ion drift velocity w_i and the average kinetic ion energy $\overline{\varepsilon}$ instead of (4.96)

$$w_i = \left[\frac{(2-\kappa)eE}{MN_a\sigma_{res}(v_0)v_0^\kappa}\right]^{\frac{1}{2-\kappa}}\frac{\Gamma\left(\frac{2}{2-\kappa}\right)}{\Gamma\left(\frac{1}{2-\kappa}\right)},$$
$$\overline{\varepsilon} = \frac{M}{2}\left[\frac{(2-\kappa)eE}{MN_a\sigma_{res}(v_0)v_0^\kappa}\right]^{\frac{2}{2-\kappa}}\frac{\Gamma\left(\frac{3}{2-\kappa}\right)}{\Gamma\left(\frac{1}{2-\kappa}\right)}. \tag{4.97}$$

Formula (4.97) allows us to take into account a weak velocity dependence of the charge exchange cross section in the expressions of the ion drift velocity and the average ion energy as the limit $\kappa \to 0$. As a result, in the limit of high field strengths (4.96) takes the form

$$w_i = \sqrt{\frac{2eE}{\pi MN_a\sigma_{res}\left(1.3\sqrt{\frac{eE\lambda}{M}}\right)}}, \quad \overline{\varepsilon} = \frac{eE}{2N_a\sigma_{res}\left(1.4\sqrt{\frac{eE\lambda}{M}}\right)}. \tag{4.98}$$

The above formulas for the ion drift velocity in the limits of low and high electric field strengths allow one to obtain the following expression in an intermediate range of electric field strengths: as [54, 56, 64]

$$w_{\mathrm{i}} = \sqrt{\frac{2T}{M}} \cdot \frac{0.48\beta}{(1 + 0.22\beta^{3/2})^{1/3}} \,, \tag{4.99}$$

where β is the solution of the equation

$$\beta = \frac{eE}{2TN_{\mathrm{a}}\sigma_{\mathrm{res}}\left[\sqrt{\frac{2T}{M}}\,(4.5 + 1.6\beta)\right]} \tag{4.100}$$

and the cross section of resonant charge exchange depends weakly on this parameter β. Figure 4.8 gives the dependence of the reduced ion drift velocity on β, and Figure 4.9 shows the mobility of atomic argon ions in argon at low and medium electric field strengths, where the mobilities according to (4.99) and (4.100) are compared with experimental data [21, 65, 66].

Let us use the above formulas for a simple estimation, namely, let us find the reduced electric field strength at which the average energy of the He^+ ion in helium is twice that at zero field. Because the average kinetic energy in directions

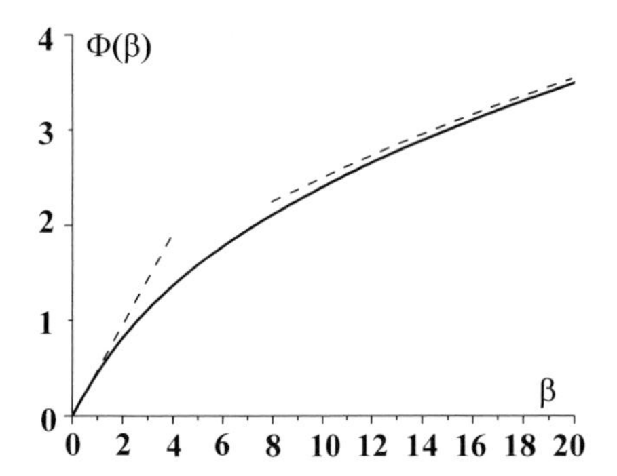

Figure 4.8 The ion drift velocity as a function of the reduced electric field strength in the case of the resonant charge exchange process. Dotted lines correspond to the limits of low and high electric field strength.

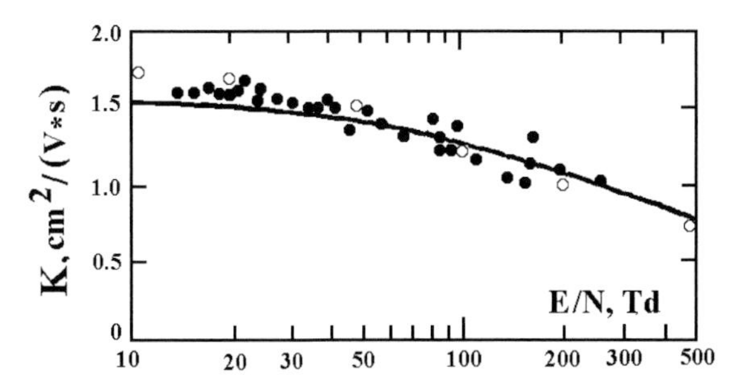

Figure 4.9 The mobility of Ar^+ ions in argon. Theory: formulas (4.99) and (4.100) [54, 56, 64] – open circles. Experiment: closed circles – [65], solid curve – [21, 66].

transverse to the field is identical in both cases, the kinetic energy of ions in the field direction varies from $T/2$ to $2T$ in these cases, and in the second case, according to (4.96) we have $eE\lambda = 4T$. Taking $T = 300\,\text{K}$ and $\sigma_{res} = 43\,\text{Å}^2$ at this temperature, we find $E/N_a = 40\,\text{Td}$. Note that the electric field strength of the electron energy doubling is 0.07 Td according to (3.39) for the regime of high electron density. From this it follows that in some range of electric field strengths the energy of plasma electrons exceeds significantly the energy of plasma ions and neutral particles.

4.4.3
Diffusion of Atomic Ions in Gases in External Fields

At low electric field strengths the diffusion coefficient of ions is expressed through the ion mobility by the Einstein relation [13, 15, 20] according to (4.38). Formula (4.90) gives the diffusion coefficient of ions in a foreign gas when ion–atom polarization interaction dominates. In the same manner we have for the diffusion coefficient of ions in a parent gas without external fields

$$D_i = 0.34\sqrt{\frac{T}{M}}\frac{1}{N_a\sigma_{res}(\sqrt{9T/M})}\,. \tag{4.101}$$

The argument in the cross section σ_{res} of resonant charge exchange indicates the collision velocity in the laboratory frame of reference at which this cross section is taken. The indicated velocity corresponds to the ion energy in the laboratory frame of reference of 4.5 T. If we express the diffusion coefficient in this formula in square centimeters per second, take the normal number density of atoms $N_a = 2.69\times 10^{19}\,\text{cm}^{-3}$, give the gas temperature T in Kelvin, represent the ion mass in atomic mass units ($1.66\times 10^{-24}\,\text{g}$), and express the cross section of resonant charge exchange σ_{res} in units of $10^{-15}\,\text{cm}^2$, this formula takes the form

$$K = 34.6\sqrt{\frac{T}{300}}\frac{1}{\sigma_{res}(2.1v_T)}\,. \tag{4.102}$$

For determination of the ion diffusion coefficient in a gas in an external electric field we use the kinetic equation for the ion distribution function (3.4) that has the form

$$(\mathbf{v}-\mathbf{w})\nabla f + \frac{e\mathbf{E}}{M}\frac{\partial f}{\partial \mathbf{v}} = I_{col}(f)\,,$$

and the ion diffusion coefficient in this gas D is defined as

$$\mathbf{j}_i = -D_i\nabla N_i\,.$$

Here $\mathbf{j}_i$ is the ion current density due to the gradient of the ion number density, which is small at a distance of the order of the mean free path of ions in a gas.

Hence, we expand the ion distribution function $f(\mathbf{v})$ over this small parameter, representing it in the form [67]

$$f(\mathbf{v}) = f_0(\mathbf{v}) - \Phi(\mathbf{v})\nabla \ln N_\mathrm{i} ,$$

and the nonperturbed distribution function $f_0(\mathbf{v})$ satisfies the equation

$$\frac{e\mathbf{E}}{M}\frac{\partial f_0}{\partial \mathbf{v}} = I_\mathrm{col}(f_0) .$$

The normalization condition for the distribution function also gives

$$\int \Phi(\mathbf{v}) d\mathbf{v} = 0 .$$

Comparing the expression for the ion current $\mathbf{j}_\mathrm{i} = \int \mathbf{v} f(\mathbf{v}) d\mathbf{v}$ with the definition of the ion diffusion coefficient, we obtain

$$D_z = \frac{1}{N_\mathrm{i}} \int (v_z - w_z)\Phi(\mathbf{v}) d\mathbf{v} , \tag{4.103}$$

where z is the direction of the ion density gradient. Substituting the above expansion of the ion distribution function into the kinetic equation, we obtain the following equation for the addition part of the distribution function

$$(v_z - w_z) f_0 = \frac{eE_z}{M}\frac{d\Phi}{dv_z} + I_\mathrm{col}(\Phi) .$$

We now obtain an integral relation for the distribution function addition $\Phi(\mathbf{v})$ analogous to relation (4.18) by multiplying the equation for Φ by v_z and integrating it over ion velocities. Using the normalization condition for $\Phi(\mathbf{v})$, we obtain

$$\left(\overline{v_z^2} - w_z^2\right) N_\mathrm{i} = \frac{m_\mathrm{a}}{M + m_\mathrm{a}} \int g_z g\sigma^*(g)\Phi(\mathbf{v})\varphi(v_a) d\mathbf{v} d\mathbf{v}_\mathrm{a} ,$$

where we used the notation in (4.18). This equation allows us to find the diffusion coefficient in the case $\sigma^*(g) \sim 1/g$, which gives

$$D_z = \frac{M}{\mu\nu}\left(\overline{v_z^2} - w_z^2\right) ,$$

where the collision rate $\nu = N_\mathrm{a} g\sigma^*(g)$ is assumed to be independent of the collision velocity. One can see that for the isotropic distribution function the diffusion coefficients of ions in different directions are identical, whereas they are different for an anisotropic distribution function in the frame of reference where ions are motionless.

We now determine the diffusion coefficient of atomic ions in a parent gas for the direction along the electric field when the electric field strength is high. Accounting for the character of the resonant charge process and assuming the mean free path of ions λ to be independent of the ion velocity, we represent the kinetic equation for the ion distribution function in the form

$$\frac{eE}{M}\frac{\partial f}{\partial v_z} = \frac{1}{\lambda}\delta(v_z)\int v_z' f(v_z')dv_z' - \frac{v_z}{\lambda} f(v_z) ,$$

where the z-axis is directed along the field. Using the reduced ion velocity

$$u = \left(\frac{M}{eE\lambda}\right)^{1/2} v_z ,$$

one can rewrite this equation as

$$\frac{2\lambda}{\pi w_{\rm i}} N_{\rm i}\left(u - \sqrt{\frac{2}{\pi}}\right)\exp\left(-\frac{u^2}{2}\right)\eta(u) = \frac{d\Phi}{du} + u\Phi - \delta(u)\int u'\Phi(u')du' ,$$

and the ion drift velocity is $w = (2eE\lambda/\pi M)^{1/2}$ in accordance with (4.96). Solving this equation, we obtain

$$\Phi = \frac{2\lambda}{\pi w} N_{\rm i}\exp\left(-\frac{u^2}{2}\right)\left(\frac{u^2}{2} - \sqrt{\frac{2}{\pi}} + C\right)\eta(u) .$$

The constant C in this expression follows from the normalization condition $\int \Phi(u)du = 0$, which gives $C = 2/\pi - 1/2$. For the longitudinal diffusion coefficient of atomic ions in a parent gas in a strong electric field this gives

$$D_\parallel = \frac{1}{N_{\rm i}}\int (v_z - w_{\rm i})\Phi\, dv_z = \lambda w_{\rm i}\left(\frac{2}{\pi} - \frac{1}{2}\right) = 0.137\lambda w_{\rm i} . \tag{4.104}$$

The transverse diffusion coefficient of atomic ions in a parent gas can be found because the distribution function in the direction perpendicular to the field coincides with the Maxwell distribution function for atoms. The variables in the kinetic equation are divided for motion along the field and for motion perpendicular to it, and the diffusion coefficient in the direction transverse to the field is [67]

$$D_\perp = \frac{T\lambda}{Mw} , \tag{4.105}$$

and the ratio of the longitudinal and transverse diffusion coefficients is equal in this case:

$$\frac{D_\parallel}{D_\perp} = 0.137\frac{Mw_{\rm i}^2}{T} = 0.087\frac{eE\lambda}{T} , \tag{4.106}$$

which is large in a strong electric field.

4.4.4
Conversion of Ions During Drift in an Electric Field

Usually different types of ions may exist in a gas when ions drift in an external electric field. Hence, in the course of ion drift, transitions between different types of ions are possible, and because the mobilities are different for different types of ions, these transitions influence the time taken by ions to travel a given distance, in particular, for their motion in a drift or gas discharge chamber. We first consider this problem for inert gases, where along with atomic ions A^+ molecular ions A_2^+ may also be present. Figure 4.10 shows the dependence on the reduced electric field strength for the mobility of helium ions moving in helium [68], and Figure 4.11 compares the mobilities of atomic and molecular ions in argon [21]. The number of types of ions in the molecular case may be higher. As examples,

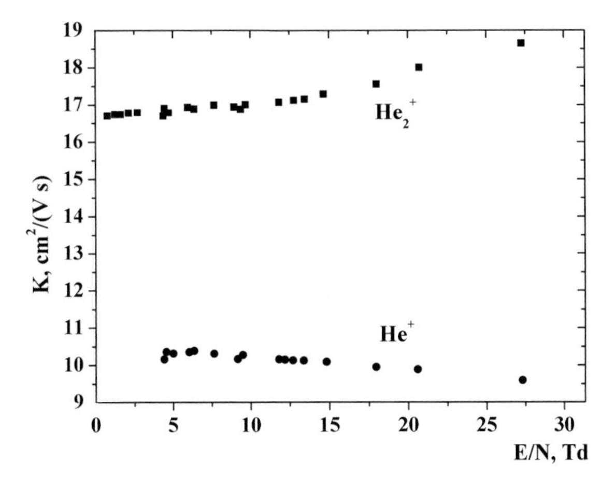

Figure 4.10 The mobility of atomic and molecular helium ions in helium as a function of the reduced electric field strength [68].

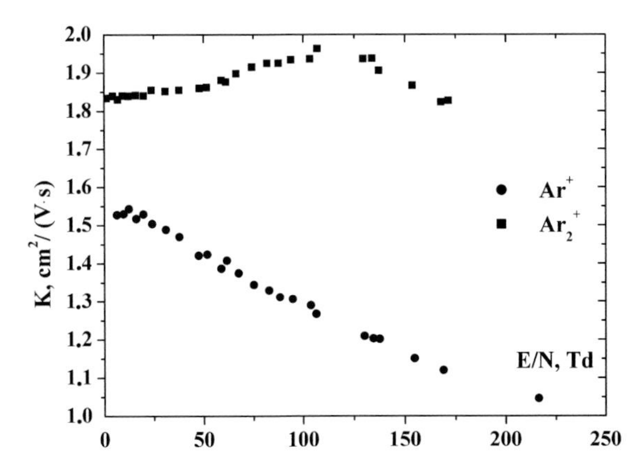

Figure 4.11 The mobility of atomic and molecular argon ions in argon [21].

Figure 4.12 gives the dependence on the electric field strengths for the ion drift velocity in the hydrogen case [69], and Figure 4.13 shows the dependencies on the reduced electric strength for the mobility of hydrogen and deuterium ions in parent gases [70]. In addition, Figure 4.14 represents the same dependencies for the mobilities of nitrogen ions in nitrogen [71, 72]. In all the cases, different types of ions are characterized by different mobilities and may be separated by the mobility.

The relative concentration of ions of different types is determined by the equilibrium constant K for the process of transition between these ions. In the simple case of ions of inert gases at room temperature, we have atomic A^+ and molecular A_2^+ ions with an equilibrium between them according to the scheme

$$A^+ + A \leftrightarrow A_2^+ .$$

Correspondingly, the equilibrium constant K for this transition is given by (1.55) and has the following form:

$$K = \frac{[A^+][A]}{[A_2^+]} = \frac{g_a g_+}{g_{mol}} \left(\frac{\mu T}{2\pi\hbar^2}\right)^{3/2} \frac{B}{T} \left[1 - \exp\left(-\frac{\hbar\omega}{T}\right)\right] \exp\left(-\frac{D}{T}\right) , \quad (4.107)$$

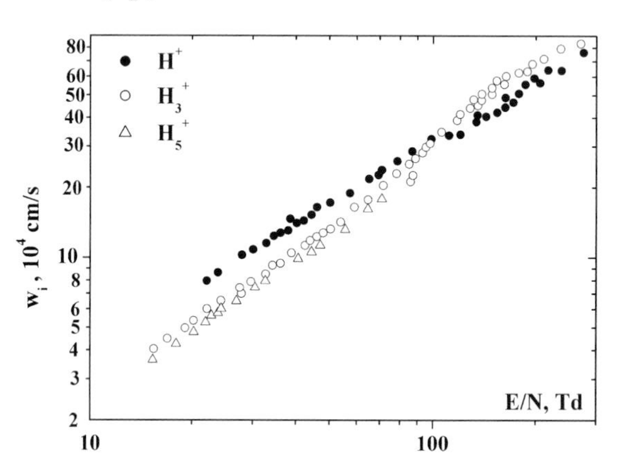

Figure 4.12 The drift velocity of positive hydrogen ions in hydrogen [69].

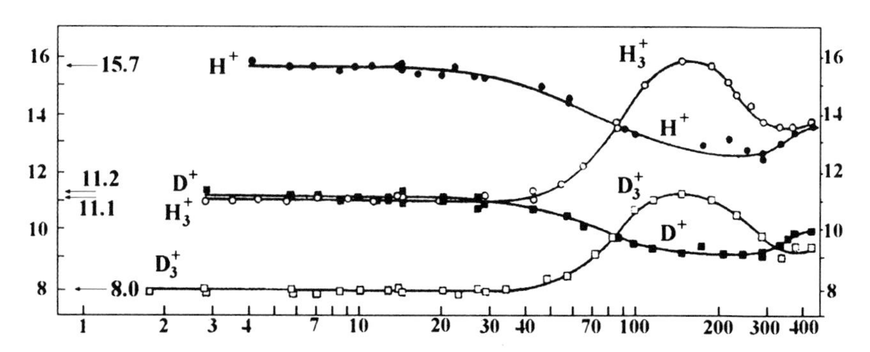

Figure 4.13 The mobility of H^+ and H_3^+ ions in hydrogen and D^+ and D_3^+ ions in deuterium [70].

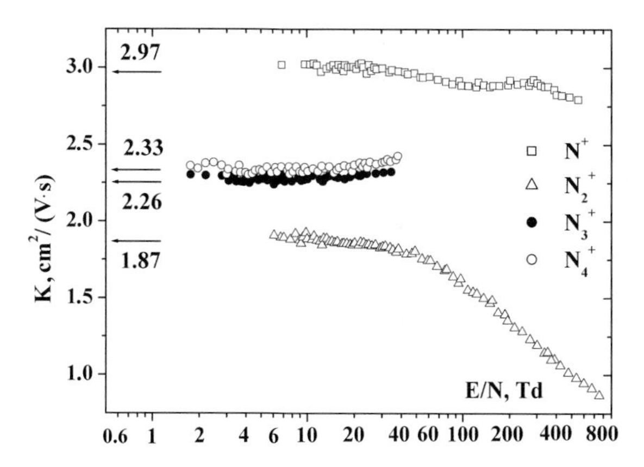

Figure 4.14 The mobility of nitrogen ions in nitrogen [71, 72]. Arrows indicate the mobilities of ions at zero electric field strength.

where $[X]$ is the number density of particles X, g_a, g_+, and g_{mol} are the statistical weights with respect to the electronic state for atoms, atomic ions, and molecular ions respectively, $\hbar\omega$ and D are the vibrational energy and the dissociation energy for the molecular ion, respectively, and B is the rotational constant for the molecular ion; we assume the temperature T expressed in energy units to be large compared with the rotational constant B, and we assume the two nuclei of the molecular ions to be different isotopes. As a demonstration, we give in Table 4.14 [34] these parameters for molecular helium and argon ions.

If we substitute these data into (4.107), in the argon case we obtain

$$K(T) = C \exp\left(-\frac{\Delta H}{T}\right) ,$$

with the enthalpy $\Delta H = 1.33\,\text{eV}$ for a pressure range $p \sim 1\,\text{Torr}$, and $C = 1.8 \times 10^{-11}\,\text{Torr}^{-1}$. This means that the equilibrium number densities for atomic and molecular argon ions are equal at pressure $p = 1\,\text{Torr}$ and temperature $T = 624\,\text{K}$ or at pressure $p = 10\,\text{Torr}$ and temperature $T = 688\,\text{K}$. We find that molecular ions dominate at low temperatures under equilibrium conditions, whereas atomic ions dominate at high temperatures. Usually ions are formed in a gas as a result of electron–atom collisions. At low temperatures they are converted into molecular ions as a result of three body collisions, and the basic ion type de-

Table 4.14 The parameters of the ground electronic state of He_2^+ and Ar_2^+ ions [34].

Molecular ion	He_2^+	Ar_2^+
D_e, eV	2.47	1.23
$\hbar\omega_e$, cm^{-1}	1698	309
$\hbar\omega_e x_e$, cm^{-1}	35.3	1.66
B_e, cm^{-1}	7.21	0.143

pends on the ratio between a typical time of ion location in a plasma and a typical time for transition between atomic and molecular ions.

Let us consider an example where atomic helium ions are formed at the initial time at the beginning of the drift chamber and pass through it under action of a low electric field in a time τ_{dr}. In the course of the drift, atomic ions are converted partially into molecular ions. Since molecular ions move faster, we obtain a broadened signal at the collector. We take the mobilities of atomic and molecular helium ions in helium to be 10.4 and 16.7 $cm^2/(V\,s)$ according to Tables 4.11 and 4.13, so their ratio is approximately 1.6. Figure 4.15 shows the time dependence for the reduced ion current $i(t/\tau_{dr})$ in this case, where the drift time τ_{dr} relates to atomic helium ions, and the current is normalized to 1:

$$\int i\left(\frac{t}{\tau_{dr}}\right)\frac{dt}{\tau_{dr}} = 1 \, .$$

In addition, we account for the broadening of the current in time as a result of longitudinal diffusion of ions in a gas. Thus, we have three typical times in this case, the drift time τ_{dr} of atomic ions through the drift chamber, the time for conversion τ_{con} of atomic ions into molecular ions, and the diffusion time τ_D for atomic ions defined according to (4.33) $L^2 = 2D_i\tau_D$, where D_i is the longitudinal diffusion coefficient for atomic ions and L is the distance traveled by ions in the drift chamber. Since the drift time is $\tau_{dr} = L/w$, where w is the ion drift velocity, we obtain $\xi = \tau_D^2/\tau_{dr}^2 = Lw/2D$. For definiteness, we use in Figure 4.15 $\xi = 100$ and $\tau_{dr}/\tau_{con} = 0.3$. As is seen, a change of the type of ion leads to a change in the character of ion drift.

One can see that this effect is stronger, the larger the ratio of the mobilities for atomic and molecular ions. At low temperatures these values are determined by the

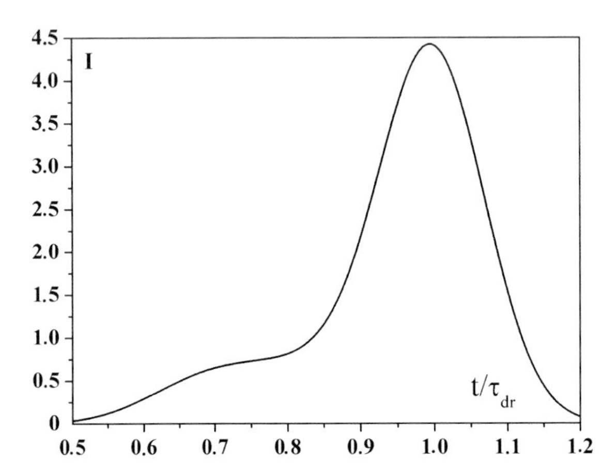

Figure 4.15 Variation in time of the current i of helium ions passing through a drift chamber under the action of an electric field if He^+ ions are formed at $t = 0$ at the beginning of this chamber. The ratio of the drift time of atomic ions τ_{dr} to the diffusion time τ_D of atomic ions in helium is $\tau_{dr}/\tau_D = 10$, and the time for conversion τ_{con} of atomic ions in molecular ions is $\tau_{con} = 3\tau_{dr}$.

ion–atom polarization scattering that is connected with atom capture by an ion, so the difference between the mobilities for atomic and molecular ions is not large at low temperatures. In contrast, at high temperatures or electric field strengths, the mobility of an atomic ion is determined by the resonant charge exchange process, and the cross section of this process is almost constant if the temperature or electric field strength increases, whereas the mobility of a molecular ion is determined by the ion–molecule polarization interaction, and the cross section of this process decreases with increasing temperature or electric field strength. Therefore, the difference in the mobilities of molecular and atomic ions increases with increasing temperature or electric field strength. Figure 4.16 represents the temperature dependence of the mobility of molecular helium ions in helium according to experiments [59, 60]. Comparing this dependence with that for atomic helium ions given in Figure 4.7, one can see that at a temperature of 100 K the mobilities of atomic and molecular ions are similar, whereas at a temperature of 700 K the mobility of molecular ions is about twice that of atomic ions.

One more aspect of transition between atomic and molecular ions in a plasma of inert gases is connected with a different rate of electron–ion recombination for atomic and molecular ions. Because molecular ions partake in dissociative recombination, and recombination of atomic ions results from the three body process, the rate of recombination of electrons and ions in these two cases differs dramatically. Therefore, the rate of electron–ion recombination in a plasma of inert gases may be determined by conversion of atomic ions in molecular ones.

Competition between processes in molecular gases involving different types of ions is more complicated than that in a plasma of inert gases. In particular, if H_2^+ ions are formed as a result of ionization of hydrogen molecules in a hydrogen gas, these molecular ions partake in the following ion–molecule reaction:

$$H_2^+ + H_2 \to H_3^+ + H + 1.7\,\text{eV}\,. \tag{4.108}$$

The rate constant for this process at room temperature is $2.1 \times 10^{-9}\,\text{cm}^3/\text{s}$ and in the case of the process $D_2^+ + D_2$, the rate constant for D_3^+ formation is $1.6 \times 10^{-9}\,\text{cm}^3/\text{s}$ [73]. These rate constants are similar to the rate constants for polarization capture of ions which follow from (2.40) for the cross section of ion–molecule polarization capture and are similar to those obtained in other experiments. From process (4.108) it follows that H_3^+ is the basic ion type in hydrogen at

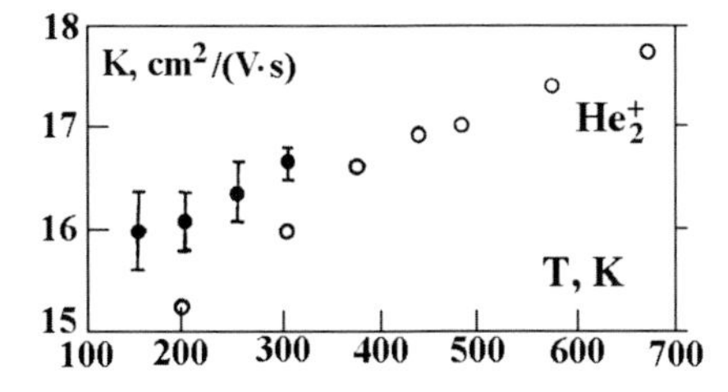

Figure 4.16 The temperature dependence of the mobility of molecular helium ions in helium according to experiments [59] (open circles) and [60] (filled circles).

room temperature. At higher gas temperatures or electric field strengths protons may be present in hydrogen along with H_3^+ [69], whereas at low temperatures ions with an odd number of nuclei, that is, H_5^+ and H_7^+, may be present. Figure 4.17 shows the equilibrium constants for the processes $H_3^+ + H_2 \leftrightarrow H_5^+$, $D_3^+ + D_2 \leftrightarrow D_5^+$, $H_5^+ + H_2 \leftrightarrow H_7^+$, and $D_5^+ + D_2 \leftrightarrow D_7^+$ obtained on the basis of experiments [74–76]. In particular, from [76] it follows that under conditions of thermodynamic equilibrium at a pressure of 1 Torr the number densities of H_3^+ and H_5^+ ions in hydrogen are equal at a temperature of 240 K, for D_3^+ and D_5^+ ions in deuterium the temperature at which the number densities of these ions are equal is 236 K, for H_5^+ and H_7^+ ions in hydrogen this temperature is 115 K, and for D_5^+ and D_7^+ ions in deuterium the temperature is 104 K. At a pressure of 10 Torr the temperatures are 287, 279, 136, and 122 K, respectively.

Establishment of thermodynamic equilibrium between ions of different types results from conversion of simple ions into complex ones in three body collisions. In particular, Figure 4.18 shows the rate constants for the three body process for formation of H_5^+ from H_3^+ according to experiments [77–79]. As is seen, the rate constant increases dramatically with decreasing temperature, and at room temperature it is according to [77] $4.5 \times 10^{-31}\,\mathrm{cm^6/s}$ for formation of H_5^+ and $6.5 \times 10^{-31}\,\mathrm{cm^6/s}$ for formation of D_5^+. One can see that in spite of the possibility of formation of various types of ions, a restricted number of ion types exists under certain conditions.

Processes of dissociative recombination of electrons and molecular ions may be of importance for the existence of ions of a certain type in real ionized gases. In the hydrogen case, the recombination coefficient of electrons and H_3^+ ions is $2.3 \times 10^{-7}\,\mathrm{cm^3/s}$ at room temperature [80], and for H_5^+ ions at 205 K the recombination coefficient is $3.6 \times 10^{-6}\,\mathrm{cm^3/s}$ [80]. The recombination coefficient measured for H_3^+ and D_3^+ ions in the temperature range 95–300 K is independent of the temperature [81], whereas beam measurements give a remarkable dependence of the cross section σ_{rec} of dissociative recombination on the electron energy ε, which is $\sigma_{rec} \sim \varepsilon^{-n}$, with $n = 1.0$–1.3, for H_3^+ and D_3^+ ions in the energy range 0.01–1 eV [82]. It is quite possible that the different temperature and energy depen-

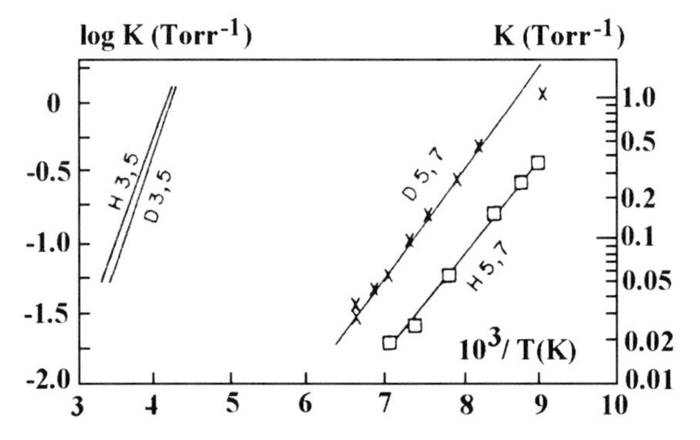

Figure 4.17 The equilibrium constants for processes $H_3^+ + H_2 \leftrightarrow H_5^+$ (H3,5), $D_3^+ + D_2 \leftrightarrow D_5^+$ (D3,5), $H_5^+ + H_2 \leftrightarrow H_7^+$ (H5,7), and $D_5^+ + D_2 \leftrightarrow D_7^+$ (D5,7) [76].

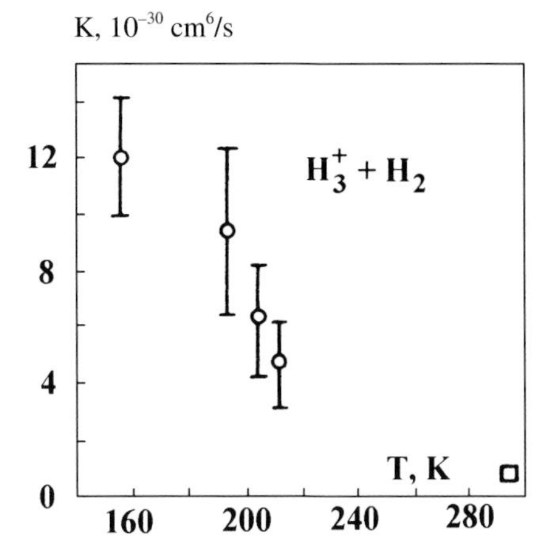

Figure 4.18 The rate constants for the three body process $H_3^+ + H_2 \leftrightarrow H_5^+$ according to experimental data: open square – [77, 78], open circles – [79].

dencies for the recombination coefficient are due to the presence of vibrationally excited ions in the beam. Next, the cross section σ_{rec} of dissociative recombination at electron energy $\varepsilon = 0.01\,\mathrm{eV}$ is $(1.1 \pm 0.1) \times 10^{-13}\,\mathrm{cm}^2$ for H_3^+ ions and $(0.5 \pm 0.1) \times 10^{-13}\,\mathrm{cm}^2$ for D_3^+ ions according to measurements [83, 84]. Note that processes of dissociative recombination involving molecular ions are of importance in a stationary plasma and may establish therein the number density of electrons and ions.

We briefly consider conversion between different types of ions in nitrogen. If in the first stage of nitrogen ionization N_2^+ ions are formed, because of a high dissociation energy of the nitrogen molecule only N_4^+ ions may be formed from the initial ion in nitrogen at room temperature. Nevertheless, along with these ions, N^+ and N_3^+ ions are also observed in nitrogen ionized by an external source. As an example, Figure 4.19 gives the pressure dependence for nitrogen ions formed in nitrogen if ions are created by an electron beam [85]. Note that transition from nitrogen ions with an even number of nuclei to ions with an odd number of nuclei is based on the process [86]

$$N_2^+(^4\Sigma_u^+) + N_2 \to N_3^+ + N\,,$$

which includes electron excited state $N_2^+(^4\Sigma_u^+)$ of the nitrogen molecule. Hence, nitrogen ions with an odd number of nuclei are formed in a source where ionization proceeds by particles (e.g., by electrons or photons) whose energy exceeds remarkably the ionization potential of the nitrogen molecule, as takes place under the conditions in Figure 4.19. Next, an increase of the nitrogen pressure must lead to an increase of the concentration of more complex ions, which follows from Figure 4.19.

Note also that transitions between nitrogen ion pairs proceed weakly compared with transitions between ion types inside one pair, so the total concentration of ions

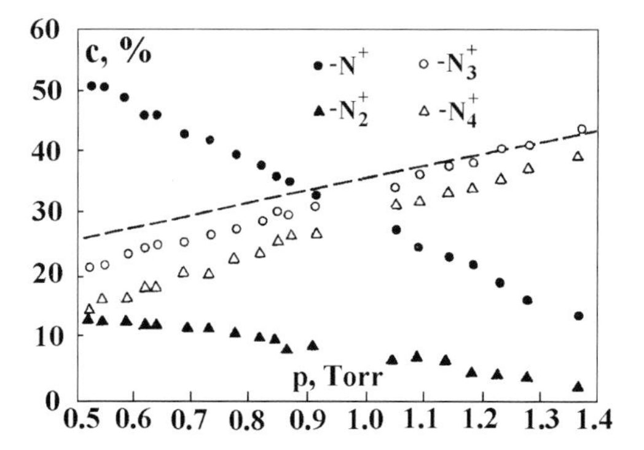

Figure 4.19 The concentration of nitrogen ions of various types in nitrogen [85].

with odd numbers of nuclei or with even numbers of nuclei must be conserved in the course of pressure variation. The total concentration of ions with even numbers of nuclei is given in Figure 4.19 by a hatch line, and as is seen, this statement is roughly fulfilled. In reality for nitrogen ions which are drifting in nitrogen in an external electric field, weak transitions between the above-mentioned pairs of nitrogen ions take place. In particular, transitions between N_2^+ and N_3^+ ions were reported in [87], and transitions between N_2^+ and N_3^+ ions were reported in [88]. Next, like the previous cases, the rate of establishment of thermodynamic equilibrium between different types of ions is determined by association of simple ions with a nitrogen molecule in a three body collision. Figure 4.20 represents the rate constants for these processes at room temperature if nitrogen ions are in nitrogen in an external electric field [89].

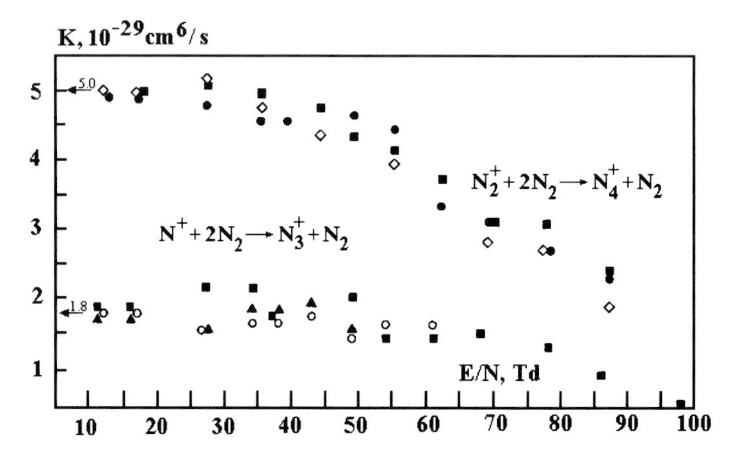

Figure 4.20 The rate constants for the three body process of conversion of nitrogen ions in nitrogen [89].

4.4.5
Mobility and Diffusion of Large Clusters in a Gas

In considering the behavior of a large cluster in an ionized gas, we use the liquid drop model, that is, we model a cluster by a spherical ball whose radius r_0 is connected to the number of its atoms by (2.16). We assume the cluster is large enough so that its radius is large compared with the distance of action of atomic forces, and then the diffusion cross section of atom–cluster collision is given by (2.18). Next, we consider the kinetic and diffusion regimes of cluster drift depending on the relation between the cluster radius r_0 and the mean free path of atoms λ. Namely, the kinetic regime corresponds to the criterion $\lambda \gg r_0$ or

$$r_0 \ll (N_a \sigma_g)^{-1} , \tag{4.109}$$

where N_a is the number density of atoms and σ_g is the gas-kinetic cross section. In particular, in atmospheric air the boundary between these regimes corresponds to a cluster radius $r_0 \sim 100$ nm, and in reality both cases may be realized.

We start form the diffusion regime if many atoms simultaneously collide with the cluster surface. The resistive force for a cluster of radius r_0 that is moving with velocity w is given in this case by the Stokes formula (4.52). Then we obtain the mobility K of a cluster of charge e and the cluster diffusion coefficient D in a gas on the basis of the Einstein relation (4.38) [90]:

$$K = \frac{\mathbf{w}}{\mathbf{E}} = \frac{e}{6\pi r_0 \eta} = \frac{K_0}{n^{1/3}} , \quad K_0 = \frac{e}{6\pi r_W \eta} ;$$
$$D = \frac{KT}{e} = \frac{T}{6\pi r_0 \eta} = \frac{D_0}{n^{1/3}} , \quad D_0 = \frac{T}{6\pi r_W \eta} , \quad \lambda \ll r_0 . \tag{4.110}$$

According to these formulas, the dependence on the number n of cluster atoms for the diffusion coefficient and mobility is $D, K \propto n^{-1/3}$. Next, the only cluster parameter in (4.110) for the cluster mobility and diffusion coefficient is the cluster radius r_0. Table 4.15 contains values of the reduced diffusion coefficients for metal clusters in some gases.

We now determine the resistive force that acts on a moving cluster in a gas in the kinetic regime of cluster drift. In this regime a cluster can collide at each moment with one atom only. The momentum transferred in atom–cluster collision is $m_a \mathbf{g}(1 - \cos\vartheta)$, where m_a is the atom mass, ϑ is the scattering angle in the center-of-mass frame of reference, and $\mathbf{g} = \mathbf{v} - \mathbf{w}$ is the collision velocity, where $\mathbf{v}$ is the atom velocity and $\mathbf{w}$ is the cluster velocity. Because the cluster mass M is large compared with the atom mass m_a, the center-of-mass frame of reference coincides practically with the frame of reference where a cluster is motionless. Next, we have a Maxwell velocity distribution function for atoms that has the form

$$\varphi(v) = N_a \left(\frac{m_a}{2\pi T}\right)^{3/2} \exp\left(-\frac{m_a v^2}{2T}\right) ,$$

Table 4.15 The diffusion coefficient D_0 of metal clusters in accordance with (4.110) reduced to the normal number density of atoms. The diffusion coefficients are given in units of $10^{-3}\ \mathrm{cm^2/s}$ and correspond to the gas temperature $T = 600$ K [91, 92].

Cluster, gas	He	Ne	Ar	Kr	Xe	H_2	N_2
Ti	0.84	0.52	0.68	0.58	0.60	1.8	0.89
Fe	0.95	0.59	0.77	0.65	0.68	2.1	1.0
Co	0.97	0.60	0.78	0.66	0.69	2.1	1.0
Ni	0.97	0.60	0.78	0.66	0.69	2.1	1.0
Cu	0.95	0.59	0.77	0.65	0.68	2.1	1.0
Zn	0.88	0.55	0.71	0.61	0.63	1.9	0.94
Sr	0.82	0.51	0.66	0.56	0.58	1.8	0.87
Mo	0.87	0.54	0.70	0.60	0.62	1.9	0.93
Pd	0.87	0.54	0.70	0.60	0.62	1.9	0.93
Ag	0.84	0.52	0.68	0.58	0.60	1.8	0.89
Ba	0.55	0.34	0.44	0.38	0.39	1.2	0.58
Ta	0.83	0.52	0.67	0.57	0.59	1.8	0.88
W	0.87	0.54	0.70	0.59	0.62	1.9	0.92
Re	0.88	0.55	0.71	0.61	0.63	1.9	0.94
Ir	0.88	0.54	0.71	0.60	0.63	1.9	0.93
Pt	0.89	0.55	0.72	0.61	0.64	1.9	0.94
Au	0.85	0.45	0.68	0.58	0.60	1.8	0.90
Tl	0.72	0.44	0.58	0.50	0.52	1.6	0.77
Pb	0.71	0.49	0.57	0.49	0.51	1.6	0.75
U	0.79	0.49	0.64	0.54	0.56	1.7	0.84
Pu	0.82	0.51	0.66	0.56	0.59	1.8	0.87

and clusters do not influence the distribution function for atoms. From this we have for the force that acts on a moving cluster in the kinetic regime of its drift

$$\mathbf{F} = \int m_a \mathbf{g}(1 - \cos\vartheta)\varphi(v) g d\sigma d\mathbf{v} = m\sigma^* \int \mathbf{g}\varphi(v) g d\mathbf{v} = \frac{m_a}{\lambda}\langle \mathbf{g}g\rangle\,,$$
$$\lambda = \frac{1}{N_a\sigma^*}\,, \tag{4.111}$$

where $d\sigma$ is the differential cross section of atom–cluster scattering, the diffusion cross section σ^* of atom–cluster collision is $\sigma^* = \pi r_0^2$ and is given by (2.18), and an average is made over the Maxwell distribution function for atoms.

Taking the cluster velocity **w** and force **F** acting on a cluster to be directed along the z-axis, we assume $w \ll v$ and perform expansion over this small parameter.

This gives

$$\langle \mathbf{g}g \rangle = \frac{1}{N} \int \mathbf{g} g \varphi(v) d\mathbf{v} = \mathbf{k} \frac{1}{N} \int g_z g w_z \frac{\partial \varphi(g)}{\partial g_z} d\mathbf{g} = -\mathbf{w} \left\langle \frac{m_a g_z^2 g}{T} \right\rangle = -\mathbf{w} \frac{m_a}{3T} \langle g^3 \rangle ,$$

where $\mathbf{k}$ is the unit vector along the direction of the cluster velocity. From this we obtain for the resistive force in the kinetic regime when a cluster moves in a gas with velocity w [93]

$$\mathbf{F} = -\mathbf{w} \cdot \frac{m_a^2}{3T\lambda} \langle v^3 \rangle = -\mathbf{w} \cdot \frac{8\sqrt{2\pi m_a T}}{3} \cdot N r_0^2 . \tag{4.112}$$

Accounting for the connection between the cluster velocity $\mathbf{w}$ and the electric field strength $\mathbf{E}$

$$\mathbf{w} = K\mathbf{E} ,$$

and taking the resistive force to be $\mathbf{F} = e\mathbf{E}$ for a cluster of charge e moving in an electric field of strength E, we obtain on the basis of (4.112) in the kinetic regime of cluster drift [94]

$$K = \frac{3e}{8\sqrt{2\pi m_a T} N r_0^2} = \frac{K_0}{n^{2/3}} , \quad K_0 = \frac{3e}{8\sqrt{2\pi m_a T} N r_W^2} . \tag{4.113}$$

This expression coincides with the first Chapman–Enskog approximation (4.21).

From this on the basis of the Einstein relation (4.38) we obtain for the cluster diffusion coefficient D in the kinetic regime of cluster drift [94]

$$D = \frac{3\sqrt{T}}{8\sqrt{2\pi m_a} N r_0^2} = \frac{D_0}{n^{2/3}} , \quad D_0 = \frac{3\sqrt{T}}{8\sqrt{2\pi m_a} N r_W^2} , \quad \lambda \gg r , \tag{4.114}$$

which also corresponds to the first Chapman–Enskog approximation (4.53). Table 4.16 gives the values of the reduced diffusion coefficients of metal clusters in some gases in accordance with (4.114) in the kinetic regime of cluster drift.

Let us compose the ratio between the cluster diffusion coefficient D_{kin} (4.114) for the kinetic regime of cluster movement in a gas and the cluster diffusion coefficient D_{dif} for the diffusion regime of cluster movement according to (4.110) using (4.59) for the viscosity coefficient. We have

$$\frac{D_{dif}}{D_{kin}} = \frac{45\pi}{64\sqrt{2}} \frac{\lambda}{r_0} = 1.56\text{Kn} , \quad \text{Kn} = \frac{\lambda}{r_0} , \tag{4.115}$$

where Kn is the Knudsen number. For the cluster diffusion coefficient this expression combines both the kinetic and diffusion regimes of cluster drift:

$$D = D_{dif}(1 + 1.56\text{Kn}) . \tag{4.116}$$

Table 4.16 The diffusion coefficient of metal clusters in a gas D_0 in the kinetic regime of cluster drift in accordance with (4.114). The diffusion coefficients are reduced to the normal density of atoms, are expressed in square centimeters per second, and correspond to gas temperature $T = 600$ K.

Cluster, gas	He	Ne	Ar	Kr	Xe	H_2	N_2
Ti	2.2	1.0	0.71	0.49	0.39	3.2	0.85
Fe	2.9	1.3	0.91	0.63	0.50	4.2	1.1
Co	3.0	1.3	0.95	0.65	0.52	4.0	1.1
Ni	3.0	1.3	0.95	0.65	0.52	4.0	1.1
Cu	2.9	1.3	0.91	0.63	0.50	4.2	1.1
Zn	2.5	1.1	0.79	0.54	0.43	3.5	0.94
Sr	2.1	0.94	0.66	0.46	0.37	3.0	0.80
Mo	2.4	1.1	0.77	0.53	0.42	3.4	0.92
Pd	2.4	1.1	0.77	0.53	0.42	3.4	0.92
Ag	2.2	1.0	0.71	0.49	0.39	3.2	0.85
Ba	0.96	0.43	0.30	0.21	0.17	1.4	0.36
Ta	2.2	0.98	0.70	0.48	0.38	3.1	0.83
W	2.4	1.1	0.76	0.52	0.42	3.4	0.90
Re	2.5	1.1	0.79	0.54	0.43	3.5	0.94
Ir	2.4	1.1	0.78	0.54	0.43	3.4	0.93
Pt	2.5	1.1	0.80	0.55	0.44	3.5	0.95
Au	2.3	1.0	0.72	0.50	0.40	3.2	0.86
Tl	1.7	0.74	0.53	0.36	0.29	2.3	0.63
Pb	1.6	0.71	0.51	0.35	0.28	2.2	0.60
U	2.0	0.88	0.63	0.43	0.34	2.8	0.75
Pu	2.1	0.96	0.68	0.47	0.37	3.0	0.81

In the limit of small Knudsen numbers (Kn $\ll$ 1) this formula is converted into (4.114), and in the limit of large Knudsen numbers (Kn $\gg$ 1) it is transformed into (4.110). In particular, if water clusters (drops) of radius r_0 move in air with temperature $T = 300$ K, (4.116) takes the form

$$D = \frac{k_*}{r_0}\left(1 + \frac{1}{N r_0 s_0}\right) ,$$

where N is the number density of air molecules, and the other parameters in this formula are $k_* = 1.2 \times 10^{-11}\,\text{cm}^3/\text{s}$ and $s_0 = 4.3 \times 10^{-19}\,\text{cm}^2$.

Let us determine the boundary number density of atoms or molecules N_* when the contributions to the cluster diffusion coefficient due to the kinetic and diffusion regimes are identical. Then we have from (4.116)

$$N_* r_0 = \frac{1.56}{\sigma_g} .$$

Table 4.17 The boundary reduced cluster radius for transition between the kinetic and diffusion regimes of cluster drift for gases at room temperature.

Gas	He	Ne	Ar	Kr	Xe	H_2	N_2	O_2
$N_* r_0$, $10^{14}\,\mathrm{cm}^{-2}$	8.0	5.8	2.8	2.2	1.5	5.1	2.7	2.7

Table 4.17 gives the values of the boundary number density of atoms on the basis of this formula, and the values of the gas-kinetic cross sections σ_g are taken from Table 4.5. Note that the boundary cluster radius does not depend on the cluster material.

4.4.6
Ambipolar Diffusion

Violation of plasma quasineutrality creates strong electric fields within the plasma. The fields between electrons and ions are associated with attractive forces, tending to move them together. The specific regime of plasma expansion in a gas results from a higher mobility of electrons than ions. But separation of electrons and ions in a gas creates an electric field that slows the electrons and accelerates the ions. This establishes a self-consistent regime of plasma motion called ambipolar diffusion and is typical for gas discharge [95, 96]. We shall examine this regime and establish the conditions necessary to achieve it.

We consider transport of electrons and ions in a weakly ionized gas in which the number densities of electrons and ions are similar and are small compared with the number density of atoms. In this case fluxes of electrons $\mathbf{j}_e$ and ions $\mathbf{j}_i$ are given by

$$\mathbf{j}_e = -K_e N_e \mathbf{E}_a - D_e \nabla N_e\,, \quad \mathbf{j}_i = K_i N_i \mathbf{E}_a - D_i \nabla N_i\,. \tag{4.117}$$

Here N_e and N_i are the number densities of electrons and ions, K_e and D_e are the mobility and the diffusion coefficient of electrons, K_i and D_i are the mobility and diffusion coefficients of ions, $\mathbf{E}_a$ is the electric field strength due to separation of electron and ion charges, and we take into account that the force from this electric field acts on electrons and ions in different directions. Since the quasineutrality of this plasma is conserved in the course of its evolution, the electron and ion fluxes are equal: $\mathbf{j}_e = \mathbf{j}_i$. From this it follows that because $K_e \gg K_i$ and $D_e \gg D_i$, in scales of electron values we have $\mathbf{j}_e = 0$. From this we find the electric field strength created by a charge separation:

$$\mathbf{E}_a = \frac{D_e}{K_e} \cdot \frac{\nabla N_e}{N_e}\,. \tag{4.118}$$

Taking the plasma flux as

$$\mathbf{j} = \mathbf{j}_i = -D_a \nabla N \tag{4.119}$$

and defining in this manner the coefficient D_a of ambipolar diffusion and accounting for the plasma quasineutrality, one finds for the coefficient of ambipolar diffusion $N_e = N_i = N$

$$D_a = D_i + \frac{D_e}{K_e} K_i \,. \tag{4.120}$$

In the case of a Maxwell distribution function for electrons and ions that corresponds to the regime of a high number density of electrons, this formula has the form

$$D_a = D_i \left(1 + \frac{T_a}{T_i}\right) , \tag{4.121}$$

where T_a and T_i are the temperatures of electrons and ions, and the effective electron temperature due to ambipolar diffusion is

$$T_a = \frac{e D_e}{d w_e / d E_a} , \tag{4.122}$$

where w_e is the electron drift velocity under the action of the ambipolar electric field strength that occurs due to charge separation because of ambipolar diffusion.

If external fields are absent and electrons are characterized by the Maxwell distribution function with a gas temperature, we have

$$T_a = T \,, \quad D_a = 2 D_i \,.$$

In the case of the Maxwell distribution function for electrons with electron temperature T_e, the coefficient of ambipolar diffusion is given by

$$D_a = D_i \left(1 + \frac{T_e}{T_i}\right) . \tag{4.123}$$

If a plasma in located in an electric field and ambipolar diffusion proceeds in the field direction z, we have for the total electric field in a plasma $E = E_z + E_a$, where E_z is an external electric field, and the effective temperature of ambipolar diffusion is

$$T_a^{\parallel} = \frac{e D_e}{d w_e / d E} , \tag{4.124}$$

where $T_a^{\parallel}$ accounts for the direction of an ambipolar field along an external field, and D_e is the coefficient of longitudinal diffusion of electrons in an external electric field. This type of ambipolar diffusion is realized in gas discharge near the anode. Note that in a general case this ambipolar temperature T_a differs from the Townsend characteristic electron temperature T_{ef} defined according to (4.82):

$$T_{ef} = \frac{e D_e}{K_e} = \frac{e D_e E}{w_e} \,.$$

In particular, in the helium case and for strong electric field strengths, when a typical electron energy exceeds the thermal atom energy significantly, we have $w_e \propto \sqrt{E}$ accounting for the cross section of electron–atom collision being independent of the collision velocity. For the ambipolar temperature this gives $T_a^{\parallel} = 2T_{ef}$.

We now consider ambipolar diffusion in the regime of a low number density of electrons for a gas discharge plasma that is located in a cylindrical discharge tube, and ambipolar diffusion of the plasma leads to plasma motion to the walls. Correspondingly, the ambipolar electric field E_a that is responsible for plasma motion to the walls is directed perpendicular to the discharge electric field. For the total electric field strength E this gives

$$E = \sqrt{E_z^2 + E_a^2} \,,$$

where E_z is the electric field strength of the discharge that is directed along the z-axis of a discharge tube. From this we have for the ambipolar temperature $T_a^{\perp}$ when ambipolar diffusion is directed perpendicular to an external electric field

$$T_a^{\perp} = \frac{e D_e}{w_a / E_a} = \frac{e D_e \sqrt{E^2 - E_z^2}}{w_e^2(E) - w_e^2(E_z)} = T_{ef} \sqrt{\frac{K_e}{\partial w_e / \partial E}} \,, \tag{4.125}$$

where T_{ef} is given by (4.82). As is seen, we have different ambipolar temperatures for electron drift along the electric field and perpendicular to it in the case of high electric field strengths and in the case of a nonlinear dependence for the electron drift velocity as a function of the electric field strength. In particular, returning to the helium case, we find $T_a^{\perp} = T_{ef}\sqrt{2}$. Note that this type of ambipolar diffusion is of importance for plasma drift to the walls of a discharge tube.

To demonstrate the characteristics of ambipolar diffusion in real gases, we evaluate below the ambipolar diffusion coefficient of helium at room temperature and an electric field strength of 1 and 10 Td. At room temperature and low electric field strengths He_2^+ is the basic type of ion, and its mobility in helium in the limit of low electric field strengths is $17\,\text{cm}^2/(\text{V s})$ at the normal number density of atoms according to the data in Table 4.11. This gives drift velocities of ions of 4.6×10^3 and 4.6×10^4 cm/s at the electric field strengths indicated if we consider the limit of low electric field strengths. Since the thermal velocity of helium atoms $v_T = \sqrt{8T/\pi m_a}$ is 1.3×10^5 cm/s, the limit of low electric field strengths is suitable for ions. The effective Townsend temperature T_{ef} defined by (4.82) is correspondingly 3.7 and 7.7 eV according to measurements [37]. From this it follows that the parameter T_{ef}/T is 140 and 300 at the above-mentioned electric field strengths. Accounting for $w_E \propto E^{1/2}$ in the helium case, we find the ratio $D_a/D_i = T_a/T$ is 200 and 420 at the indicated electric field strengths. This testifies to the specific character of ambipolar diffusion in atomic (and molecular) gases. Since the elastic electron–atom (or electron–molecule) scattering dominates over a wide energy range, and exchange of energy in a single collision is small, an electron may have enough energy at electric field strengths, where ions have the thermal energy. Then the second term in (4.120) exceeds the first term significantly, and $D_a \gg D_i$.

Let us find the criterion of the ambipolar diffusion regime for plasma propagation. We denote by L a typical distance over which plasma parameters vary remarkably, and this distance exceeds significantly the mean free path of charged particles in the gas. From the Poisson equation we have for a typical difference between the electron and ion number densities

$$\Delta N \equiv |N_i - N_e| \sim \frac{E}{4\pi e L} .$$

The estimate for the electric field strength resulting from charge separation has the form

$$\frac{\Delta N}{N} \sim \left(\frac{r_D}{L}\right)^2 ,$$

where r_D is the Debye–Hückel radius (1.13). From this it follows that quasineutrality of this plasma $\Delta N \ll N$ corresponds to the criterion $r_D \ll L$, that is, the criterion of plasma existence.

4.4.7
Double Layer

When Langmuir [97, 98] identified plasma as a specific type of gas, he was guided by gas discharge and divided the gas inside a gas discharge tube into two parts, the plasma and the plasma sheath. The plasma is a uniform quasineutral gas including electrons and ions, and according to Langmuir [97], "we shall use the name plasma to describe this region containing balanced charges of ions and electrons." The plasma sheath is formed near electrodes and according to the Langmuir analysis of the positive column of a mercury arc in 1923 [99], "Electrons are repelled from the negative electrode while positive ions are drawn towards it. Around each negative electrode there is thus a sheath of definite thickness containing only positive ions and neutral atoms."

Thus, we have that if a uniform quasineutral plasma involving electrons and ions is formed in some spatial region, the plasma becomes nonneutral at its boundary. In the case of gas discharge these boundaries relate to the electrodes and walls of the gas discharge chamber, and these regions are called the plasma sheath or double layer. As an example of the plasma type we analyze the plasma of the positive column of gas discharge within the framework of the Schottky model [100]. Then electrons and ions are formed in the positive column as a result of collision of electrons with gas atoms, and the loss of charged particles results from ambipolar diffusion of the plasma to the walls, and the ionization balance has the form

$$\frac{D_a}{\rho_0^2} \sim \nu_{ion} = N_a k_{ion} ,$$

where D_a is the coefficient of ambipolar diffusion, ρ_0 is the tube radius, ν is the ionization rate, N_a is the number density of gas atoms, k_{ion} is the rate constant for ionization in electron–atom collisions, and the mean free path of charged particles is small compared with the tube radius. This is the equation for the longitudinal electric field strength which determines the ionization rate compensated for by the departure of electrons and ions to the walls.

We now analyze the character of motion of electrons and ions to the walls. Electrons and ions attach to the walls and recombine there, and we have that the electron j_e and ion j_i fluxes to the walls are identical:

$$j_e = j_i \,. \tag{4.126}$$

To fulfill this, it is necessary that the electric field at the walls precludes approach of electrons to the walls, and this electric field arises if the walls are charged negatively. Next, this electric field is damped by the Debye screening by the plasma as we approach to the walls, and for definiteness we take the following relations between plasma parameters:

$$\rho_0 \gg \lambda \gg r_D \,,$$

where λ is the mean free path of electrons and ions and r_D is the Debye–Hückel radius. Under these conditions we have that the processes under consideration proceed in a flat layer near the walls and plasma quasineutrality occurs near the walls, where the field is absent. Then the action of the walls reduces to the potential U near the walls, which repels the electrons.

Thus, introducing the electron T_e and ion T_i temperatures, we have for the electron and ion fluxes to the walls

$$j_e = N_e \sqrt{\frac{T_e}{2\pi m_e}} \,, \quad j_i = N_i \sqrt{\frac{T_i}{2\pi m_i}} \,, \tag{4.127}$$

where N_e and N_i are the electron and ion number densities near the walls and m_e and m_i are the electron and ion masses. Equality of the electron and ion fluxes is governed by the wall electric potential U with respect to the plasma, which gives

$$N_e = N_i \exp\left(-\frac{eU}{T_e}\right) \,,$$

which gives the following expression for the wall potential:

$$eU = \frac{T_e}{2} \ln\left(\frac{T_e m_i}{T_i m_e}\right) \,. \tag{4.128}$$

Thus, the plasma sheath is formed at the plasma boundary, and its parameters may be different depending on the parameters of the ionized gas near the plasma boundary. One more effect may follow from the nature of the double layer where electron and ion fluxes toward the plasma boundary originate, because of interaction of these fluxes with a surrounding plasma, and this interaction may lead to instability of the double layer [101, 102].

4.4.8
Electrophoresis

A weakly ionized gas can be used to achieve the separation of isotopes and elements in a mixture of gases, by making use of the different currents for different types of charged atomic particles. As such an example, we consider electrophoresis in a gas discharge plasma. This phenomenon corresponds to a partial separation of the components of the system [103]. We take a gas in a cylindrical tube to consist of two components, a buffer gas (e.g., helium) and an admixture of easily ionized atoms (e.g., mercury, cadmium, zinc). The admixture atoms will give rise to an ionic component because of the low ionization potential of these atoms. The number densities of the components satisfy the inequalities $N \gg N_a \gg N_i$, where N is the number density of the atoms of the buffer gas, N_a is the number density of admixture atoms, and N_i is the number density of ions.

Because the ion current arises from the admixture ions, we have the balance equation

$$-D\frac{dN_a}{dx} + w_i N_i = 0$$

for the admixture, where D is the diffusion coefficient of the admixture atoms in the buffer gas, w_i is the ion drift velocity, and the x-axis is directed along the tube axis. Assuming the ion diffusion coefficient D_i and the diffusion coefficient D in a gas for admixture atoms to be the same order of magnitude $D_i \sim D$, and using the Einstein relation (4.38) for a typical size $L = (d \ln N_a/dx)^{-1}$ responsible for the gradient of the admixture number density, we find that

$$L \sim \frac{T}{eEc_i} ,$$

and admixture atoms are located in a region of size L. In this expression, $c_i = N_i/N_a$ is the degree of ionization of the admixture, T is the temperature of atoms or ions, and E is the electric field strength. If the size L is small compared with the discharge tube length, the admixture gas is gathered near the cathode.

Under electrophoresis conditions, a discharge plasma is nonuniform, and the glowing of this plasma due to radiation from the excited atoms of the admixture is concentrated near the cathode; this is of importance for gas lasers [104]. Electrophoresis is established for a time $\tau \sim L/(c_i w)$ after switching off the discharge. In fact, in a region near the anode, the number density of admixture atoms and ions decreases. An increase of the rate constant for ionization is required, so the electric field strength in this region must be increased. Thus, electrophoresis can change the parameters of a gas discharge.

4.4.9
Recombination of Positive and Negative Ions in Gases

The analysis of motion of ions in a gas in an external electric field allows us to describe the character of recombination of positive and negative ions in a dense gas. In this case, approach of ions with charges of different signs results from the action of the field of a given ion on the ion of the opposite sign, and approach of ions is braked due to gas resistance through ion collisions with gas atoms. As a result, the mobilities of ions under the action of the field of another ion give the rate of ion–ion recombination.

Indeed, if each ion has charge $\pm e$ and the distance between ions is R, the electric field of each ion at the location of another ion is $E = \pm e/R^2$, and this field causes ion approach, which is determined by the drift velocity:

$$w = E(K_+ + K_-) = \frac{eE(K_+ + K_-)}{R^2} .$$

This is valid for a dense gas if the criterion $R \gg \lambda$ holds true, and λ is the mean free path of ions in a gas.

To determine the decay rate for positive ions due to recombination, imagine a sphere of radius R around the positive ion, and compute the number of negative ions entering this sphere per unit time. This is given by the product of the surface area of the sphere, $4\pi R^2$, and the negative ion flux, $j = N_- w = N_- e(K_+ + K_-)/R^2$. From this we have the balance equation for the number density of positive ions is

$$\frac{dN_+}{dt} = -N_+ N_- 4\pi e(K_+ + K_-) .$$

Comparing this with the definition of the recombination coefficient, we derive for this coefficient the Langevin formula [105]:

$$\alpha = 4\pi e(K_+ + K_-) . \tag{4.129}$$

This formula holds true for a dense gas, and we now combine all the limiting cases for recombination of positive and negative ions. Below we summarize various cases of recombination of positive and negative ions in a gas.

Figure 4.21 gives the dependence of the recombination coefficient for positive and negative ions in a gas on the number density N of gas atoms, and we divide the range of the atom number density into three groups. At low number densities recombination of positive and negative ions in a gas has the pairwise character (2.98), and in accordance with (2.100) the value of the pairwise recombination coefficient in this range of densities is $\alpha_1 \geq \hbar^2/\left(m_e^2 \mu T\right)^{1/2}$, since $R_0 \gg a_0$. Here m_e is the electron mass, μ is the reduced mass of the ions, T is the gas (ion) temperature, and $a_0 = \hbar^2/(m_e e^2)$ is the Bohr radius. The three body process of ion–ion recombination (2.90) corresponds to region 2, and the recombination coefficient for positive and negative ions in a gas is determined on the basis of (2.91), $\alpha_2 \sim (Ne^6/T^3)(\beta e^2/\mu)^{1/2}$, where β is the polarizability of the gas atom, N is the number density of gas atoms, and μ is the reduced mass of the ion and atom. From this one can estimate the boundary number density N_1 of atoms for transition from region 1 to region 2 in Figure 4.21:

$$N_1 \sim a_0 \left(\frac{T}{e^2}\right)^{5/2} \beta^{-1/2} .$$

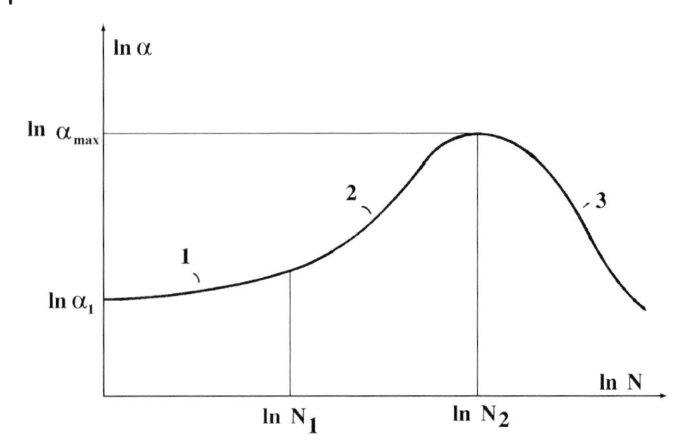

Figure 4.21 The recombination coefficient of positive and negative ions as a function of the gas density. 1 – region of pairwise ion–ion recombination; 2 – region of three body recombination of positive and negative ions in a gas; 3 – region of high pressures.

Region 3 in Figure 4.21 is described by the Langevin theory, and (4.129) gives the order-of-magnitude estimate $\alpha_3 \sim e/(N\sqrt{\beta\mu})$ for the recombination coefficient in this region if we assume that the polarization interaction potential determining the ion–atom scattering process acts. The transition between regions 2 and 3 is associated with the number density

$$N_2 \sim T^{3/2}/(e^3\sqrt{\beta}) ,$$

and this number density corresponds to the largest recombination coefficient, $\alpha_{\max} \sim e^4\mu^{-1/2}T^{-3/2}$. The relevant physical process is pairwise ion–ion recombination if the cross section corresponds to elastic scattering of particles experiencing the Coulomb interaction. Thus, the maximum recombination coefficient has the same order of magnitude as the elastic rate constant of ions.

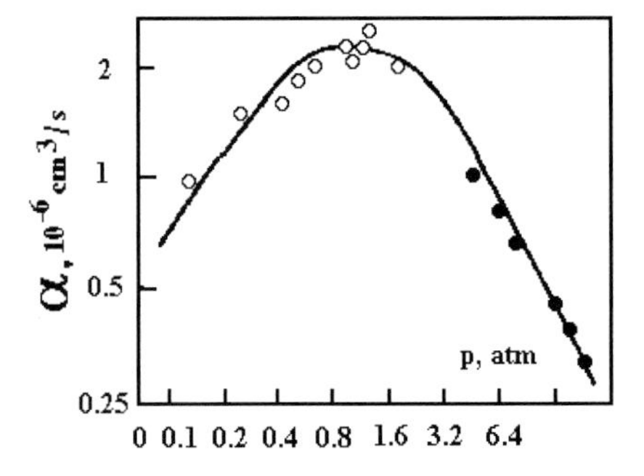

Figure 4.22 The recombination coefficient of positive and negative ions in air as a function of pressure. The curve is an approximation of experimental data given by open circles [106] and closed circles [107].

Typical values for the quantities indicated in Figure 4.21 for ions produced in air at room temperature are $N_1 \sim 10^{17}\,\mathrm{cm}^{-3}$, $N_2 \sim 10^{20}\,\mathrm{cm}^{-3}$, $\alpha_1 \sim (10^{-9}\text{–}10^{-7})\,\mathrm{cm}^3/\mathrm{s}$, and $\alpha_{\max} \sim 10^{-6}\,\mathrm{cm}^3/\mathrm{s}$. Figure 4.22 shows measured rate constants [106, 107] for recombination of positive and negative ions in air as a function of pressure. The maximum of the recombination coefficient can be seen to occur at atmospheric pressure.

4.5 Plasma in a Magnetic Field

4.5.1 Electron Hydrodynamics in a Gas in an External Field

We consider below the behavior of electrons in a gas in external fields. For simplicity, this analysis will be based on hydrodynamics of electrons in a gas in external fields, which corresponds to the tau approximation for the electron–atom collision integral (3.5) in the regime of a low electron number density, when one can ignore electron–electron collisions. In the tau approximation, the kinetic equation (3.4) for the distribution function for electrons in a gas in an external field has the form

$$\frac{\partial f}{\partial t} + \frac{\mathbf{F}}{m_e} \cdot \frac{\partial f}{\partial \mathbf{v}} = -\frac{f - f_0}{\tau} ,$$

where $\mathbf{F}$ is the force acting on the electron from external fields, m_e is the electron mass, $1/\tau = N_a v \sigma_{ea}$ is the rate of electron–atom collisions, N_a is the number density of atoms, σ_{ea} is the cross section for electron–atom collisions, and the collision time τ is assumed to be independent of the electron velocity. We take the force $\mathbf{F}$ acting on an electron in a general form:

$$\mathbf{F} = -e\mathbf{E}\exp(-i\omega t) - \frac{e}{c}[\mathbf{v} \times \mathbf{H}] ,$$

that is, the external field consists of a harmonic electric field and a constant magnetic field. In the standard method, by multiplication of the kinetic equation by the electron velocity and by subsequent integration over electron velocities, we obtain the equation of electron motion in external fields:

$$m_e \frac{d\mathbf{w}_e}{dt} = -m_e \frac{\mathbf{w}_e}{\tau} - e\mathbf{E}\exp(-i\omega t) - \frac{e}{c}[\mathbf{w}_e \times \mathbf{H}] , \tag{4.130}$$

where $\mathbf{w}_e$ is the electron drift velocity. As is seen, electron–atom collisions give rise to a frictional force $m_e \mathbf{w}_e/\tau$.

Let us write the motion equation in terms of its components. If the magnetic field $\mathbf{H}$ is directed along the z-axis and the electric field direction is in the xz plane ($\mathbf{E} = \mathbf{i}E_x + \mathbf{k}E_z$), (4.130) leads to the following set of equations:

$$\frac{dw_x}{dt} + \frac{w_x}{\tau} = a_x e^{i\omega t} + \omega_H w_y \,, \quad \frac{dw_y}{dt} + \frac{w_y}{\tau} = -\omega_H w_x \,,$$
$$\frac{dw_z}{dt} + \frac{w_z}{\tau} = a_z e^{i\omega t} \,, \tag{4.131}$$

where the Larmor frequency is $\omega_H = eH/(m_e c)$ and $a_x = -eE_x/m_e$, $a_z = -eE_z/m_e$. This set of equations has the following steady-state solution that is independent of the initial conditions:

$$w_x = \frac{\tau(1+i\omega\tau)a_x e^{i\omega t}}{1+\left(\omega_H^2-\omega^2\right)\tau^2+2i\omega\tau} \,, \quad w_y = \frac{\omega_H \tau^2 a_x e^{i\omega t}}{1+\left(\omega_H^2-\omega^2\right)\tau^2+2i\omega\tau} \,,$$
$$w_z = \frac{\tau a_z e^{i\omega t}}{1+i\omega\tau} \,. \tag{4.132}$$

Expressions (4.132) for the electron drift velocity allow one to determine the conductivity tensor $\Sigma_{\alpha\beta}$ that is introduced in terms of a generalized Ohm's law that establishes the connection between components the current density vector i_α and the electric field strength vector E_β in the form

$$i_\alpha = \Sigma_{\alpha\beta} E_\beta \,,$$

where $\Sigma_{\alpha\beta}$ is the conductivity tensor. This equation is valid at low electric field strengths, so the electron drift velocity is small compared with electron thermal velocity. Because the electron current density is

$$i_\alpha = eN_e w_e \,,$$

we have the general expression for the conductivity tensor of an ionized gas:

$$\Sigma_{\alpha\beta} = \Sigma_0 \begin{pmatrix} \frac{1+i\omega\tau}{1+\left(\omega_H^2-\omega^2\right)\tau^2+2i\omega\tau} & \frac{\omega_H\tau}{1+\left(\omega_H^2-\omega^2\right)\tau^2+2i\omega\tau} & 0 \\ -\frac{\omega_H\tau}{1+\left(\omega_H^2-\omega^2\right)\tau^2+2i\omega\tau} & \frac{1+i\omega\tau}{1+\left(\omega_H^2-\omega^2\right)\tau^2+2i\omega\tau} & 0 \\ 0 & 0 & \frac{1}{1+i\omega\tau} \end{pmatrix} . \tag{4.133}$$

where $\Sigma_0 = N_e e^2 \tau / m_e$ is the plasma conductivity in a constant electric field for the tau approximation in accordance with (3.42). This expression may be used for the analysis of the plasma interaction with external fields.

4.5.2
Hall Effect

In the absence of a magnetic field, the plasma conductivity is a scalar quantity. When a magnetic field is applied to a weakly ionized gas, the conductivity acquires a tensor character. This means that an electric current can occur in directions in which the electric field component is zero. If we consider the case where a constant electric field strength is perpendicular to a magnetic field, then (4.133) yields ($\omega = 0$)

$$\Sigma_{xx} = \Sigma_{yy} = \Sigma_0 \frac{1}{1+\omega_H^2\tau^2} \,, \quad \Sigma_{yx} = -\Sigma_{xy} = \Sigma_0 \frac{\omega_H \tau}{1+\omega_H^2\tau^2} \,. \tag{4.134}$$

In the limit case $\omega_H \tau \gg 1$, the total current is directed perpendicular to both the electric and the magnetic fields. In this case the plasma conductivity and electric current do not depend on the collision time because the change of the direction of electron motion is determined by the electron rotation in a magnetic field. In this case Ohm's law has the form

$$i_y = ecN_e \frac{E_x}{H} .$$

We consider the case $\omega_H \tau \ll M/m_e$, where M is the atom mass when the variation of the electron energy is small during the rotation period in the magnetic field. If the transverse electric current does not reach the plasma boundary, this results in a separation of charges that in turn creates an electric field that slows and eventually stops the electrons. This gives rise to an electric current in the direction perpendicular to the electric and magnetic fields that is the Hall effect [108].

We now consider the kinetic description of the Hall effect. By analogy with the case when an electron is moving in a gas in an electric field and the electron distribution function has the form (3.10), let us represent the electron distribution function as

$$f(\mathbf{v}) = f_0(v) + v_x f_1(v) + v_y f_2(v) .$$

The electric and magnetic fields are taken to be constant and mutually perpendicular, with the electric field along the x-axis and the magnetic field along the z-axis. By analogy with the case of the constant electric field, when the electron drift velocity is given by (3.26), for the electron distribution functions we now have

$$v f_1 = \frac{a v}{(\nu^2 + \omega_H^2)} \frac{d f_0}{d v} , \quad v f_2 = \frac{a \omega_H}{(\nu^2 + \omega_H^2)} \frac{d f_0}{d v} , \tag{4.135}$$

where $a = eE/m_e$, and $\nu = N_a v \sigma^*_{ea}$ is the rate of electron collisions with atoms. These equations lead to the following expressions for the components of the electron drift velocity:

$$w_x = \frac{eE}{3m_e} \left\langle \frac{1}{v^2} \frac{d}{dv} \left(\frac{\nu v^2}{\nu^3 + \omega_H^2} \right) \right\rangle , \quad w_y = \frac{eE}{3m_e} \left\langle \frac{1}{v^2} \frac{d}{dv} \left(\frac{\omega_H v^3}{\nu^2 + \omega_H^2} \right) \right\rangle . \tag{4.136}$$

In the limit $\omega_H \ll \nu$, the first of these expressions transforms into (3.26).

In the case when the rate of electron–atom collisions $\nu = 1/\tau$ is independent of the electron velocity or for a hydrodynamic description of electron motion in crossed electric and magnetic field we have in the limit $\omega_H \tau \gg 1$

$$w_y = -\frac{eE_x}{m_e \omega_H} = -c\frac{E_x}{H} , \quad w_x = \frac{w_y}{\omega_H \tau} . \tag{4.137}$$

As is seen, in the limit $\omega_H \tau \gg 1$ an electron (or a charged particle) is moving in crossed fields in the direction that is perpendicular to the field directions, and the electron drift velocity is proportional to the electric field strength and inversely proportional to the magnetic field strength. One can represent the drift velocity **w** of a charged particle in crossed electric **E** and magnetic **H** fields in the form

$$\mathbf{w} = c\frac{[\mathbf{E} \times \mathbf{H}]}{H^2} . \tag{4.138}$$

Let us find the electron temperature when an ionized gas is located in crossed electric and magnetic fields. We use the balance equation for the electron energy that has the form

$$eEw_x = \int \frac{m_e v^2}{2} I_{ea}(f_0) d\mathbf{v} .$$

Using (4.136) for the electron drift velocity and (3.18) for the collision integral, we obtain by analogy with the derivation of (3.36)

$$T_e - T = \frac{Ma^2}{3} \frac{\langle v^2 \nu / (\nu^2 + \omega_H^2) \rangle}{\langle v^2 \nu \rangle} . \tag{4.139}$$

Evidently, in the limit of low magnetic fields $\omega_H \ll \nu$ this formula is transformed into (3.36). In the case $\nu = \text{const}$, this expression gives

$$T_e - T = \frac{Ma^2}{3(\nu^2 + \omega_H^2)} , \tag{4.140}$$

and in the limit of strong magnetic field strengths $\omega_H \gg \nu$ we have from this

$$T_e - T = \frac{Ma^2}{3\omega_H^2} = \frac{Mc^2E^2}{3H^2} . \tag{4.141}$$

We now examine the case where a weakly ionized gas moves with an average velocity u in a transverse magnetic field of strength H. Then an electric field of strength $E' = Hu/c$ exists in the fixed frame of axes, where c is the light velocity. This field creates an electric current that is used for the production of electric energy in magnetohydrodynamic generators [109–111]. The energy released in a plasma under the action of this electric current corresponds to transformation of the flow energy of a gas into electric and heat energies. In consequence, this process leads to a deceleration of the gas flow and a decrease of the average gas velocity. In addition, the generation of an electric field causes an increase in the electron temperature as given by (4.141). The maximum increase in the electron temperature corresponds to the limit $\omega_H \gg \nu$. In this limit, (4.141) becomes

$$T_e - T = \frac{Ma^2}{3\omega_H^2} = \frac{1}{3}Mu^2 . \tag{4.142}$$

This value follows from (3.36) if electrons are in a gas in an induced electric field of strength $E' = Hu/c$.

4.5.3
Cyclotron Resonance

One more interesting effect takes place when electrons are in a gas in a strong magnetic field $\omega_H \tau \gg 1$ and in an alternating electric field of high frequency $\omega\tau \gg 1$. Then the behavior of electrons may have a resonant character. Indeed, according to (4.133) for the conductivity tensor, we have for its components in the directions parallel $\Sigma_{\parallel}$ and perpendicular $\Sigma_{\perp}$ to the magnetic field

$$\Sigma_{\parallel} = \Sigma_0 \frac{1 + i\omega\tau}{1 + \left(\omega_H^2 - \omega^2\right)\tau^2 + 2i\omega\tau}\,,$$
$$\Sigma_{\perp} = \Sigma_0 \frac{\omega_H \tau}{1 + \left(\omega_H^2 - \omega^2\right)\tau^2 + 2i\omega\tau}\,. \tag{4.143}$$

The conductivity is seen to have a resonance (so-called cyclotron resonance) at $\omega = \omega_H$ of the width $\Delta\omega \sim 1/\tau$, and the components of the conductivity tensor are $\Sigma_{\parallel} = i\Sigma_{\perp} = \Sigma_0/2$ at the resonance.

The cyclotron resonance has a simple physical explanation. In a magnetic field an electron travels in a circular orbit with cyclotron frequency ω_H, and if a harmonic electric field is applied in the plane of the circular orbit, the electron continuously receives energy from the field if the frequency of the electric fields coincides with the electron cyclotron frequency. Similar to motion in a constant electric field, the electron is accelerated until it collides with atoms. Hence, in both cases the conductivities are of the same order of magnitude and are expressed in terms of the rate $1/\tau \ll \omega$ of collisions between the electron and atoms. If the field frequency ω differs from the cyclotron frequency ω_H, the conductivity is considerably lower, since the conditions of interaction between the electron and the field are not optimal.

We can analyze the cyclotron resonance from the standpoint of energy absorption. The power per unit volume absorbed by a plasma is $p = \mathbf{j}\cdot\mathbf{E}$. If we take the x-axis to be in the direction of the electric field, then the specific power absorbed by the plasma is

$$p = \left(\Sigma_{xx} + \Sigma_{xx}^*\right)\frac{E^2}{2}$$
$$= \frac{p_0}{2}\left[\frac{1 - i\omega\tau}{1 + \left(\omega_H^2 - \omega^2\right)\tau^2 - 2i\omega\tau} + \frac{1 + i\omega\tau}{1 + \left(\omega_H^2 - \omega^2\right)\tau^2 + 2i\omega\tau}\right], \tag{4.144}$$

where $p_0 = \Sigma_0 E^2$ is the specific absorbed power for the constant electric field. Absorption of energy by a plasma is accompanied subsequently by electron–atom collisions that lead to energy transfer to the atoms.

In the region of the cyclotron resonance, with $\omega\tau \gg 1$, $\omega_H\tau \gg 1$, and $(\omega - \omega_H)\tau \sim 1$, (4.144) yields

$$p = \frac{p_0}{2}\left[\frac{1}{1 + \left(\omega_H^2 - \omega^2\right)\tau^2}\right]. \tag{4.145}$$

One can see that the resonant absorbed power is less than half of what is absorbed in a constant electric field. That the same order of magnitude is obtained for these values is explained by the related character of the electron motion in these cases.

4.5.4
Motion of Charged Particles in a Nonuniform Magnetic Field

We now analyze the behavior of a charged particle in a weakly nonuniform magnetic field. We consider a magnetic trap of axial symmetry with a higher magnetic field strength near its ends, as shown in Figure 4.23. In this case a charged particle may be reflected from the ends with a higher magnetic field and become locked in the trap. A weak nonuniformity of the magnetic field is governed by the criterion

$$r_{\mathrm{L}} \frac{\partial \ln H_z}{\partial z} \ll 1 , \quad r_{\mathrm{L}} \frac{\partial \ln H_\rho}{\partial \rho} \ll 1 . \tag{4.146}$$

A charged particle is moving along a helical trajectory near the axis. In addition, we assume that the presence of charged particles does not influence the character of the magnetic field because of their small density, and the spatial variation of the magnetic field is subjected to the Maxwell equation $\operatorname{div} \mathbf{H} = 0$. Because of the axial symmetry, $\operatorname{div} \mathbf{H} = 0$ can be written as

$$\frac{1}{\rho} \frac{\partial}{\partial \rho} (\rho H_\rho) + \frac{\partial H_z}{\partial z} = 0 , \tag{4.147}$$

where $H_\rho = 0$ at the axis. Near the axis, (4.146) gives

$$H_\rho = -\frac{\rho}{2} \left(\frac{\partial H_z}{\partial z} \right)_{\rho=0} .$$

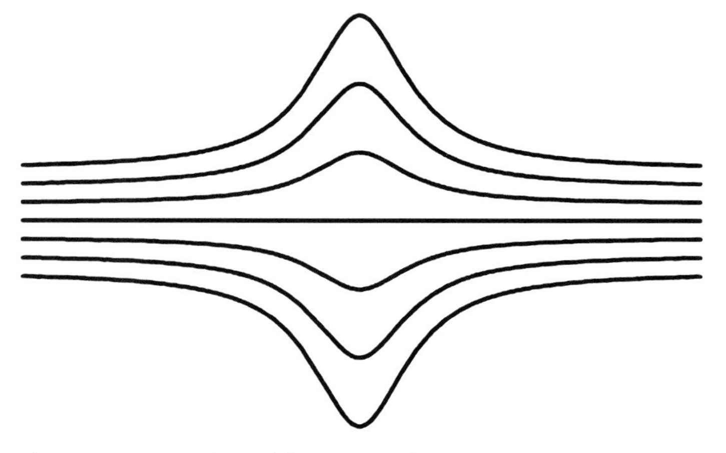

Figure 4.23 Magnetic lines of force in a simple magnetic trap with axial symmetry. Charged particles may be locked in this trap.

Because the magnetic field is directed along the axis z, the force acting on a charged particle is given by

$$F_z = -\frac{e}{c} v_\tau H_\rho = \frac{v_\tau \rho}{2c} \frac{\partial H_z}{\partial z} .$$

Let us introduce the particle magnetic moment μ of a moving particle of charge e in the usual way as

$$\mu = \frac{IS}{c} , \tag{4.148}$$

where I is the particle current, S is the area enclosed by its trajectory, and c is the velocity of light. Under the given conditions we have $I = e\omega_H/(2\pi)$ and $S = \pi r_L^2$, where r_L is the Larmor radius of the particle. Hence, we have

$$\mu = \frac{e\omega_H r_L^2}{2c} = \frac{m v_\tau^2}{2H} ,$$

where $v_\tau = r_L \omega_H$ is the tangential component of the velocity of the charged particle, and the Larmor frequency is $\omega_H = eH/mc$. Near the axis, where $H_z \gg H_\rho$, we obtain

$$\mu H_z = \frac{m v_\tau^2}{2} . \tag{4.149}$$

The force acting on the particle in the magnetic field direction is given by

$$F_z = -\mu \frac{\partial H_z}{\partial z} .$$

The minus sign means that the force is in the direction of decreasing magnetic field.

Let us prove that the magnetic moment of the charged particle is an integral of the motion, that is, it is a conserved quantity. One can analyze the particle motion along a magnetic line of force when it is averaged over gyrations. The equation of motion along the magnetic field gives $m dv_z/dt = F_z = -\mu dH_z/dz$, and since $v_z = dz/dt$, it follows from this that $d(mv_z^2/2) = -\mu dH_z$. From the energy conservation condition for the particle, we have accounting for (4.149)

$$\frac{d}{dt}\left(\frac{mv_\tau^2}{2} + \frac{mv_z^2}{2}\right) = 0 = \frac{d}{dt}\left(\mu H_z + \frac{mv_z^2}{2}\right)$$
$$= \frac{d}{dt}(\mu H_z) - \mu \frac{dH_z}{dt} = H_z \frac{d\mu}{dt} = 0 .$$

This yields the equation for the magnetic momentum of the particle motion,

$$\frac{d\mu}{dt} = 0 , \tag{4.150}$$

so the magnetic moment is conserved during the motion of a charged particle in a weakly nonuniform magnetic field. It should be noted that in deriving this formula, we assume $H_z \gg H_\rho$. Therefore, in analyzing the motion of a charged particle in a weakly nonuniform magnetic field, for a given spatial region we choose the magnetic field to be directed almost along the z-axis .

Let us analyze the motion of a charged particle in a weakly nonuniform magnetic field along the magnetic field direction, which is taken as the z-axis. A particle is moving along a helical trajectory, and the motion equation has the form

$$m_e \frac{dv_z}{dt} = F_z = -\mu \frac{\partial H_z}{\partial z} = \frac{\varepsilon_\tau}{H} \frac{\partial H_z}{\partial z} ,$$

where ε_τ is the transverse kinetic energy of the charged particle, and we use the definition (4.149) of the longitudinal force and the value of the magnetic momentum of the charged particle. Multiplying the motion equation by the particle velocity along the magnetic field v_z, we reduce this equation to the form

$$\frac{d\varepsilon_z}{dt} = -\frac{\varepsilon_\tau}{H} \frac{dH_z}{dt} .$$

On the basis of the conservation of the total electron kinetic energy $\varepsilon_z + \varepsilon_\tau = \text{const}$ of the charged particle, we obtain from this equation

$$\frac{\varepsilon_\tau}{H} = \text{const} . \qquad (4.151)$$

Thus, the magnetic field gradient leads to exchange between the the kinetic energies of the charged particle in the longitudinal and transverse directions. If the particle is moving in the direction of a magnetic field increase, the Larmor radius increases in the course of this motion, as shown in Figure 4.24. Note that (4.151) is analogous to (4.150). Indeed, the magnetic moment of an electron that is moving along a circular trajectory is

$$\mu = \frac{IS}{c} = \frac{\varepsilon_\tau}{H} ,$$

where the electric current is $I = e\omega_H/(2\pi)$, and the area for this current is $S = \pi r_L^2$, where r_L is the Larmor radius. Therefore, (4.151) corresponds to conservation of the electron magnetic moment (4.149) in the course of exchange between the transverse and longitudinal electron kinetic energies of the particle.

Another example of the behavior of a charged particle in a weakly nonuniform magnetic field is its drift due to a magnetic field gradient. Indeed, let the magnetic field be directed along the z-axis and be decreased weakly in the y direction. A

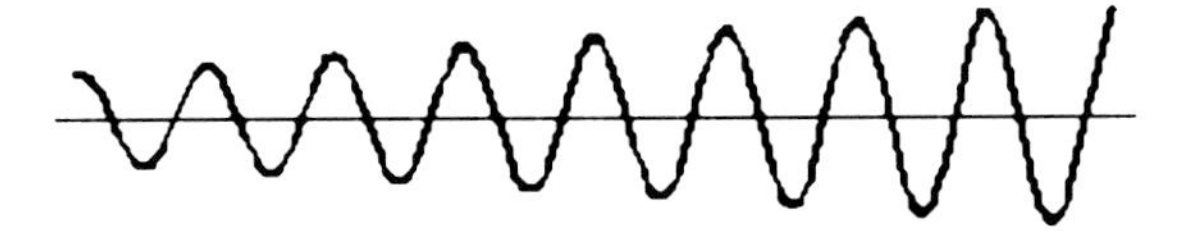

Figure 4.24 Trajectory of a charged particle moving in a weakly nonuniform magnetic field along the magnetic field. Because of a variation of Larmor radius of the particle, its drift takes place in accordance with (4.151).

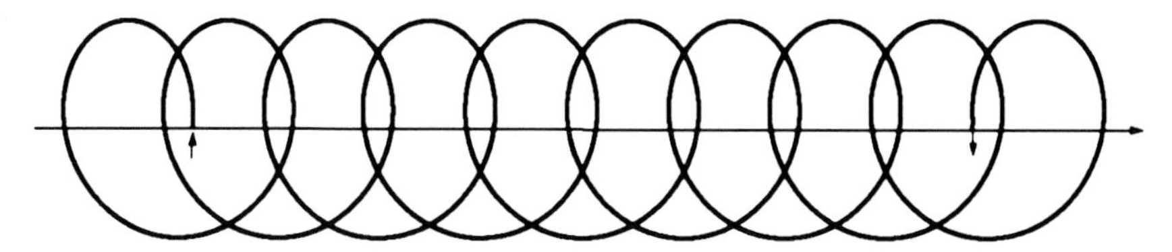

Figure 4.25 Trajectory of a charged particle moving along the gradient of the magnetic field that is directed perpendicular to the magnetic field.

charged particle rotates clockwise in the xy plane, so the upper Larmor radius of particle rotation is larger than the lower one. As a result, a particle moves in the x direction, as shown in Figure 4.25.

Let us rewrite (4.138) in the form

$$\mathbf{w} = c\frac{[\mathbf{F} \times \mathbf{H}]}{emH^2},$$

where $\mathbf{F}$ is the force acting on a charged particle of charge e and mass m. Because $\mathbf{F} = -\mu\nabla H$ and $\mu = mv^2/2H$, we obtain

$$\mathbf{w} = \frac{v^2}{2\omega_H}\frac{[\nabla\mathbf{H} \times \mathbf{H}]}{H^2}. \tag{4.152}$$

Hence, if the magnetic field is directed along the z-axis and it varies in the y direction, a charged particle is moving in the x direction. From this it follows that if the magnetic field strength has a maximum at the z-axis and decreases in the ρ direction, the particle moves along the z-axis with a helical trajectory.

4.5.5 Magnetic Traps

We can now analyze the motion of a charged particle in a magnetic trap. If the magnetic field has the form given in Figure 4.26 and the particle is moving from the region of a weak magnetic field, the region of increasing magnetic field acts on the charged particle like a magnetic mirror, and the particle reflects from this region and returns. Indeed, as the magnetic field increases, some of the particle's kinetic energy $mv_\tau^2/2$ that corresponds to the particle's motion in the direction perpendicular to the magnetic field increases proportionally. If this value becomes as large as the total initial kinetic energy, motion of the particle along the z-axis stops, and the particle starts to move in the opposite direction. This is the principle of a magnetic mirror in which a particle moves along a magnetic line of force between a region with a weak magnetic field and a region with a strong magnetic field. We take $H_{\min}$ to be the minimum magnetic field and $H_{\max}$ to be the maximum magnetic field along this magnetic line of force; θ is the angle between the particle velocity and magnetic field line at the field minimum. Then if this angle exceeds θ_m, given by the relation

$$\sin^2\theta_m = H_{\min}/H_{\max}, \tag{4.153}$$

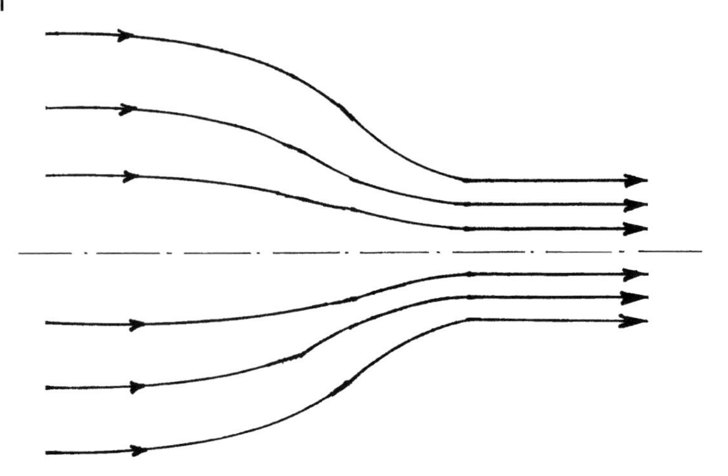

Figure 4.26 Magnetic mirror for a charged particle that moves along a varying magnetic field and reflects from a region with a high magnetic field.

the particle reflects from the region of the strong magnetic field and is trapped in a bounded space. The above principle is the basis of various magnetic traps. It acts for both positively and negatively charged particles, so this arrangement works for both electrons and protons.

This principle is used in magnetic traps, and a simple magnetic trap is represented in Figure 4.26. If a charged particle is located in such a trap, and since (4.151) holds true in the course of particle motion, the particle reflects from trap ends of a high magnetic field and is locked in this magnetic trap. Then the lifetime of a charged particle in this magnetic trap is determined by its collisions with other particles (or by the action of other fields), and it is large compared with the time of one passage between trap ends.

Note that in this type of a trap, magnetic walls reflect charged particles. In a magnetic trap of another type, a charged particle moves along a closed trajectory under the action of the magnetic field, and the potential well for the particle is created by the magnetic field. Figure 4.27 represents the magnetic trap of magnetron discharge. In this case the magnetic field is created by two coaxial magnets with different directions of the poles, and the maximum magnetic field has the form of a ring over these magnets with the same axis. Stationary magnets allow one to reach a magnetic field strength of 100 G, and we will use this value for subsequent estimations. The Larmor frequency of electrons in this magnetic field is $\omega_H = 1.8 \times 10^9\,\mathrm{s}^{-1}$, and this value must exceed the rate of electron collisions with atoms in this trap.

Let us make some estimations for the magnetic trap of magnetron discharge. An electron drifts under the action of the magnetic and electric field moving over a ring, and the electron drift velocity w_ρ along the ring is given by (4.138), $w_\rho = cE_z/H$, where E_z is the projection of the electric field strength on the z-axis of this system. We have for the depth of the potential well $U_{\max} = \mu H$, and the magnetic

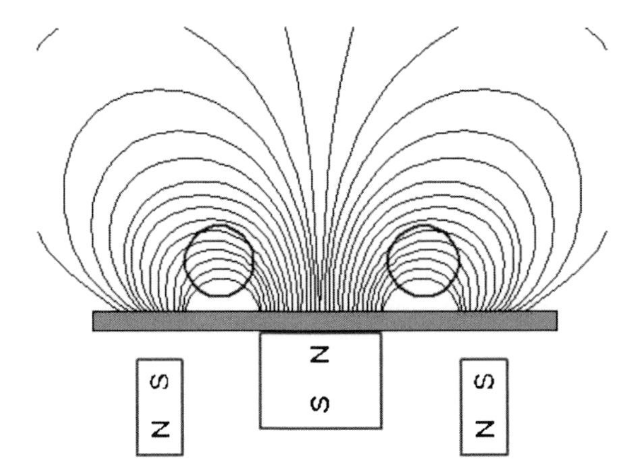

Figure 4.27 Magnetic lines of force in magnetron discharge. As a result of summation of magnetic fields of two magnets of axial symmetry, the total magnetic field has a maximum of circular shape. Electrons may be captured by this trap and do not partake in the ionization processes at the cathode.

moment μ for this electron motion is

$$\mu = \frac{IS}{c} = \frac{1}{c}\frac{ew_\rho}{2\pi r}\pi r^2 = \frac{eE_z r}{2H} ,$$

where $I = ew_\rho/(2\pi r)$ is the current created by this electron, r is the radius of the ring where the magnetic field has a maximum, and we use (4.138) for the electron drift velocity. For the depth $U_{\max}$ of the potential well due to action of the magnetic field this gives

$$U_{\max} = \frac{eE_z r}{2} . \tag{4.154}$$

For definiteness, we make an estimate for magnetron discharge in helium, being guided by the electron temperature $T_e = 3\,\text{eV}$. The criterion $\omega_H \gg \nu$ gives in this case $N_a \ll 3 \times 10^{16}\,\text{cm}^{-3}$ for the number density of helium atoms, which corresponds to a helium pressure of $p \ll 1\,\text{Torr}$ if the helium temperature is of the order of room temperature. Next, the ratio of the depth of the potential well $U_{\max}$ to the electron temperature T_e is

$$\frac{U_{\max}}{T_e} = \frac{3}{2}\frac{eH}{Mc}\frac{r_0}{c}\frac{H}{E} \gg 1 ,$$

and taking typical parameters of the magnetic trap for magnetron discharge, $H = 100\,\text{G}$ and $r = 3\,\text{cm}$, we obtain from $U_{\max} \sim T_e$ that $E_z \sim 2\,\text{V/cm}$. Correspondingly, on the basis of these parameters according to (4.141) we obtain $T_e \approx 3\,\text{eV}$ for the electron temperature.

We note the principal difference between the magnetic mirror shown in Figure 4.26 and the magnetic trap of magnetron discharge. There is a low plasma density in a magnetic mirror, and a charged particle travels between its ends (magnetic taps) many times. Let us make some estimations for the magnetic trap of magnetron discharge. We take for definiteness the helium pressure in magnetron discharge to be 0.1 Torr, which corresponds to a number density of helium atoms at room temperature of $N_a = 3 \times 10^{15}\,\text{cm}^{-3}$ and leads to the rate of electron–atom scattering of $\nu = 2 \times 10^8\,\text{s}^{-1}$. It is small compared with the Larmor frequency for electrons $\omega_H \approx 2 \times 10^9\,\text{s}^{-1}$, and hence electrons are magnetized, that is, they are captured by the magnetic trap. But they have a small lifetime in the magnetic trap. Indeed, the thermal electron velocity is $v_T \sim 10^8\,\text{cm/s}$, and the mean free path of electrons $\lambda \sim 0.5\,\text{cm}$ is small compared with the ring radius. In addition, the electron drift velocity in crossed electric and magnetic fields is in this case $v_\rho \approx 2 \times 10^6\,\text{cm/s}$ and is lower that the thermal electron velocity. Correspondingly, its shift under the action of a field is relatively small, approximately 0.01 cm, between neighboring electron collisions with atoms. Nevertheless, this magnetic trap increases the number density of electrons by the factor $\exp(U_{max}/T_e)$ in comparison with regions outside the action of the magnetic field. This is of importance for magnetron discharge, because a heightened plasma density is created in the magnetic trap region, and intense bombardment of the cathode by ions starts from this region.

4.5.6
Charge Particles in the Earth's Magnetic Field

The Earth's magnetic field is determined by internal Earth currents [112–114] which create the magnetic moment of the Earth, whose value is $M = 7.8 \times 10^{19}$ $\text{G}\,\text{m}^3$. The magnetic field strength $\mathbf{H}$ at distance R from the center of this magnetic dipole is

$$\mathbf{H} = \frac{3(\mathbf{MR})\mathbf{R} - \mathbf{M}R^2}{R^5}\,, \tag{4.155}$$

and Figure 4.28 shows the positions of magnetic lines of force for the Earth. As is seen, these lines are parallel to the Earth's surface near the equator, and the magnetic field strength at the equator near the Earth's surface is $H_0 = 0.31\,\text{G}$. On the basis of this formula we have for the magnetic field strength H at distance R from the Earth's center

$$H = H_0 \left(\frac{R_\oplus}{R}\right)^3 \sqrt{\frac{5 - 3\cos 2\theta}{2}}\,.$$

We assume here the Earth to be a ball of radius $R_\oplus = 6370\,\text{km}$, and θ is the latitude of a given point. Note that (4.155) in terms of components of the magnetic field vector $\mathbf{H}$ has the form

$$H_R = -\frac{2M\sin\theta}{R^3}\,, \quad H_\theta = \frac{M\cos\theta}{R^3}\,.$$

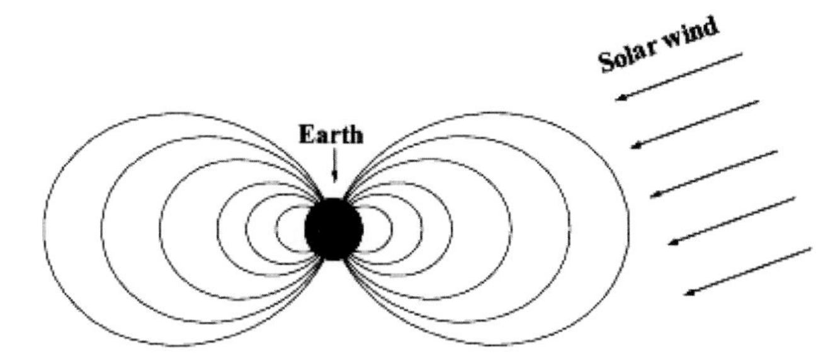

Figure 4.28 Magnetic lines of force near the Earth, which is a magnetic dipole.

This model of the Earth's magnetic field due to a magnetic dipole, as well as the ball model of Earth, is enough for the analysis of magnetic effects for the Earth's atmosphere.

The magnetic field of the Earth may prevent charged particles, both electrons and protons, penetrating the Earth's atmosphere. We assume a charged particle is moving along a magnetic line of force, as shown in Figure 4.29. Let us find at what altitudes this takes place. In considering magnetic traps for charged particles, we assume the Larmor frequency ω_H to be large compared with the rate ν of collisions between charged particles and atmospheric nitrogen molecules. For definiteness, we consider the motion of fast protons in the Earth's atmosphere, and take for estimation the gas-kinetic cross section $\sigma = 3 \times 10^{-15}\,\text{cm}^2$ as the cross section of proton–molecule collision. We have $\omega_H = 3 \times 10^3\,\text{Hz}$ for the Larmor frequency of protons in the magnetic field of the Earth near its surface, and the criterion $\omega_H \gg \nu$ holds true if the number density of molecules is less than $N_0 = 2 \times 10^{10}\,\text{cm}^{-3}$. This occurs at altitudes above 140 km. Thus, the character of proton motion in the Earth's atmosphere in accordance with Figure 4.29 takes place in the upper atmosphere.

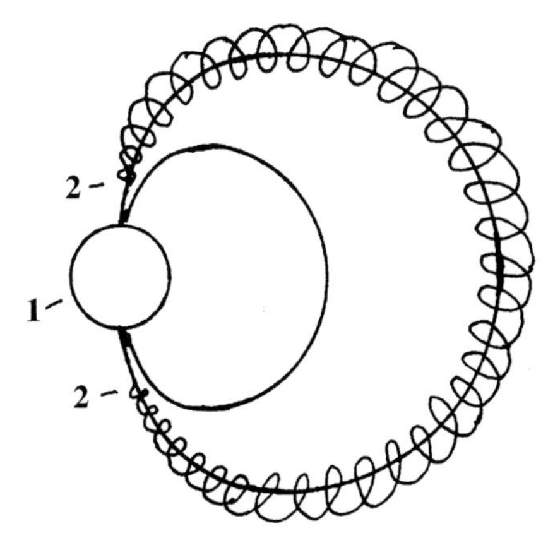

Figure 4.29 Trajectory of a charged particle along a magnetic line of force which is captured by the magnetic field of the Earth: 1 – the Earth; 2 – points of reflection of the particle.

Let us make simple estimations for capture of fast protons by the Earth's magnetic field when protons are moving as shown in Figure 4.29. For definiteness, we take a beam of protons of energy 100 keV that crosses magnetic lines of force at latitudes corresponding to the equator. This beam is captured by the magnetic field at distances R from the Earth where the Larmor radius r_L for protons is comparable to this distance. Because the proton velocity is $v_p = 4 \times 10^8$ cm/s at this energy, we obtain from the above criterion

$$R \sim \sqrt{\frac{\omega_0 R_\oplus^3}{v_p}} \sim 70 R_\oplus ,$$

where $\omega_0 = 3 \times 10^3$ Hz is the Larmor frequency at the equator near the Earth's surface and $R_\oplus$ is the Earth's radius. Being captured, protons are moving along an appropriate magnetic line of force and approach the Earth near its poles, where the Larmor radius of protons is $r_L \sim 1$ km. The protons may be reflected in regions of high magnetic field strength, as shown in Figure 4.29 or may penetrate the Earth's atmosphere depending on the angle with the magnetic line of force when they are captured by the Earth's magnetic field. If protons penetrate the Earth's atmosphere in the polar region, they their energy is partially used in excitation of atmospheric atoms and molecules and causes glowing of the atmosphere. In this manner auroras occur that are observed at latitudes near the Earth poles.

Along with the character of motion of charged particles in the Earth's atmosphere shown in Figure 4.28, when charged particles move along a helical trajectory, particle motion along an elliptic trajectory is possible if the particle energy is high enough. There are so-called radiation belts where fast electrons and protons are captured by the Earth's magnetic field. These regions are located at a distance of several Earth radii from the Earth's surface, and captured fast electrons and protons may have an energy of several million electronvolts. Although along with the Earth's magnetic field other factors provide the stability of the radiation belts, we now estimate the depth of the potential well for a captured electron. Using (4.148), we have for the maximum depth of the magnetic potential well U_{max} for an electron

$$U_{max} = \mu H = U_0 \left(\frac{R_\oplus}{R}\right)^4 do , \quad U_0 = e\omega_0 \cdot \pi R_\oplus^2 = 1.7 \times 10^{10}\ \text{MeV} .$$

Here R is the distance from the Earth's surface, do is the solid angle related to the electron trajectory, so that the area $S = \pi R^2 do$ is located inside the electron trajectory, and $\omega_0 = 5.5 \times 10^6\ \text{s}^{-1}$ is the Larmor frequency for an electron located near the equator and the Earth's surface. As is seen, the depth of the magnetic well in this case is large enough to trap electrons. Because of their larger mass, protons have a longer lifetime in the Earth's magnetic trap than do electrons, and hence the number of captured protons is greater than that for electrons if this lifetime is determined by collisions with atmospheric atoms. In addition, owing to the influence of the solar wind, this magnetic trap acts more effectively on the side of the Earth opposite the Sun.

4.5.7
High-Conductivity Plasma in a Magnetic Field

A plasma with high conductivity will have electrons with velocities considerably greater than the velocities of the ions. Then the electric current is created by electrons and is given by

$$\mathbf{j}_e = -eN_e\mathbf{w}_e \,,$$

where $\mathbf{w}_e$ is the drift velocity of the electrons and N_e is their number density. If the motion occurs in a magnetic field, an additional electric field is produced in the laboratory frame of axes, whose strength is

$$\mathbf{E}' = \frac{1}{c}[\mathbf{w}_e \times \mathbf{H}] = -\frac{1}{ecN_e}[\mathbf{j} \times \mathbf{H}] \,. \tag{4.156}$$

This field acts on the electrons, giving rise to an additional force acting on the entire plasma. The force per unit volume of the plasma is

$$e\mathbf{E}'N_e = -\frac{1}{c}[\mathbf{j} \times \mathbf{H}] \,.$$

If the plasma conductivity is sufficiently high, its response to the electric field (4.156) will result in the movement of electrons. This movement will continue until separation of the electrons and ions gives rise to an internal electric field in the plasma,

$$\mathbf{E} = -\frac{1}{c}[\mathbf{w}_e \times \mathbf{H}] \,, \tag{4.157}$$

which will compensate for the field (4.156). We insert (4.157) into the Maxwell equation $-(\partial\mathbf{H}/\partial t) = c \operatorname{curl} \mathbf{E}$, which yields

$$\frac{\partial \mathbf{H}}{\partial t} = \operatorname{curl}[\mathbf{w}_e \times H] \,. \tag{4.158}$$

We shall analyze the variation of the magnetic field and the plasma motion when the electric current is due to electrons and the plasma conductivity is high. We transform (4.158) by writing $\operatorname{curl}(\mathbf{w}_e \times \mathbf{H}) = \mathbf{w}_e \operatorname{div} \mathbf{H} + (\mathbf{H} \cdot \nabla)\mathbf{w}_e - (\mathbf{w}_e \cdot \nabla)\mathbf{H} - \mathbf{H} \operatorname{div} \mathbf{w}_e$. Using the Maxwell equation $\operatorname{div} \mathbf{H} = 0$, and taking the expression for $\operatorname{div} \mathbf{w}_e$ from the continuity equation for electrons, we obtain

$$\frac{\partial \mathbf{H}}{\partial t} - \frac{\mathbf{H}}{N_e}\frac{\partial N_e}{dt} + (\mathbf{w}_e \cdot \nabla)\mathbf{H} - \frac{\mathbf{H}}{N_e}(\mathbf{w}_e \cdot \nabla)N_e = (\mathbf{H} \cdot \nabla)\mathbf{w}_e \,.$$

We divide this equation by N_e and find that

$$\frac{d}{dt}\left(\frac{\mathbf{H}}{N_e}\right) = \left(\frac{\mathbf{H}}{N_e} \cdot \nabla\right)\mathbf{w}_e \,, \tag{4.159}$$

where

$$\frac{d}{dt}\left(\frac{\mathbf{H}}{N_e}\right) = \frac{\partial}{\partial t}\left(\frac{\mathbf{H}}{N_e}\right) + (\mathbf{w}_e \cdot \nabla)\frac{\mathbf{H}}{N_e}$$

is the derivative at a point that moves with the plasma.

To analyze the motion of an element of plasma volume with length $d\mathbf{l}$ and cross section $d\mathbf{s}$ containing $N_e d\mathbf{s} d\mathbf{l}$ electrons, we assume at first that the vector $d\mathbf{l}$ is parallel to the magnetic field $\mathbf{H}$, so the magnetic flux through this elementary plasma volume is $H ds$. If the plasma velocity at one end of the segment $d\mathbf{l}$ is $\mathbf{w}_e$, then at its other end the velocity is $\mathbf{w}_e + (d\mathbf{l} \cdot \nabla)\mathbf{w}_e$, so the variation of the segment length during a small time interval δt is $\delta t (d\mathbf{l} \cdot \nabla)\mathbf{w}_e$. Hence, the length of the segment satisfies the equation

$$\frac{d}{dt}(d\mathbf{l}) = (d\mathbf{l} \cdot \nabla)\mathbf{w}_e \ ,$$

which is identical to (4.159). From this it follows, first, that in the course of plasma evolution the segment $d\mathbf{l}$ has the same direction as the magnetic field and, second, that the length of the plasma element remains proportional to the quantity H/N_e, that is, the magnetic flux through this plasma element does not vary with time during the plasma motion [115]. Thus, the magnetic lines of force are frozen into the plasma [116], that is, their direction is such that the plasma electrons travel along these lines. This takes place in the case when the plasma conductivity is high. Interaction of charged particle flow and magnetic fields leads to specific forms of the plasma flow which are realized in both laboratory devices and cosmic and solar plasmas [109, 117–119].

To find the steady-state motion of a high-conductivity plasma, we start with (4.157), which gives the force on each plasma electron as

$$\mathbf{F} = -e\mathbf{E} = \frac{e}{c}[\mathbf{w}_e \times \mathbf{H}] = -\frac{1}{cN_e}[\mathbf{j} \times \mathbf{H}] \ .$$

Inserting into the expression for the force the current density $\mathbf{j} = (c/4\pi) \cdot \mathrm{curl}\, \mathbf{H}$, we obtain

$$\mathbf{F} = -\frac{1}{cN_e}[\mathbf{j} \times \mathbf{H}] = \frac{1}{4\pi N_e}[\mathbf{H} \times \mathrm{curl}\, \mathbf{H}] = \frac{1}{4\pi N_e}\left[\frac{1}{2}\nabla H^2 - (\mathbf{H} \cdot \nabla)\mathbf{H}\right] . \quad (4.160)$$

We now substitute (4.160) into the Euler equation and assume that the drift velocity of the electrons is considerably smaller than their thermal velocity. Hence, we can ignore the term $(\mathbf{w}_e \cdot \nabla)\mathbf{w}_e$ compared with the term $\nabla p/(m N_e)$, and obtain

$$\nabla\left(p + \frac{H^2}{8\pi}\right) - \frac{(\mathbf{H} \cdot \nabla)\mathbf{H}}{4\pi} = 0 \ . \quad (4.161)$$

The quantity

$$p_H = \frac{H^2}{8\pi} \quad (4.162)$$

is the magnetic field pressure or magnetic pressure; it is the pressure that the magnetic field exerts on the plasma.

4.5.8
Pinch Effect

We next analyze shrinkage under the action of a magnetic field for a cylindrical plasma column maintained by a direct current [120, 121]. Here the magnetic lines of force are cylinders, and because of the axial symmetry, (4.161) for the direction perpendicular to the field and current has the form

$$\nabla \left(p + \frac{H^2}{8\pi} \right) = 0 \,. \tag{4.163}$$

The solution of this equation shows that the total pressure $p + H^2/(8\pi)$, which is the sum of the gas-kinetic pressure and the magnetic field pressure, is independent of the transverse coordinate. Let the radius of the plasma column be a and the current in it be I, so the magnetic field at the surface of the column is $H = 2I/(ca)$. The total pressure outside the column near its surface is equal to the magnetic field pressure $I^2/(2\pi c^2 a^2)$, and the total pressure inside the plasma column is equal to the gas-kinetic pressure p. Equating these two pressures, we find the radius of the plasma column to be [121]

$$a = \frac{I}{c\sqrt{2\pi p}} \,. \tag{4.164}$$

An increase in the current of the plasma column is accompanied by a corresponding increase in the magnetic field, which gives rise to a contraction of the plasma column. This phenomenon is called the pinch effect, and the state of the plasma column itself is known as z-pinch.

4.5.9
Reconnection of Magnetic Lines of Force

In a cold plasma of high conductivity, magnetic lines of force are frozen in the plasma. This means that internal magnetic fields support electric currents inside the plasma. But plasma motion and the interaction of currents may cause a short circuit of some currents. This leads to an instability called the reconnection of magnetic lines of force [122]. As a result of this process, the energy of the magnetic fields is transformed into plasma energy in an explosive process that generates plasma fluxes. This phenomenon is observed in solar plasmas. Various solar plasma structures, such as spicules and prominence, result from this phenomenon.

We consider first a simple example of this phenomenon, when there are two antiparallel currents of amplitude I located a distance $2a$ apart, and with length $l \gg a$. We assume these currents have transverse dimensions that are small compared with a. Taking the direction of the currents to be along the z-axis, taking the plane of the currents to be the xz plane, and placing the origin of the coordinates in the middle of the currents, we find the magnetic field strength at distances $r \ll l$

from the origin to be

$$H_x = H_0 y \left[\frac{1}{(x-a)^2 + y^2} + \frac{1}{(x+a)^2 + y^2} \right],$$

$$H_y = -H_0 \left[\frac{x-a}{(x-a)^2 + y^2} + \frac{x+a}{(x+a)^2 + y^2} \right],$$

where $H_0 = 2I/(ca)$. From this it follows that the energy released from reconnection of these currents can be estimated to be

$$\varepsilon = \int \frac{H^2}{8\pi} d\mathbf{r} \sim \frac{I^2 l}{c^2}.$$

This means that the energy released per unit length of the conductors is constant near them.

This estimate can be used for understanding the general properties of a turbulent plasma of high conductivity in a magnetic field. The plasma is characterized by a typical drift velocity v of the electrons, and a typical length Δr over which this velocity varies. The energy of this plasma is contained both in the plasma motion and in its magnetic fields, which are of the order of $H \sim eN_e v/(c\Delta r)$, where N_e is the number density of electrons. As a result of reconnection of magnetic lines of force and reconnection of currents, transformation of the plasma magnetic energy into energy of plasma motion takes place, followed by an inverse transformation. In the end, these forms of energy will be transformed into heat. A typical time for reconnection is $\tau \sim \Delta r/v$. This plasma can be supported by external fields. Figure 4.30 gives an example of reconnection of magnetic lines of force. Released energy is transferred to a plasma and generates intense plasma flows. This phenomenon is of importance for the Sun's plasma. The resulting plasma flow inside the Sun creates a shock wave that is responsible for generation of X-rays. The plasma flows outside the Sun generate a hot plasma in the Sun's corona.

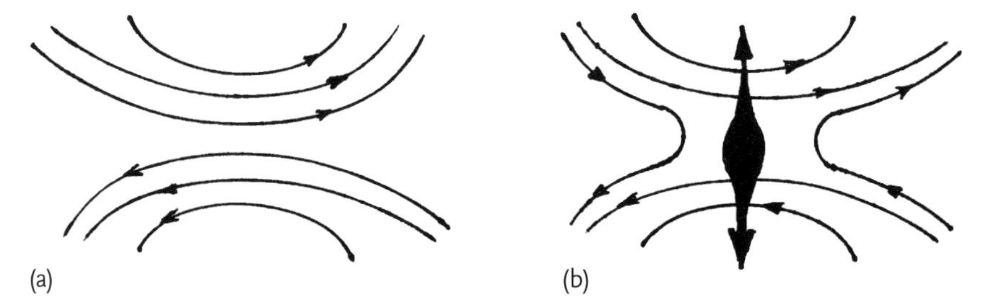

Figure 4.30 Reconnection of magnetic lines of force: (a) before reconnection; (b) after reconnection. A dark plasma which produces a pulse as a result of reconnection of magnetic lines of force moves in two opposite directions perpendicular to these lines.

References

1 Eckert, E.R.G. and Drake, R.M. (1972) *Analysis of Heat and Mass Transfer*, McGraw-Hill, New York.
2 Wetty, J.R. (1976) *Fundamentals of Momentum, Heat and Mass Transport*, John Wiley & Sons, Inc., New York.
3 Cengel, Y.A. (2003) *Heat and Mass Transfer. A Practical Approach*, McGraw-Hill, Boston.
4 Euler, L. (1757) *Mem. Acad. Sci. Berl.*, **11**, 217.
5 Euler, L. (1757) *Opera Omnia* Ser. 2, **12**, 2.
6 Ladyzhenskaya, O. (1969) *The Mathematical Theory of Viscous Incompressible Flows*, Gordon and Breach, New York.
7 Landau, L.D. and Lifshitz, E.M. (1980) *Statistical Physics*, vol. 1, Pergamon Press, Oxford.
8 Faber, T.E. (1995) *Fluid Dynamics for Physicists*, Cambridge University Press, New York.
9 Navier, C.L.M.H. (1822) *Mem. Acad. R. Sci. Inst. Fr.*, **6**, 389.
10 Stokes, G.G. (1845) *Trans. Camb. Philos. Soc.*, **8**, 287.
11 Chapman, S. and Cowling, T.G. (1952) *The Mathematical Theory of Non-uniform Gases*, Cambridge University Press, Cambridge.
12 Ferziger, J.H. and Kaper, H.G. (1972) *Mathematical Theory of Transport Processes in Gases*, North Holland, Amsterdam.
13 Einstein, A. (1905) *Ann. Phys.*, **17**, 549.
14 Smoluchowski, M. (1906) *Ann. Phys.*, **21**, 756.
15 Einstein, A. (1906) *Ann. Phys.*, **19**, 371.
16 Einstein, A. (1908) *Z. Electrochem.*, **14**, 235.
17 Nernst, W. (1888) *Z. Phys. Chem.*, **2**, 613.
18 Townsend, J.S. and Bailey, V.A. (1899) *Philos. Trans. A* **193**, 129.
19 Townsend, J.S. and Bailey, V.A. (1900) *Philos. Trans. A*, **195**, 259.
20 Huxley, L.G.H. and Crompton, R.W. (1973) *The Diffusion and Drift of Electrons in Gases*, John Wiley & Sons, Inc., New York.
21 McDaniel, E.W. and Mason, E.A. (1973) *The Mobility and Diffusion of Ions in Gases*, John Wiley & Sons, Inc., New York.
22 Stokes, G.G. (1851) *Trans. Camb. Philos. Soc.* **9**(II), 8.
23 Landau, L.D. and Lifshitz, E.M. (1986) *Fluid Mechanics*, Pergamon Press, Oxford.
24 Chapman, S. (1916) *Philos. Trans. R. Soc. Lond. A*, **216**, 279.
25 Chapman, S. (1917) *Philos. Trans. R. Soc. Lond. A*, **217**, 115.
26 Enskog, D. (1921) *Arkiv Mat. Astron. Fys.*, **16**, 1.
27 Smirnov, B.M. (1982) *Sov. Phys.-Phys. Usp.*, **138**, 519.
28 Palkina, L.A., Smirnov, B.M., and Chibisov, M.I. (1969) *Zh. Exp. Teor. Fiz.*, **56**, 340.
29 Palkina, L.A. and Smirnov, B.M. (1974) *High Temp.*, **12**, 37.
30 Eletskii, A.V., Palkina, L.A., and Smirnov, B.M. (1975) *Transport Phenomena in a Weakly Ionized Plasma*, Atomizdat, Moscow, in Russian.
31 Vargaftic, N.B. (1972) *Reference Book on Thermophysical Properties of Gases and Liquids*, Nauka, Moscow, in Russian.
32 Zwakhals, C.J. and Reus, K.W. (1980) *Physica C*, **100**, 231.
33 Zhiglinskii, A.G. (ed) (1994) *Reference Data on Elementary Processes Involving Atoms, Ions, Electrons and Photons*, St. Petersburg University Press, St. Petersburg, in Russian.
34 Smirnov, B.M. (2008) *Reference Data on Atomic Physics and Atomic Processes*, Springer, Heidelberg.
35 Vargaftic, N.B., Filipov, L.N., Tarismanov, A.A., and Totzkii, E.E. (1990) *Reference Book on Thermophysical Properties of Liquids and Gases*, Energoatomizdat, Moscow, in Russian.
36 Kesten, J. *et al.* (1984) *J. Chem. Phys. Ref. Data*, **13**, 229.
37 Dutton, J. (1975) *J. Phys. Chem. Ref. Data*, **4**, 577.
38 Townsend, J.S. (1908) *Proc. R. Soc. A*, **80**, 207.
39 Townsend, J.S. (1908) *Proc. R. Soc. A*, **81**, 464.

40 Sena, L.A. (1946) *Zh. Exp. Teor. Fiz.*, **16**, 734.

41 Sena, L.A. (1948) *Collisions of Electrons and Ions with Atoms*, Gostekhizdat, Leningrad, in Russian.

42 Dalgarno, A. (1958) *Philos. Trans. A*, **250**, 428.

43 Smirnov, B.M. (1974) *Excited Atoms and Ions in Plasma*, Atomizdat, Moscow, in Russian.

44 Tyndall, A.M. (1938) *The Mobility of Positive ions in Gases*, Cambridge University Press, Cambridge.

45 Mason, E.A. and McDaniel, E.W. (1988) *Transport Properties of Ions in Gases*, John Wiley & Sons, Inc., New York.

46 Crompton, R.W. and Elford, M.T. (1964) *Proc. R. Soc.*, **74**, 497.

47 Ellis, H.W., Pai, R.Y., McDaniel, E.W., Mason, E.A., and Viehland, L.A. (1976) *Atom. Data Nucl. Data Tables*, **17**, 177.

48 Ellis, H.W., McDaniel, E.W., Albritton, D.L., Viehland, L.A., Lin, S.L., and Mason, E.A. (1978) *Atom. Data Nucl. Data Tables*, **22**, 179.

49 Ellis, H.W., Trackston, M.G., McDaniel, E.W., and Mason, E.A. (1984) *Atom. Data Nucl. Data Tables*, **31**, 113.

50 Viehland, L.A. and Mason, E.A. (1995) *Atom. Data Nucl. Data Tables*, **60**, 37.

51 Radzig, A.A. and Smirnov, B.M. (1984) in *Chemistry of Plasma*, (ed. B.M. Smirnov), **11**, 170, in Russian.

52 Holstein, T. (1952) *J. Chem. Phys.*, **56**, 832.

53 Yu.P. Mordvinov, Smirnov, B.M. (1965) *Zh. Exp. Teor. Fiz.*, **48**, 133.

54 Smirnov, B.M. (1968) *Atomic Collisions and Elementary Processes in Plasma*, Atomizdat, Moscow, in Russian.

55 Kihara, T. (1953) *Rev. Mod. Phys.*, **25**, 944.

56 Smirnov, B.M. (2003) in *Reviews of Plasma Physics*, vol. 23, (ed. V.D. Shafranov), Kluwer Academic, New York, p. 209.

57 Dalgarno, A., McDowell, M.R.C., and Williams, A. (1958) *Philos. Trans. A*, **250**, 411.

58 Chanin, L.M. and Biondi, M.A. (1957) *Phys. Rev.*, **106**, 473.

59 Orient, O.J. (1967) *Can. J. Phys.*, **45**, 3915.

60 Patterson, P.L. (1970) *Phys. Rev. A*, **2**, 1154.

61 Courville, G.E. and Biondi, M.A. (1962) *J. Chem. Phys.*, **37**, 616.

62 Gerber, R.A. and Gusinov, M.A. (1971) *Phys. Rev. A*, **4**, 2027.

63 Sena, L.A. (1939) *Zh. Exp. Teor. Fiz.*, **9**, 1320.

64 Smirnov, B.M. (1968) *Dokl. Acad. Nauk SSSR*, **181**, 61.

65 Rakshit, A.B. and Warneck, P. (1980) *J. Chem. Phys.*, **73**, 2673.

66 Beaty, E.C. (1961) *Proc. 5 Int. Conf. Phen. Ioniz. Gases*, vol. 1, Munich, p. 183.

67 Smirnov, B.M. (1981) *Physics of Weakly Ionized Gases*, Mir, Moscow.

68 Beaty, E.C. and Patterson, P.L. (1965) *Phys. Rev. A*, **137**, 346.

69 Saporoshenko, M. (1965) *Phys. Rev. A*, **139**, 349.

70 Miller, T.M. *et al.* (1968) *Phys. Rev.*, **173**, 115.

71 McKnight, L.G., McAffee, K.B., and Sipler, D.P. (1967) *Phys. Rev.*, **164**, 62.

72 Moseley, J.T. *et al.* (1969) *Phys. Rev.*, **178**, 240.

73 Bowers, M.T. *et al.* (1969) *J. Chem. Phys.*, **50**, 4787.

74 Bennett, S.L. and Field, F.H. (1972) *J. Am. Chem. Soc.*, **94**, 8669.

75 Hiraoka, K. and Kebarle, P. (1974) *J. Chem. Phys.*, **62**, 2267.

76 Beuhler, R.J., Ehrenson, S., and Friedman, L. (1983) *J. Chem. Phys.*, **79**, 5982.

77 Arifov, U.A. *et al.* (1971) *High Energy Chem.*, **5**, 81.

78 Elford, M.T. and Milloy, H.B. (1974) *Aust. J. Phys.*, **27**, 795.

79 Johnsen, R., Huang, C.M., and Biondi, M.A. (1976) *J. Chem. Phys.*, **65**, 1539.

80 Leu, M.T., Johnsen, R., and Biondi, M.A. (1973) *Phys. Rev. A*, **8**, 413.

81 Adams, N.G., Smith, D., and Agle, E. (1984) *J. Chem. Phys.*, **81**, 1778.

82 Mitchell, J.B.A. *et al.* (1984) *J. Phys. B*, **17**, L909.

83 Mul, P.M. *et al.* (1977) *J. Chem.Phys.*, **67**, 373.

84 McGowan, J.W. *et al.* (1979) *Phys. Rev. Lett.*, **42**, 413.

85 Saporoshenko, M. (1965) *Phys. Rev. A*, **139**, 352.

86 Čermak, V. and Herman, Z. (1965) *Collect. Czech. Chem. Commun.*, **30**, 1343.

87 Bian, W.S. (1965) *J. Chem. Phys.*, **42**, 1261.

88 Shanin, M.M. (1965) *J. Chem. Phys.*, **43**, 1798.

89 Veatch, G.E. and Oskam, H.J. (1970) *Phys. Rev. A*, **2A**, 1422.

90 Smirnov, B.M. (2001) *Physics of Ionized Gases*, John Wiley & Sons, Inc., New York.

91 Smirnov, B.M. (1999) *Clusters and Small Particles in Gases and Plasmas*, Springer, New York.

92 Smirnov, B.M. (2010) *Cluster Processes in Gases and Plasmas*, Wiley-VCH Verlag GmbH.

93 Lifshits, E.M. and Pitaevskii, L.P. (1981) *Physical Kinetics*, Pergamon Press, Oxford.

94 Smirnov, B.M. (1997) *Phys. Usp.*, **40**, 1117.

95 Franklin, R.N. (1977) *Plasma Phenomena in Gas Discharges*, Oxford University Press, Oxford.

96 Raizer, Yu.P. (1991) *Gas Discharge Physics*, Springer, Berlin.

97 Langmuir, I. (1928) *Proc. Natl. Acad. Sci. USA*, **14**, 628.

98 Tonks, L. and Langmuir, I. (1929) *Phys. Rev.*, **33**, 195.

99 Langmuir, I. (1923) *Science*, **58**, 290.

100 Schottky, W. (1924) *Phys. Z.*, **25**, 345, 635.

101 Bohm, D. and Gross, E.P. (1949) *Phys. Rev.*, **75**, 1851.

102 Iizuka, S., Saeki, K., Sato, N., and Hatta, Y. (1979) *Phys. Rev. Lett.*, **43**, 1404.

103 Druyvesteyn, M.J. (1935) *Physica*, **2**, 255.

104 Chebotarev, G.D., Prutsakov, O.O., and Latush, E.L. (2003) *J. Russ. Laser Res.*, **24**, 37.

105 Langevin, P. (1905) *Ann. Chem. Phys.*, **8**, 245.

106 Sayers, J. (1938) *Proc. R. Soc. A*, **169**, 83.

107 Mächler, W. (1964) *Z. Phys.*, **104**, 1.

108 Hall, E. (1879) *Am. J. Math.*, **2**, 287.

109 Kulikovskiy, A.G. and Lyubimov, G.A. (1965) *Magnetohydrodynamics*, Addison-Wesley, Reading.

110 Womac, G.J. (1969) *MHD Power Generation*, Chapman and Hall, London.

111 Rosu, R.J. (1987) *Magnetohydrodynamic Energy Conversion*, Hemisphere, Washington.

112 Chapman, S. and Bartels, J. (1940) *Geomagnetism*, Oxford University Press, Oxford.

113 Stacey, R.D. (1977) *Physics of Earth*, John Wiley & Sons, Inc., New York.

114 Merill, R.T., McElhinny, M.W. (1983) *The Earth Magnetic Field: Its History, Origin and Planetary Perspective*, Academic Press, London.

115 Landau, L.D., Lifshitz, E.M. (1960) *Electrodynamics of Continuous Media*, Pergamon Press, Oxford.

116 Alfvèn, H. (1942) *Nature*, **150**, 495.

117 Hughes, W.F., Young, F.J. (1966) *The Electromagnetodynamics of Fluids*, John Wiley & Sons, Inc., New York.

118 Biskamp, D. (1993) *Nonlinear Magnetohydrodynamics*, Cambridge University Press, Cambridge.

119 Lorrain, P., Lorrain, F., Houle, S. (2006) *Magneto-fluid Dynamics: Fundamentals and Case Studies of Natural Phenomena*, Springer, New York.

120 Northupp, E.F. (1934) *Phys. Rev.*, **24**, 474.

121 Benett, W.H. (1934) *Phys. Rev.*, **45**, 890.

122 Kadomtsev, B.B. (1988) *Cooperative Phenomena in Plasma*, Nauka, Moscow, in Russian,English version: Shafranov, V.D. (ed) (2001) *Rev. Plasma Phys.*, **22**, 1.

5
Waves, Instabilities, and Structures in Excited and Ionized Gases

5.1
Instabilities of Excited Gases

5.1.1
Convective Instability of Gases

Currents in a plasma lead to its heating, and this may cause new forms of gas motion in accord with general laws of hydrodynamics [1–4]. In the first stage of transformation of a quiet plasma into a moving plasma, where the motion supports the heat transport, they may be considered as an instability of a motionless plasma. The simple form of gas or plasma motion that increases the heat transport compared with heat transport in a motionless gas is convection, which consists in the movement of streams of hot gas to a cooler region and streams of cold gas to a warmer region. The conditions for the development of this heat transport mechanism [1] and the convective movement arising for a motionless gas will be considered below.

To find the stability conditions for a gas with a temperature gradient dT/dz oriented in the same direction as an external force [1], we assume the gas to be in equilibrium with this external force. If the external force is directed along the negative z-axis, the number density of gas molecules decreases with increasing z. We consider two elements of gas of the same volume located a distance dz from each other, and calculate the energy required to exchange them in terms of their positions. The mass of the element with the smaller z coordinate is larger by ΔM than that of the element with the larger coordinate, so the work required for the exchange is $\Delta M g dz$, where g is the force per unit mass exerted by an external field. We ignore heat conductivity in this operation. The gas element initially at larger z has its temperature increased by ΔT as a result of this exchange, and the element initially at smaller z is cooled by this amount. Hence, the heat energy taken from a gas as a result of the transposition of the two elements is $C_p \Delta M dT = c_p \Delta M dT/m$, where C_p is the heat capacity of the gas per unit mass, c_p is the heat capacity per molecule, and m is the molecular mass. The instability criterion states that the above exchange of gas elements is energetically advantageous, and has the

Fundamentals of Ionized Gases, First Edition. Boris M. Smirnov.

form [1]

$$\frac{dT}{dz} > \frac{mg}{c_p} . \tag{5.1}$$

In particular, for atmospheric air subjected to the action of the Earth's gravitational field, this relation gives ($g = 10^3$ cm/s^2, $c_p = 7/2$)

$$\frac{dT}{dz} > 10\ \text{K/km} . \tag{5.2}$$

Criterion (5.1) is a necessary condition for the development of convection, but if it is satisfied that does not mean that convection will necessarily occur. Indeed, if we transfer an element of gas from one point to another, it is necessary to overcome a resistive force that is proportional to the displacement velocity. The displacement work is proportional to the velocity of motion of the gas element. If this velocity is low, the gas element exchanges energy with the surrounding gas in the course of motion by means of thermal conductivity of the gas, and this exchange is greater the more slowly the displacement proceeds. From these arguments it follows that the gas viscosity and thermal conductivity coefficients can determine the nature of the development of convection. We shall develop a formal criterion for this process.

5.1.2
Rayleigh Problem

If a motionless gas is unstable, small perturbations are able to cause a slow movement of the gas, which corresponds to convection. Our goal is to find the threshold for this process and to analyze its character. We begin with the simplest problem of this type, known as the Rayleigh problem [5]. The configuration of this problem is that a gas, located in a gap of length L between two infinite plates, is subjected to an external field. The temperature of the lower plate is T_1, the temperature of the upper plate is T_2, and the indices are assigned such that $T_1 > T_2$.

We can write the gas parameters as the sum of two terms, where the first term refers to the gas at rest and the second term corresponds to a small perturbation due to the convective gas motion. Hence, the number density of gas molecules is $N + N'$, the gas pressure is $p_0 + p'$, the gas temperature is $T + T'$, and the gas velocity $\mathbf{w}$ is zero in the absence of convection. When we insert the parameters in this form into the stationary equations for continuity (4.1), momentum transport (4.6), and heat transport (4.27), the zeroth-order approximation yields

$$\nabla p_0 = -\mathbf{F}N , \quad \Delta T = 0 , \quad \mathbf{w} = 0 ,$$

where $\mathbf{F}$ is the force acting on a single gas molecule. The first-order approximation for these equations gives

$$\frac{\nabla (p_0 + p')}{N + N'} + \frac{\eta \Delta \mathbf{w}}{N + N'} + \mathbf{F} = 0 , \quad w_z \frac{(T_2 - T_1)}{L} = \frac{\kappa}{c_V N} \Delta T' . \tag{5.3}$$

The parameters in the present problem are used in the last equation. The z-axis is taken to be perpendicular to the plates.

We transform the first term in the first equation in (5.3) with first-order accuracy, and obtain

$$\frac{\nabla(p_0 + p')}{N + N'} = \frac{\nabla p_0}{N} + \frac{\nabla p'}{N} - \frac{\nabla p_0}{N}\frac{N}{N'} = \mathbf{F}\left(1 - \frac{N'}{N}\right) + \frac{\nabla p'}{N}\,.$$

According to the equation of state (4.12) for a gas, $N = p/T$, we have $N' = (\partial N/\partial T)_p T' = -NT'/T$. Inserting this relation into the second equation in (5.3), we can write the set of equations in the form

$$\operatorname{div}\mathbf{w} = 0\,, \quad -\frac{\nabla p'}{N} - \mathbf{F}\frac{T'}{T} - \frac{\eta\Delta\mathbf{w}}{N} = 0\,, \quad w_z = \frac{\kappa L}{c_V N(T_2 - T_1)}\Delta T'\,. \quad (5.4)$$

We can reduce the set of equations (5.4) to an equation of one variable. For this goal we first apply the div operator to the second equation in (5.3) and take into account the first equation in (5.3). Then we have

$$\frac{\Delta p'}{N} - \frac{F\partial T'}{T\partial z} = 0\,. \quad (5.5)$$

Here we assume that $T_1 - T_2 \ll T_1$. Therefore, the unperturbed gas parameters do not vary very much within the gas volume. We can ignore their variation and assume the unperturbed gas parameters to be spatially constant.

We take w_z from the third equation in (5.4) and insert it into the z component of the second equation. Applying the operator Δ to the result, we obtain

$$\frac{1}{N}\frac{\partial}{\partial z}\Delta p' - \frac{F\Delta T'}{T} + \frac{\eta\kappa L}{c_V N^2(T_2 - T_1)}\Delta^2 T' = 0\,.$$

Using the relation (5.5) between $\Delta p'$ and T', we obtain [1]

$$\Delta^3 T' = -\frac{\mathrm{Ra}}{L^4}\left(\Delta - \frac{\partial^2}{\partial z^2}\right)T'\,, \quad (5.6)$$

where the dimensionless combination of parameters

$$\mathrm{Ra} = \frac{(T_1 - T_2)c_V F N^2 L^3}{\eta\kappa T} \quad (5.7)$$

is called the Rayleigh number.

The Rayleigh number is fundamental for the problem we are considering. We can rewrite it in a form that conveys clearly its physical meaning. Introducing the kinetic viscosity $\nu = \eta/\rho = \eta/(Nm)$, where ρ is the gas density, the thermal diffusivity coefficient is $\chi = \kappa/(Nc_V)$, and the force per unit mass is $g = F/m$, the Rayleigh number then takes the form

$$\mathrm{Ra} = \frac{T_1 - T_2}{T}\frac{gL^3}{\nu\chi}\,. \quad (5.8)$$

This expression is given in terms of the primary physical parameters that determine the development of convection: the relative difference of temperatures $(T_1 - T_2)/T_1$, the specific force of the field g, a typical system size L, and also the transport coefficients taking into account the types of interaction of gas flows in the course of convection. Note that we use the specific heat capacity at constant volume c_V because of the condition that exists in our investigation. If equilibrium is maintained instead at a fixed external pressure, it is necessary to use the specific heat capacity at constant pressure in the above equations.

Equation (5.6) shows that the Rayleigh number determines the possibility of the development of convection. For instance, in the Rayleigh problem the boundary conditions at the plates are $T' = 0$ and $w_z = 0$. Also, the tangential forces $\eta(\partial w_x/\partial z)$ and $\eta(\partial w_y/\partial z)$ are zero at the plates. Differentiating the equation $\operatorname{div} \mathbf{w} = 0$ with respect to z and using the conditions for the tangential forces, we find that at the plates $\partial^2 w_z/\partial z^2 = 0$. Hence, we have the boundary conditions

$$T' = 0 \,, \quad w_z = 0 \,, \quad \frac{\partial^2 w_z}{\partial z^2} = 0 \,.$$

We denote the coordinate of the lower plate by $z = 0$ and the coordinate of the upper plate by $z = L$. A general solution of (5.6) with the stated boundary conditions at $z = 0$ can be expressed as

$$T' = C \exp[i(k_x x + k_y y)] \sin k_z z \,. \tag{5.9}$$

The boundary condition $T' = 0$ at $z = L$ gives $k_z L = \pi n$, where n is an integer. Inserting the solution (5.9) into (5.6), we obtain

$$\mathrm{Ra} = \frac{(k^2 L^2 + \pi^2 n^2)^3}{k^2 L^2} \,, \tag{5.10}$$

where $k^2 = k_x^2 + k_y^2$. The solution (5.9) satisfies all boundary conditions.

Equation (5.10) shows that convection can occur for values of the Rayleigh number not less than $\mathrm{Ra}_{\min}$, where $\mathrm{Ra}_{\min}$ refers to $n = 1$ and $k_{\min} = \pi/(L\sqrt{2})$. The numerical value of $\mathrm{Ra}_{\min}$ is [6]

$$\mathrm{Ra}_{\min} = 27\pi^4/4 = 658 \,.$$

The magnitude of $\mathrm{Ra}_{\min}$ can vary widely depending on the geometry of the problem and the boundary conditions, but in all cases the Rayleigh number is a measure of the possibility of convection.

5.1.3
Convective Movement of Gases

To gain some insight into the nature of convective motion, we consider the simple case of motion of a gas in the xz plane. Inserting solution (5.9) into the equation $\operatorname{div} \mathbf{w} = \partial w_x/\partial x + \partial w_z/\partial z = 0$, the components of the gas velocity are

$$w_z = w_0 \cos(kx) \sin \frac{\pi n z}{L} \,, \qquad w_x = -\frac{\pi n}{kL} w_0 \sin(kx) \cos \frac{\pi n z}{L} \,, \tag{5.11}$$

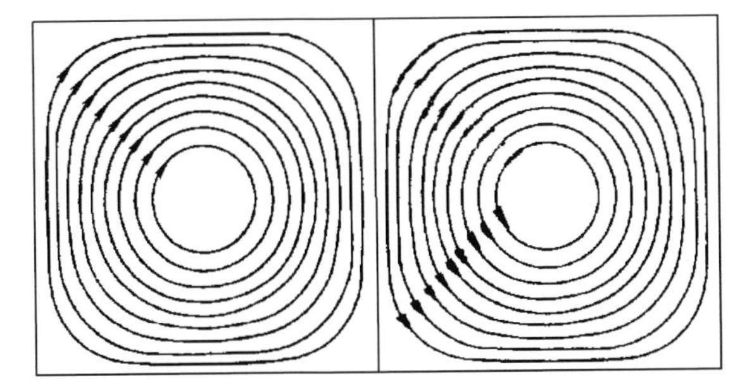

Figure 5.1 The paths of the gas elements in the Rayleigh problem for $n = 1$ and $k = \pi/L$.

where n is an integer and the amplitude of the gas velocity w_0 is assumed to be small compared with the corresponding parameters of the gas at rest. In particular, w_0 is small compared with the thermal velocity of the gas molecules.

The equations of motion for an element of the gas are $dx/dt = w_x$ and $dz/dt = w_z$, where relations (5.11) give the components of the gas velocity. We obtain $dx/dz = -[\pi n/(kL)]\tan(kx)\cot(\pi nz/L)$. This equation describes the path of the gas element. The solution of the equation is

$$\sin(kx)\sin\left(\frac{\pi nz}{L}\right) = C\,, \tag{5.12}$$

where C is a constant determined by the initial conditions. This constant is bounded by -1 and $+1$; its actual value depends on the initial position of the gas element. Of special significance are the lines at which $C = 0$. These lines are given by

$$z = \frac{Lp_1}{n}\,, \quad x = \frac{Lp_2}{n}\,, \tag{5.13}$$

where p_1 and p_2 are nonnegative integers. The straight lines determined by (5.13) divide the gas into cells. Molecules inside such a cell can travel only within this cell and cannot leave it. Indeed, (5.13) shows that the component of the gas velocity directed perpendicular to the cell boundary is zero; that is, the gas cannot cross the boundary between the cells. These cells are known as Benard cells [7–9].

Formula (5.13) shows that inside each cell the gas elements travel along closed paths around the cell center, where the gas is at rest. Figure 5.1 shows the path of the elements of gas in the Rayleigh problem for $n = 1$ and $k = \pi/L$, corresponding to the Rayleigh number $\text{Ra} = 8\pi^4 = 779$. In the Rayleigh problem, the Benard cells are pyramids with regular polygons as bases; in a general case, these cells can have a more complicated structure.

5.1.4
Convective Heat Transport

Convection is a more effective mechanism of heat transport than is thermal conduction. We can illustrate this by examining convective heat transport for the

Rayleigh problem. There will be a boundary layer of thickness δ formed near the walls, within which a transition occurs from zero fluid velocity at the wall itself to the motion occurring in the bulk of the fluid. The thickness of the boundary layer is determined by the viscosity of the gas, and the heat transport in the boundary layer is accomplished by thermal conduction, so the heat flux can be estimated to be $q = -\kappa\nabla T \sim \kappa(T_1 - T_2)/\delta$. Applying the Navier–Stokes equation (4.15) to the boundary region, one can estimate its thickness. This equation describes a continuous transition from the walls to the bulk of the gas flow. We now add the second term in the Navier–Stokes equation, $m(\mathbf{w}\cdot\nabla)\mathbf{w}$, which cannot be ignored here, to the expression in (5.3). An order-of-magnitude comparison of separate terms in the z-component of the resulting equation yields $mw_z^2/\delta \sim F(T_1 - T_2)/T \sim \eta w_z/N\delta^2$. Hence, we find that the boundary layer thickness is

$$\delta \sim \left[\frac{\eta^2 T}{N^2 F m^3 (T_1 - T_2)}\right]^{1/3} . \tag{5.14}$$

In the context of the Rayleigh problem, we can compare the heat flux transported by a gas due to convection (q) and that transported due to thermal conduction (q_{cond}). The thermal heat flux is $q_{\text{cond}} = \kappa(T_1 - T_2)/L$, and the ratio of the fluxes is

$$\frac{q}{q_{\text{cond}}} \sim \frac{L}{\delta} \sim \left[\frac{N^2 F m L^3 (T_1 - T_2)}{\eta^2 T}\right]^{1/3} \sim \text{Gr}^{1/3} . \tag{5.15}$$

Here Gr, the Grashof number, is the following dimensionless combination of parameters:

$$\text{Gr} = \frac{N^2 F m L^3 (T_1 - T_2)}{\eta^2 T} = \frac{\Delta T}{T}\frac{g L^3}{\nu^2} . \tag{5.16}$$

A comparison of the definitions of the Rayleigh number (5.7) and the Grashof number (5.16) gives their ratio as

$$\frac{\text{Ra}}{\text{Gr}} = \frac{c_V \eta}{m\kappa} = \text{Pr} = \frac{\nu}{\chi} .$$

where Pr is the Prandtl number.

The continuity equation (4.1), the equation of momentum transport (4.6), and the equation of heat transport (4.27) are valid not only for a gas, but also for a liquid. Therefore, the results we obtain are also applicable to liquids. However, a gas does have some distinctive features. For example, estimates (4.40) and (4.13) show that for a gas the ratio $\eta/(m\kappa)$ is of order unity. Furthermore, the specific heat capacity c_V of a single molecule is also close to 1. Hence, the Rayleigh number has the same order of magnitude as the Grashof number for a gas. Since convection develops at high Rayleigh numbers, we find that for convection $\text{Gr} \gg 1$. Therefore, according to the ratio in (5.15), we find that heat transport via convection is considerably more effective than heat transport in a motionless gas via thermal conduction.

The ratio (5.15) between the convective and conductive heat fluxes was derived for an external force directed perpendicular to the boundary layer. We can derive the corresponding condition for an external force directed parallel to the boundary layer. We take the z-axis to be normal to the boundary layer and the external force to be directed along the x-axis. Then, to the second equation of the set of equations (5.3), we add the term $m(\mathbf{w} \cdot \nabla)\mathbf{w}$ and compare the x components of this equation. This comparison yields

$$\frac{m w_x^2}{L} \sim \frac{F(T_1 - T_2)}{T} \sim \frac{\eta}{N} \cdot \frac{w_x}{\delta^2} \,. \tag{5.17}$$

Equation (5.17) gives the estimate

$$\delta \sim \left[\frac{\eta^2 T L}{N^2 F m^3 (T_1 - T_2)} \right]^{1/4}$$

for the thickness of the boundary layer. Hence, the ratio of the heat flux q due to convection and the flux q_{cond} due to thermal conduction is

$$\frac{q}{q_{\text{cond}}} \sim \frac{L}{\delta} \sim \mathrm{Gr}^{1/4} \,. \tag{5.18}$$

The ratio of the heat fluxes in this case is seen to be different from that when the external force is perpendicular to the boundary layer (see (5.15)). However, the convective heat flux in this case is still considerably larger than the heat flux due to thermal conduction in a motionless gas.

5.1.5 **Instability of Convective Motion**

We find that the convective motion of an ionized gas (as of any gas or liquid) occurs as a Rayleigh–Taylor instability that results in convective gas movement along simple closed trajectories for a simple geometry of its boundaries, and the space is divided in Benard cells, so gas trajectories (or Taylor vortices) do not intersect the boundaries of these cells. But this motion becomes more complicated with increasing Rayleigh number [10–12], and new types of convective motion develop when the Rayleigh and Grashof numbers become sufficiently large. The orderly convective motion becomes disturbed, and this disturbance increases until the stability of the convective motion of a gas is entirely disrupted, giving rise to disordered and turbulent flow of the gas. This will happen even if the gas is contained in a stationary enclosure. To analyze the development of turbulent gas flow we consider once again the Rayleigh problem: a gas at rest between two parallel and infinite planes maintained at different constant temperatures is subjected to an external force. We shall analyze the convective motion of a gas described by (5.11), and corresponding to sufficiently high Rayleigh numbers with $n \geq 2$. In this case there can develop simultaneously at least two different types of convection.

Figure 5.2 shows two types of convective motion for the Rayleigh number $\mathrm{Ra} = 108\pi^4$ corresponding to the wave number $k_1 = 9.4/L$ for $n = 1$ and to $k_2 = 4.7/L$ for $n = 2$. To analyze this example using these parameters, we combine the solutions so that the gas flows corresponding to $n = 1$ and to $n = 2$ travel in the same

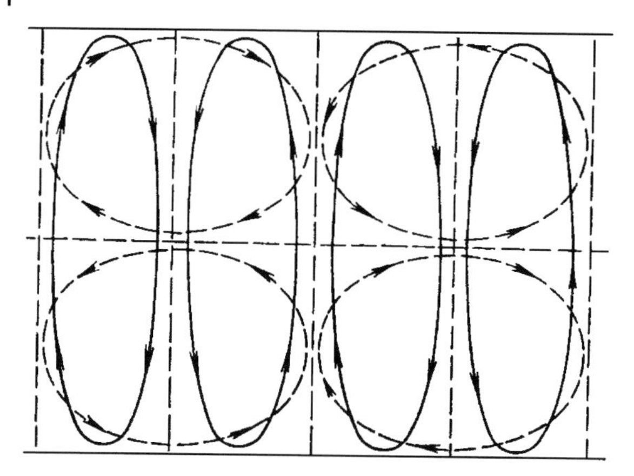

Figure 5.2 The types of convective motion in the Rayleigh problem for Ra $= 108\pi^4$. The mixing of the gas fluxes traveling in opposite directions finally results in random gas motion, or turbulence. Regions of countercurrents are marked.

direction in some region of the gas volume. Then in other regions these flows must move in opposite directions. The existence of two solutions with opposite directions of the gas flow does not mean that the ordered flow of a gas is disturbed. A combination of two solutions is itself a solution. For opposite gas flows, a combination means that at some points the gas is motionless. Nevertheless, the fact that an increase of the Rayleigh number gives rise to new types of solutions means that the convective flow can become turbulent. Assume that there is an ordered convective flow in the system corresponding to one of the solutions. Then a small perturbation in one of the regions of the gas volume gives rise to different types of flow. At the boundary of this region two opposite gas flows meet, so the kinetic energy of motion of the gas is transformed into thermal energy of the gas. This results in disordered motion. The development of turbulence changes altogether the character of heat transport.

We can estimate the thermal conductivity coefficient in a plasma or hot gas with developed turbulence. For example, turbulent flow can occur in a column of hot gas resulting from the passage of an electric current through it, and the temperature evolution of this column is of interest. We add the force of an external field directed perpendicular to the temperature gradient (as in the case of lightning in air), so the transport of heat occurs over small distances. We assume that a typical transverse size r of the column is large compared with a typical size l of small vortices, as determined by (5.8) and (5.10), giving for atmospheric air

$$\mathrm{Ra} = \frac{\Delta T}{T}\frac{g l^3}{\nu\chi} = \frac{l}{r}\frac{g l^3}{\nu\chi} \sim 10^3 \,.$$

In this case the thermal conductivity coefficient is given by estimate (4.40), $\kappa \sim N v l$, where l is the mean free path for vortices and v is a typical velocity in the fluxes. The value of v is obtained from the Navier–Stokes equation (4.15) as $v \sim (g l)^{1/2}$. The Reynolds number $\mathrm{Re} = v l/\nu$ for these motions can be estimated from the above expression for the Rayleigh number, yielding

$$\mathrm{Ra} = \frac{l}{r}\frac{\nu}{\chi}\mathrm{Re}^2 \sim 10^3 .$$

Since $l \ll r$ and $\nu \sim \chi$, it follows that $\mathrm{Re} \gg 1$. Thus, at large values of the Rayleigh number when several types of motion of gaseous flows are possible, the movement of the gas is characterized by large values of the Reynolds number. As these values increase, the gas motion tends toward disorder, and turbulent motion develops.

5.1.6
Thermal Explosion

We shall now consider the other type of instability of a motionless gas, which is called thermal instability or thermal explosion. It occurs in a gas experiencing heat transport by way of thermal conduction, where the heat release is determined by processes (e.g., chemical) whose rate depends strongly on the temperature. There is a limiting temperature where heat release is so rapid that thermal conductivity processes cannot transport the heat released, and then an instability occurs, which is the thermal explosion [13, 14]. As a result of this instability, the internal energy of the system is transformed into heat, which leads to a new regime of heat transport. We use the Zeldovich approach [15, 16] in the analysis of this problem.

We shall analyze this instability within the framework of the geometry of the Rayleigh problem. Gas is located in a gap between two infinite plates with a distance L between them. The wall temperature is T_w. We take the z-axis as perpendicular to the walls, with $z = 0$ in the middle of the gap, so the coordinates of walls are $z = \pm L/2$. We introduce the specific power of heat release $f(T)$ as the power per unit volume and use the Arrhenius law [17],

$$f(T) = A \exp\left(-\frac{E_\mathrm{a}}{T}\right) , \tag{5.19}$$

for the temperature dependence of this value, where E_a is the activation energy of the heat release process. This dependence of the rate of heat release is identical to that for the chemical process and represents a strong temperature dependence because $E_\mathrm{a} \gg T$. This dependence is used widely for chemical processes [18, 19].

To find the temperature distribution inside a gap in the absence of the thermal instability, we note that (4.41) for the transport of heat has the form

$$\kappa \frac{d^2 T}{dz^2} + f(T) = 0 .$$

We introduce a new variable $X = E_\mathrm{a}(T - T_0)/T_0^2$, where T_0 is the gas temperature in the center of the gap, and obtain the equation

$$\frac{d^2 X}{dz^2} - B \exp(-X) = 0 ,$$

where $B = E_a A \exp(-E_a/T_0)/(T_0^2 \kappa)$. Solving this equation with the boundary conditions $X(0) = 0$, $dX(0)/dz = 0$ (the second condition follows from the symmetry $X(z) = X(-z)$ in this problem), we have

$$X = 2 \ln \cosh z \ .$$

The temperature difference between the center of the gap and the walls is

$$\Delta T \equiv T_0 - T_w = \frac{2T_0^2}{E_a} \ln \cosh \left[\frac{L}{2} \sqrt{\frac{AE_a}{2T_0^2 \kappa}} \exp\left(-\frac{E_a}{2T_0} \right) \right] . \tag{5.20}$$

To analyze this expression, we refer to Figure 5.3, illustrating the dependence on T_0 for the left-hand and right-hand sides of this equation (curves 1 and 2, respectively) at a given T_w. The intersection of these curves yields the center temperature T_0. The right-hand side of the equation does not depend on the temperature of the walls and depends strongly on T_0. Therefore, it is possible that curves 1 and 2 do not intersect. That would mean there is not a stationary solution of the problem. The physical implication of this result is that thermal conduction cannot suffice to remove the heat released inside the gas. This leads to a continuing increase of the temperature, and thermal instability occurs.

To find the threshold of the thermal instability corresponding to curve 1′ in Figure 5.3, we establish the common tangency point of the curves describing the left-hand and right-hand sides of (5.20). The derivatives of the two sides are equal when

$$\Delta T = \frac{2T_0^2}{E_a} \ln \cosh y \ , \quad 1 = y \tanh y \ , \tag{5.21}$$

where

$$y = \frac{L}{2} \sqrt{\frac{AE_a}{2T_0^2 \kappa}} \exp\left(-\frac{E_a}{2T_0} \right) .$$

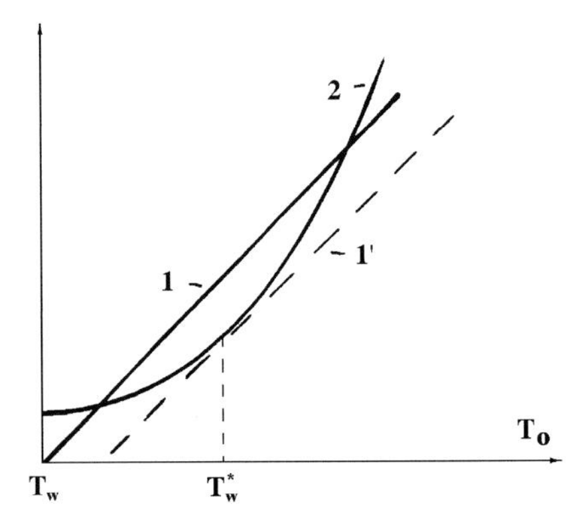

Figure 5.3 The dependence for the right-hand and left-hand sides of (5.20) on the temperature in the center. The wall temperature T_w^* corresponds to the threshold of the thermal instability.

The solution of the second equation in (5.21) is $y = 1.2$, so

$$\frac{AE_a L^2}{T_0^2 \kappa} \exp\left(-\frac{E_a}{T_0}\right) = 11.5\,, \quad \Delta T = 1.19\frac{T_0^2}{E_a}\,. \tag{5.22}$$

From this, it follows that the connection between the specific powers of heat release is

$$f(T_w) = f(T_0)\exp -1.19 = 0.30 f(T_0)\,.$$

This gives the following connection between the parameters for heat release at the walls at the threshold of thermal instability:

$$\frac{L^2 AE_a}{T_w^2 \kappa} \exp\left(-\frac{E_a}{T_w}\right) = 3.5\,. \tag{5.23}$$

Although we referred to the Arrhenius law for the temperature dependence of the rate of heat release, in actuality we used only the implication that this dependence is strong. We can, therefore, rewrite the condition for the threshold of the thermal instability in the form

$$\frac{L^2}{\kappa}\left|\frac{df(T_w)}{dT_w}\right| = 3.5\,. \tag{5.24}$$

The criterion for a sharp peak in the power involved in the heat release is

$$T_w\left|\frac{df(T_w)}{dT_w}\right| \gg 1\,. \tag{5.25}$$

Relation (5.24) has a simple physical meaning. It relates the rate of the heat release process to the rate of heat transport. If their ratio exceeds a particular value of the order of unity, then thermal instability develops.

5.1.7 Thermal Waves

The development of thermal instability can lead to the formation of a thermal wave. This takes place if the energy in some internal degree of freedom is significantly in excess of its equilibrium value. It can occur in a chemically active gas in combustion processes, and the theory of propagation of thermal waves [20–23] used below was developed for these processes. But the thermal explosion that is accompanied by propagation of a thermal wave has a universal character and may be employed also for problems involving ionized gases. Below we will be guided by a thermal wave propagating in a nonequilibrium molecular gas where the vibration temperature exceeds significantly the gas translation temperature, as occurs in molecular gas lasers, in particular, CO_2 and CO gas lasers. In these cases the development of thermal instability leads to a vibrational relaxation that is accompanied by equalizing of the translational and vibrational temperatures and proceeds in the form of a propagating thermal wave. Below we will be guided by this problem.

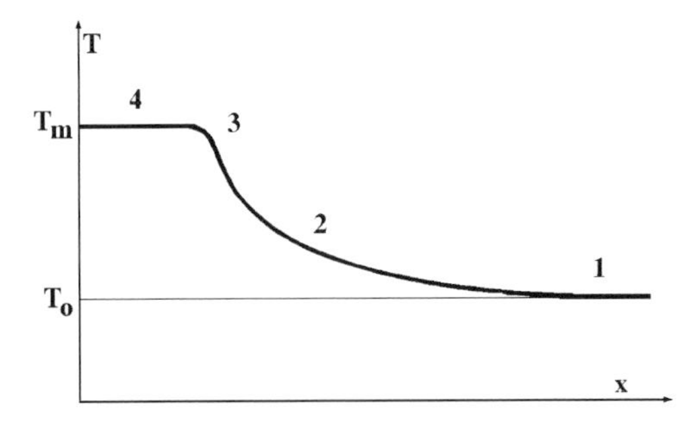

Figure 5.4 The temperature distribution in a gas in which a thermal wave propagates.

Figure 5.4 shows the temperature distribution in a gas upon propagation of a thermal wave. Region 1 has the initial gas temperature. The wave has not yet reached this region at the time represented in the figure. The temperature rise observed in region 2 is due to heat transport from hotter regions. The temperature in region 3 is close to the maximum. Processes that release heat occur in this region. We use the strong temperature dependence to establish the specific power of heat release, which according to (5.19) has the form

$$f(T) = f(T_{\mathrm{m}}) \exp\left[-\alpha(T_{\mathrm{m}} - T)\right] , \quad \alpha = \frac{E_{\mathrm{a}}}{T_{\mathrm{m}}^2} . \tag{5.26}$$

Here T_{m} is the final gas temperature, determined by the internal energy of the gas. Region 4 in Figure 5.4 is located after the passage of the thermal wave. Thermodynamic equilibrium among the relevant degrees of freedom has been established in this region.

To calculate the parameters of the thermal wave, we assume the usual connection between spatial coordinates and the time dependence of propagating waves. That is, the temperature is taken to have the functional dependence $T = T(x - ut)$, where x is the direction of wave propagation and u is the velocity of the thermal wave. With this functional form, the heat balance equation (4.41) becomes

$$u \frac{dT}{dx} + \chi \frac{d^2 T}{dx^2} + \frac{f(T)}{c_p N} = 0 . \tag{5.27}$$

Our goal is to determine the eigenvalue u of this equation. To accomplish this, we employ the Zeldovich approximation method [15, 16], which uses a sharply peaked temperature dependence for the rate of heat release. We introduce the function $Z(T) = -dT/dx$, so for $T_0 < T < T_{\mathrm{m}}$ we have $Z(T) \geq 0$. Because of the equation

$$\frac{d^2 T}{dx^2} = \frac{d}{dx}\left(\frac{dT}{dx}\right) = \frac{Z dZ}{dT} ,$$

equation (5.27) takes the form

$$-uZ + \chi \frac{Z dZ}{dT} + \frac{f(T)}{c_p N} = 0 . \tag{5.28}$$

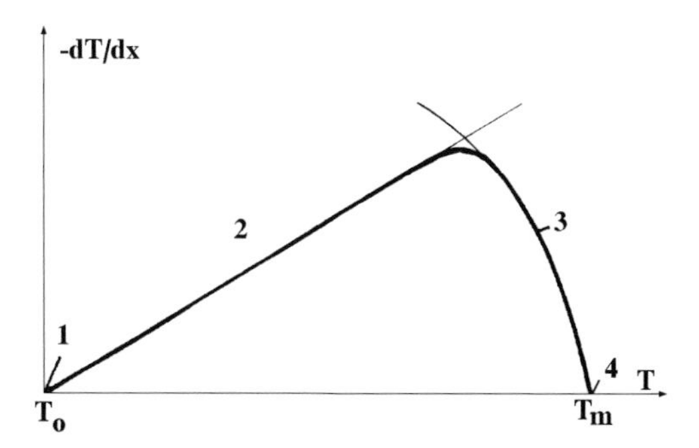

Figure 5.5 The solution of (5.28) for various regions of the thermal wave giving rise to Figure 5.4.

The solution of this equation for the regions defined in Figure 5.4 is illustrated in Figure 5.5. Before the thermal wave (region 1) and after it (region 4), we have $Z = 0$. In region 2 there is no heat release, and one can ignore the last term in (5.28). This yields

$$Z = \frac{u(T - T_0)}{\chi} . \tag{5.29}$$

Ignoring the first term in (5.28), we obtain

$$Z = \sqrt{\frac{2}{c_p N \chi} \int_T^{T_m} f(T) d T} \tag{5.30}$$

for region 3 in Figure 5.5. Equations (5.29) and (5.30) are not strictly joined at the interface between regions 2 and 3 because there is no interval where one can ignore both terms in (5.28). But because of the very strong dependence of $f(T)$ on T, there is only a narrow temperature region where it is impossible to ignore one of these terms. This fact allows us to connect solution (5.29) with (5.30) and find the velocity of the thermal wave.

Introducing the temperature T_* that corresponds to the maximum of $Z(T)$, (5.28) gives

$$Z_{\max} = Z(T_*) = \frac{f(T_*)}{u c_p N} .$$

Equation (5.28) for temperatures near T_* then takes the form

$$\frac{d Z}{d T} = \frac{u}{\chi} \left[1 - \frac{f(T)}{f(T_*)} \right] .$$

Solving this equation and taking into account (5.26), we have

$$Z = \frac{u}{\chi}(T - T_0) - \frac{u}{\chi \alpha} \exp[\alpha(T - T_*)] . \tag{5.31}$$

The boundary condition for this equation is that it should coincide with (5.29) far from the maximum Z. In this region one can ignore the third term in (5.28), which is the approximation that leads to (5.29). This is valid under the condition

$$\alpha(T_* - T_0) \gg 1 . \tag{5.32}$$

Now we seek the solution of (5.28) at $T > T_*$. Near the maximum of $Z(T)$, Z is given by (5.31), so (5.28) has the form

$$\frac{dZ}{dT} = \frac{Z_{\max}}{T_* - T_0}\{1 - \exp[\alpha(T - T_*)]\} .$$

This equation is valid in the region where the second term in (5.31) is significantly smaller than the first one. According to condition (5.32), this condition is fulfilled at temperatures where $\exp[\alpha(T - T_*)] \gg 1$. Therefore, because of the exponential dependence of the second term in (5.28), there is a temperature region where the second term in (5.28) gives a scant contribution to Z, but nevertheless determines its derivative. This property will be used below. On the basis of the above information and the dependence given in (5.26), we find that the solutions of (5.28) at $T > T_*$ are

$$Z = \sqrt{\frac{2f(T_m)}{c_p N\chi\alpha}}\sqrt{1 - \exp[\alpha(T - T_m)]} ,$$

$$\frac{dZ}{dT} = \sqrt{\frac{2f(T_m)}{c_p N\chi\alpha}}\frac{\exp[\alpha(T - T_m)]}{\sqrt{1 - \exp[\alpha(T - T_m)]}} .$$

Comparing the expressions for dZ/dT near the maximum, one can see that they can be connected if condition (5.32) is satisfied. Then solution (5.30) is valid in the temperature region $T_* < T \leq T_m$ up to temperatures near the maximum of Z. Connecting values of dZ/dT in regions where $\alpha(T - T_*) \gg 1$ and where $\alpha(T_m - T) \gg 1$, we obtain

$$Z_{\max} = \sqrt{\frac{2f(T_m)}{c_p N\chi\alpha}}(T_* - T_m)\exp[\alpha(T_* - T_m)] .$$

Next, from (5.31) it follows that $Z_{\max} = Z(T_*) = f(T_*)/(u c_p N)$. Comparing these expressions, one can find the temperature T_* corresponding to the maximum of $Z(T)$ and hence to the velocity of the thermal wave. We obtain

$$\frac{\alpha(T_* - T_0)}{2} = \exp[\alpha(T_m - T_*)] , \tag{5.33}$$

$$u = \frac{T_m}{T_m - T_0}\sqrt{\frac{2\chi f(T_m)}{c_p N E_a}} . \tag{5.34}$$

We used (5.26) for α. Relation (5.33) together with condition (5.32) gives $T_m - T_* \ll T_* - T_0$. This was taken into account in (5.32).

Formula (5.34) corresponds to the dependence (5.26) for the rate of heat release near the maximum. In a typical case at $T = T_m$ all the "fuel" is used, and $f(T_m) = 0$. Then all the above arguments are valid, because the primary portion of the heat release takes place at temperatures T, where $\alpha(T_m - T) \sim 1$, that is, where $\alpha(T - T_*) \gg 1$. Then, using the new form of the function $f(T)$ near the maximum of $Z(T)$, we transform (5.34) for the velocity of the thermal wave into the form

$$u = \frac{1}{T_m - T_0}\sqrt{\frac{2\chi}{c_p N}\int_T^{T_m} f(T)dT} \,. \tag{5.35}$$

This formula is called the Zeldovich formula.

We can analyze the problem from another standpoint. We take an expression for $Z(T)$ such that, in the appropriate limits, it would agree with (5.29) and (5.30). The simplest expression of this type has the form

$$Z = \frac{u}{\chi}(T - T_0)\left\{1 - \exp[-\alpha(T_m - T)]\right\} \,.$$

If we insert this into (5.28), $f(T)$ is given by

$$\frac{f(T)}{c_p N} = uZ - \frac{\chi}{2}\frac{d}{dT}Z^2 = \frac{u^2}{\chi}(T - T_0)\sqrt{1 - \exp -\alpha(T_m - T)}$$
$$\times\left[1 - \sqrt{1 - \exp(-\alpha(T_m - T))}\right] + \frac{\alpha u^2}{2\chi}(T - T_0)^2 \exp[-\alpha(T_m - T)] \,.$$

In the region $\alpha(T - T_0) \gg 1$, the first term is small compared with the second one and one can ignore it. Then the comparison of this expression with that in (5.26) in the temperature region $\alpha(T_m - T) \sim 1$ and $T_m - T \ll T_m - T_0$ gives the velocity of the thermal wave as

$$u = \frac{T_m}{T_m - T_0}\sqrt{\frac{2\chi}{E_a}\frac{f(T_m)}{c_p N}} \,.$$

where $\alpha = E_a / T_m^2$. This equation agrees exactly with (5.34) because of the identical assumptions used for construction of the solution in both cases.

One can use this method for the alternative case when the function $f(T)$ has an exponential dependence far from T_m, and goes to zero at $T = T_m$. For example, we take the approximate dependence

$$f(T) = A(T_m - T)\exp[-\alpha(T_m - T)] \,.$$

An approximate solution of (5.28) constructed on the basis of (5.29) and (5.30) has the form

$$Z = \frac{u}{\chi}(T - T_0)\sqrt{1 - e^{-\alpha(T_m - T)}[\alpha(T_m - T) + 1]} \,.$$

Substituting this into (5.28), we have

$$\frac{f(T)}{c_p N} = uZ - \frac{\chi}{2}\frac{dZ^2}{dT} = \frac{u^2}{\chi}(T - T_0)\sqrt{1 - e^{-\alpha(T_m - T)}}$$
$$\times \left[1 - \sqrt{1 - e^{-\alpha(T_m - T)}}\right] + \frac{\alpha u^2}{2\chi}(T - T_0)^2 e^{-\alpha(T_m - T)} .$$

In the region $\alpha(T - T_0) \gg 1$, where the heat release is essential, the first term is small compared with the second one. Then comparing this expression with the approximate dependence assumed above for $f(T)$, we find the velocity of the thermal wave $\alpha = E_a/T_m^2$ to be

$$u = \frac{T_m^2}{E_a(T_m - T_0)}\sqrt{\frac{2\chi A}{c_p N}} .$$

This result is in agreement with the Zeldovich formula (5.35) for the dependence employed for $f(T)$. The above analysis shows that the Zeldovich formula for the velocity of a thermal wave is valid under the condition $\alpha(T_m - T_0) \gg 1$.

5.1.8
Thermal Waves of Vibrational Relaxation

We shall apply the above results to the analysis of illustrative physical processes. First we consider a thermal wave of vibrational relaxation that can propagate in an excited molecular gas that is not in equilibrium. In particular, this process can occur in molecular lasers, where it can result in the quenching of laser generation. We consider the case where the number density of excited molecules is considerably greater than the equilibrium density. Vibrational relaxation of excited molecules causes the gas temperature to increase and the relaxation process accelerates. There will be a level of excitation and a temperature at which thermal instability develops, leading to the establishment of a new thermodynamic equilibrium between excited and nonexcited molecules.

The balance equation for the number density of excited molecules N_* has the form

$$\frac{\partial N_*}{\partial t} = D\Delta N_* - N N_* k(T) ,$$

where N is the total number density of molecules, and where we assume $N \gg N_*$. D is the diffusion coefficient for excited molecules in a gas, and $k(T)$ is the rate constant for vibrational relaxation. Taking into account the usual dependence of traveling wave parameters $N_*(x, t) = N_*(x - ut)$, where u is the velocity of the thermal wave, we transform the above equation into the form

$$u\frac{dN_*}{dx} + D\frac{d^2 N_*}{dx^2} - N_* N k(T) = 0 . \qquad (5.36)$$

In front of the thermal wave we have $N_* = N_{max}$, and after the wave we have $N_* = 0$. That is, we are assuming the equilibrium number density of excited molecules to be small compared with the initial density. We introduce the mean energy $\Delta\varepsilon$ released in a single vibrational relaxation event. Then the difference between the gas temperatures after T_m and before T_0 the thermal relaxation wave is

$$T_\mathrm{m} - T_0 = \frac{N_0 \Delta\varepsilon}{N c_p} ,$$

where N_0 is the initial number density of excited molecules.

The heat balance equation (5.27) now has the form

$$u \frac{dT}{dx} + \chi \frac{d^2 T}{dx^2} - \frac{\Delta\varepsilon N_* k(T)}{c_p} = 0 . \qquad (5.37)$$

The wave velocity can be obtained from the simultaneous analysis of (5.36) and (5.37). The simplest case occurs when $D = \chi$. Then, both balance equations are identical, and the relation between the gas temperature and the number density of excited molecules is

$$T_\mathrm{m} - T = \frac{N_* \Delta\varepsilon}{N c_p} . \qquad (5.38)$$

We have only one balance equation in this case. Comparing it with (5.27), we have $f(T)/(c_p N) = (T_\mathrm{m} - T) N k(T)$. On the basis of the Zeldovich formula we find the wave velocity to be

$$u = \frac{T_\mathrm{m}^2}{E_\mathrm{a}(T_\mathrm{m} - T_0)} \sqrt{\frac{2\chi}{\tau(T_\mathrm{m})}} , \qquad (5.39)$$

where $\tau(T_\mathrm{m}) = 1/[N k(T_\mathrm{m})]$ is a typical time for vibrational relaxation at temperature T_m. Because of the assumption $\alpha(T_\mathrm{m} - T_0) \gg 1$ and the dependence (5.26) for the rate constant for vibrational relaxation, relation (5.39) with the assumptions employed gives

$$u \ll \sqrt{\frac{\chi}{\tau(T_\mathrm{m})}} .$$

We shall now examine the propagation of a vibrational relaxation thermal wave for limiting relations between D and χ. We analyze first the case $D \gg \chi$. Figure 5.6a shows the distribution of the number density of excited molecules N_* and of the gas temperature T along the wave. We note that the centers of these two distributions coincide. This reflects the fact that the excited molecules that are quenched at a given time introduce heat into the gas.

We can analyze the balance equation (5.36) for the number density of excited molecules in a simple fashion by ignoring thermal conductivity processes. We obtain the temperature distribution in the form of a step as shown in (Figure 5.6a). At $x > 0$ the vibrational relaxation is weak, and the solution of (5.36) in this region

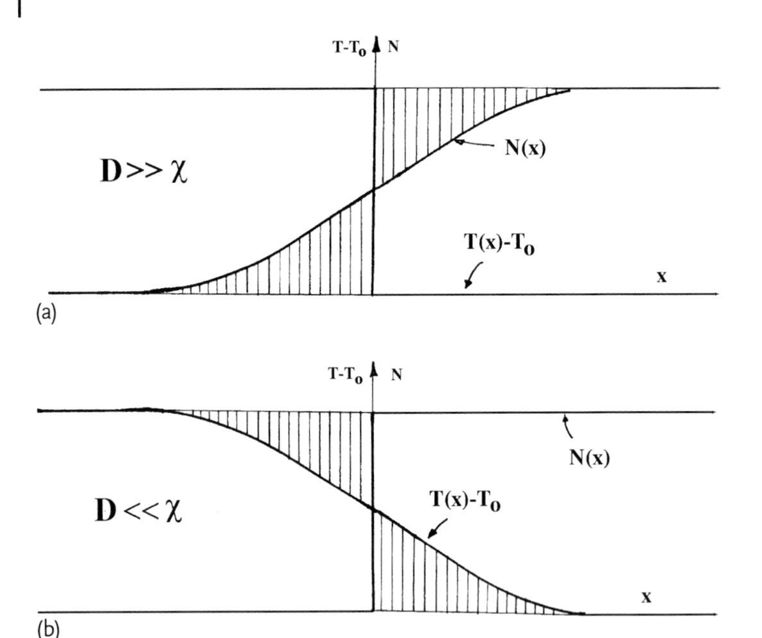

Figure 5.6 The distribution of the gas temperature and the number density of excited molecules in the thermal wave of vibrational relaxation for different limiting ratios between the diffusion coefficient of excited molecules D and the thermal diffusivity coefficient χ of the gas: (a) $D \gg \chi$ and (b) $\chi \gg D$

has the form $N_* = N_{\max} - (N_{\max} - N_0)\exp(-ux/D)$ if $x > 0$, where N_0 is the number density of the excited molecules at $x = 0$ and N_0 is the integration constant. In the region $x < 0$ vibrational relaxation is of importance, but the gas temperature is constant. This leads to

$$N_* = N_0 \exp(\alpha x)\,, \quad x < 0\,, \quad \alpha = \sqrt{\left(\frac{u}{2D}\right)^2 + \frac{1}{D\tau}} - \frac{u}{2D}\,,$$

where $\tau = 1/[Nk(T_m)]$. The transition region, where the gas temperature is not at its maximum, but the vibrational relaxation is essential, is narrow under the conditions considered. Hence, at $x = 0$ the above expressions must give the same results both for the number density of excited molecules and for their derivatives. We obtain

$$\alpha = \frac{u}{D}\,, \quad u = \sqrt{\frac{2D}{\tau T}} = \sqrt{2DNk(T_m)}\,, \quad D \gg \chi\,. \tag{5.40}$$

In this case the propagation of the thermal wave of vibrational relaxation is governed by the diffusion of excited molecules in a hot region where vibrational relaxation takes place. Hence, the wave velocity is of the order of $u \sim \sqrt{D/\tau}$ in accordance with (5.40). The width of the front of the thermal wave is estimated as $\Delta x \sim \sqrt{D\tau}$.

The opposite limiting case $\chi \gg D$ is shown in Figure 5.6b, illustrating the distribution of the gas temperature and number density of excited molecules along the thermal wave for this case. Because at $x > 0$ the rate of vibrational relaxation is low, and at $x < 0$ excited molecules are absent, one can ignore the last term in (5.37). If we assume the transition region to be small, (5.37) leads to the result

$$T(x) = T_0 + (T_m - T_0)\exp\left[-u\frac{(x + x_0)}{\chi}\right], \quad x > x_0 ; \quad T(x) = T_m , \quad x < x_0 ,$$

where $-x_0$ is the back boundary of the thermal wave. This value can be determined from the condition that the positions of the centers for the gas temperature distributions and the number density of molecules are coincident; that is, the areas of the shaded regions in Figure 5.6b must be the same. This yields $x_0 = \chi/u$ and

$$T(0) = T_r = T_m\left(1 - \frac{1}{e}\right) + \frac{T_0}{e} , \tag{5.41}$$

where e is the base of Naperian logarithms. The wave velocity is determined by the Zeldovich formula (5.35), where $f(T) = \Delta\varepsilon_* N k(T)$ and N_* is the step function. This allows one to take T_r as the upper limit of integration in (5.35), leading to the result

$$u = \frac{1}{T_r - T_0}\sqrt{\frac{2\chi f(T_r)}{c_p N\alpha}} .$$

This formula, with (5.38) and (5.41), yields

$$u = \sqrt{\frac{2e}{e-1}}\frac{T_r}{\sqrt{E_a(T_m - T_0)}}\sqrt{\frac{\chi}{\tau(T)}} , \tag{5.42}$$

where $\tau(T_r) = [N k(T_r)]^{-1}$ and $\alpha = E_a/T_r^2$. Since $\alpha(T_r - T_0) \gg 1$, we conclude that

$$u \ll \sqrt{\frac{\chi}{k(T)}} .$$

In this case the wave velocity is slow compared with that for the case $D = \chi$ because the vibrational relaxation process proceeds at lower temperatures and lasts longer than in the case $D = \chi$. Summing up the above results, we point out that the vibrational relaxation thermal wave is created by the processes of diffusion of excited molecules in a gas, by thermal conductivity of the gas, and by vibrational relaxation of excited molecules. Hence, the wave velocity depends on D and χ and on a typical time τ for vibrational relaxation.

5.1.9
Ozone Decomposition Through Thermal Waves

The process of propagation of a thermal wave results from competition between the processes of heat release and heat transport, where the temperature dependence for the heat release is sharper than for heat transport. This character of heat processes is of universal character and may be applied to some processes in ionized gases. We considered above a thermal wave of vibrational relaxation in a nonequilibrium molecular gas, and below we consider one more process of this type, which results in decomposition of ozone in air or other gases and proceeds in the form of a thermal wave. The propagation of the thermal wave results from chemical processes whose basic stages are

$$O_3 + M \overset{k_{dis}}{\rightarrow} O_2 + O + M \,, \quad O + O_2 + M \overset{K}{\rightarrow} O_3 + M \,, \quad O + O_3 \overset{k_1}{\rightarrow} 2O_2 \quad (5.43)$$

where M is a gas molecule and the relevant rate constants for the processes are given above the arrows. If these processes proceed in air, the temperature T_m after the thermal wave is connected to the initial gas temperature T_0 by

$$T_m = T_0 + 48c \,, \quad (5.44)$$

where the temperatures are expressed in Kelvin and c is the ozone concentration in air expressed as a percentage.

On the basis of scheme (5.43), we obtain the set of balance equations

$$\frac{d[O_3]}{dt} = -k_{dis}[M][O_3] + K[O][O_2][M] - k_1[O][O_3] \,,$$
$$\frac{d[O]}{dt} = k_{dis}[M][O_3] - K[O][O_2][M] - k_1[O][O_3] \,, \quad (5.45)$$

where [X] is the number density of particles X. Estimates show that at gas pressure $p \leq 1$ atm and $T_m > 500$ K we have $K[O_2][M] \ll k_1[O_3]$, that is, the second term on the right-hand side of each equation in (5.45) is smaller than the third one. In addition, we know that $[O] \ll [O_3]$, that is, $d[O]/dt \ll d[O_3]/dt$. This gives $d[O]/dt = 0$ and $[O] = k_{dis}[M]/k_1$. Using this result in the first equation in (5.45), we obtain

$$\frac{d[O_3]}{dt} = -2k_{dis}[M][O_3] \,.$$

Then, with (5.36) and (5.37), we obtain for the thermal wave

$$u\frac{d[O_3]}{dx} + D\frac{d^2[O_3]}{dx^2} - 2k_{dis}[M][O_3] = 0 \,,$$
$$u\frac{dT}{dx} + \chi\frac{d^2T}{dx^2} + \frac{2}{c_p}\Delta\varepsilon k_{dis}[O_3] = 0 \,, \quad (5.46)$$

where $\Delta\varepsilon = 1.5$ eV is the energy released from the decomposition of one ozone molecule.

We can now substitute numerical parameters of the above processes for a thermal wave in air at atmospheric pressure, namely, $D = 0.16\,cm^2/s$ and $\chi = 0.22\,cm^2/s$. These quantities are almost equal numerically and have similar temperature dependence, so we take them to be equal and given by

$$D = \chi = \frac{0.19}{p}\left(\frac{T}{300}\right)^{1.78} .$$

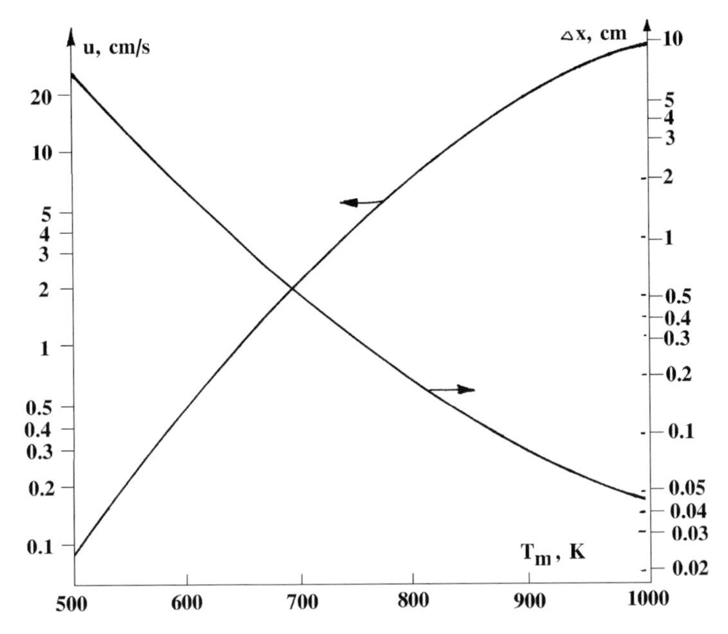

Figure 5.7 The velocity of the thermal wave of ozone decomposition (u) and the width of the wave front (Δx) as a function of the final temperature.

Here D and χ are expressed in square centimeters per second, the air pressure p is given in atmospheres, and the temperature is expressed in Kelvin. Next, we employ the expression $k_{\mathrm{dis}} = 1.0 \times 10^{-9}\,\mathrm{cm^3/s}\exp(-11\,600/T)$ for the dissociation rate constant. We can observe how equating D and χ simplifies the problem. Then (5.39) gives the thermal wave velocity

$$u = \frac{1.3\,T_{\mathrm{m}}^{2.39}}{T_{\mathrm{m}} - T_0}\exp\left(-\frac{5800}{T_{\mathrm{m}}}\right),$$

where the initial (T_0) and final (T_{m}) air temperatures are expressed in Kelvin, and the thermal wave velocity is in centimeters per second. Figure 5.7 illustrates the temperature dependence for the wave velocity with $T_0 = 300$ K. Note that it does not depend on the air pressure.

The width of the wave front can be characterized by the $\Delta x = (T_{\mathrm{m}} - T_0)/(dT/dx)_{\max}$, where the maximum temperature gradient is $(dT/dx)_{\max} = u(T_* - T_0)/\chi$ and the temperature T_* is determined by (5.33). Figure 5.7 gives this value as a function of the temperature. From the data in Figure 5.7, one can see that the thermal wave velocity is small compared with the sound velocity. This means that the propagation of a thermal wave is a slow process in contrast to that for a shock wave.

5.2 Waves in Ionized Gases

5.2.1 Acoustic Oscillations

Oscillations and noise in a plasma play a much greater role than in an ordinary gas because of the long-range nature of charged particle interactions. If a plasma is not uniform and is subjected to external fields, a wide variety of oscillations can occur [24, 25]. Under some conditions, these oscillations can become greatly amplified. Then the plasma oscillations affect basic plasma parameters and properties. Below we analyze the simplest types of oscillations in a gas and in a plasma. The natural vibrations of a gas are acoustic vibrations, that is, waves of alternating compressions and rarefactions that propagate in gases. We shall analyze these waves with the goal of finding the relationship between the frequency ω of the oscillation and the wavelength λ, which is connected to the wave vector $\boldsymbol{k}$ by $|\boldsymbol{k}| = 2\pi/\lambda$. It is customary to refer to the amplitude $k = |\boldsymbol{k}|$ as the wave number.

In our analysis, we assume the oscillation amplitudes to be small. Thus, any macroscopic parameter of the system can be expressed as

$$A = A_0 + \sum_{\omega} A'_{\omega} \exp[i(kx - \omega t)] , \tag{5.47}$$

where A_0 is an unperturbed parameter (in the absence of oscillations), A'_{ω} is the amplitude of the oscillations, ω is the oscillation frequency, and k is the appropriate wave number. The wave propagates along the x-axis. Since the oscillation amplitude is small, an oscillation of a given frequency does not depend on oscillations with other frequencies. In other words, there is no coupling between waves of different frequencies when the amplitudes are small. Therefore, one need retain only the leading term in sum (5.47) and can express the macroscopic parameter A in the form

$$A = A_0 + A' \exp[i(kx - \omega t)] . \tag{5.48}$$

To analyze acoustic oscillations in a gas, we can apply (5.48) to the number density of gas atoms (or molecules) N, the gas pressure p, and the mean gas velocity w, and take the unperturbed gas to be at rest ($w_0 = 0$). Using the continuity equation (4.1) and ignoring terms with squared oscillation amplitudes, we obtain

$$\omega N' = k N_0 w' . \tag{5.49}$$

The gas velocity $\mathbf{w}$ is directed along the wave vector $\boldsymbol{k}$ for an acoustic wave (a longitudinal oscillation). Similarly, the Euler equation (4.15) in the linear approximation leads to

$$\omega w' = \frac{k}{m N_0} p' , \tag{5.50}$$

where m is the mass of the particles of the gas.

Equations (5.49) and (5.50) connect the oscillation frequency ω and the wave number by the relation

$$\omega = c_s k \,, \tag{5.51}$$

where the speed of sound c_s is

$$c_s = \sqrt{\frac{p'}{m N'}} = \sqrt{\frac{1}{m}\frac{\partial p}{\partial N}} \,. \tag{5.52}$$

An equation of the type of (5.51) that connects the frequency ω of the wave with its wave number k is called a dispersion relation. We see that here the group velocity of sound propagation $\partial\omega/\partial k$ is the same as the phase velocity ω/k and does not depend on the sound frequency.

To find the sound velocity, it is necessary to know the connection between variations of the gas pressure and density in the acoustic wave. For long waves, the regions of compression and rarefaction do not exchange energy during wave propagation. Hence, this process is adiabatic, and the parameters of the acoustic wave satisfy the adiabatic equation

$$p N^{-\gamma} = \text{const} \,, \tag{5.53}$$

where $\gamma = c_p/c_V$ is the adiabatic exponent. That is, c_p is the specific heat capacity at constant pressure, and c_V is the specific heat capacity at constant volume. On the basis of expansion (5.48), the wave parameters are related by

$$\frac{p'}{p_0} = \gamma \frac{N'}{N_0} \,.$$

Because of the state equation (4.11), $p = NT$, where T is the gas temperature, we obtain $p'/N' = \partial p/\partial N = \gamma T$. Thus, the dispersion relation (5.51) yields

$$\omega = \sqrt{\frac{\gamma T}{m}} k \,. \tag{5.54}$$

The sound velocity is seen to be of the order of the thermal velocity of gas particles. Figure 5.8 demonstrates this dependence for acoustic oscillations in air.

The dispersion relation (5.54) is valid under adiabatic conditions in the wave if a typical time τ for heat transport in the wave is short compared with the period of oscillations $1/\omega$. Assuming the heat transport in the wave to be due to thermal conductivity, we have $\tau \sim r^2/\chi \sim (\chi k^2)^{-1}$, where the distance r is of the order of the wavelength and χ is the thermal diffusivity coefficient. From this we obtain the adiabatic criterion

$$\omega \ll \frac{c_s^2}{\chi} \tag{5.55}$$

for the wave, where c_s is the wave velocity. For example, in the case of air under standard conditions ($p = 1\,\text{atm}$), condition (5.55) has the form $\omega \ll 5 \times 10^9\,\text{s}^{-1}$.

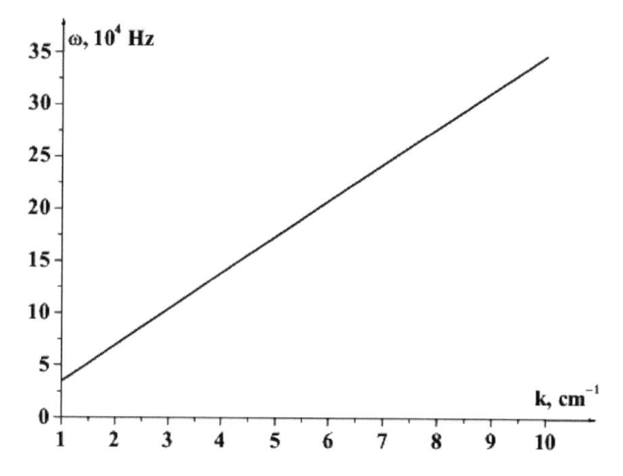

Figure 5.8 Dispersion relation (5.54) for air at $T = 300$ K in the centimeter range of wavelengths.

Because $\chi = \kappa/(c_p N) \sim v_T/\lambda$ and $c_s \sim v_T$, where λ is the free path length of gas molecules and v_T is a typical thermal velocity of the molecules, condition (5.55) may be rewritten in the form

$$\lambda k \ll 1 .$$

That is, the wavelength of the oscillations is long compared with the mean free path of the molecules of the gas.

5.2.2 Plasma Oscillations

Plasma oscillations which result from long-range interaction of electrons as a light charged plasma component are weakly excited oscillations in ionized gases [26–28]. Plasma oscillations are the simplest oscillations for a plasma, and the frequency of such oscillations of an infinite wavelength is given by (1.13). Below we derive the dispersion relation for these oscillations of finite wavelengths. Then we assume the ions to be at rest and their charges to be uniformly distributed over the plasma volume. Like to the case of acoustic oscillations in a gas, we shall derive the dispersion relation for the plasma waves from the continuity equation (4.1), the Euler equation (4.6), and the adiabatic condition (5.55) for the wave. Moreover, we take into account that the motion of electrons produces an electric field owing to the disturbance of the plasma quasineutrality. The electric field term is introduced into the Euler equation (4.6), and the electric field strength is determined by the Poisson equation (1.4).

As with the deduction of the dispersion relation for acoustic oscillations, we assume that the parameters of the oscillating plasma can be written in the form (5.48), and in the absence of oscillations both the mean electron velocity w

and the electric field strength E are zero. Hence, we obtain

$$-i\omega N_e' + ikN_0 w' = 0\,, \quad -i\omega w' + i\frac{kp'}{m_e N_0} + \frac{eE'}{m_e} = 0\,,$$
$$\frac{p'}{p_0} = \gamma\frac{N_e'}{N_0}\,, \quad ikE' = -4\pi e N\,. \tag{5.56}$$

Here k and ω are the wave number and the frequency of the plasma oscillations, N_0 is the mean number density of charged particles, $p_0 = N_0 m_e \langle v_x^2\rangle$ is the electron gas pressure in the absence of oscillations, m_e is the electron mass, v_x is the electron velocity component in the direction of oscillations, and the angle brackets denote averaging over electron velocities. N_e', w', p', and E' in (5.56) are the oscillation amplitudes for the electron number density, mean velocity, pressure, and electric field strength, respectively.

Eliminating the oscillation amplitudes from the system of equations in (5.56), we obtain the dispersion relation for plasma oscillations;

$$\omega^2 = \omega_p^2 + \gamma\langle v_x^2\rangle k^2\,, \tag{5.57}$$

where $\omega_p = \sqrt{4\pi N_0 e^2/m_e}$ is the plasma frequency defined by (1.13). Plasma oscillations or Langmuir oscillations are longitudinal, in contrast to electromagnetic oscillations. Hence, the electric field due to plasma waves is directed along the wave vector. This fact was used in deducing the set of equations in (5.56).

The dispersion relation (5.57) is valid for adiabatic propagation of plasma oscillations. If heat transport is due to thermal conductivity of electrons, the adiabatic condition takes the form $\omega\tau \sim \omega/(\chi k^2) \gg 1$ (compare this with criterion (5.55)), where ω is the frequency of oscillations, $\tau \sim (\chi k^2)^{-1}$ is a typical time for heat transport in the wave, χ is the electron thermal diffusion coefficient, and k is the wave number of the wave. Since $\chi \sim v_e\lambda$, $\omega \sim \omega_p \sim v_e/r_D$, where v_e is the mean electron velocity, λ is the electron mean free path, and r_D is the Debye–Hückel radius given by (1.9). The adiabatic condition yields

$$\lambda r_D k^2 \ll 1\,.$$

If this inequality is reversed, then isothermal conditions in the wave are fulfilled. In this case the adiabatic parameter γ in the dispersion relation (5.57) must be replaced by the coefficient 3/2 since for the pressure we use the equation of the gaseous state, $p = N/T$. As a result, we have in the isothermal case the following dispersion relation for plasma oscillations instead of (5.57) [29, 30]:

$$\omega^2 = \omega_p^2 + 3k^2\frac{T_e}{m_e}\,. \tag{5.58}$$

Note that because the frequency of plasma oscillations is much greater than the inverse of a typical time interval between electron–atom collisions, we have $\omega_p \gg N_a v_e \sigma_{ea} \sim v_e/\lambda$. From this it follows that $\lambda \gg r_D$.

5.2.3
Ion Sound

Since an ionized gas contains two types of charged particles, electrons and ions, there are two types of plasma oscillations due to electrons and ions [28]. Above we considered fast oscillations, in which ions as a slow plasma component do not partake, and now we analyze the oscillations due to the motion of ions in a homogeneous plasma. The special character of these oscillations is due to the large mass of ions. This stands in contrast to the small mass of electrons that enables them to follow the plasma field, so the plasma remains quasineutral on average:

$$N_e = N_i \, .$$

Moreover, the electrons have time to redistribute themselves in response to the electric field in the plasma. Then the Boltzmann equilibrium is established, and the electron number density is given by the Boltzmann formula (1.43):

$$N_e = N_0 \exp\left(\frac{e\varphi}{T_e}\right) \approx N_0 \left(1 + \frac{e\varphi}{T_e}\right) ,$$

where φ is the electric potential due to the oscillations and T_e is the electron temperature. These properties of the electron oscillations allows us to express the amplitude of oscillations of the ion number density as

$$N_i' = N_0 \frac{e\varphi}{T_e} \, . \tag{5.59}$$

We can now introduce the equation of motion for ions. The continuity equation (4.1) has the form $\partial N_i/\partial t + \partial(N_i w_i)/\partial x = 0$ and gives

$$\omega N_i' = k N_0 w_i \, , \tag{5.60}$$

where ω is the frequency, k is the wave number, and w_i is the mean ion velocity due to the oscillations. Here we assume the usual harmonic dependence (5.48) for oscillation parameters. The equation of motion for ions due to the electric field of the wave has the form

$$m_i \frac{d\mathbf{w}_i}{dt} = e\mathbf{E} = -e\nabla\varphi \, .$$

Taking into account the harmonic dependence (5.48) on the spatial coordinates and time, we have

$$m_i \omega w_i = e k \varphi \, . \tag{5.61}$$

Eliminating the oscillation amplitudes of N_i, φ, and w_i in the set of equations (5.59), (5.60), and (5.61), we obtain the dispersion relation

$$\omega = k \sqrt{\frac{T_e}{m_i}} \tag{5.62}$$

connecting the frequency and wave number. These oscillations caused by ion motion are known as ion sound. As with plasma oscillations, ion sound is a longitudinal wave, that is, the wave vector $\boldsymbol{k}$ is parallel to the oscillating electric field vector $\mathbf{E}$. The dispersion relation for ion sound is similar to that for acoustic waves. This is because both types of oscillations are characterized by a short-range interaction. In the case of ion sound, the interaction is short-ranged because the electric field of the propagating wave is shielded by the plasma. This shielding is effective if the wavelength of the ion sound is considerably greater than the Debye–Hückel radius for the plasma where the sound propagation occurs, that is, $k r_D \ll 1$. Dispersion relation (5.62) is valid if this condition is fulfilled.

To find the dispersion relation for ion sound in a general case, we start with the Poisson equation for the plasma field in the form

$$\frac{d^2\varphi}{dx^2} = 4\pi e(N_e - N_i) .$$

In the case of long-wave oscillations treated above, we took the left-hand side of this equation to be zero. Now, using the harmonic dependence of wave parameters on the coordinates and time, we obtain $-k^2\varphi$ for the left-hand side of this equation. Taking $N_e' = N_0(1 + e\varphi/T_e)$ in the right-hand side of this Poisson equation, we obtain

$$N_i' = N_0 \frac{e\varphi}{T_e}\left(1 + \frac{k^2 T_e}{4\pi N_0 e^2}\right) .$$

In the treatment of long-wave oscillations, we ignored the second term in the parentheses. Hence, dispersion relation (5.62) is now replaced by [28]

$$\omega = k\sqrt{\frac{T_e}{m_i}}\sqrt{1 + \frac{k^2 T_e}{4\pi N_0 e^2}} . \tag{5.63}$$

This dispersion relation transforms into (5.62) in the limit $k r_D \ll 1$ when the oscillations are determined by short-range interactions in the plasma. In the opposite limit $k r_D \gg 1$, we get

$$\omega = \sqrt{\frac{4\pi N_0 e^2}{m_i}} . \tag{5.64}$$

In this case a long-range interaction in the plasma is of importance, and from the form of the dispersion relation, we see that ion oscillations are similar to plasma oscillations.

5.2.4
Magnetohydrodynamic Waves

New types of oscillations arise in a plasma subjected to a magnetic field. We consider the simplest oscillations of this type in a high-conductivity plasma. In this case the magnetic lines of force are frozen in the plasma [31], and a change in the plasma current causes a change in the magnetic lines of force, which acts in opposition to this current and creates oscillations of the plasma together with the magnetic field [32–34]. Such oscillations are called magnetohydrodynamic waves.

For magnetohydrodynamic waves with wavelengths less than the radius of curvature of the magnetic lines of force, we have

$$\frac{1}{k} \ll \left| \frac{H}{\nabla H} \right| , \tag{5.65}$$

where H is the magnetic field strength. Then one can consider the magnetic lines of force to be straight lines. We construct a simple model of oscillations of a high-conductivity plasma, where the magnetic lines of force are frozen in the plasma. The displacement of the magnetic lines of force causes a plasma displacement, and because of the plasma elasticity, these motions are oscillations. The velocity of propagation of this oscillation is, according to dispersion relation (5.52), given by $c = \sqrt{\partial p/\partial \rho}$, where p is the pressure, $\rho = MN$ is the plasma density, and M is the ion mass. Because the pressure of a cold plasma is equal to the magnetic pressure $p = H^2/(8\pi)$, we have

$$c = \sqrt{\frac{H \partial H/\partial N}{4\pi M}}$$

for the velocity of wave propagation. Since the magnetic lines of force are frozen in the plasma, $\partial H/\partial N = H/N$. This gives [32–34]

$$c = c_{\mathrm{A}} = \frac{H}{\sqrt{4\pi MN}} \tag{5.66}$$

for the velocity of these waves. c_{A} is called the Alfvén speed.

The oscillations being examined may be of two types depending on the direction of wave propagation, as shown in Figure 5.9 [35]. If the wave propagates along the magnetic lines of force, it is called an Alfvén wave or magnetohydrodynamic wave. This wave is analogous to a wave propagating along an elastic string. The other wave type propagates perpendicular to the magnetic lines of force. Then the vibration of one magnetic line of force causes the vibration of a neighboring line. Such waves are called magnetic sound. The dispersion relation for both types of oscillations has the form

$$\omega = c_{\mathrm{A}} k .$$

The oscillations in a magnetic field are of importance for the Sun's plasma [36]. There are various types of waves in a plasma in a magnetic field [37].

5.2.5 Propagation of Electromagnetic Waves in a Plasma

We shall now derive the dispersion relation for an electromagnetic wave propagating in a plasma. Plasma motion due to an electromagnetic field influences the wave

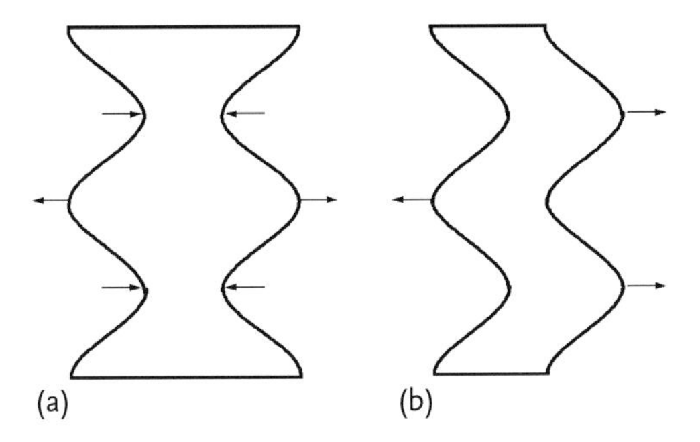

Figure 5.9 Two types of magnetohydrodynamic waves: (a) magnetic sound; (b) Alfvén waves.

parameters, and therefore the plasma behavior establishes the dispersion relation for the electromagnetic wave. We employ Maxwell's equations [38, 39] as they were represented by Heaviside [40] for the electromagnetic wave:

$$\operatorname{curl}\mathbf{E} = -\frac{1}{c}\frac{\partial \mathbf{H}}{\partial t}\,, \quad \operatorname{curl}\mathbf{H} = \frac{4\pi}{c}\mathbf{j} - \frac{1}{c}\frac{\partial \mathbf{E}}{\partial t}\,. \tag{5.67}$$

Here **E** and **H** are the electric and magnetic fields in the electromagnetic wave, **j** is the density of the electron current produced by the action of the electromagnetic wave, and c is the light velocity. Applying the curl operator to the first equation in (5.67) and the operator $-(1/c)(\partial/\partial t)$ to the second equation, and then eliminating the magnetic field from the resulting equations, we obtain

$$\nabla \operatorname{div}\mathbf{E} - \Delta\mathbf{E} + \frac{4\pi}{c^2}\frac{\partial \mathbf{j}}{\partial t} - \frac{1}{c^2}\frac{\partial^2 \mathbf{E}}{\partial t^2} = 0\,.$$

We assume the plasma to be quasineutral, so $\operatorname{div}\mathbf{E} = 0$. The electric current is due to motion of the electrons, so $\mathbf{j} = -eN_0\mathbf{w}$, where N_0 is the average number density of electrons and **w** is the electron velocity due to the action of the electromagnetic field. The equation of motion for the electrons is $m_e d\mathbf{w}/dt = -e\mathbf{E}$, which leads to the relation

$$\frac{\partial \mathbf{j}}{\partial t} = -eN_0\frac{d\mathbf{w}}{dt} = \frac{e^2 N_0}{m_e}\mathbf{E}\,.$$

Hence, we obtain

$$\Delta\mathbf{E} - \frac{\omega_p^2}{c^2}\mathbf{E} + \frac{1}{c^2}\frac{\partial^2 \mathbf{E}}{\partial t^2} = 0 \tag{5.68}$$

for the electric field of the electromagnetic wave, where ω_p is the plasma frequency. Writing the electric field strength in the form (5.48) and substituting it into the above equation, we obtain the dispersion relation for the electromagnetic wave [41]:

$$\omega^2 = \omega_p^2 + c^2 k^2\,. \tag{5.69}$$

If the plasma density is low ($N_e \to 0, \omega_p \to 0$), the dispersion relation agrees with that for an electromagnetic wave propagating in a vacuum, $\omega = kc$. Equation (5.69) shows that electromagnetic waves do not propagate in a plasma if their frequencies are lower than the plasma frequency ω_p. A characteristic damping distance for such waves is of the order of $c/\sqrt{\omega_\mathrm{p}^2 - \omega^2}$ according to dispersion relation (5.69).

5.2.6 The Faraday Effect in a Plasma

The Faraday effect manifests itself as a rotation of the polarization vector of an electromagnetic wave propagating in a medium in an external magnetic field. The interaction of an electromagnetic wave with the medium results in electric currents being induced in this medium by the electromagnetic wave, and these currents act on the propagation of the electromagnetic wave. If this medium is subjected to a magnetic field, different interactions occur for waves with left-handed as compared with right-handed circular polarization. Hence, electromagnetic waves with different circular polarizations propagate with different velocities, and propagation of electromagnetic waves with plane polarization is accompanied by rotation of the polarization vector of the electromagnetic wave. This effect was discovered by Faraday in 1845 [42, 43] and is known by his name. The Faraday effect has been observed in different media [44–48], and the specific characteristic of a plasma is the interaction between an electromagnetic wave with a gas of free electrons. The Faraday effect in a plasma will be analyzed below.

We consider an electromagnetic wave in a plasma propagating along the z-axis while being subjected to an external magnetic field. The wave and the constant magnetic field $\mathbf{H}$ are in the same direction. We treat a frequency regime such that we can ignore ion currents compared with electron currents. Hence, one can ignore motion of the ions. The electron velocity under the action of the field is given by (4.132). The electric field strengths of the electromagnetic wave corresponding to right-handed (subscript $+$) and left-handed (subscript $-$) circular polarization are given by

$$E_+ = (E_x + i E_y)e^{-i\omega t}\,, \quad E_- = (E_x - i E_y)e^{-i\omega t}\,.$$

Using the criterion $\omega\tau \ll 1$ for a plasma without collisions, we obtain on the basis of expressions (4.132) for the electron drift velocities the following current densities:

$$j_+ = -e N_\mathrm{e}(w_x + i w_y) = \frac{i N_\mathrm{e} e^2 E_+}{m_\mathrm{e}(\omega + \omega_H)} = \frac{i\omega_\mathrm{p}^2 E_+}{4\pi(\omega + \omega_H)}\,,$$
$$j_- = -e N_\mathrm{e}(w_x - i w_y) = \frac{i\omega_\mathrm{p}^2 E_-}{4\pi(\omega - \omega_H)}\,.$$

If we employ in (5.68) the harmonic dependence (5.48) on time and spatial coordinates of wave parameters, we have

$$k^2\mathbf{E} - \frac{4\pi i\omega\mathbf{j}}{c^2} - \frac{\omega^2\mathbf{E}}{c^2} = 0\,. \tag{5.70}$$

In the case the electromagnetic wave propagating along the direction of the magnetic field on the basis of the above expressions for the density of the electron currents, we obtain the dispersion relations for the electromagnetic waves with different circular polarizations in the form

$$k_+^2 c^2 + \frac{\omega_+ \omega_p^2}{\omega_+ + \omega_H} - \omega_+^2 = 0\,, \quad k_-^2 c^2 + \frac{\omega_- \omega_p^2}{\omega_- - \omega_H} - \omega_-^2 = 0\,, \qquad (5.71)$$

where subscripts + and − refer to right-handed and to left-handed circular polarizations, respectively. On the basis of these dispersion relations, we can analyze the propagation of an electromagnetic wave of frequency ω in a plasma in an external magnetic field. At $z = 0$ we take the wave to be polarized along the direction of the x-axis, so $\mathbf{E} = \mathbf{i}E\exp(-i\omega t)$, and we introduce the unit vectors $\mathbf{i}$ and $\mathbf{j}$ along the x-axis and the y-axis, respectively. The electric field of this electromagnetic wave in the plasma is

$$\mathbf{E} = \mathbf{i}E_x + \mathbf{j}E_y = \frac{\mathbf{i}}{2}(E_+ + E_-) + \frac{\mathbf{j}}{2i}(E_+ - E_-)\,.$$

We use the boundary condition

$$E_+ = E_0 e^{i(k_+ z - i\omega t)}\,, \quad E_- = E_0 e^{i(k_- z - i\omega t)}\,.$$

Introducing $k = (k_+ + k_-)/2$ and $\Delta k = k_+ - k_-$, we obtain the result

$$\mathbf{E} = E_0 e^{i(kz - \omega t)}\left(\mathbf{i}\cos\frac{\Delta k z}{2} + \mathbf{j}\sin\frac{\Delta k z}{2}\right)\,. \qquad (5.72)$$

From dispersion relations (5.71) it follows that

$$k_+^2 = \frac{\omega^2}{c^2}\left[1 - \frac{\omega_p^2}{\omega(\omega + \omega_H)}\right]\,, \quad k_-^2 = \frac{\omega^2}{c^2}\left[1 - \frac{\omega_p^2}{\omega(\omega - \omega_H)}\right]\,. \qquad (5.73)$$

If we assume the inequalities $\Delta k \ll k$ and $\omega_H \ll \omega$, relations (5.73) yield

$$\Delta k = \frac{\omega_H}{c}\frac{\omega_p^2}{\omega^2}\,. \qquad (5.74)$$

This result establishes the rotation of the polarization vector during propagation of the electromagnetic wave in a plasma. The angle φ for the rotation of the polarization is proportional to the distance z of propagation. This is a general property of the Faraday effect. In the limiting case $\omega \gg \omega_p$ and $\omega \gg \omega_H$, we have

$$\frac{\partial \varphi}{\partial z} = \frac{\Delta k}{2} = \frac{\omega_H}{c}\frac{\omega_p^2}{2\omega^2}\,.$$

For a numerical example we note that for maximum laboratory magnetic fields $H \sim 10^4$ G the first factor ω_H/c is about $10\,\mathrm{cm}^{-1}$, so for these plasma conditions the Faraday effect is detectable for propagation distances of the order of 1 cm.

From the above results it follows that the Faraday effect is strong in the region of the cyclotron resonance $\omega \approx \omega_H$. Then a strong interaction occurs between the plasma and the electromagnetic wave with left-handed polarization. In particular, it is possible for the electromagnetic wave with left-handed polarization to be absorbed, but for the wave with right-handed circular polarization to pass freely through the plasma. Then the Faraday effect can be detected at small distances.

5.2.7
Whistlers

Insertion of a magnetic field into a plasma leads to a large variety of new types of oscillations in it. We considered above magnetohydrodynamic waves and magnetic sound, both of which are governed by elastic magnetic properties of a cold plasma. In addition to these phenomena, a magnetic field can produce electron and ion cyclotron waves that correspond to rotation of electrons and ions in the magnetic field. Mixing of these oscillations with plasma oscillations, ion sound, and electromagnetic waves creates many types of hybrid waves in a plasma. As an example of this, we now consider waves that are a mixture of electron cyclotron and electromagnetic waves. These waves are called whistlers and are observed as atmospheric electromagnetic waves of low frequency (in the frequency interval 300–30 000 Hz). These waves are a consequence of lightning in the upper atmosphere and propagate along magnetic lines of force. They can approach the magnetosphere boundary and then reflect from it. Therefore, whistlers are used for exploration of the Earth's magnetosphere up to distances of 5–10 Earth radii. The whistler frequency is low compared with the electron cyclotron frequency $\omega_H = eH/(m_e c) \sim 10^7$ Hz, and it is high compared with the ion cyclotron frequency $\omega_{iH} = eH/(Mc) \sim 10^2$–$10^3$ Hz (M is the ion mass). Below we consider whistlers as electromagnetic waves of frequency $\omega \ll \omega_H$ that propagate in a plasma in the presence of a constant magnetic field.

We employ relation (5.48) to give the oscillatory parameters of a monochromatic electromagnetic wave. Then (5.68) gives

$$k^2\mathbf{E} - \boldsymbol{k}(\mathbf{k}\cdot\mathbf{E}) - i\mathbf{j}\frac{4\pi\omega}{c^2} = 0 \tag{5.75}$$

where $\omega \ll kc$. We state the current density of electrons in the form $\mathbf{j} = -eN_e\mathbf{w}$, where N_e is the electron number density. The electron drift velocity follows from the electron equation of motion (4.157), which, when $\nu \ll \omega \ll \omega_H$, has the form $e\mathbf{E}/m_e = -\omega_H(\mathbf{w}\times\mathbf{h})$, where $\mathbf{h}$ is the unit vector directed along the magnetic field. Substituting this into equation (5.75), we obtain the dispersion relation

$$k^2(\mathbf{j}\times\mathbf{h}) - \boldsymbol{k}[\boldsymbol{k}\cdot(\mathbf{j}\times\mathbf{h})] - i\mathbf{j}\frac{\omega\omega_p^2}{\omega_H c^2} = 0\,, \tag{5.76}$$

where $\omega_p = \sqrt{4\pi N_0 e^2/m_e}$ is the plasma frequency in accordance with (1.13). We introduce a coordinate system such that the z-axis is parallel to the external magnetic field (along the unit vector **h**) and the xz plane contains the wave vector. The x and y components of this equation are

$$-\frac{i\omega\omega_{\rm p}^2}{\omega_H c^2} j_x + \left(k^2 - k_x^2\right) j_y = 0\,, \quad -k^2 j_x - \frac{i\omega\omega_{\rm p}^2}{\omega_H c^2} j_y\,. \tag{5.77}$$

The determinant of this system of equations must be zero, which leads to the dispersion relation

$$\omega = \frac{\omega_H c^2 k k_z}{\omega_{\rm p}^2} = \frac{\omega_H c^2 k^2 \cos\vartheta}{\omega_{\rm p}^2}\,. \tag{5.78}$$

Here ϑ is the angle between the direction of wave propagation and the external magnetic field.

Usually whistlers are considered as electromagnetic waves excited by atmospheric lightning and propagating in a magnetized plasma of the atmosphere with frequencies below the electron plasma frequency and electron cyclotron frequency. In this form whistlers were observed in the Earth's atmosphere long ago [49, 50], and their nature was understood in the early stage of plasma physics [51]. As is seen, the whistler frequency is considerably higher than the frequency of Alfvén waves and magnetic sound. In particular, if the whistler propagates along the magnetic field, dispersion relation (5.78) gives $\omega = \omega_{\rm A}^2/\omega_{{\rm i}H}$, where $\omega_{\rm A}$ is the frequency of the Alfvén wave and $\omega_{{\rm i}H}$ is the ion cyclotron frequency. Since we assume $\omega \gg \omega_{{\rm i}H}$, this implies that

$$\omega \gg \omega_{\rm A} \gg \omega_{{\rm i}H}\,. \tag{5.79}$$

In addition, because $\omega \ll \omega_H$, the condition $\omega \ll kc$ leads to the inequalities

$$\omega \ll kc \ll \omega_{\rm p}\,. \tag{5.80}$$

Whistlers are determined entirely by the motion of electrons. To examine the nature of these waves, we note first that the electron motion and the resultant current in the magnetized plasma give rise to an electric field according to (4.157). This electric field, in turn, leads to an electron current according to (5.75). In the end, the whistler oscillations are generated. Note that because the dispersion relation has the dependence $\omega \sim k^2$, the group velocity of these oscillations, $v_{\rm g} = \partial\omega/\partial k \sim \sqrt{\omega}$, increases with the wave frequency. This leads to an identifying characteristic of the received signal as a short-time signal with a wide band of frequencies. Decrease of the tone of such a signal with time is the reason for the name "whistler". This character of whistlers as electromagnetic signals allowed these waves to be detected long ago [49, 50] as a result of atmospheric excitation by lightning and allows them to be separated from other types of waves. This gave the possibility to ascertain reliably the nature of whistlers [51].

The polarization of a whistler propagating along the magnetic field can be found from dispersion relation (5.77) together with dispersion relation (5.78). The result is

$$j_y = i j_x\,, \quad j_x = -i j_y\,. \tag{5.81}$$

From this it follows that the wave has circular polarization. This wave propagating along the magnetic field therefore has a helical structure. The direction of rotation of wave polarization is the same as the direction of electron rotation. The development of such a wave can be described as follows. Suppose electrons in a certain region possess a velocity perpendicular to the magnetic field. This electron motion gives rise to an electric field and compels electrons to circulate in the plane perpendicular to the magnetic field. This perturbation is transferred to the neighboring regions with a phase delay. Such a wave is known as a helicon wave.

5.3 Plasma Instabilities

5.3.1 Damping of Plasma Oscillations in Ionized Gases

Interaction of electrons and atoms leads to damping of plasma oscillations because electron–atom collisions shift the phase of the electron vibration and change the character of collective interaction of electrons in plasma oscillations. We shall take this fact into account below, and include it in the dispersion relation for the plasma oscillations. To obtain this relation, we use (4.9) instead of (4.6) as the equation for the average electron momentum. Then the second equation in set (5.56) is transformed into

$$-i\omega w' + i\frac{kp'}{m_e N_0} + \frac{eE'}{m_e} = \frac{w'}{\tau}\,, \tag{5.82}$$

and the remaining equations of this set are unchanged. Here τ is the characteristic time for elastic electron–atom collisions.

Replacing the first equation in system (5.56) by (5.82), we obtain the dispersion relation in the form

$$\omega = \sqrt{\omega_p^2 + \gamma \left\langle v_x^2 \right\rangle k^2} - \frac{i}{\tau} \tag{5.83}$$

instead of (5.57). This dispersion relation requires the condition

$$\omega\tau \gg 1\,. \tag{5.84}$$

Substituting dispersion relation (5.83) into (5.48), we find that the wave amplitude decreases with time as $\exp(-t/\tau)$, where this decrease is due to the scattering of electrons by atoms of the gas. The condition for existence of plasma oscillations is such that the characteristic time of the wave damping must be considerably higher than the oscillation period; namely, inequality (5.84) must hold. The frequency of collisions between electrons and atoms is $1/\tau \sim N\sigma v$, where N is the atom number density, v is a typical velocity of the electrons, and σ is the cross section for electron–atom collisions. Assuming this cross section to be of the order of a gas-kinetic cross section, the mean electron energy to be approximately 1 eV, and the frequency of plasma oscillations to be given by (1.12), we find that condition (5.84) is equivalent to the criterion

$$\frac{N_e^{1/2}}{N} \gg 10^{-12}\,\text{cm}^{3/2}\,.$$

This shows that in some gas discharge plasmas the conditions for the existence of plasma oscillations are not fulfilled.

We now consider the mixture of plasma oscillations with electromagnetic waves that is given by dispersion relation (5.69). We now insert damping in (5.68) that follows from the Maxwell equations with inclusion of electron currents. Introducing the dependence of the electric field strength on time in the form $E \sim \exp(ikx - i\omega t)$, we obtain in the limit $\omega\tau \ll 1$ the following dispersion equation:

$$\omega^2 = c^2k^2 + \omega_p^2\left(1 - \frac{i}{\omega\tau}\right),$$

which in the limit $\omega \ll \omega_p$ when a wave is damped in a plasma gives

$$\omega = -i\frac{k^2c^2}{\omega_p^2\tau}\,. \tag{5.85}$$

This dispersion relation allows us to analyze the character of penetration into a plasma. Defining the dependence of the signal amplitude on the distance x from the plasma boundary in the form $\exp(-x/\Delta)$, we have the depth of penetration of an electromagnetic wave in a plasma:

$$\Delta = \frac{1}{Imk} = \frac{c}{\omega_p\sqrt{2\omega\tau}}\,. \tag{5.86}$$

As is seen, the penetration depth becomes infinite in the limit of low frequencies, until this depth attains the Debye–Hückel radius. This phenomenon is the normal skin effect.

5.3.2
Interaction Between Plasma Oscillations and Electrons

The above damping mechanism for plasma oscillations is due to electron–atom collisions, since these collisions cause a phase shift of the oscillations of the colliding electron, leading to the damping of the oscillations. Now we consider another mechanism for interaction of electrons with waves. Electrons can be captured by waves (see Figure 5.10), and then a strong interaction takes place between these electrons and the waves. To analyze this process in detail, we note that in the frame of reference where the wave is at rest, a captured electron travels between the potential "walls" of the wave. Reflecting from one wall, an electron exchanges energy with the wave. If the electron velocity in this frame of reference is u, the energy is

$$\Delta\varepsilon = \frac{1}{2}m_e(v_p + u)^2 - \frac{1}{2}m_e(v_p - u)^2 = 2m_e v_p\,,$$

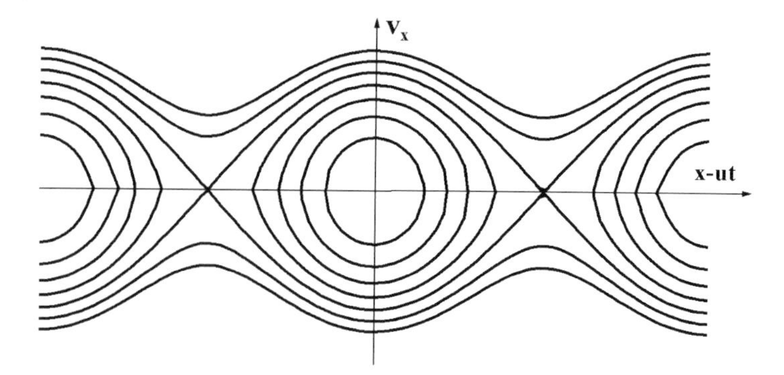

Figure 5.10 Trajectories of electrons in their phase space when electrons interact with plasma oscillations. Captured electrons have closed trajectories.

where $v_p = \omega/k$ is the phase velocity of the wave. The characteristic velocity of captured electrons in the frame of reference being considered is $u \sim (e\varphi/m_e)^{1/2}$, where φ is the amplitude of the scalar potential of the wave. We assume that collisions of a captured electron with other plasma electrons occur often enough that the captured electron can reflect only once from the potential wall of the wave. The electron obtains energy from the plasma electrons and escapes from the potential well. This means that the rate of collisions with plasma electrons is greater than the frequency of oscillations of the captured electron in the potential well of the wave. The rate of electron–electron collisions follows from (2.38) and is of the order of $N_e e^4 T_e^{3/2} m_e^{-1/2} \ln \Lambda$, and the frequency of oscillations of captured electrons is of the order of $\sqrt{e\varphi k/m_e} \sim \sqrt{eE'k/m_e} \sim \sqrt{N'e^2/m_e}$, where k is the wave number of the oscillation, and φ, E', and N' are the corresponding parameters of the oscillation. This gives the condition

$$\frac{N'}{N_e} \ll \frac{N_e e^6}{T_e^3} \tag{5.87}$$

for the interaction between captured electrons and the wave.

Assuming this criterion to be satisfied, we now find the character of the energy exchange between the wave and plasma electrons. In the first approximation, the electron distribution function is not altered by the interaction with the wave, and it is necessary to compare the number of electrons with velocity $v_p + u$ that transfer energy to the wave and the number of electrons with velocity $v_p - u$ that take energy from the wave, where u is a positive quantity. The number of captured electrons is proportional to the electron distribution function $f(v)$. Hence, the wave gives its energy to electrons and is damped if $f(v_p + u)$ is larger than $f(v_p - u)$. This means that the wave is damped when

$$\left[\frac{\partial f}{\partial v_x}\right]_{v_x = v_p} < 0 \,. \tag{5.88}$$

Here v_x is the component of the electron velocity in the direction of the wave propagation, and the derivative is taken for the electron velocity being equal to the phase

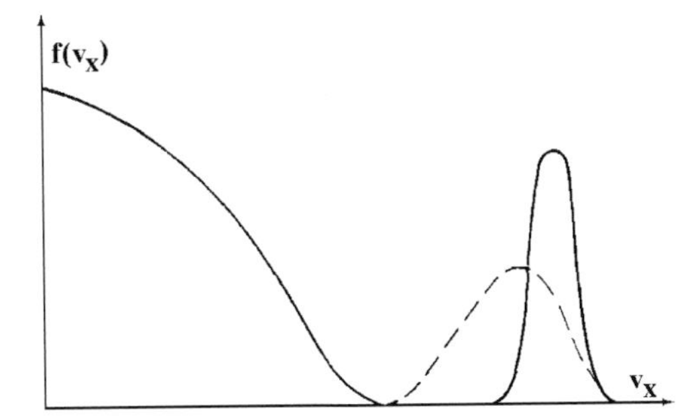

Figure 5.11 The velocity distribution function for plasma and beam electrons when an electron beam is injected in a plasma. The solid curve corresponds to the initial electron distribution in the plasma and beam. Ranges with a positive derivation are unstable, and development of the instability transforms the distribution function into that indicated by the dotted line.

velocity v_p of the wave. When condition (5.88) is not satisfied, the wave takes energy from the electrons and its amplitude increases.

In deriving condition (5.88) for wave damping, we stated that condition (5.87) is satisfied when the field of the wave does not affect the electron distribution function. However, criterion (5.88) holds also when condition (5.87) is not satisfied and the electron distribution function is altered owing to the interaction with the wave. Interaction with the wave tends to equalize the number densities of electrons with velocities $v_p + u$ and $v_p - u$ and, hence, it decreases the derivative $\partial f/\partial v_x$ at $v_x = v_p$, albeit without a change in its sign. That is, criterion (5.88) remains valid even when the interaction with the wave alters the electron distribution function.

Condition (5.88) is satisfied for the Maxwell electron energy distribution function or if it is simply a monotonically decreasing function, and oscillations in such a plasma decrease with time owing to the interaction with electrons. However, when an electron beam is injected into the plasma producing the electron distribution function in Figure 5.11, plasma oscillations in this system will be amplified, getting the necessary energy from the electrons. The interaction between plasma oscillations and electrons will reduce the velocity of the electrons until the electron distribution function again becomes monotonically decreasing.

Amplification of the oscillations can cause their amplitude to increase rapidly with time. If the oscillation amplitude is small and the oscillations do not alter the electron distribution function, that is, if condition (5.87) is satisfied, then the oscillation amplitude will increase exponentially. When particles of the plasma transfer their energy to the plasma wave, thus amplifying it, such a plasma state is termed an unstable state.

5.3.3 Attenuation Factor for Waves in Plasmas

We introduce into relation (5.48) the attenuation factor γ for the waves, so the oscillation amplitude varies as $\exp(-\gamma t)$. We can estimate the attenuation factor when the wave does not affect the electron distribution in the plasma, so condition (5.87) remains satisfied. Variation with time of the energy density W of the plasma oscillations can be estimated to be

$$\frac{dW}{dt} \sim \nu \int_{-u_0}^{u_0} f(v) \Delta\varepsilon du \,. \tag{5.89}$$

Here $\nu \sim u_0 k$ is the oscillation frequency for an electron captured in the potential well of the wave, $u_0 = (2e\varphi/m_e)^{1/2}$, φ is the amplitude of the oscillation potential, and $\Delta\varepsilon = 2m_e v_p u_0$ is the maximum change in the electron energy when the direction of the electron motion is reversed. (We take into account only the interaction between the wave and captured electrons.) The right-hand side of (5.89) may be estimated as

$$\nu \frac{\partial f}{\partial v_x} u_0 \Delta\varepsilon u_0 \sim u_0 k \frac{\partial f}{\partial v_x} u_0 m_e v_p u_0 \cdot u_0 \sim \frac{\partial f}{\partial v_x} \frac{e^2\omega}{m_e k_2} W \,,$$

where we use the relations $v_p = \omega/k$ and $W \sim (E')^2 \sim \varphi^2/k^2 (E'$ is the amplitude of the electric field of the wave). From the definition of the attenuation factor γ of the plasma wave, we have $dW/dt = -\gamma W$, so we obtain the estimate

$$\gamma \sim \frac{e^2\omega}{m_e k^2} \frac{\partial f}{\partial v_x} \,. \tag{5.90}$$

Attenuation occurs when condition (5.88) is satisfied. The attenuation factor (5.90), which is due to the interaction between the charged particles and the wave, is known as the Landau damping factor.

The condition for the existence of oscillations is

$$\gamma \ll \omega \,. \tag{5.91}$$

We can examine the implications of condition (5.91) for plasma oscillations, and we can also investigate ion sound, assuming a Maxwell distribution function for the particles. Taking $\omega \sim \omega_p$, we obtain

$$k r_D \ll 1 \tag{5.92}$$

for plasma oscillations, where r_D is the Debye–Hückel radius (1.7). When this condition is fulfilled, the phase velocity of the wave is considerably greater than the thermal velocity, so the electrons captured by the wave are in the tail of the distribution function.

When the ion sound propagates in a plasma in which the temperatures of electrons and ions are identical, the phase velocity of the ion sound is of the order of the thermal velocity of the ions, and the attenuation factor is of the order of the wave frequency. Therefore, ion sound can propagate only in plasmas in which the electron temperature T_e is considerably higher than the ion temperature T_i, that is, when we have

$$T_e \gg T_i\,. \tag{5.93}$$

5.3.4
Beam-Plasma Instability

Assume that an electron beam penetrates into a plasma, where the velocity of the electrons in the beam is much higher than the thermal velocity of the plasma electrons and the number density N_b of the electrons in the beam is considerably lower than the number density N_0 of plasma electrons. Deceleration of the electron beam can occur owing to the scattering of electrons of the beam by electrons and ions of the plasma. There is, however, another mechanism for deceleration of the electron beam: beam instability. In the early days of plasma physics this phenomenon was known as the Langmuir paradox [26–28]. It was revealed in these investigations that when electrons are ejected from a cathode surface in the form of a beam, the average energy of ejected electrons becomes equal to the thermal energy of the electrons of the gas discharge plasma into which this electron beam penetrates, at small distances from the cathode as compared with the mean free path of the electrons. Because it was assumed in that era that energy exchange between beam and plasma electrons results only from collisions because oscillations of the plasma were weak, the observed effect was considered to be a paradox. In actuality, interaction of the beam electrons with plasma electrons proceeds more effectively through collective degrees of freedom of the beam-plasma system than through collisions due to the resonant character of this interaction, and this explains the effect under consideration [29, 30, 52, 53]. This interaction can be succinctly described as follows. Suppose plasma oscillations are generated in a plasma. Interacting with electrons of the beam and taking energy from them, these oscillations are amplified. Thus, some of the energy of the electron beam is transformed into the energy of plasma oscillations and remains in the plasma. This energy may subsequently be transferred to other degrees of freedom of the plasma.

One can estimate the amplification of the plasma oscillations in the above scenario assuming the oscillation amplitude to be small, and taking the temperatures of the electrons in the plasma and in the beam to be zero. Hence, the pressure of the electrons in the plasma and beam is zero. Applying the continuity equation (4.1) and the Euler equation (4.6) to the plasma electrons, we derive equations for the amplitudes of the plasma parameters that follow from the two first equations of set (5.91) with $p' = 0$. Elimination of the electron velocity w' in the wave in these equations yields

$$N_e' = -i\frac{kE'}{m_e\omega^2}N_0\,. \tag{5.94}$$

One can obtain the expression for the amplitude of oscillations of the electron number density N_b' in the beam in a similar way by writing the electron number density in the beam as $N_b + N'_b \exp[i(kx - \omega t)]$ and the velocity of the electrons in the beam as $u + w_b \exp[i(kx - \omega t)]$, where the x-axis is parallel to the velocity of the beam and N_b and u are, respectively, the electron number density and the electron velocity in the unperturbed beam. Then we have

$$N'_b = -\frac{ikE'N_b}{m_e(\omega - ku)^2} . \tag{5.95}$$

As was done in arriving at the last equation of system (5.56), the Poisson equation (3.2) yields

$$ikE' = -4\pi e(N'_e + N'_b) \tag{5.96}$$

for the amplitude of the system's parameters. Eliminating the amplitudes N', N'_b, and E' from the set of equations (5.94), (5.95), and (5.96), we obtain the dispersion relation

$$\frac{\omega_p^2}{\omega^2} + \frac{\omega_p^2}{(\omega - ku)^2} \cdot \frac{N_b}{N_0} = 1 . \tag{5.97}$$

Here $\omega_p = (4\pi N_0 e^2/m_e)^{1/2}$ is the frequency of plasma oscillations in accordance with (1.12). When the number density of the beam electrons is zero ($N_b = 0$), dispersion relation (5.97) reduces to (5.57), where the electron temperature is taken to be zero. The strongest interaction between the beam and plasma occurs when the phase velocity of the plasma waves ω/k is equal to the velocity of the electron beam. Let us analyze this case. Since the number density of the beam electrons N_b is small compared with the number density N_0 of the plasma electrons, the frequency of the plasma oscillations is close to the plasma frequency ω_p. Hence, we shall consider waves with wave number $k = \omega/u$, which have the most effective interaction with the electron beam. We write the frequency of these oscillations as $\omega = \omega_p + \delta$ and insert it into dispersion relation (5.97). Expanding the result in a series in terms of the small parameter δ/ω_p, we obtain

$$\delta = \omega_p \left(\frac{N_b}{2N_0}\right)^{1/3} \exp\left(\frac{2\pi i n}{3}\right) ,$$

where n is an integer. One can see that $\delta/\omega_p \sim (N_b/N_0)^{1/3} \ll 1$. That is, the expansion over the small parameter δ/ω_p is valid.

If the imaginary component of the frequency (which is equal to the imaginary component of δ) is negative, the wave is attenuated; if it is positive, the wave is amplified. The maximum value of the amplification factor is given by ($n = 1$)

$$-\gamma = \frac{\sqrt{3}}{2} \left(\frac{N_b}{N_0}\right)^{1/3} \omega_p = 0.69\omega_p \left(\frac{N_b}{N_0}\right)^{1/3} . \tag{5.98}$$

The amplitude N'_b varies with time as $\exp(\gamma t)$; this result is valid if the plasma oscillations do not affect the properties of the plasma. This type of instability is known as beam-plasma instability. As a result of this instability, the distribution function for the beam electrons expands (see Figure 5.11), and the energy surplus is transferred to plasma oscillations. In reality, various factors influence the character and rate of a beam of charged particles propagating in a plasma [54].

5.3.5 Buneman Instability

We now consider instability of another type that develops if the mean velocity of the electrons differs from the mean velocity of the ions. We formulate the problem by taking all the plasma ions to be at rest and all the electrons to be traveling with velocity u with respect to the ions. The plasma is quasineutral, that is, the number densities of the electrons and ions are equal. Our goal is to determine the maximum amplification factor of the plasma oscillations. The electron beam is decelerated owing to the transfer of energy from the beam to the plasma oscillations [55].

With this formulation, the problem is equivalent to the previous case of interaction of an electron beam and a plasma. In both problems an electron beam is penetrating a plasma, so the dispersion relation can be derived in a similar way. Denoting the ion mass as M and taking into account the equality of ion and electron number densities, we obtain the dispersion relation

$$\frac{m_e}{M}\frac{\omega_p^2}{\omega^2} + \frac{\omega_p^2}{(\omega - ku)^2} = 1 \tag{5.99}$$

instead of (5.97). If we let the ratio m_e/M go to zero, we obtain the dispersion relation $\omega = \omega_p + ku$. Hence, one can write the frequency of the plasma oscillations as

$$\omega = \omega_p + ku + \delta \,.$$

Substituting this frequency into dispersion relation (5.99) and expanding the result in a power series in terms of the small parameter δ/ω_p, we obtain

$$\frac{2\delta}{\omega_p} = \frac{m_e}{M}\frac{\omega_p^2}{(\omega_p + ku + \delta)^2} \,.$$

The electron beam has the strongest interaction with the wave whose wave number is $k = -\omega_p/u$. For this wave we have

$$\delta = \left(\frac{m_e}{M}\right)^{1/3} \omega_p \exp\left(\frac{2\pi i n}{3}\right) ,$$

where n is an integer. The highest amplification factor corresponds to $n = 1$ and is given by

$$-\gamma = Im\delta = \frac{\sqrt{3}}{2}\left(\frac{m_e}{2M}\right)^{1/3} \omega_p = 0.69\omega_p \left(\frac{m_e}{M}\right)^{1/3} . \tag{5.100}$$

Note that the frequency of oscillations is the order of the amplification factor. This type of instability of the electron beam due to interaction with plasma ions is known as Buneman instability [55]. This instability can exist in an ionospheric plasma [56] and in a hot or nonequilibrium plasma [57].

5.3.6
Hydrodynamic Instabilities

The instabilities discussed above are so-called kinetic instabilities, for which the amplification of oscillations is due to the differences in the character of the motion of various groups of particles [59, 60]. The development of oscillations ultimately results in a change of the velocity distribution function for the charged particles of the plasma. Another class of instabilities is known as hydrodynamic instabilities. The development of hydrodynamic instabilities involves a displacement of the plasma regions and results, finally, in variation of the spatial configuration of the plasma. We shall analyze the simplest type of hydrodynamic instability, pinch instability.

We examine the stability of the pinch configuration with respect to variation of a current radius at a given point where the current axial symmetry is conserved at such a variation. The corresponding instability is called a sausage instability. We have to find under what conditions an accidental distortion of the pinch will not develop further. Let us assume that the distortion of the pinch results only in a slight curving of the magnetic lines of force; that is, the radius of curvature of the magnetic lines of force is considerably larger than the radius of the pinch. According to (4.163), the relation [58]

$$p + \frac{H^2}{8\pi} = \text{const}$$

is satisfied in the plasma region.

We must analyze the variation of the parameters of the pinch due to variation of its radius. The total current and magnetic flux through the cross section of the pinch will be conserved. The electric current is $I_z = caH_\varphi/2$, where a is the pinch radius and H_φ is the axial magnetic field strength. The condition $\delta I_z = 0$ yields $(\delta a/a) + (\delta H_\varphi/H_\varphi) = 0$, where δa is the variation of the pinch radius, and δH_φ is the variation of the axial magnetic field at the pinch surface outside the plasma. The longitudinal magnetic field is frozen into the plasma, so a displacement of plasma elements does not change the magnetic flux through them. The condition for conservation of the magnetic flux, $\Phi_z = \pi a^2 H_z$, yields $(2\delta a/a) + (\delta H_z/H_z) = 0$, where H_z is the longitudinal external magnetic field as it exists inside the plasma. Hence, $(\delta H_z/H_z = 2(\delta H_\varphi/H_\varphi)$. The variation of the magnetic field pressure inside the plasma is $\delta[H_z^2/(8\pi)] = H_z \delta H_z/(4\pi)$, and the variation of the magnetic pressure outside the plasma is $H_\varphi \delta H_\varphi/(4\pi) = H_\varphi^2 \delta H_z/(8\pi H_z)$. It can be seen that if

$$H_z^2 \geq \frac{H_\varphi^2}{2} \tag{5.101}$$

holds true, the additional internal magnetic field pressure produced by the distortion of the pinch is smaller than the additional external magnetic field pressure. When condition (5.101) is satisfied, the pinch is stable with respect to displacements of the sausage type.

5.4 Nonlinear Phenomena in Plasmas

5.4.1 The Lighthill Criterion

Plasma instabilities cause the development of certain types of oscillations. The subsequent evolution of these oscillations is determined by interactions between the oscillations and the plasma itself. The strength of these interactions depends on the wave amplitude. Hence, if a plasma instability develops, we deal with nonlinear plasma phenomena. But the role of nonlinear plasma processes is not restricted to the development of instabilities [61]. Nonlinear processes affect the nature of wave propagation in a plasma. In fact, interaction between elementary waves acts through the interaction of these waves with the plasma. Therefore, nonlinear phenomena determine both the profile and the character of the propagation of all waves in a plasma except those of low intensity. Along with this, nonlinear phenomena create a new plasma state [54, 62–64].

To study the development of perturbations in a plasma, we examine a perturbation in the form of a one-dimensional wave packet consisting of waves encompassing a narrow range of wave numbers ($\Delta k \ll k$). The perturbation amplitude at point x is

$$a(x,t) = \sum_k a(k) \exp(ikx - i\omega t) , \tag{5.102}$$

where $a(k)$ is the amplitude of the wave with wave number k. We take into account the wave dispersion, expressible as [65]

$$\begin{aligned} \omega(k) &= \omega(k_0) + \frac{\partial \omega(k_0)}{\partial k}(k-k_0) + \frac{\partial^2 \omega(k_0)}{\partial k^2}(k-k_0)^2 \\ &= \omega_0 + v_g(k-k_0) + \frac{1}{2}\frac{\partial v_g(k_0)}{\partial k}(k-k_0)^2 . \end{aligned} \tag{5.103}$$

Here k_0 is the mean wave number of the wave packet and v_g is the group velocity of the wave. Since wave number values are restricted to an interval of width Δk near k_0, then according to (5.102) the perturbation is initially concentrated in a spatial region of extent $\Delta x \sim 1/\Delta k$. As the wave packet evolves, it diverges owing to the different group velocities of individual waves. The initial wave packet, which has a size of the order of $1/\Delta k$, diverges on a timescale given by $\tau \sim (\Delta k^2)^{-1}(\partial v_g/\partial k)^{-1}$. That is, the wave dispersion usually leads to increasing spatial extension of the wave as a function of time.

The interaction of waves of different k values influences the behavior of this process. In particular, it can even lead to compression of the waves. To assess this possibility, we take into account the dependence

$$\omega = \omega_0 - \alpha E^2 \tag{5.104}$$

of the wave frequency on its amplitude, where E is a (small) field amplitude and ω_0 is the frequency of this wave. Inserting expansions (5.103) and (5.104) into (5.102), we obtain

$$a(x,t) = \sum_k a(k) \times \exp\left[i(k-k_0)(x-x_0) - ik_0x_0 - i(k-k_0)^2\frac{\partial v_g}{\partial k}t - i\alpha E^2(x)t\right] \quad (5.105)$$

for the wave packet amplitude. From (5.105) it follows that nonlinear wave interactions can lead to modulation of the wave packet. With certain types of modulation the wave packet may decay into separate bunches, or it may be compressed into a solitary wave – a soliton. This phenomenon is known as modulation instability.

Formula (5.105) shows that the compression of a wave packet or its transformation into separate bunches can take place only if the last two terms in the exponent in this equation have opposite signs. Only in this case can a nonlinear interaction compensate for the usual divergence of the wave packet. Therefore, the modulation instability can only occur if the following criterion holds true:

$$\alpha\frac{\partial v_g}{\partial k} < 0 . \quad (5.106)$$

This condition is called the Lighthill criterion [66].

5.4.2 The Korteweg–de Vries Equation

The above analysis shows that wave dispersion leads to divergence of the wave packet. If this divergence is weak, a weak nonlinearity is able to change its character. We now consider an example of such behavior when a wave is characterized by low dispersion and nonlinearity. Consider the propagation of long-wavelength waves in a medium in which the dispersion relation has the form

$$\omega = v_g k(1 - r_0^2 k^2) , \quad r_0 k \ll 1 . \quad (5.107)$$

This form describes a variety of oscillations, with sound and ion sound as examples. We shall obtain below a nonlinear equation that describes such waves.

We take as the starting point the Euler equation (4.6) for the velocity of particles in a longitudinal wave that has the form

$$\frac{\partial v}{\partial t} + v\frac{\partial v}{\partial x} - \frac{F}{m} = 0 . \quad (5.108)$$

Here $v(x,t)$ is the particle (gas atom or gas molecule) velocity in a longitudinal wave which propagates along the x-axis, F is the force per atomic particle, and m is the mass of the atomic particle. Within the framework of a linear approximation, one can write the particle velocity in the form $v = v_g + v'$, where v_g is the group velocity and v' is the particle velocity in the frame of reference where the wave is at rest. Because $v' \ll v_g$, we have

$$\frac{\partial v'}{\partial t} + v_g \frac{\partial v'}{\partial x} - \frac{F}{m} = 0$$

in the linear approximation, where the last term is a linear operator with respect to v'. In the harmonic approximation we have $v' \sim \exp(ikx - i\omega t)$. We seek the form of the operator F/m that leads to dispersion relation (5.107) in this approach. This gives

$$\frac{\partial v'}{\partial t} + v_g \left(\frac{\partial v'}{\partial x} + r_0^2 \frac{\partial^3 v'}{\partial x^3} \right) = 0 \,.$$

The last term takes into consideration a weak dispersion of long-wave oscillations. To account for nonlinearity of these waves, we analyze the second term of this equation. In the linear approach we replace the particle velocity v by the group velocity v_g. Returning this term to its initial form, that is, taking into account weak nonlinear effects, we obtain

$$\frac{\partial v}{\partial t} + v \frac{\partial v}{\partial x} + v_g r_0^2 \frac{\partial^3 v}{\partial x^3} = 0 \,. \tag{5.109}$$

This equation is called the Korteweg–de Vries equation [67] and was originally obtained in the analysis of wave propagation processes in shallow water.

The Korteweg–de Vries equation accounts for nonlinearity and weak dispersion simultaneously, and therefore serves as a convenient model equation for the analysis of nonlinear dissipative processes. As applied to plasmas, it describes propagation of long-wavelength waves in a plasma for which the dispersion relation is given by (5.107).

5.4.3 Solitons

The Korteweg–de Vries equation has solutions that describe a class of solitary waves. These waves are called solitons and conserve their form in time. It follows from dispersion relation (5.107) that short waves propagate more slowly than long waves, but nonlinear effects compensate for the spreading of the wave. We now show that this property holds true for waves that are described by the Korteweg–de Vries equation. We consider a wave of velocity u. The spatial and temporal dependence for the particle velocity has the form $v = f(x - ut)$. This gives $\partial v/\partial t = -u \partial v/\partial x$, and the Korteweg–de Vries equation is transformed to the form

$$(v - u)\frac{dv}{dx} + v_g r_0^2 \frac{d^3 v}{dx^3} = 0 \,. \tag{5.110}$$

Assuming the perturbation to be zero at large distances from the wave, that is $v = 0$ and $d^2v/dx^2 = 0$ at $x \to \infty$, this equation reduces to

$$v_g r_0^2 \frac{d^2 v}{dx^2} = uv - \frac{v^2}{2} ,$$

with the solution

$$v = 3u \cosh^{-2}\left(\frac{x}{2r_0}\sqrt{\frac{u}{v_g}}\right) . \tag{5.111}$$

The wave described by (5.111) is concentrated in a limited spatial region and does not diverge in time. The wave becomes narrower with an increase of its amplitude, that is, the wave is concentrated in a restricted region, and its spatial width is inversely proportional to the square root of the wave amplitude.

In other words, the Korteweg–de Vries equation has stationary solutions describing a nonspreading solitary wave, or soliton. The amplitude a and extension $1/\alpha$ of solitons are such that the value a/α^2 does not depend on the wave amplitude. If the initial perturbation is relatively small, evolution of the wave packet leads to formation of one solitary wave. If the amplitude of an initial perturbation is relatively large, this perturbation in the course of evolution of the system will split into several solitons. Thus, solitons are not only stable steady-state perturbations in the system, but they can also play a role in the evolution of some perturbations in a nonlinear dispersive medium.

5.4.4 Langmuir Solitons

The occurrence and propagation of solitons is associated with an electric field that exists in a plasma as the result of a wave process, and this field confines the perturbation to a restricted region. This can be demonstrated using the example of plasma oscillations. Denoting by $E(x,t)$ the electric field strength of the plasma oscillations, we have

$$W(x) = \frac{\overline{E^2}}{8\pi}$$

for their energy density, where the bar means a time average. Assuming the equality of electron and ion temperatures, $T_e = T_i = T$, the pressure of a quasineutral plasma is $p = 2N_e T$. The plasma pressure is established with a sound velocity that is greater than the velocity of propagation of long-wave oscillations. Then, because of the uniformity of plasma pressure at all points of the plasma, we have

$$2N(x)T + W(x) = 2N_0 T , \tag{5.112}$$

where N_0 is the number density of charged particles at large distances where the plasma oscillations are absent, and the plasma temperature is assumed to be constant in space.

The dispersion relation for plasma oscillations (5.57) has the form

$$\omega^2 = \omega_p^2 \left(1 - \frac{\overline{E^2}}{16\pi N_0 T}\right) + \gamma \left\langle v_x^2 \right\rangle k^2 \tag{5.113}$$

when (5.112) is taken into account, where ω_p is the plasma frequency in the absence of fields. In (5.113) we used (1.12) for the plasma frequency.

We can write (5.113) in a more convenient form. Take the electric field strength of the wave in the form $E = E_0 \cos \omega t$, so $\overline{E^2} = E_0^2/2$. Using $\overline{v_x^2} = T/m_e$ for electrons and introducing the Debye–Hückel radius r_D according to (1.7), we represent (5.113) in the form

$$\omega^2 = \omega_p^2 \left(1 - \frac{E_0^2}{32\pi N_0 T} + 2\gamma r_D^2 k^2\right) . \tag{5.114}$$

The first term on the right-hand side of this expression is considerably larger than the other two terms.

One can see that dispersion relation (5.114) satisfies the Lighthill criterion (5.106) because in this case we have

$$\alpha \frac{\partial v_g}{\partial k} = -\frac{\gamma \omega_p}{32\pi m_e N_0} < 0 .$$

Thus, nonlinear Langmuir oscillations can form a soliton. Dispersion relation (5.114) shows that if the energy density of plasma oscillations is high enough (so that the second term of (5.114) is larger than the third one), the oscillations cannot exist far from the soliton. The oscillations create a potential well in the plasma and are enclosed in this well. They can propagate in the plasma together with the well and occupy a restricted spatial region. The size of the potential well (or the soliton size) decreases with increase of the energy density of the plasma oscillations. Because $r_D k \ll 1$, the solitons are formed when the energy density of the oscillations is small compared with the specific thermal energy of charged particles of the plasma. Thus, this analysis demonstrates the tendency to form solitons from long-wave plasma oscillations. However, the analysis used does not allow one to study the evolution of oscillations at large amplitudes. Nevertheless, this exhibits the conditions for the existence of the Langmuir soliton that were realized experimentally [68–72] and that can exist in a solar plasma [73].

5.4.5
Nonlinear Ion Sound

We now analyze nonlinear ion sound when the nonlinearity is large [74]. For this purpose we use the Euler equation (4.6), the continuity equation (4.1), and the Poisson equation (1.4), which, in the linear approach, leads to dispersion relation (5.62) for ion sound. When we use these equations without any linear approximation, they can be written as

$$\frac{\partial v_i}{\partial t} + v_i \frac{\partial v_i}{\partial x} + \frac{e}{m_i}\frac{\partial \varphi}{\partial x} = 0 , \quad \frac{\partial N_i}{\partial t} + \frac{\partial (N_i v_i)}{\partial x} = 0 , \quad \frac{\partial^2 \varphi}{\partial x^2} = 4\pi e(N_e - N_i) . \tag{5.115}$$

Here m_i is the ion mass, v_i is the velocity of ions in the wave, φ is the electric potential of the wave, and N_e and N_i are the number densities of electrons and ions, respectively. As discussed above, electrons are in equilibrium with the field owing to their high mobility. Therefore, the Boltzmann distribution applies to the electrons, so $N_e = N_0 \exp(e\varphi/T_e)$, where N_0 is the mean number density of charged particles and T_e is the electron temperature.

We now analyze the motion of ions in the field of a steady-state wave when the plasma parameters v_i, N_i, and φ depend on time and the spatial coordinate as $f(x - ut)$, where u is the velocity of the wave. Then the set of equations (5.115) can be rewritten as

$$(v_i - u)\frac{dv_i}{dx} + \frac{e}{m_i} \cdot \frac{d\varphi}{dx} = 0\,, \quad \frac{d}{dx}[N_i(v_i - u)] = 0\,,$$
$$\frac{d^2\varphi}{dx^2} = 4\pi e\left[N_0 \exp\left(\frac{e\varphi}{T_e}\right) - N_i\right]. \tag{5.116}$$

The perturbation is assumed to be zero far from the wave, so $N_i = N_0$, $v_i = 0$ and $\varphi = 0$ at $x \to \infty$. The first two equations of set (5.116) then give

$$\frac{v_i^2}{2} - uv_i + \frac{e\varphi}{m_i} = 0\,, \quad N_i = N_0\frac{u}{u - v_i}\,.$$

The second equation implies that $v_i < u$ because $N_i > 0$. This means that $\varphi \geq 0$ in the first equation, that is, the electric potential of this ion acoustic wave, is always positive. The first equation leads to $v_i = u - \sqrt{u^2 - 2e\varphi/m_i}$, and the second equation gives $N_i = N_0 u(u^2 - 2e\varphi/m_i)^{-1/2}$. Substituting these expressions into the last equation of system (5.116), we obtain

$$\frac{d^2\varphi}{dx^2} = 4\pi e N_0\left[\exp\left(\frac{e\varphi}{T_e}\right) - \frac{u}{\sqrt{u^2 - 2e\varphi/m_i}}\right] \tag{5.117}$$

for the ion sound. This equation describes the potential of the electric field for the nonlinear ion acoustic wave. It is in the form of an equation of motion for a particle if φ is regarded as the coordinate and x as time. A general property of this type of equation is that multiplication by the integration factor $d\varphi/dx$ makes it possible to perform one integration immediately. Equation (5.117) then becomes

$$\frac{1}{2}\left(\frac{d\varphi}{dx}\right)^2 - 4\pi N_0 T_e \exp\left(\frac{e\varphi}{T_e}\right) - 4\pi N_0 m_i u\sqrt{u^2 - \frac{2e\varphi}{m_i}} = \text{const}\,.$$

Assuming that at large distances from the wave the potential φ and the electric field strength of the wave $-d\varphi/dx$ are zero, we can evaluate the constant of integration and obtain

$$\frac{1}{2}\left(\frac{d\varphi}{dx}\right)^2 + 4\pi N_0 T_e \left[1 - \exp\left(\frac{e\varphi}{T_e}\right)\right] + 4\pi N_0 m_i u \left(u - \sqrt{u^2 - \frac{2e\varphi}{m_i}}\right) = 0\,. \tag{5.118}$$

This solution describes a solitary wave because, according to the boundary conditions, the perturbation tends to zero at large distances. The solution allows one to determine the shape of the soliton and the relation between its parameters for various amplitudes of the wave. We first determine from (5.118) the relation between the maximum potential of the wave $\varphi_{\max}$ and its velocity u. We can do this by putting $d\varphi/dx = 0$ and $\varphi = \varphi_{\max}$ into (5.118). Introducing the reduced variables $\zeta = e\varphi_{\max}/T_e$ and $\eta = mu^2/(2T_e)$, one can rewrite (5.118) as

$$1 - \exp\zeta + 2\eta\left(1 - \sqrt{1 - \frac{\zeta}{\eta}}\right) = 0\,. \tag{5.119}$$

The solution of this equation is given in Figure 5.12. The limiting cases of this equation are instructive. For a small amplitude of the ion acoustic wave, $\zeta \to 0$, (5.119) gives $\eta = 1/2$. This yields the phase velocity $u = (T_e/m_i)^{1/2}$ for this wave, corresponding to dispersion relation (5.62) for ion sound of small amplitude. For large-amplitude waves, (5.119) implies $\zeta = \eta$, so the equation for ζ takes the form

$$1 - \exp\zeta + 2\zeta = 0\,. \tag{5.120}$$

This equation yields $\zeta = 1.26$, so $e\varphi_{\max} = 1.26T_e$ and $u = 1.58(T_e/m_i)^{1/2}$. For larger wave amplitudes the electric potential in the wave center becomes too large to admit solutions of (5.119), and ions are reflected from the crest of the wave. As a result, part of the wave reverses and the wave separates into parts. Thus, a solitary

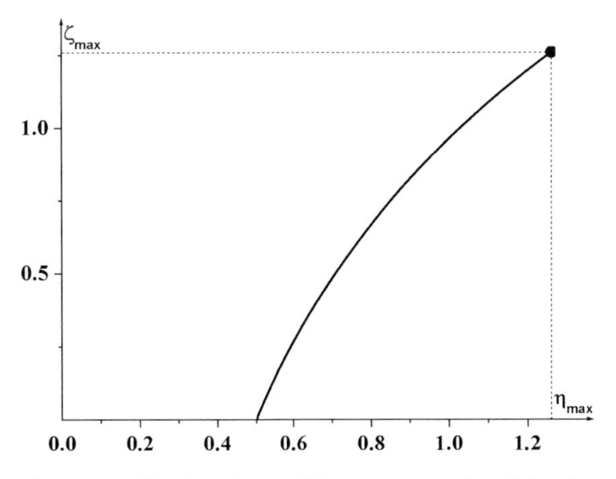

Figure 5.12 The dependence of the maximum reduced electric potential $\zeta_{\max} = e\varphi_{\max}/T_e$ on the reduced kinetic energy $\eta_{\max} = m_i u^2/2T_e$ at this maximum.

ion acoustic wave exists only in a restricted range of wave amplitudes and velocities. Exceeding the limiting amplitude leads to a wave decaying into separate waves.

5.4.6
Parametric Instability

Nonlinear phenomena are responsible for interaction between different modes of oscillation. A possible consequence of this interaction is the decay of a wave into two waves. Since the wave amplitude depends on time and spatial coordinates by the harmonic dependency $\exp(i\boldsymbol{k}\cdot\mathbf{r} - i\omega t)$, such a decay corresponds to fulfilling the relations

$$\omega_0 = \omega_1 + \omega_2 , \quad \boldsymbol{k}_0 = \boldsymbol{k}_1 + \boldsymbol{k}_2 , \tag{5.121}$$

where subscript 0 relates to the parameters of the initial wave and subscripts 1 and 2 refer to the decay waves. This instability is called parametric instability. We consider below an example of this instability in which a plasma oscillation decays into a plasma oscillation of a lower frequency and an ion acoustic wave (ion sound).

The electric field of the initial plasma oscillation is

$$\mathbf{E} = \mathbf{E}_0 \cos(k_0 x - \omega_0 t) ,$$

where x is the direction of propagation. In the zero approximation we assume the electric field amplitude E_0 and other wave parameters to be real values. The equation of motion for electrons $m_e dv_0/dt = -eE$ yields the electron velocity $v_0 = u_0 \cos(k_0 x - \omega_0 t)$, where $u_0 = eE_0/(m_e\omega_0)$.

Let another plasma wave and the ion sound wave be excited in the system simultaneously with the initial plasma oscillation, and let their amplitudes be small compared with the amplitude of the initial oscillation. Consider the temporal development of these waves taking into account their interaction with each other and with the initial oscillation. Since ion velocities are much lower than electron velocities, one can analyze these waves separately. The equation of motion and the continuity equation for ions are

$$m_i \frac{dv_i}{dt} = eE , \quad \frac{\partial N_i'}{\partial t} + N_0 \frac{\partial v_i}{\partial x} = 0 .$$

Here m_i is the ion mass, v_i is the ion velocity, N_0 is the equilibrium number density of ions, N_i' is the perturbation of the ion number density due to the oscillation, and E is the electric field due to the oscillations. Elimination of the ion velocity from these equations yields

$$\frac{\partial^2 N_i'}{\partial t^2} + \frac{eN_0}{m_i}\frac{\partial E}{\partial x} = 0 . \tag{5.122}$$

One can find the electric field strength from the equation of motion for the electrons by averaging over fast oscillations. The one-dimensional Euler equation (4.6) can be rewritten for electrons as

$$\frac{\partial v_e}{\partial t} + v_e \frac{\partial v_e}{\partial x} + \frac{1}{m_e N} \frac{\partial p_e}{\partial x} + \frac{eE}{m_e} = 0 , \tag{5.123}$$

where $p_e = NT_e$ is the the electron gas pressure, T_e is the electron temperature, and N_e is the electron number density. After averaging over fast oscillations, we find the first term in this equation is zero. We write the electron velocity as $v_e = v_0 + v_e'$, where v_e' is the electron velocity due to the small-amplitude plasma wave. Then we have

$$\overline{v_e \frac{\partial v_e}{\partial x}} = \frac{1}{2} \frac{\partial \overline{v_e^2}}{\partial x} = \frac{1}{2} \frac{\partial}{\partial x} \overline{(v_0 + v_e')^2} = \frac{\partial}{\partial x} \overline{(v_0 v_e')} ,$$

where the bar denotes averaging over fast oscillations. We assume that $T_e = \text{const}$, corresponding to a high rate of energy exchange in the ion acoustic wave. During the ion motion, the electron number density has a relaxation time for maintaining the quasineutrality of the plasma. Therefore, after averaging, we find the deviation of the electron number density from equilibrium is the same as that for ions, and the third term in the Euler equation (4.6) becomes

$$\frac{1}{m_e N} \frac{\partial p_e}{\partial x} = \frac{T_e}{m_e N_0} \frac{\partial N_i'}{\partial x} \quad (N \approx N_0) .$$

The Euler equation after averaging is transformed into

$$\frac{\partial}{\partial x} \overline{(v_0 v_e')} + \frac{T_e}{m_e N_0} \frac{\partial N_i'}{\partial x} + \frac{eE}{m_e} = 0 .$$

Substituting the electric field derived from this equation into (5.122), we obtain

$$\frac{\partial^2 N_i'}{\partial t^2} - \frac{T_e}{m_i} \frac{\partial^2 N_i'}{\partial x^2} - \frac{m_e N_0}{m_i} \frac{\partial^2}{\partial x^2} \overline{(v_0 v_e')} = 0 . \tag{5.124}$$

If we ignore the last term in (5.124) and assume harmonic dependence of the ion density on time and spatial coordinates, we obtain dispersion relation (5.62) for ion sound: $\omega = c_s k$, $c_s = \sqrt{T/m_i}$. To take into account interaction between ion sound and plasma oscillations, we have to trace the motion of electrons in the field of the small-amplitude plasma wave. To do this, we shall use the Maxwell equation for the electric field of the small-amplitude wave, assuming the magnetic field to be zero. This gives $\partial E'/\partial t + 4\pi j' = 0$. Here E' is the electric field of the wave and j' is the electric current density generated by it.

For simplicity, we shall ignore thermal motion of the electrons, since it has only a small effect on the oscillation frequency. Therefore, when writing the expression for the current density, we can ignore the variation of the electron number density due to the electron pressure of the plasma wave. We assume the electron number density to have the form $N_e + N_i'$, where N_e includes the equilibrium electron number density and its variation under the action of the initial plasma wave and N_i' is the variation of the ion number density owing to the motion of ions. Accordingly, the electron velocity is $v_0 + v'_e$, where v_0 is the electron velocity due to the initial plasma wave and v'_e is the electron velocity due to the small-amplitude plasma wave. The current density due to the small-amplitude plasma wave is then

$$j' = -e(N_e + N'_i)(v_0 + v'_e) + eN_e v_0 = -eN_e v'_e - eN'_i v_0 ,$$

where we ignore second-order terms. The Maxwell equation $\partial E'/\partial t + 4\pi j' = 0$ can now be rewritten as

$$\frac{\partial E'}{\partial t} - 4\pi e N'_i v_0 - 4\pi e N_e v'_e = 0 .$$

The electron equation of motion is

$$m_e \frac{dv'_e}{dt} = -eE' .$$

Eliminating E' from these equations, we obtain

$$\frac{\partial^2 v'_e}{\partial t^2} + \omega_p^2 v'_e + \frac{N'_i}{N_e}\omega_p^2 v_0 = 0 \tag{5.125}$$

as the equation for the electron velocity due to the small-amplitude plasma wave. Here $\omega_p = \sqrt{4\pi N_e e^2/m_e}$ is the frequency of plasma oscillations ignoring the thermal motion of electrons. One can see that if we do not take into account the interaction between the small-amplitude plasma wave and the initial plasma oscillation and the ion acoustic wave (that is, if we ignore the last term in (5.125)), then the assumptions used here give the small-amplitude plasma wave frequency as being equal to the plasma frequency.

Let us represent the solutions of (5.124) and (5.125) in the form

$$v_0 = u_0 \cos(k_0 x - \omega_0 t) , \quad v_e = a\cos(k_e x - \omega_e t) , \quad N'_i = bN_0 \cos(k_i x - \omega_i t) ,$$

where a and b are slowly varying oscillation amplitudes, ω_e and k_e are the frequency and the wave number of the small-amplitude plasma wave, ω_i and k_i are the frequency and the wave number of the ion sound, and N_0 is the equilibrium number density of the charged particles. (We assume that $N_e = N_0$ in (5.125).) Since the oscillation amplitudes vary slowly, the temporal and spatial dependencies are identical, so we find from (5.124) and (5.125) that

$$\omega_0 = \omega_e + \omega_i ; \quad k_0 = k_e + k_i , \tag{5.126}$$

as in (5.121). This condition is similar to that for parametric resonance of coupled oscillators, and therefore the instability that we analyze is termed a parametric instability. In addition, because of the nature of this process, this instability is called modulation instability. This instability has a universal character and is suitable both for plasma oscillations and for electromagnetic waves [75, 76].

Taking into account a slow variation of the oscillation amplitudes and condition (5.126), we find from (5.124) and (5.125) [35]

$$\frac{\partial a}{\partial t} = b\frac{\omega_e u_0}{4} , \quad \frac{\partial b}{\partial t} = -a\frac{m_e k_i u_0}{4m_i\omega_i}$$

for the oscillation amplitudes. Solution of these equations shows that the oscillations increase with the dependence $a, b \sim \exp(\gamma t)$, with

$$\gamma = \frac{1}{4}\sqrt{\frac{m_e \omega_e}{m_i \omega_i}} u_0 k_i = \frac{1}{4}\sqrt{\frac{m_e \omega_e}{m_i \omega_i}} \cdot \frac{e E_0}{m_e \omega_0} k_i \, . \tag{5.127}$$

Thus, the initial plasma wave is unstable. It can decay into a plasma wave of a lower frequency and the ion sound. This instability is also known as decay instability. The exponential growth parameter of the new wave is proportional to the amplitude of the decaying wave.

5.5 Ionization Instabilities and Plasma Structures

5.5.1 Drift Waves

A laboratory ionized gas is usually maintained by an external electric field that generates an electric current and causes ionization of the gas. Electrons are the principal plasma component in the sense that electrons contribute most of the total electric current, and formation of new charged particles is due to collisions of electrons with gas atoms or molecules. Perturbations of the electron number density are thus of central importance to the properties of the plasma.

To examine the simplest form of perturbations in the plasma, we begin with the continuity equation (4.1)

$$\frac{\partial N_e}{\partial t} + \operatorname{div} \mathbf{j} = 0$$

for electrons, and we assume that the electron flux $\mathbf{j}$ is determined solely by the electron drift in an external electric field. The current is then $\mathbf{j} = \mathbf{w} N_e$, where $\mathbf{w}$ is the electron drift velocity. Using expansion (5.48) for a perturbation to the electron number density, we obtain the dispersion relation

$$\omega = k w \, . \tag{5.128}$$

That is, there is a wave associated with the perturbation that propagates together with the electric current. In other words, the perturbation moves together with the electric current. Such waves which transfer plasma perturbation by an electric current are drift waves.

Damping or amplification of these waves can occur by several mechanisms. One such mechanism is the diffusive motion of the electrons. The electron flux is then $\mathbf{j} = \mathbf{w} N_e - D \nabla N_e$, where D is the diffusion coefficient for electrons in a gas, and the dispersion relation that follows from the continuity equation for electrons has the form

$$\omega = k w - i D k^2 \, . \tag{5.129}$$

The diffusion of the electrons is seen to lead to damping of the drift wave because the dependence of electron parameters on time has the form $\exp(-i\omega t)$. As a result of electron diffusion, the perturbation region increases in size and the perturbation dissipates.

The behavior of the drift wave depends on the type of ionization processes in the gas. Under certain conditions, the interaction of drift waves with the ionization process of gas atoms by electron impact leads to amplification of the drift waves. This phenomenon is called ionization instability. Various types of ionization instability can occur depending on the particular properties of a plasma and the processes within it. The development of ionization instability leads to the formation of structures in the plasma that will be considered below.

5.5.2
Ionization Instability from Thermal Effects

A common causal mechanism for ionization instability is to be found in the positive column of an arc, where the instability arises from thermal processes. The equilibrium electron number density depends strongly on the plasma temperature, whereas the temperature dependence of the heat transport coefficient is weak. At high intensities, thermal instability occurs because the heat transport mechanism due to atom thermal conductivity is unable to provide heat release. This instability may lead to contraction of the plasma that increases heat release or it can cause the formation of new types of structures in the plasma. A simple example of this phenomenon arises when the heat transport in a cylindrical discharge tube depends upon thermal conductivity. The heat balance equation in this case has the form

$$\frac{1}{\rho}\frac{d}{d\rho}\left[\rho\kappa(\rho)\frac{dT}{d\rho}\right] + p(\rho) = 0\,, \tag{5.130}$$

where ρ is the distance from the tube axis, κ is the thermal conductivity coefficient, and $p(\rho) = iE$ is the specific power of heat release, where i is the current density and E is the electric field strength. We assume that the rate of heat release is strongly dependent on the local plasma temperature, and is a function only of this temperature.

This model is descriptive of arc discharges at high currents and high pressures. If the gas discharge plasma is in equilibrium, the electron and gas temperatures are similar, and the electron number density can be estimated from the Saha formula (1.52) to behave as $N_e \sim \exp[-J/(2T)]$, where J is the ionization potential of the gas atoms, and it is usually true that $T \ll J$. The specific power of heat release is $p = iE$, where the electric field strength E does not vary over the discharge cross section, and the variation of the current density i is proportional to N_e. The specific power of heat release in an arc discharge is determined not only entirely by the temperature, but also has a strong dependence on the temperature. In this case the equilibrium between ionization and recombination is maintained at each point in the plasma, that is, local ionization equilibrium is supported in the plasma.

We introduce the new variables $z = \rho^2/r_0^2$ and $\theta = (T - T_\mathrm{w})/T_\mathrm{w}$ into (5.130), where r_0 is the tube radius and T_w is the wall temperature. Then (5.130) can be written in the form

$$\frac{d}{dz}\left(z\frac{d\theta}{dz}\right) + A\exp(b\theta) = 0\,, \tag{5.131}$$

where $A = r_0^2 p(T_\mathrm{w})/[4T_\mathrm{w}\kappa(T_\mathrm{w})]$, $b = T_\mathrm{w} d\ln p(T_\mathrm{w})/dT_\mathrm{w}$, and in this case $b \gg 1$. Because of the strong dependence of p on T, we can assume that the thermal conductivity coefficient κ is independent of the temperature. The solution of (5.131) is given by the Fock formula:

$$\theta = \frac{1}{b}\ln\frac{2\gamma}{Ab(1+\gamma z)^2}\,, \tag{5.132}$$

where γ is a parameter to be determined by boundary conditions. One of the boundary conditions of (5.131) is $\theta(0) = 0$, following from the definition of this function. This gives the relation for γ that

$$2\gamma = Ab(1+\gamma)^2\,. \tag{5.133}$$

Since γ is real, we have $Ab \leq 1/2$. If this condition is not satisfied, then (5.131) has no real solution. This means that thermal instability arises because heat extraction is not compensated for by heat release. The instability then leads to a different discharge regime where the current occurs in a narrow region near the tube center.

The threshold for this instability corresponds to $\gamma = 1$ and $Ab = 1/2$, that is, it has its onset when $p(T_0) = 4p(T_\mathrm{w})$, where T_0 is the temperature at the axis. The instability threshold at $Ab = 1/2$ corresponds to

$$\left[\frac{dp(T)}{dT}\right]_{T_\mathrm{w}} = \frac{2\kappa(T_\mathrm{w})}{r_0^2}\,. \tag{5.134}$$

If this instability leads to contraction of the plasma, these relations make it possible to estimate the radius ρ_0 for the new contracted regime of the plasma. Formula (5.134) then gives

$$\rho_0^2 \sim \kappa(T_0)\left[\frac{dp(T)}{dT}\right]^{-1}_{T=T_0}\,, \tag{5.135}$$

where T_0 is the temperature at the axis.

5.5.3 Ionization Wave in a Photoresonant Plasma

In considering a photoresonant plasma in Section 3.3.9, we found two regimes of plasma to exist. In the regime of low intensities of incident resonant radiation being absorbed in a plasma thickness of approximately $1/k_0$ (for definiteness, we assume the frequency of incident radiation is close to the center of the spectral line for the radiative transition, and k_0 is the absorption coefficient at the line center), this radiation flux is equal to the flux of scattering isotropic radiation that returns back through the plasma boundary, and this balance is determined by the degree of atom excitation, so the excitation temperature is given by (3.107). Correspondingly, the number density of excited atoms, as well as the flux of resonant radiation inside the plasma, drops exponentially with distance from the plasma boundary, and a typical depth of penetration of resonant radiation is approximately $1/k_0$.

Another mechanism of interaction between incident resonant radiation and a plasma takes place at high intensities of radiation and results in a chain of processes which lead to full plasma ionization. These processes establish the electron temperature according to (3.114), and this temperature is independent of the radiation intensity. But incident radiation propagates in this regime in the form of an ionization wave, and penetration of resonant radiation inside the plasma leads to its full ionization. As a result, the plasma becomes transparent, and radiation penetrates into deeper plasma layers.

Evidently, the front width for this ionization wave is approximately $1/k_0$, and the velocity w of propagation of this wave follows from the balance equation for the energy flux I according to $I = wNJ$, where we assume the thermal electron energy of approximately T_e to be small compared with the atomic ionization potential J, N is the number density of atoms of an ionized gas that is transformed subsequently into the fully ionized plasma, and the energy flux is $I = \hbar\omega j$, where j is the flux of incident photons. Thus, we have

$$w = \frac{\hbar\omega j}{JN} . \tag{5.136}$$

As follows from this formula, one can reach a high velocity of propagation of the ionization wave in reality. For example, for an alkali metal plasma (Tables 3.3 and 3.4) at a vapor pressure of 1 Torr the light speed is attained at radiation intensity $I \sim 10^9\,\mathrm{W/cm^2}$. Of course, in this case we are dealing with a nonrelativistic character of interaction between incident radiation and an ionized gas, and we will be guided by lower intensities of incident radiation, but note that in reality radiation intensities are available which exceed the above value by several orders of magnitude.

We assume that a transparent channel of a fully ionized plasma is conserved in the course of propagation of resonant radiation. But this channel expands after gas ionization because of gas heating. This creates an acoustic wave, that is, a photoresonant plasma is a source of acoustic oscillations. Next, because of nonuniformity of a formed plasma, atoms from external regions penetrate inside the transparent channel and are excited by resonant radiation. Let the radius r_0 of a radiation beam be large compared with the mean free path of atoms in this ionized gas. Then the atom flux inside a transparent channel results from diffusion of atoms in a plasma, and a typical time for its occupation is $\tau \sim r_0^2/D$, where D is the diffusion coefficient of atoms in the plasma. Hence, in this regime of propagation of resonant radiation, it penetrates inside a plasma to a depth L that is estimated from the relation

$$\frac{L^2}{D} \sim \frac{I r_0^2}{J N D} \,. \tag{5.137}$$

Being guided by the reduced diffusion coefficient of atoms in a plasma, $DN \sim 10^{19}\,\mathrm{cm^{-1}\,s^{-1}}$ according to the data in Tables 4.2–4.4, we find that radiation penetrates to a depth $L \sim 10\,\mathrm{cm}$, if the radiation power is $I r_0^2 \sim 100\,\mathrm{W}$. This shows that conditions for passage of photoresonant radiation through an ionized gas are available in reality.

5.5.4
Ionization Instability of a Plasma in a Magnetic Field

We now examine the behavior of a dense plasma in crossed electric and magnetic fields at low electric currents. In this problem, local ionization equilibrium exists everywhere within the plasma, so the Saha relation for the electron number density is valid, but heat transport processes are not essential. We can examine a time for evolution of a perturbation of the electron number density. The electron energy per unit volume is $W = 3 N_e T_e/2$, and the balance equation for this quantity has, according to (3.35), the form

$$\frac{d W_e}{dt} = -e E w_e N_e - 3\frac{m_e}{M}(T_e - T)\nu N_e \,. \tag{5.138}$$

Here we account for the negative direction of the electric field, m_e and M are the electron and atom masses, T_e and T are the electron and atom temperatures, and for simplicity we assume that the electron–atom collision rate $\nu = N_a v \sigma^*_{ea}$ is independent of the electron velocity, N_a is the atom number density, σ^*_{ea} is the diffusion cross section of electron–atom collisions, and v is the electron velocity.

Because of the local ionization equilibrium, we have $N_e \sim \exp(-J/2T_e)$, where J is the atom ionization potential. This gives the relation

$$\frac{N_e'}{N_e} = \frac{T_e'}{T_e}\frac{J}{2T_e}$$

between perturbation values of the electron number density N_e' and temperature T_e', so we can conclude that

$$\frac{T_e'}{T_e} \ll \frac{N_e'}{N_e} \,.$$

On the basis of this inequality, the term $T_e^{-1} d T_e/dt$ can be ignored compared with the term $N_e^{-1} d N_e/dt$ on the left-hand side of (5.138). This yields the result

$$\frac{d N_e}{dt} = -\gamma N_e' \,, \quad \gamma = \frac{4 T_e}{J}\frac{m_e}{M}\nu \,. \tag{5.139}$$

Thus, the ionization perturbation under consideration has a timescale for damping that is much longer than a typical time for electron–atom collisions.

In considering this plasma in an external magnetic field, we employ the usual geometry, where the electric field is directed along the x-axis and the magnetic field is directed along the z-axis. Then the electron drift velocity $\mathbf{w}$ in crossed electric and magnetic fields is given by (4.136), and in the case when the rate of electron–atom collisions is independent of the electron velocity $\nu = \text{const}$, it is convenient to reduce the indicated formulas to the following vector form:

$$\mathbf{w} = -\frac{e\mathbf{E}\nu}{m_e(\omega_H^2 + \nu^2)} + \frac{e\mathbf{E} \times \omega_H}{m_e(\omega_H^2 + \nu^2)}, \tag{5.140}$$

where the vector $\boldsymbol{\omega}_H = e\mathbf{H}/(m_e c)$ is parallel to the magnetic field.

The physical framework for this problem is such that electrodes collect the electron current in the x direction; hence, there is no electric current along the y-axis. The electron drift velocity along the x-axis creates the electric field along the y-axis, where the strength of this electric field is

$$\mathbf{E} = -\frac{m_e \nu}{e}\mathbf{w} - \frac{m_e}{e}\omega_H \times \mathbf{w}, \tag{5.141}$$

as follows from (5.140).

A perturbation in the electron number density causes a perturbation of the electric field strength and the electron drift velocity. We can determine these perturbations from (5.138), which is now

$$\frac{d W_e}{dt} = -eEwN_e' - eE'wN_e + \left(\frac{d W_e}{dt}\right)_T, \tag{5.142}$$

where $(d W_e/dt)_T = -3N_e T_e(m_e/M)$ is established by energy changes in electron–atom collisions, and was evaluated above.

The relationship between perturbations w' and N_e' can be derived from the continuity equation (4.1) for electrons. This can be used in the steady-state form $\text{div}(N_e\mathbf{w}) = 0$ since the instability develops slowly. Writing the dependence on $\mathbf{r}$ in the form $\exp(i\boldsymbol{k} \cdot \mathbf{r})$, where $\boldsymbol{k}$ is the wave vector, we obtain

$$N_e(\mathbf{w}' \cdot \boldsymbol{k}) + N_e'(\mathbf{w} \cdot \boldsymbol{k}) = 0. \tag{5.143}$$

We can now apply (5.140) to eliminate the perturbed drift velocity of the electrons from this expression. First, we find the direction of $\mathbf{w}'$. Since the perturbation develops slowly, we can write $\mathbf{E}' = -\nabla\varphi'$, where φ' is the perturbation of the electric potential. As before, we assume $\varphi' \sim \exp(i\boldsymbol{k} \cdot \mathbf{r})$ and obtain $\mathbf{E}' = -i\boldsymbol{k}\varphi'$, that is, the vectors $\mathbf{E}'$ and $\boldsymbol{k}$ are either parallel or antiparallel. Then, using the relationship (5.140) between vectors $\mathbf{w}'$ and $\mathbf{E}'$, we obtain

$$\mathbf{w}' = \text{const} \cdot \left[\boldsymbol{k} - \frac{\boldsymbol{k} \times \boldsymbol{\omega}_H}{\nu}\right].$$

Multiplying vector $\mathbf{w}'$ by itself, we evaluate the constant in the above expression, and obtain

$$\mathbf{w}' = \pm w' \frac{\nu\boldsymbol{k} - (\boldsymbol{k} \times \boldsymbol{\omega}_H)}{k\sqrt{\omega_H^2 + \nu^2}}.$$

Substituting this expression into the relationship derived from the continuity equation yields

$$\frac{w'}{w} = \pm \frac{N'_e}{N_e} \sqrt{\omega_H^2 + \nu^2} \cos\left(\frac{\alpha}{\nu}\right) , \tag{5.144}$$

where $\cos\alpha = k_x/k$, with α the angle between vectors $\mathbf{w}$ and $\boldsymbol{k}$. Substitution of this into (5.142) gives, with the help of (5.139),

$$\frac{dW'}{dt} = \frac{d}{dt}\left(\frac{3N_e T_e}{2}\right) = N'_e m_e w^2 \nu \left(\frac{\omega_H \sin 2\alpha}{\nu} - \cos^2\alpha - \frac{T_e}{2J}\right) . \tag{5.145}$$

If the term on the right-hand side of this equation is positive, any random deviation of the electron number density from its equilibrium value continues to increase, that is, instability develops. This term has a maximum when $\tan 2\alpha = -\omega_H/\nu$. For this direction of vector $\boldsymbol{k}$, (5.145) has the form

$$\frac{dW'}{dt} = N'_e m_e w^2 \nu \left(\frac{\sqrt{\omega_H^2 + \nu^2}}{\nu} - 1 - \frac{T_e}{2J}\right) . \tag{5.146}$$

From this it follows that this instability has a threshold, stated by $\omega_H/\nu \geq (T_e/J)^{1/2}$. When the ratio ω_H/ν is large, ionization instability develops for perturbations propagating at an angle $\alpha = 45°$ to the direction of the current. If this ratio is small, the most unstable perturbations propagate in a direction almost perpendicular to the current.

5.5.5
Attachment Instability of a Molecular Gas

Several different instabilities are possible during the formation of negative ions by attachment of electrons to molecules. This process is related to the formation of positive ions by loss of electrons, and to recombination events. All of these processes can create ionization structures and waves in a plasma. As an example, we shall consider attachment as it occurs in the plasma of an excimer laser, where a strong dependence of the attachment rate for electrons on the vibrational state of the molecule may lead to attachment instability. Then vibrational excitation of molecules causes acceleration of the electron attachment process and may produce a decrease in the electric current.

In an excited molecular gas (e.g., HCl) the attachment rate constant increases with an increase of the vibrational temperature. The positions of electronic states, as illustrated in Figure 5.13, determine the properties of this process. It proceeds according to the following scheme:

$$AB + e \to (AB^-)^{**} , \quad (AB^-)^{**} \to A + B^- , \quad (AB^-)^{**} \to AB^* + e . \tag{5.147}$$

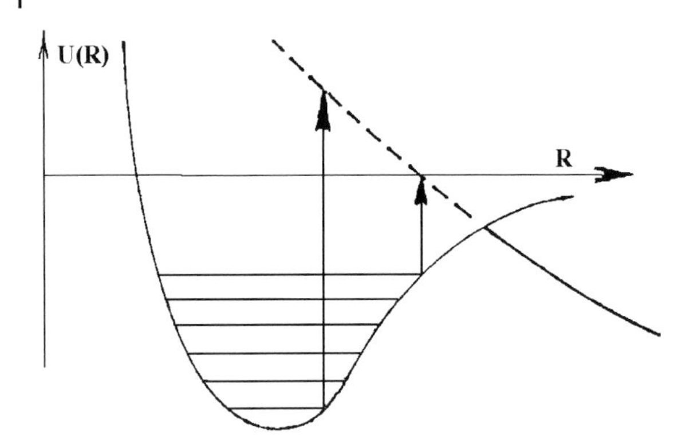

Figure 5.13 Electronic states which determine the electron attachment process involving vibrationally excited molecular states.

Here $(AB^-)^{**}$ denotes an autodetaching state of the negative ion and AB^* denotes a vibrational excited state of the molecule. The probability of attachment after formation of the autodetaching state $(AB^-)^{**}$ is proportional to

$$\exp\left[-\frac{\int_{R_c}^{R} \Gamma(R) dR}{\hbar v}\right], \qquad (5.148)$$

where $\Gamma(R)$ is the width of the autodetachment level, v is the relative velocity of the nuclei, R is the distance of the electron captured into the autodetachment level, and R_c is the distance at which the states intersect. The exponential factor is the survival probability of the autodetaching state, and it is usually quite small. Hence, the greater the distance at which the electron is captured, the higher the probability for formation of the negative ion. Because the capture distance increases with an increase of the molecular vibrational temperature, the attachment rate constant also increases with an increase of the vibrational temperature.

We can apply these features to the analysis of the balance equations for the electron number density and vibrational temperature T_v, which have the form

$$\frac{\partial N_e}{\partial t} = -\nu_{at} N_e + \nu_{ion} N_e, \quad \frac{\partial T_v}{\partial t} = N_e k_{ex} \hbar\omega - M_{rel}.$$

Here $\nu_{at} = N k_{at}$ is the electron attachment frequency, (k_{at} is the rate constant for this process, N is the number density of molecules), ν_{ion} is the ionization frequency, k_{ex} is the rate constant for the vibrational excitation of molecules by electron impact, $\hbar\omega$ is the vibrational energy of the molecule, and M_{rel} is the vibrational relaxation rate. Linearizing these equations, assuming that perturbations of N_e and T_v vary with time as $\exp(-i\omega t)$, and taking $\nu_{at} = \nu_{ion}$ for an unperturbed state, we find that the process being studied is oscillatory in nature if $d\nu_{at}/dT_v > 0$. The frequency is

$$\omega_E^2 = k_{ex} N_e \hbar\omega \frac{d\nu_{at}}{dT_v}, \qquad (5.149)$$

and the oscillations are known as Eletskii oscillations [77].

One can combine these oscillations and drift waves by inserting the electron drift into the balance equation for the electron number density. Then instead of dispersion relations (5.128) and (5.149), we obtain the following dispersion relation:

$$\omega = \frac{kw}{2} + \sqrt{\frac{k^2 w^2}{4} + \omega_E^2}\,, \tag{5.150}$$

which transforms to (5.128) or (5.149) in the appropriate limits.

5.5.6 Current Convective Instability

Just as there are a variety of mechanisms that can alter the manner in which unbound electrons can be introduced into and removed from a plasma, so also there are different ways in which ionization instabilities can develop. In addition to the above-described mechanisms for destabilizing a plasma, we now analyze one more case that involves a plasma in a strong magnetic field. The configuration we explore is one in which electric current flows in a cylindrical tube so that the electric and magnetic fields are directed along the axis of this plasma column. The magnetic field is so strong that electrons and ions in the plasma are magnetized, that is, the Larmor radii of electrons and ions are small compared with the mean free paths for these particles. Since charged particles recombine on the walls of the tube, the plasma is uniform across the diameter of the tube. These are the conditions in which an ionization instability called the current convective instability may occur [78, 79].

In the environment described, electrons will rotate with respect to ions in any cross section of the plasma column. This gives rise to an azimuthal electric field that enhances this motion. The development of the instability is slow if plasma quasineutrality is conserved. If a small inhomogeneity of the plasma is present in the direction of the current, then the plasma density is different in neighboring cross sections of the column. Since the current in the plasma is conserved, a decrease in the plasma density in any cross section of the column must be compensated for by an additional electric field.

Now let us assume an "oblique" perturbation of the plasma density that is compensated for by an electric field directed at an angle to the column axis. Then the azimuthal component of the electric field will produce rotation of the electrons and ions. This will enhance a suitably directed perturbation, leading to an instability. Let us derive a dispersion relation for such perturbations. The plasma is quasineutral and magnetized, so the velocity of electrons and ions is

$$\mathbf{w} = c\frac{\mathbf{E} \times \mathbf{H}_0}{H_0^2}\,, \tag{5.151}$$

where $\mathbf{H}_0$ is the magnetic field and $\mathbf{E}$ is the electric field of the wave. The magnetic field of the wave can be ignored since the perturbation is slow. For the same reason, the electric field can be described by the potential φ, where $\mathbf{E} = -\nabla\varphi$.

The current density is $i_z = \Sigma E_z$, where the z-axis is along the axis of the column and Σ is the plasma conductivity. The electric field is the sum of the external

(E_0) and wave fields. Since the wave parameters are proportional to $\exp(i\mathbf{kr} - i\omega t)$, we have $E_z = E_0 - ik_z\varphi$. The plasma conductivity is proportional to the electron number density, so $\Sigma = \Sigma_0 + \Sigma' = \Sigma_0(1 + N_e'/N_0)$, where Σ_0 is the plasma conductivity, N_0 is the electron number density in the absence of perturbations, and Σ' and N_e' are the corresponding perturbed parameters. From the above expressions, the condition for conservation of the current density in the field direction is

$$-\Sigma_0 ik_z\varphi + \Sigma' E_0 = 0 ,$$

or

$$-ik_z\varphi + \frac{N_e'}{N_0}E_0 = 0 . \tag{5.152}$$

The continuity equation (4.1) for electrons is

$$\frac{\partial N_e}{\partial t} + \operatorname{div}(N_e\mathbf{w}) - D_a\frac{\partial^2 N_e}{\partial z^2} = 0 ,$$

where D_a is the ambipolar diffusion coefficient for the plasma. We have ignored the diffusive flux of electrons perpendicular to the magnetic field, because the magnetic field strength is high and this flux is small. Assuming a harmonic dependence of the perturbed parameters on time and spatial coordinates, we can rewrite this equation to first order as

$$(-i\omega + k_z^2 D_a)N_e' + w_x\frac{\partial N_0}{\partial x} = 0 , \tag{5.153}$$

where the x-axis is in the direction of the maximum gradient of the equilibrium number density.

Equations (5.151) and (5.152) give

$$w_x = \frac{cE_y}{H_0} = -c\frac{ik_y\varphi}{H_0} = \frac{k_y}{k_z}\frac{cE_0}{H_0}\frac{N_e'}{N_0} .$$

Substituting this into dispersion relation (5.153), we obtain

$$i\omega = k_z^2 D_a + \frac{k_y}{k_z}\frac{cE_0}{H_0 L} , \tag{5.154}$$

where $1/L = -d(\ln N_0)/dx$. One can see that instability ($\text{Im}\,\omega < 0$) will develop if the ratio k_y/k_z has an appropriate sign and value [78]. The instability has a threshold with respect to the electric field. The magnetic field must be high enough to meet the conditions described above.

Equation (5.154) shows that the optimum condition for this instability is such that the cyclotron frequency for ions is of the order of the collision frequency of ions with gas particles. The threshold for this instability is connected to the diffusion of charged particles. If the diffusion rate is sufficiently high, perturbations of the plasma are damped by the diffusion.

5.5.7
Electric Domain

One of a variety of steady-state structures that can be formed by nonlinear processes in a plasma is the electric domain – a perturbation of the electron number density and the electric field in a gas discharge plasma that propagates with the electron current. The physical picture of this phenomenon is as follows. The external segment of an electric circuit that passes through a gas discharge has a large resistance that maintains a constant electric current density in the discharge. This constant current is

$$i_e = -ewN_e = \text{const}\,, \tag{5.155}$$

where w is the electron drift velocity and N_e is the electron number density. The dependence of the electron drift velocity on the electric field strength has the behavior shown in Figure 5.14, where two different electric current regimes are possible at some values of the electric field strength. This behavior can preserve a perturbation that propagates together with the current, meaning that it is a drift wave. Such a perturbation is identified as the electric domain. Although this plasma structure was discovered in a semiconductor plasma [80–83], it is of importance for a gas discharge plasma [35].

To describe the electric domain, we use the continuity equation (4.1) and the Poisson equation (1.4), which here take the forms

$$\frac{\partial N_e}{\partial t} + \frac{\partial j_e}{\partial x} = 0\,, \quad \frac{\partial E}{\partial x} = 4\pi e(N_0 - N_e)\,, \tag{5.156}$$

where j_e is the electron flux, E is the electric field strength, and N_0 is the equilibrium number density of electrons. We first treat these equations in a linear approximation so as to analyze the electric domain at low intensity. The electron flux can be written

$$j_e = -N_e w - D\frac{\partial N_e}{\partial x}\,, \tag{5.157}$$

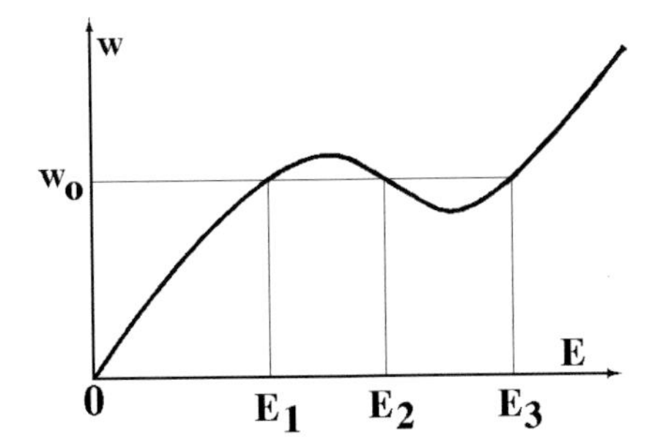

Figure 5.14 A nonmonotonic dependence of the electron drift velocity on the electric field strength that can lead to the formation of a domain.

where D is the diffusion coefficient of electrons in a plasma, and the minus sign in the first term accounts for the direction of the electron current. Taking E and N_e to exhibit harmonic behavior as expressed by (5.48), we can write

$$E = E_0 + E' e^{i(kx-\omega t)}, \quad N_e = N_0 + N' e^{i(kx-\omega t)}.$$

We then obtain the dispersion relation

$$\omega = -kw - iDk^2 - i \cdot 4\pi N_0 e \frac{dw}{dE} \tag{5.158}$$

on the basis of (5.156) and (5.157) using the connection $\partial w/\partial x = (dw/dE) \times (dE/dx)$. Dispersion relation (5.158) describes drift waves (5.129) that move together with the electron current. The diffusion process removes electrons from the perturbation zone, thus causing damping of these waves. But if $dw/dE < 0$, long drift waves with $Dk^2 < i4\pi N_0 e dw/dE$ can develop.

We now study a nonlinear electric domain. In a nonsteady interval of the electric field strengths in Figure 5.14, (5.157) has the form

$$j_0 = -N_0 w(E_2) = -N_e w(E) - D\frac{\partial N_e}{\partial x}.$$

We should add to this equation the Poisson equation (1.4) or (5.156). Eliminating the electron number density between these equations, we obtain

$$D\frac{d^2 E}{dx^2} = w(E)\frac{dE}{dx} - 4\pi e N_0[w(E) - w(E_2)]. \tag{5.159}$$

Equation (5.159) describes the behavior of the electric field strength in the electric domain.

When we assume that diffusion plays a secondary role, and ignore diffusion in the first stage of the analysis, (5.159) takes the form

$$\frac{dE}{dx} = 4\pi e N_0 \left[\frac{w(E_2)}{w(E)} - 1\right]. \tag{5.160}$$

We solve this equation with the boundary condition $E = E_2$ at $x = 0$. The solution is shown in Figure 5.15a. As follows from (5.160), $E(x)$ increases with x until $E_2 < E < E_3$, where $w(E_3) = w(E_2)$. At $E = E_3$ we have $dE/dx = 0$, so for subsequent values of x we obtain $E = E_3$. Thus, this solution describes the conversion of the system from the unstable state E_2 to the stable state with $E = E_3$.

This transition means that the change of the discharge regime as a result of the perturbation leads to an increase of the electric field strength up to E_3. This would require a variation of the discharge voltage, which is impossible because the discharge voltage is maintained by the external voltage. Therefore, the variation of the electric field strength from E_2 to E_3 is a perturbation that takes place in a limited region of the plasma. A return to the initial value of the field occurs as a result of diffusion, leading to decay of the perturbation. Hence, a typical size of the back

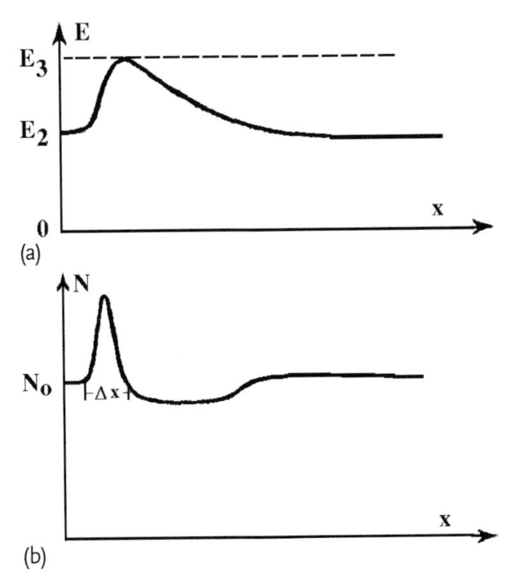

Figure 5.15 The distribution of the electric field (a) and the electron number density (b) in the electric domain.

boundary zone of the electric domain is of the order of D/w. The forward boundary zone size can be estimated from (5.160), and is $\Delta x = \Delta E/(4\pi e N_0 \Delta w)$, where $\Delta E = E_3 - E_2$ and $\Delta w = w(E_2) - w_{\min}$ (see Figure 5.14). A concomitant of the distribution of the electric field strength in the electric domain is the distribution of the electron number density that is shown in Figure 5.15b and follows from the Poisson equation (5.156).

5.5.8
Striations as Ionization Waves

Striations or strates are plasma structures that are observed in a positive column of a gas discharge of low or moderate pressure [84–89]. The column is striped in an alternation of luminous and dark regions. Striations have a rich history of study, and a detailed description of some aspects of this phenomenon was made in the nineteenth century [90–94]. Thomson [95, 96] was the first to describe the oscillations of the positive column of a gas discharge, and his equations were improved by Druuyvesteyn [97, 98]. These oscillations may be considered as standing waves – stationary striations. Striations can be at rest or move to electrodes from the positive column. In addition, striations may be self-induced oscillations or may be excited by an external source [99–101]. In both cases they characterize wave properties of a gas discharge plasma. Striations exist only in certain ranges of currents and pressures, although this range may be wide enough. For example, in hydrogen, striations exist in the pressure range which varies within the limits of eight orders of magnitude and in the range of discharge currents that covers four orders of magnitude [101]. Striations have different properties for atomic and for molecular gases.

Striations are ionization waves and may be considered as plasma structures resulting from an instability of ionization and recombination processes or as nonlinear ionization waves [61, 102]. From this standpoint, striations exist if a stratified plasma structure is more favorable than the uniform plasma distribution, that is, the electric voltage for a stratified structure of a gas discharge plasma at a given electric current corresponds to a lower voltage for gas discharge than that for a uniform plasma [103]. Then in considering striations as an ionization instability [85, 86, 104–106], one can understand the conditions under which these ionization waves exist.

It is clear that if the ionization rate is independent of the electron number density, we obtain a stable uniform plasma, and for existence of an ionization instability the above dependence is necessary. We now consider ionization waves in a plasma resulting from stepwise ionization of atoms. This mechanism corresponds to low-pressure glow discharges in inert gases, where metastable atoms are of importance for ionization processes, and below we will be guided by this case. Hence, from the standpoint of the classification of striation types [107–109], we analyze below the range of low discharge currents. Along with ionization processes, we consider the ambipolar diffusion that leads to loss of electrons and provides the stabilization of a gas discharge plasma. The balance equation for the electron number density is

$$\frac{\partial N_e}{\partial t} + D_a \frac{\partial^2 N_e}{\partial x^2} = Z N_e \,, \tag{5.161}$$

where D_a is the ambipolar diffusion coefficient, the x-axis is along the axis of the discharge tube, and Z is the sum of the rates of generation and loss of electrons. Assuming that the ionization wave has a small amplitude, we have the usual harmonic dependence $N_e = N_0 + N' \exp(ikx - i\omega t)$, where $N' \ll N_0$. When we introduce the effective electron temperature $T_e = T_0 + T' \exp(ikx - i\omega t)$ and insert these relations into (5.159), we obtain the dispersion relation [110]

$$\omega = -i D_a k^2 + i Z_T N_e \frac{\partial T_e}{\partial N_e} + i Z_N N_e \tag{5.162}$$

for the ionization wave, where we have used the notation $Z_N = \partial Z/\partial N_e$ and $Z_T = \partial Z/\partial T_e$, and have taken into account that $Z = 0$ at equilibrium.

This dispersion relation was obtained by a formal procedure. It is of importance that the ionization rate Z depends on both the number density and the temperature of the electrons, and variation of one of these quantities alters the other. Depending on the connection between these quantities, waves can be either developed or damped. We give some illustrative examples. Assume that the resistance of the external circuit is large enough to conserve the electron current density $\mathbf{i}$. Then we have

$$\mathbf{i} = \Sigma \mathbf{E} + e D_a \nabla N_e = \text{const} \,.$$

This gives the electric field

$$\mathbf{E} = \frac{\mathbf{i}}{\Sigma} - \frac{e D_a \nabla N_e}{\Sigma} , \tag{5.163}$$

where Σ is the plasma conductivity and D_a is the coefficient of ambipolar diffusion that characterizes the diffusion of electrons in a plasma.

When we use the balance equation for the mean energy of electrons, and take into account elastic collisions of electrons and gas atoms, the heat balance equation for electrons has the form

$$\operatorname{div} \mathbf{q} = \mathbf{i} \cdot \mathbf{E} - N_e \nu \delta \varepsilon = 0 , \tag{5.164}$$

where $\mathbf{q}$ is the heat flux. The first term on the right-hand side of this equation accounts for heating of electrons under the action of the electric current. The second term takes into account elastic electron–atom collisions; ν is the collision rate and $\delta \varepsilon$ is the mean energy transferred to atoms in a single collision. Further, we assume that $\delta \varepsilon$ is proportional to the mean electron energy and that the collision rate is independent of the electron energy.

We assume that the heat flux q is determined by the electron thermal conductivity ($\mathbf{q} = -\kappa \nabla T_e$, where κ is the thermal conductivity coefficient of electrons) and $\Sigma \sim N_e$. Taking the values Σ and D_a to be dependent on T_e, (5.163) and (5.164) give the dispersion relation

$$k^2 \kappa T' = -\frac{i^2}{\Sigma} \frac{N'}{N_0} - \frac{i k E j D_a}{\Sigma} N' - \frac{i^2}{\Sigma} \frac{N'}{N_0} + \frac{T'}{T_0} .$$

From this it follows that

$$\frac{\partial T_e}{\partial N_e} = \frac{T'}{N'} = \frac{T_0}{N_0} \frac{(-2 - i k / k_0)}{(1 + k^2 / k_1^2)} , \tag{5.165}$$

where $k_0 = i/(e D_a N_0)$ and $k_1 = i\sqrt{\Sigma T_0 \kappa}$. On the basis of (4.40) we have the estimate for the thermal conductivity coefficient due to electrons $\kappa \sim N_0 v_e \lambda$, and from (3.42) we have $\Sigma \sim N_0 e^2 \lambda / (m_e v_e)$, where N_0 is the equilibrium electron number density and $v_e \sim \sqrt{T_e / m_e}$ is a typical thermal electron velocity. From this we conclude that k_0 and k_1 have the same order of magnitude.

We now find the conditions that will give rise to an ionization instability leading to the formation of striations. For this goal we substitute (5.165) into (5.162) and analyze the resulting expression. Since the effective electron temperature T_e is small compared with the atomic ionization potential $J/T_e \ll J$, (3.69) and (3.70) give $Z_T \sim (J/T_e^2) \nu_{ion}$, where ν_{ion} is the rate of atom ionization by electron impact. Because of the dependence of Z on the electron number density, we have $Z_N \sim \nu_{ion}/N_0$. This gives $Z_N N_0 \ll Z_T T_e$, and for long-wave oscillations ($k \ll k_0$), (5.162) and (5.165) give $\operatorname{Im} \omega < 0$, so long-wave oscillations are damped. Therefore, striations as nonlinear ionization waves can only be short-wave oscillations. Then the criterion for the existence of striations (the Rother condition [111–113]) has the form

$$Z_N > 0 . \tag{5.166}$$

There is a range of wavelengths satisfying the condition $k > k_0$, where $\mathrm{Re}\,\omega > \mathrm{Im}\,\omega$, so ionization waves – striations – can exist. These waves are developed by nonlinear interaction processes.

Consider the example of ionization instability, governed by the nonlinear stepwise ionization of atoms by electron impact. We shall assume that there is only one excited atomic state that contributes to the ionization. We denote this state by the subscript or superscript $*$. We thus have

$$Z = N_a k_{ion} + N_* k^*_{ion} - 1/\tau_D \,, \tag{5.167}$$

where N_a and N_* are the atom number densities in the ground and excited states, k_{ion} and k^*_{ion} are the rate constants for ionization of atoms in the ground and excited states caused by electron impact, and τ_D is the time for diffusion of electrons to the walls of the discharge tube. At equilibrium we have $Z = 0$. The dependence of Z on the electron number density due to ionization of excited atoms is

$$Z_N = k^*_{ion} \frac{\partial N_*}{\partial N_e} \,. \tag{5.168}$$

Using (3.66) for the number density of excited atoms, we obtain

$$N_* = N_*^B \left[1 + \frac{1}{N_e k_q \tau_D} \right]^{-1} ,$$

where N_*^B is given by the Boltzmann formula and corresponds to equilibrium between the ground and excited states. Substituting this into (5.165) and (5.166), and using the relation $n_* = 1/(k_q \tau_D)$, we have

$$Z_N = \frac{k^*_{ion} n_* N_*^B}{(N_e + n_*)^2} \,, \quad Z_T = \frac{J}{T_e^2} \left(k_{ion} N_a + \frac{k_{ion} N_*^B N_e}{N_e + n_*} \right) .$$

Inserting these expressions into (5.160) yields

$$\begin{aligned} \mathrm{Re}\,\omega &= -Im \left(Z_T N_e \frac{\partial T_e}{\partial N_e} \right) = \frac{J}{T_e^2} \left(k_{ion} N_a + \frac{k_{ion} N_*^B N_e}{N_e + n_*} \right) \frac{k/k_0}{1 + k^2/k_1^2} \,, \\ \mathrm{Im}\,\omega &= \frac{k^*_{ion} n_* N_*^B N_e}{(N_e + n_*)^2} - D_a k^2 - \frac{2J}{T_e^2} \left(k_{ion} N_a + \frac{k_{ion} N_*^B N_e}{N_e + n_*} \right) \frac{1}{1 + k^2/k_1^2} \end{aligned} \tag{5.169}$$

for the frequency and damping (or amplification) of the ionization wave, where k_0 and k_1 are defined in (5.165).

Equations (5.169) identify that range of parameters where amplification of the ionization wave ($\mathrm{Im}\,\omega > 0$) can occur. According to (3.69), we find that $k_{ion} N_a/(k^*_{ion} N_*^B) \sim J_*/J < 1$, where J and J_* are the ionization potentials of the atom in the ground and excited states. Hence, when $N_e \sim N_*$ and $k > k_1(J/T_e)$, the last term in expressions (5.169) for the growth of the ionization wave is not compensated for by the first term. Then the ionization wave can develop until it will no longer decay as a result of electron diffusion. Thus, the ionization wave can amplify in the range of wave vectors

$$k_1\sqrt{\frac{J}{T_e}} < k < \sqrt{\frac{k_{\text{ion}}\,N_*^{\text{B}}}{D_a}}\ . \tag{5.170}$$

If this condition is satisfied, ionization waves can develop, and striations arise. The plasma then exhibits a striped structure with a specific periodicity.

Considering the striations as ionization waves gives an understanding of some aspects of this problem [85, 86]. One can formulate conclusions which follow from the above analysis of striations as instabilities of ionization waves in a gas discharge plasma. For low gas discharge currents this instability proceeds only in an intermediate range of electron number densities where the ionization rate depends on the number density of electrons (the Rother condition (5.166)). Next, we have different regions in the spatial distributions of species, so the wave distribution of the electron number density is shifted with respect to the electron temperature.

5.5.9
Self-Consistent Structure of Striations

In considering striations as ionization waves, we divide a region into elements with the length of a wave for each element, so plasma parameters are repeated in each element. According to the Pupp classification [107–109], for striations of low pressure and discharge electric currents, the wavelengths of striations as oscillations in a plasma of inert gases are small compared with the mean free path electrons with respect to energy change in elastic electron–atom collisions, but are large in comparison with the mean free path of electrons with respect to electron scattering by atoms. According to the properties of waves, the electric voltage $\varphi(\lambda)$ between two points a distance of one wave period λ apart is the same

$$\varphi(\lambda) = \int_0^{\lambda} E(x)dx = \text{const}\ ,$$

where $E(x)$ is the electric field strength at point x and λ is the wavelength for striations. According to the Novak rule [116], the voltage per striation is almost constant for a given striation type, that is, this value is independent of the electric current and pressure. As a demonstration of this, Table 5.1 gives the values of the voltage per striation for three striation types (P, R, S) in gas discharge of inert gases, where these values are compared with the atomic excitation energy ΔE, the atomic ionization potential J and the ionization potential $J*$ for a metastable atom.

Our goal is to understand the nature of striations as ionization waves in an atomic gas of low density if the mean free path of electrons is small compared with a striation length (the spatial period of oscillations), but the mean free path of electrons with respect to variation of their energy as a result of elastic collisions with

Table 5.1 The electric potential (V) on one striation for three types of striations in inert gases [86]. The atomic excitation energy ΔE is given along with the atomic ionization potential in the ground J and metastable $J*$ states.

Types of striation	P	R	S	ΔE, eV	J, eV	J^*, eV
He	14.2	–	30.05	19.8	24.6	4.8
Ne	9.2	12.7	19.5	16.62	21.5	4.9
Ar	6.7	9.5	12.0	11.55	15.76	4.2

atoms exceeds the striation length significantly. Under these conditions, the variation of the electron energy as a result of the electron displacement from coordinate x_1 to coordinate x_2 along the field is [114, 115]

$$\varepsilon(x_1) + \varphi(x_1) = \varepsilon(x_2) + \varphi(x_2) ,$$

where ε is the electron energy at an indicated point and φ is the plasma electric potential at this point. Hence, in the first approximation each electron acquires a certain energy after passage of one striation.

Let us combine this with the Novak rule [116], according to which the energy acquired by an electron after passage of one striation period is constant for a given striation type, that is, it is independent of the gas pressure and discharge current. One can conclude from this that a certain energy is given to an individual electron on each striation. On this basis one can construct a certain physical picture of plasma processes with formation of striations. Indeed, if a group of electrons lose

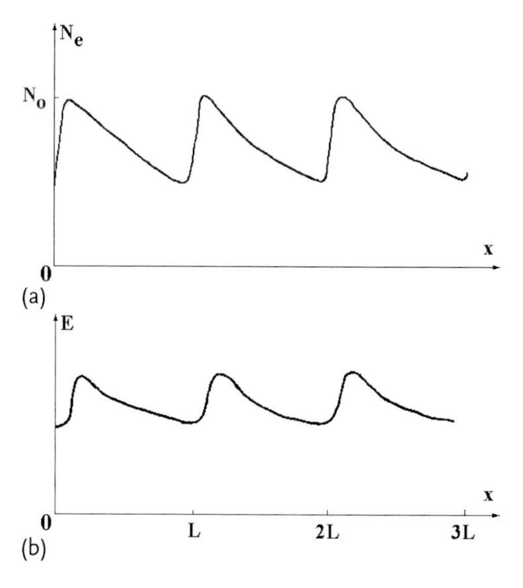

Figure 5.16 Distributions of the number density of electrons (a) and the electric field strength (b) along a gas discharge tube of small radius under the formation of striations.

excitation energy, these electrons are stopped, and their density increases in a corresponding point of space (see Figure 5.16). Then the voltage distribution in space has the form of a "Tsendin staircase", as shown in Figure 5.17 [89, 117]. Moreover, Figure 5.17 shows the bunching of the electron flux in this case. Indeed, each electron that partakes in excitation loses an identical amount of energy, and for these electrons in this approximation the electron energy distribution function is conserved on the following striation. If we take into account the energy loss in elastic electron–atom collisions, fast electrons lose more high energy than slow ones. As a result, the electron energy distribution function becomes narrower, and bunching of electrons takes place. Within the framework of this description, one can expect formation of striations in the case when two regimes of the plasma state, with striations and for the uniform plasma distribution, may be realized. Then one can consider a plasma with striations as an energetically profitable plasma state [103], that is, in this plasma state the total voltage of this plasma is optimal.

Seemingly, the Tsendin model, which is based on a stronger braking of fast electrons compared with slow ones, allows us to explain the nature of striation formation. Indeed, the electron velocity distribution function is narrow owing to bunching of the electron beam and is not broadened after passage of many striation periods. Then the energy

$$\int_0^{\lambda} eE(x)dx = \Delta\varepsilon \tag{5.171}$$

Figure 5.17 Tsendin staircase – the form of the plasma potential in striation formation and the character of bunching [89, 117].

obtained by an individual electron after passage of one period subsequently goes into atom ionization. Equation (5.171) holds in the absence of energy exchange between electrons and atoms as a result of elastic collisions, and in the absence of electron–electron collisions. Evidently, the energy $\Delta\varepsilon$ is equal to $e\varphi$, where φ is the electric voltage on one striation given in Table 5.1 for inert gases. For example, one can expect $\Delta\varepsilon$ to be equal to or close to the atomic excitation energy for striations of a certain type. Then collision of two excited atoms formed after this striation leads to ionization. Unfortunately, we do not have a direct connection between the values in Table 5.1 and the atomic excitation energies or atomic ionization potentials. This testifies also to the role of metastable atoms in ionization processes involving striations.

Nevertheless, if we overlook this contradiction, we can come to an agreement about the character of variation of striation parameters in this model with the Pekarek theory [86, 104–106]. Indeed, at the beginning of some striation, where the electrons are slow, the number density of electrons is higher than that in a region of fast electrons because their flux in the longitudinal direction is conserved. Hence, the regions of the maximum number density and the maximum energy (temperature) of electrons (or glowing maximum) are shifted in accordance with the above-mentioned theory. Note that along with ionization processes, the loss of electrons and ions takes place as a result of their ambipolar diffusion to walls. We assume there are long striations in a discharge tube if the tube radius is small compared with the striation wavelength, so the loss process proceeds uniformly along the tube.

In conclusion of this analysis, we note that the striation mechanism under consideration corresponds to low pressures in atomic gases, which allows us to ignore the energy loss of electrons in elastic collisions with atoms. Next, we ignore electron–electron collisions, so the discharge currents for this mechanism of striations are limited. Under such conditions, the energy transferred to electrons from the electric field of the discharge is consumed in excitation and ionization of atoms. Because striations exist for a nonlinear dependence of the ionization rate on the number density of electrons, the ionization process in a striated plasma has a stepwise character. From this it follows that metastable atoms are of importance for generation of striations in inert gases, and along with electron collisions with metastable atoms, diffusion of metastable atoms in a plasma may be important for formation of striations. The latter process leads to washing out of the striation wave in contrast to the above-mentioned bunching mechanism.

This model shows that two regimes of the plasma can exist. In the first, the plasma is homogeneous, and ionization of atoms by electron impact occurs throughout the plasma. In the other regime, ionization occurs in narrow ionization zones distributed periodically over the length of the discharge tube. That is, striations exist in this regime. The regime in which the discharge plasma finds itself is determined by the energy available to the system, so the characteristics of an external source of electricity are of great importance for the plasma behavior.

References

1 Landau, L.D. and Lifshits, E.M. (1959) *Fluid Mechanics*, Pergamon Press, London.
2 Batchelor, G.K. (1967) *An Introduction to Fluid Dynamics*, Cambridge University Press, Cambridge.
3 Drazin, P. and Reid, W. (1981) *Hydrodynamic Stability*, Cambridge University Press, Cambridge.
4 Acheson, D.J. (1990) *Elementary Fluid Dynamics*, Oxford University Press, Oxford.
5 Lord Rayleigh (1916) *Philos. Mag.*, **32**, 529.
6 Gershuni, G.Z. and Zhukhovitskii, E.M. (1972) *Convective Stability of Compressed Liquid*, Nauka, Moscow, in Russian.
7 Benard, H. (1900) *Rev. Gen. Sci. Pure Appl.*, **11**, 261.
8 Benard, H. (1900) *Rev. Gen. Sci. Pure Appl.*, **11**, 1309.
9 Benard, H. (1901) *Ann. Chim. Phys.*, **23**, 62.
10 Chandrasekhar, S. (1961) *Hydrodynamic and Hydromagnetic Stability*, Oxford University Press, London.
11 Koschmieder, E.L. (1993) *Benard Cells and Taylor Vortices*, Cambridge University Press, Cambridge.
12 Getting, A.V. (1998) *Rayleigh–Taylor Convection Structure and Dynamics*, World Scientific Publishing, London.
13 Nernst, W. (1918) *Z. Electrochem.*, **24**, 335.
14 Semenov, N.N. (1928) *Z. Phys.*, **48**, 571.
15 Zeldovich, Ya.B. and Frank-Kamenestkii, D.A. (1938) *Zh. Fiz. Khim.*, **12**, 100.
16 Zeldovich, Ya.B. and Frank-Kamenestkii, D.A. (1938) *Dokl. Akad. Nauk SSSR*, **19**, 693.
17 Arrhenius, S. (1889) *Z. Phys. Chem.*, **4**, 226.
18 Laidler, K.J. (1993) *The World of Physical Chemistry*, Oxford University Press, Oxford.
19 Levine, R.D. (2005) *Molecular Reaction Dynamics*, Cambridge University Press, Cambridge.
20 Zeldovich, Ya.B. (1944) *Theory of Combustion and Gas Detonation*. Izdat. AN SSSR, Moscow, in Russian.
21 Frank-Kamenestkii, D.A. (1969) *Diffusion and Heat Transfer in Chemical Kinetics*, Plenum Press, New York.
22 Kondrat'ev, V.N. and Nikitin, E.E. (1974) *Kinetics and Mechanism of Gas-Phase Reactions*, Nauka, Moscow, in Russian.
23 Zeldovich, Ya.B., Barenblatt, G.I., Librovich, V.B., and Makhviladze, G.M. (1985) *The Mathematical Theory of Combustion and Explosions*, Consultants Bureau, New York.
24 Stix, T.H. (1992) *Waves in Plasmas*, American Institute of Physics, New York.
25 Swanson, D.G. (2003) *Plasma Waves*, Institute of Physics Publishing, Bristol.
26 Langmuir, I. (1925) *Phys. Rev.*, **26**, 585.
27 Langmuir, I. (1928) *Proc. Natl. Acad. Sci. USA*, **14**, 627.
28 Tonks, L. and Langmuir, I. (1929) *Phys. Rev.*, **33**, 195.
29 Bohm, D. and Gross, E.P. (1949) *Phys. Rev.*, **75**, 1851.
30 Bohm, D. and Gross, E.P. (1949) *Phys. Rev.*, **75**, 1864.
31 Alfvén, H. (1942) *Nature*, **150**, 405.
32 Alfvén, H. (1950) *Cosmical Electrodynamics*, Oxford University Press, London.
33 Landau, L.D. and Lifshitz, E.M. (1960) *Electrodynamics of Continuous Media*, Pergamon Press, Oxford.
34 Alfvén, H. and Fälthammar, C.G. (1963) *Cosmical Electrodynamics. Fundamental Principles*, Clarendon Press, London.
35 Kadomtsev, B.B. (1988) *Cooperative Phenomena in Plasma*, Nauka, Moscow, in Russian,English version: Shafranov, V.D. (2001) *Rev. Plasma Phys.*, **22**, 1.
36 Fälthammar, C.G. (1963) *Cosmic Electrodynamics. Fundamental Principles*, Clarendon Press, London.
37 Ginzburg, V.L. and Rukhadze, A.A. (1972) *Waves in Magnetoactive Plasma*, Nauka, Moscow, in Russian.
38 Maxwell, J.C. (1865) *Philos. Trans. Phys. Soc.*, **155**, 459.

39 Maxwell, J.C. (1873) *Treatise on Electricity and Magnetism*, Oxford University, London.

40 Heaviside, O. (1884) *Electrician*, **12**, 583–586.

41 Thomson, J.J. (1928) *Proc. Phys. Soc.*, **40**, 82.

42 Faraday, M. (1846) *Philos. Trans. R. Soc.*, **136**, 1.

43 Faraday, M. (1859) *Experimental Researches in Chemistry and Physics*, Taylor and Francis, London.

44 VanVleck, J.H. (1932) *The Theory of Electric and Magnetic Susceptibilities*, Oxford University Press, New York.

45 Born, M. (1933) *Optik*, Springer, Berlin.

46 Jenkins, F.A. and White, H.E. (1976) *Fundamentals of Optics*, McGraw-Hill, New York.

47 Wood, R.W. (1988) *Physical Optics*, Optical Society of America, Washington.

48 Mansuripur, M. (1995) *The Physical Principles of Magneto-optical Recording*, Cambridge University Press, London.

49 Piece, W.H (1894) *Nature*, **49**, 554.

50 Barkhausen, H. (1919) *Phys. Z.*, **20**, 401.

51 Helliwell, R.A. (1965) *Whistlers and Related Ionospheric Phenomena*, Stanford University Press, Stanford.

52 Akhiezer, A.I. and Feinberg, Ya.B. (1951) *Zh. Eksp. Teor. Fiz.*, **21**, 1262.

53 Gabor, D., Ash, E.A., and Dracot, D. (1955) *Nature*, **178**, 916.

54 Nezlin, M.V. (1995) *Physics of Intense Beams in Plasmas*, IOP Publishing, Bristol.

55 Buneman, O. (1958) *Phys. Rev. Lett.*, **1**, 8.

56 Buneman, O. (1963) *Phys. Rev. Lett.*, **10**, 285.

57 Khalil, S.H.M. and Mohamed, B.F. (1991) *Contrib. Plasma Phys.*, **31**, 513.

58 Benett, W.H. (1934) *Phys. Rev.*, **45**, 890.

59 Mikhailowskii, A.B. (1974) *Theory of Plasma Instabilities*, Consultant Bureau, New York.

60 Mikhailovskii, A.B. (1992) *Electromagnetic Instabilities in an Inhomogeneous Plasma*, IOP Publishing, Bristol.

61 Landa, P.C. (2010) *Nonlinear Oscillations and Waves in Dynamical Systems*, Kluwer, Dordrecht.

62 Kadomtzev, B.B. (1965) *Plasma Turbulence*, Academic Press, New York.

63 Tsytovich, V.N. (1970) *Nonlinear Effects in Plasmas*, Plenum Press, New York.

64 Hasegawa, A. (1975) *Plasma Instabilities and Nonlinear Effects*, Springer, Berlin.

65 Wojaczek, K. (1962) *Beitr. Plasmaphys.*, **2**, 1.

66 Lighthill, M.J. (1965) *J. Inst. Math. Appl.*, **1**, 1.

67 Korteweg, D.J. and de Vries, G. (1895) *Philos. Mag.*, **39**, 422.

68 Stenzel, R., Kim, H.C., and Wong, A.Y. (1974) *Phys. Rev. Lett.*, **32**, 659.

69 Ikezi, H., Nishikawa, K., and Mima, K. (1974) *J. Phys. Soc. Jpn.*, **37**, 766.

70 Antipov, S.V. *et al.* (1976) *JETP Lett.*, **23**, 562.

71 Antipov, S.V. *et al.* (1977) *JETP Lett.*, **25**, 145.

72 Antipov, S.V. *et al.* (1981) *Physica D*, **3**, 311.

73 Thejappa, G. *et al.* (1999) *J. Geophys. Res.*, **104**, 279.

74 Ivanov, A.A. (1977) *Physics of Strongly Nonequilibrium Plasma*, Atomizdat, Moscow, in Russian.

75 Das, K.P. and Sihi, S. (1982) *Plasmaphysics*, **22**, 29.

76 Agrawal, G.P. (1995) *Nonlinear Fiber Optics*, Academic Press, San Diego.

77 Eletskii, A.V. (1986) *Sov. Phys. JTP*, **56**, 850.

78 Nedospasov, A.V. (1975) *Sov. Phys. Usp.*, **18**, 588.

79 Khalil, S.H.M., Sayee, Y.A., and Zaki, N.G. (1989) *Contrib. Plasma Phys.*, **29**, 605.

80 Gunn, J.B. (1963) *Solid State Commun.*, **1**, 88.

81 Gunn, J.B. (1965) *Int. Sci. Technol.*, **46**, 43.

82 Bonch-Bruevich, V.L., Zvyagin, I., and Mironov, A. (1972) *Domain Electric Instability in Semiconductors*, Nauka, Moscow, in Russian.

83 Pozhela, J. (1981) *Plasma and Current Instabilities in Semiconductors*, Pergamon Press, Oxford.

84 Oleson, N.L. and Cooper, A.W. (1968) *Adv. Electron. Electron Phys.*, **1**, 65.

85 Nedospasov, A.V. (1968) *Sov. Phys. Usp.*, **11**, 174.

86 Pekarek, L. (1968) *Sov. Phys. Usp.*, **11**, 188.

87 Garscadden, A. (1978) in *Gaseous Electronics*, (eds. M.N. Hirsh, H.J. Oskam), Academic Press, New York p. 1, 65.

88 Kolobov, V.I. (2006) *J. Phys. D*, **39**, R487.

89 Tsendin, L.T. (2010) *Phys. Usp.*, **180**, 130.

90 Wüllner, A. (1874) *Ann. Phys. Chem.*, **32**, 1.

91 Spottiswoode, W. (1875) *Proc. R. Soc.*, **23**, 455.

92 Spottiswoode, W. (1876) *Proc. R. Soc.*, **24**, 73.

93 Spottiswoode, W. (1877) *Proc. R. Soc.*, **25**, 547.

94 Spottiswoode, W. (1877) *Proc. R. Soc.*, **26**, 90.

95 Thomson, J.J. (1921) *Philos. Mag.*, **42**, 1454.

96 Thomson, J.J. and Thomson, G.P. (1933) *Conduction of Electricity through Gases*, Cambridge University Press, London.

97 Druyvesteyn, M.J. (1934) *Physica*, **1**, 273.

98 Druyvesteyn, M.J. (1934) *Physica*, **1**, 1003.

99 Zaitsev, A.A. (1951) *Dokl. Akad. Nauk SSSR*, **79**, 779, in Russian.

100 Zaitsev, A.A. (1952) *Dokl. Akad. Nauk SSSR*, **84**, 41, in Russian.

101 Klarfel'd, B.N. (1952) *Zh. Teor. Exp. Fys.*, **22**, 6.

102 Landa, P.C., Miskinova, N.A., and Ponomarev, Yu.V. (1990) *Sov. Phys. Usp.* **22**, 813.

103 Raizer, Yu.P. (1997) *Gas Discharge Physics*, Springer, Berlin.

104 Pekarek, L. (1954) *Czech. J. Phys.*, **4**, 221.

105 Pekarek, L. (1957) *Phys. Rev.*, **108**, 1371.

106 Rohlena, K., Ruzicka, T., and Pekarek, L. (1972) *Czech. J. Phys. B*, **22**, 72.

107 Pupp, W. (1932) *Phys. Z.*, **33**, 844.

108 Pupp, W. (1933) *Phys. Z.*, **34**, 756.

109 Pupp, W. (1934) *Phys. Z.*, **35**, 705.

110 Wojaczek, K. (1959) *Monatsber. Dtsch. Akad. Wiss.*, **1**, 23.

111 Rother, H. (1959) *Ann. Phys.*, **4**, 373.

112 Rother, H. (1959) *Z. Phys.*, **157**, 326.

113 Rother, H. (1960) *Ann. Phys.*, **5**, 252.

114 Bernstein, I.B. and Holstein, T. (1954) *Phys. Rev.*, **94**, 1475.

115 Tsendin, L.T. (1974) *Sov. Phys. JETP*, **39**, 805.

116 Novak, M. (1960) *Czech. J. Phys.*, **10**, 954.

117 Tsendin, L.T. (1995) *Plasma Source Sci.Technol.* **4**, 200.

6
Complex Plasmas, Including Atmospheric Plasmas

6.1
Single Cluster or Particle in an Ionized Gas

6.1.1
Characteristics of a Complex Plasma

According to its definition, a complex plasma is an ionized gas with small particles in it [1–5]. The characteristic feature of this system is that particles are sinks for electrons and ions, and hence the presence of particles in a plasma may change its properties. There are various types of complex plasma. Atmospheric air is an aerosol plasma, and aerosol particles (or aerosols) in air may be of various types and of various sizes in the range 0.01–100 μm. In particular, water aerosols are found at altitudes of a few kilometers and are responsible for electric processes in the atmosphere [6, 7]; below we consider this problem in detail. Tropospheric aerosol particles are formed near the Earth's surface as a result of natural processes and human activity. These aerosols are made up of SO_2, NH_3, NO_x, and mineral salts and other compounds that result from natural processes on the Earth's surface and may reach high altitudes. The smallest aerosols, Aitken particles [8, 9] of sizes 10–100 nm are present at high altitudes above clouds. Scattering of sunlight by Aitken particles determines the blue color of the sky, and Aitken particles are partially responsible for interaction of solar radiation with the Earth's atmosphere [6, 7]. The number density of Aitken particles at optimal altitudes is 10^2–10^4 cm^{-3} [10, 11], and they partake in atmospheric chemical reactions at high altitudes.

Let us consider the atmospheric plasma from another standpoint. The basis of this plasma is the solar wind – a flux of plasma that is emitted by the Sun's corona [12, 13]. Encountering dust generated by a cold condensed system, the solar wind forms a dusty plasma, and the number density of charged particles in this plasma exceeds by several orders of magnitude that in the solar wind. In this way the dusty plasma of Saturn's rings is formed, and the dusty plasma of comet tails results from the same processes. Magnetic fields are of importance for the properties of dusty plasmas in the solar system. The long lifetime of trapped ions near charged dust particles leads to an increased number density of ions, which for comet tails is 10^3–10^4 cm^{-3} [13–15] and exceeds that in the solar wind, whereas the

Fundamentals of Ionized Gases, First Edition. Boris M. Smirnov.

electron temperature $T_e \sim 10^4$ K [13, 15–17] corresponds to that of the solar wind. In this manner other types of dusty plasma are formed in the solar system [18–22]. In particular, interaction of the solar wind, a flux of plasma emitted by the Sun's corona [12, 13], with dust from the rings of Jupiter and Saturn leads to formation of a specific dusty plasma [23–26]. A similar type of dusty plasma corresponds to comet tails [27, 28], and the properties of this plasma are determined by interaction of comet dust with the solar wind [29]. A laboratory dusty plasma is formed in a trap, usually near the cathode of radio-frequency discharge or inside striations of glow discharge. A strong impetus for study of the laboratory dusty plasma was the discovery of ordered structures formed by particles of identical size in a trap [30–33]. This fact testifies to the self-consistency of a dusty plasma as a physical system. This allowed the laboratory dusty plasma to be considered as a specific physical object [34–38]. A detailed study of the laboratory dusty plasma paves the way for applications of this plasma [4, 39].

A cluster plasma is an ionized gas containing nanosized clusters [40–42]. This plasma differs from a dusty plasma not only by particle sizes, but also by the processes which occur this system. Indeed, clusters may grow or evaporate in a cluster plasma, that is, the nucleation and evaporation processes are of importance for a cluster plasma. Below we will be guided by a cluster plasma consisting of a buffer gas (e.g., argon) and a metal vapor with a low concentration of metal atoms. This plasma may be used for transport of heatproof metals and as a light source. We consider below some aspects of complex plasmas.

We mostly consider the behavior of an individual nanocluster or microparticle in ionized gas. We use the liquid drop model for a cluster [43] to identify it as a spherical liquid drop that is cut out from a bulk system. Within the framework of this model, the principal property of a cluster as a physical object due to magic numbers [44–51] is lost, so nanoclusters and microparticles become identical. This model corresponds to averaging a cluster property under consideration of nearby sizes, which simplifies the analysis a certain cluster property. Therefore, we consider now the behavior of individual clusters or particles in a gas. In this context, the behavior of a probe in an ionized gas [52–56] and satellites in the atmosphere [57] has an analogy with the properties of individual nanoclusters and microparticles.

6.1.2
Transport Parameters of Clusters

In considering transport phenomena involving large clusters, it is convenient to use for a cluster the liquid drop model, according to which the cluster is cut out of a bulk system and has spherical shape. Then the number densities of atoms in the cluster and the bulk liquid are identical, and the number of cluster atoms n and the cluster radius r_0 are connected by (2.16), $n = (r_0/r_W)^3$, where r_W is the Wigner–Seitz radius [58, 59], which is given by (2.17) and its values are represented in Table 2.2. For large clusters $r_0 \gg a_0$, where a_0 is a typical atomic size, the diffusion cross section of atom–cluster collisions is given by (2.18), $\sigma^* = \pi r_0^2$. We have two limiting cases for cluster transport in a gas depending on the relation between the cluster radius r_0 and the mean free path λ for atoms in a gas, and the mobility and diffusion coefficient of clusters in gas for these cases are given by (2.19), (4.113), and (4.114).

The relaxation character of a cluster in a gas is connected to the large cluster mass compared with the mass of an atom, and hence equilibrium of clusters with a gas results from many collisions with atoms of a gas. In particular, for a rareness gas (2.19) represents the motion equation for a cluster in the form

$$\frac{d\mathbf{w}}{dt} = -\nu\mathbf{w} ,$$

where $\mathbf{w}$ is the cluster velocity with respect to the gas in which it is located and ν is the rate of relaxation. If the relaxation rate for gas atoms is determined by the rate of collisions with other atoms, that is, atom relaxation is determined by a single collision, many collisions are required for a cluster to change its velocity because of the large cluster mass. Hence, the rate of cluster relaxation decreases with increasing cluster size. In particular, within the framework of the liquid cluster model, for the relaxation rate of a large cluster we have [60, 61]

$$\nu = \frac{8\sqrt{2\pi m T}}{3n m_a} \cdot N_a r_0^2 = \frac{8N_a r_W^2 \sqrt{2\pi m T}}{3m_a} \cdot n^{-1/3} , \tag{6.1}$$

where m is the mass of a gas atom, m_a is the mass of a cluster atom, and n is the number of cluster atoms.

Long relaxation of the cluster velocity if the cluster velocity differs from the mean gas velocity is determined by the large cluster mass. This takes place in gas flows with clusters where the gas velocity varies. In particular, we consider the spread case when clusters are in a gas flow in a tube of variable cross section as shown in Figure 6.1. Then the gas flow leaves the tube through an orifice and acquires a velocity c_s of the order of the sound velocity near the orifice. Clusters in this flow do not influence the character of this laminar flow, but in the case

$$\frac{\rho_0}{c_s}\nu \ll 1$$

the drift velocity of clusters near the orifice is low compared with the flow velocity c_s. In particular, in the case of a cylindrical tube and a conic exit, as shown in

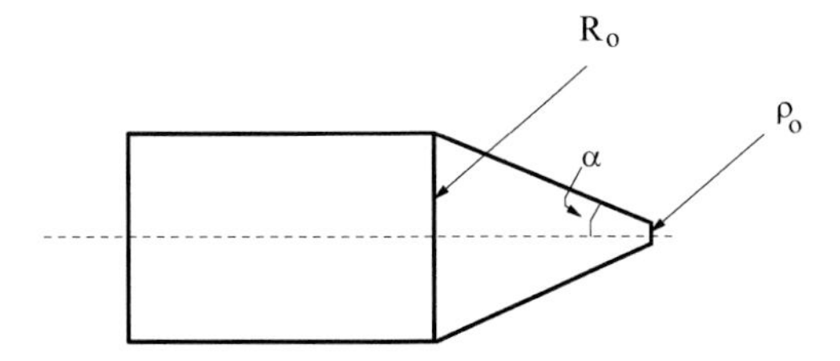

Figure 6.1 The geometry of a cylindrical tube with a conic exit through which a gas with clusters flows. R_0 is the radius of the cylindrical tube, ρ_0 is the orifice radius, and α is the conic angle.

Figure 6.1, the drift velocity of clusters w near the orifice is [60, 61]

$$w = 2.68 c_s (\nu\tau)^{2/3}\,, \quad \tau = \frac{\rho_0}{3c_s \tan\alpha}\,, \tag{6.2}$$

where the relaxation rate ν is given by (6.1), τ is the time for variation of the flow velocity, and α is the conic angle.

We consider also falling of a cluster or a particle under the action of their weight. Taking the case of a dense gas (6.5) and equalizing the weight to the resistive force that is given by the Stokes formula, we obtain the free-fall velocity w for a spherical particle of radius r_0:

$$w = \frac{2\rho g r_0^2}{9\eta}\,, \tag{6.3}$$

where g is the free-fall acceleration and η is the gas viscosity. In particular, for a water particle in air at atmospheric pressure and temperature $T = 300\,\text{K}$, this formula may be represented in the form

$$w = 0.0118\frac{r_0^2}{a^2}\,, \tag{6.4}$$

where the velocity of the falling particle is given in centimeters per second and $a = 1\,\mu\text{m}$. This formula holds true at small values of the Reynolds number $\text{Re} = vr/v \ll 1$, where $v = \eta/\rho_g$ is the kinematic viscosity of the gas, and ρ_g is the gas density. In the case of water drops in air this criterion is $r_0 \ll 30\,\mu\text{m}$.

6.1.3
Charging of Particles in a Dense Ionized Gas

When a cluster is in an ionized gas, electrons and ions may attach to its surface, which leads to cluster charging and also produces a sink for electrons and ions of an ionized gas. In considering the process of cluster charging in an ionized gas, we will model clusters by a macroscopic spherical liquid drop (Section 2.1.3), so its radius is large compared with the distance of interaction between the cluster surface and atomic particles, and both large nanoclusters and microparticles are identical in this consideration. Since a cluster is charged, a self-consistent field is formed near it that influences the equilibrium cluster charge and the distribution of electrons and ions near the cluster. We consider below charging processes for an isolated cluster in an ionized gas that determine the cluster charge and the self-consistent field near it.

For a dense buffer gas, when the criterion

$$r_0 \gg \lambda \tag{6.5}$$

holds true (λ is the mean free path of ions in a gas), we have the following expression for the ion flux $J_+(R)$ toward the cluster, which is the total number of positive ions crossing a sphere of a radius R around a cluster per unit time:

$$J_+ = 4\pi R^2 \left(-D_+ \frac{dN_+}{dR} + K_+ E N_+ \right) , \tag{6.6}$$

where we ignore the cluster field screening by a surrounding plasma, so the electric field strength acting on an ion from the charged cluster is $E = Ze/R^2$, $N_+(R)$ is the current number density of positive ions with electron charge e, D_+ and K_+ are the diffusion coefficient and the mobility of positive ions, and $-|Z|e$ is the negative cluster charge. The first term in this expression corresponds to diffusive motion, the second term relates to ion drift.

If outside the cluster there is no recombination involving positive ions, (6.6) may be considered as the equation for the ion number density in this region. Assuming that a positive ion transfers its charge to the cluster as a result of contact with the cluster surface, we have the boundary condition $N(r_0) = 0$. Another boundary condition far from the cluster is $N(\infty) = N_0$, (N_0 is the equilibrium number density of atomic ions in an ionized gas). The latter means that attachment of electrons and ions to clusters does not violate the ionization balance in this plasma. Assuming the electric field of the cluster to be relatively small, we use the Einstein relation $K_+ = eD_+/T$ between the ion mobility K_+ and diffusion coefficient D_+, where T is the ion temperature. This leads to the following equation for the number density $N_+(R)$ of positive ions:

$$J_+ = -4\pi R^2 D_+ e \left(\frac{dN_+}{dR} - \frac{Ze^2 N_+}{TR^2} \right) .$$

Solution of this equation with the boundary condition $N_+(r_0) = 0$ gives

$$N_+(R) = \frac{J_+}{4\pi D_+} \int_{r_0}^{r} \frac{dR'}{(R')^2} \exp\left(\frac{Ze^2}{TR'} - \frac{Ze^2}{TR} \right)$$
$$= \frac{J_+ T}{4\pi D_+ Ze^2} \left[\exp\left(\frac{Ze^2}{Tr_0} - \frac{Ze^2}{TR} \right) - 1 \right] .$$

Applying the boundary condition $N_+(\infty) = N_0$ far from the cluster leads to the Fuks formula [62] for the total rate of ion attachment to the cluster surface:

$$J_+ = \frac{4\pi D_+ N_0 Z e^2}{T \left\{ \exp\left[Ze^2/(Tr_0) \right] - 1 \right\}} . \tag{6.7}$$

In the limiting case $Z \to 0$, the Fuks formula (6.7) is transformed into the Smoluchowski formula for the diffusive flux J_0 of neutral atomic particles to the surface of an absorbed sphere of radius r_0 [63]:

$$J_0 = 4\pi D_+ N_0 r_0 . \tag{6.8}$$

We now consider the case of different signs of the particle and ion, in particular, the positive ion current goes to a negatively charged particle. Then replacing

$Z \to -Z$ in the Fuks formula (6.7), we obtain for the rate of attachment of positively charged ions (electrons) to a negatively charged cluster

$$J_- = -\frac{4\pi D_- N_0 Z e^2}{T \cdot \left[1 - \exp\left(-\frac{Ze^2}{Tr_0}\right)\right]} . \tag{6.9}$$

In the limit $Ze^2/(r_0 T) \gg 1$ this formula is transformed into the Langevin formula [64]:

$$J_- = \frac{4\pi Z e^2 D_-}{T} = 4\pi Z e K_- N_0 . \tag{6.10}$$

Let us find the charge Z of a cluster in an ionized gas with positive ions and electrons as charged atomic particles. Equalizing the rates of attachment of positive ions and electrons to the cluster surface, we obtain the following expression for the equilibrium cluster charge:

$$Z = \frac{r_0 T}{e^2} \ln \frac{D_+ N_+}{D_- N_-} = \frac{r_0 T}{e^2} \ln \frac{K_+ N_+}{K_- N_-} , \tag{6.11}$$

where N_+, and N_- are the number densities of positive ions and electrons far from the cluster. In a quasineutral plasma, $N_+ = N_-$, the currents of positive ions and electrons given by (6.7) and (6.9) are equal. For the cluster equilibrium charge this yields

$$Z = \frac{r_0 T}{e^2} \ln \frac{D_+}{D_-} = \frac{r_0 T}{e^2} \ln \frac{K_+}{K_-} . \tag{6.12}$$

We also consider the case of a nonequilibrium plasma consisting of electrons and positive ions if $T_e \gg T_i$. Then (6.11) takes the form

$$Z = -\frac{r_0 T_e}{e^2} \ln \frac{K_e}{K_i} ,$$

where K_e and K_i are the mobilities of electrons and ions in a buffer gas. Note that in a gas discharge plasma the electron distribution function differs from the Maxwell one, and the electron temperature is not a characteristic of a Maxwell distribution of electrons, but it is introduced as $T_e = e D_e / K_e$.

In deriving the Fuks formula, we assume the electric field of the cluster $E = Ze/r_0^2$ is weak near the cluster surface, which gives

$$e E \lambda \sim \frac{Ze^2}{r_0^2} \lambda \ll T .$$

Since $Ze^2/r_0 \sim T$, this condition gives $\lambda \gg r_0$, which is the criterion for a dense ionized gas. In addition, the Fuks formula (6.7) is valid under the criterion

$$r_0 \gg \frac{e^2}{T} . \tag{6.13}$$

Because in this case the average cluster charge $|Z|$ is large, in spite of the stepwise character of attachment of electrons and ions to the cluster surface, this process may be considered as a continuous one.

Below we find the rate of charge relaxation to the average charge and the particle charge distribution function in the stationary case assuming a typical cluster charge to be large. We assume that the plasma consists of positive and negative ions with similar diffusion coefficients D_+ and D_- in a buffer gas. Introducing the mean diffusion coefficient of ions $D = (D_+ + D_-)/2$ and the difference $\Delta D = |D_+ - D_-|$, which is relatively small, $\Delta D \ll D$, we reduce (6.12) for the average particle charge to the form

$$Z_0 = \frac{r_0 T}{e^2} \cdot \frac{\Delta D}{D} .$$

The equation of cluster charging has a simple form in this case,

$$\frac{dZ}{dt} = e(J_- - J_+) = \frac{Z_0 - Z}{\tau} ,$$

and the time τ for establishment of the equilibrium cluster charge is given by

$$\frac{1}{\tau} = \frac{4\pi N_0 e^2 D}{T} = 2\pi \Sigma , \tag{6.14}$$

where $\Sigma = N_0 e(K_+ + K_-) = 2N_0 e^2 D/T$ is the conductivity coefficient of the ionized gas. The solution of the charging equation in this case at the initial condition $Z(0) = 0$ has the form

$$Z = Z_0(1 - e^{-t/\tau}) ,$$

and as is seen, a typical time for establishment of the equilibrium cluster charge does not depend on the cluster radius.

6.1.4
The Charge Distribution Function for Clusters

We now find the charge distribution function for clusters $f(Z)$ taking into account that as a result of ion attachment, the charge varies by one unit. This leads to the following kinetic equation for the charge distribution function:

$$f(Z)[J_+(Z) + J_-(Z)] = f(Z-1)J_+(Z-1) + f(Z+1)J_-(Z+1) ,$$

where $J_+(Z)$ is the rate (the probability per unit time) for the cluster charge to change from Z to $Z+1$ as a result of attachment of positive ions to a cluster and, correspondingly, $J_-(Z)$ is the rate for the cluster charge to change from Z to $Z-1$ due to attachment of negative ions to a cluster. It is convenient to reduce (6.7) and (6.9) for charging rates to the form

$$J_+(Z) = \frac{4\pi D_+ N_0 r_0 Z x}{\exp(Zx) - 1} , \quad J_-(Z) = \frac{4\pi D_- N_0 r_0 Z x}{1 - \exp(-Zx)} , \quad x = \frac{e^2}{r_0 T} .$$

Substituting these expressions into the above kinetic equation for the charge distribution function of clusters, we reduce the kinetic equation to the form

$$\frac{\left[D_+ + D_- \exp(Zx)\right] Zx}{\exp(Zx) - 1} f(Z) = \frac{D_+(Z-1)x}{\exp[(Z-1)x] - 1} f(Z-1) + \frac{D_-(Z+1)x}{\exp[(Z+1)x] - 1} f(Z+1) \,. \qquad (6.15)$$

This kinetic equation may be used for determination of the charge distribution function for clusters. We consider first the case when along with criterion (6.5), the criterion $x \gg 1$ holds true, which is inverse with respect to criterion (6.13). In this case clusters are mostly neutral, and a small number of clusters have charge $Z = \pm 1$, whereas the probability for a cluster having charge $|Z| \geq 2$ is exponentially small. Indeed, from the kinetic equation (6.15) it follows that

$$\frac{f(2)}{f(1)} = \frac{D_+(e^{2x} - 1)}{2(D_+ + D_- e^{2x})(e^x - 1)} = \frac{D_+}{2D_-} e^{-x} \,,$$

where $f(Z)$ with $Z \geq 2$ is exponentially small. The same conclusion relates to $Z \leq -2$. This allows us to restrict our attention to neutral and singly charged ions. We have $f(0) \approx 1$, and from (6.14) we have

$$\frac{f(1)}{f(0)} \approx f(1) = \frac{D_+(e^x - 1)}{(D_+ + D_- e^x)x} = \frac{D_+}{D_- x} \,,$$

$$\frac{f(-1)}{f(0)} \approx f(-1) = \frac{D_-(e^x - 1)}{(D_- + D_+ e^x)x} = \frac{D_-}{D_+ x} \,.$$

From this we find for the average charge of particles

$$\overline{Z} = f_1 - f_{-1} = \frac{D_+^2 - D_-^2}{D_+ D_- x} \,.$$

In the case $D_+ - D_- = \Delta D \ll D_+$, this formula gives $\overline{Z} = 2\Delta D/(Dx) \ll 1$.

We now consider the case when criterion (6.5) holds true along with criterion (6.13). It is convenient to introduce a new variable $z = xZ$, and because Z is a whole number and $x \ll 1$, this variable is taken to be continuous. Then we represent the kinetic equation (6.15) in the form

$$-f(z)[D_+ F(z) + D_- F(z) e^z] + f(z-x) D_+ F(z-x) + f(z+x) D_- F(z+x) e^{z+x} = 0 \,,$$

where

$$F(z) = \frac{z}{e^z - 1} \,.$$

Expanding this equation over the small parameter $x \ll 1$ and accounting for the first two expansion terms, we reduce the kinetic equation to the form

$$\frac{d}{dz}\left[\left(\frac{D_+}{D_-} - e^z\right) f(z) F(z)\right] - \frac{x}{2}\frac{d^2}{dz^2}\left[\left(\frac{D_+}{D_-} + e^z\right) f(z) F(z)\right] = 0 \,. \qquad (6.16)$$

We first consider the case $D_+ = D_-$ when the average cluster charge is zero and the distribution function is an even function of z and the kinetic equation for the charge distribution function for clusters takes the form

$$\frac{d}{dz}[z f(z)] + \frac{x}{2}\frac{d^2}{dz^2}\left[\frac{e^z+1}{e^z-1} z f(z)\right] = 0 .$$

For small values of z that determine the normalization of the charge distribution function, this kinetic equation takes the form

$$\frac{d[z f(z)]}{dz} + x\frac{d^2 f(z)}{dz^2} = 0 .$$

Solving this equation taking into account the symmetry of the charge distribution function $f(z) = f(-z)$ and the boundary condition $f(\infty) = 0$, we obtain

$$f(z) = C \exp\left(-\frac{z^2}{2x}\right) .$$

Returning to the cluster charge Z as a variable for the distribution function and normalizing the distribution function, we obtain

$$f(Z) = \sqrt{\frac{x}{2\pi}} \exp\left(-\frac{Z^2 x}{2}\right) .$$

In the general case, when $D_- \neq D_+$, we obtain for the charge distribution function for clusters

$$f(Z) = \sqrt{\frac{2\pi}{x}} \exp\left[-\frac{(Z-Z_0)^2 x}{2}\right] , \tag{6.17}$$

where Z_0 is the average cluster charge, and since $x \ll 1$, a typical particle charge $Z \gg 1$.

6.1.5
Charging of Clusters or Particles in a Rare Ionized Gas

We now consider the kinetic regime of cluster charging where criterion $r_0 \ll \lambda$ is fulfilled and the problem of cluster charging is reduced to pairwise collisions of electrons and ions with the cluster. Assuming that each contact of a colliding ion or electron with the cluster surface leads to transfer of its charge to the cluster, we find the cross sections of collisions where the distance of closest approach r_{min} of colliding particles does not exceed the cluster radius r_0. The cross section of ion contact with the cluster surface, if the cluster charge Z has the same sign as that of colliding ion, in the classical case is

$$\sigma = \pi r_0^2 \left(1 - \frac{Ze^2}{r_0 \varepsilon}\right) , \quad \varepsilon \geq \frac{Ze^2}{r_0} .$$

In the case of different signs of charges for the colliding ion and cluster, this cross section is

$$\sigma = \pi r_0^2 \left(1 + \frac{|Z|e^2}{r_0 \varepsilon}\right), \quad \varepsilon \geq 0 .$$

For a Maxwell distribution function for ions this gives for the rate constant for cluster charging when each contact leads to charge transfer and the same sign of cluster and ion charges

$$k = \langle v\sigma \rangle = k_0 \int_x^\infty t\,dt \exp(-t) \left(1 - \frac{x}{t}\right) = k_0 e^{-x} , \quad t = \frac{\varepsilon}{T} ,$$

$$x = \frac{|Z|e^2}{r_0 T} , \quad k_0 = \sqrt{\frac{8T}{\pi m_i}} \pi r_0^2 , \tag{6.18}$$

where m_i is the ion mass, k_0 is the rate constant for ion collision with a neutral cluster when this collision leads to their contact, and T is the ion temperature. For different signs of charges of colliding particles the rate constant for their contact averaged over the Maxwell distribution function for ions is given by

$$k = k_0 \int_0^\infty t\,dt \exp(-t) \left(1 + \frac{x}{t}\right) = k_0(1 + x) . \tag{6.19}$$

One can combine these formulas for the rates of cluster charging by introducing the probability ξ that the ion transfers its charge to the cluster as a result of their contact [65]:

$$J_> = \xi k_0 N_i \cdot (1 + x) , \quad J_< = \xi k_0 N_i \cdot \exp(-x) . \tag{6.20}$$

Here N_i is the number density of ions and the rates $J_>$ and $J_<$ relate to different and identical signs of ion and cluster charges.

Let us find on the basis of these rates the equilibrium cluster charge in a nonequilibrium plasma consisting of electrons and ions with different electron T_e and ion T_i temperatures. Then equalizing the rates of attachment of electrons and ions to the cluster surface, we obtain for the equilibrium cluster charge $Z = -|Z|$ of a quasineutral plasma

$$|Z| = \frac{r_0 T_e}{e^2} \ln \left[\sqrt{\frac{m_i T_e}{m_e T_i}} \left(1 + \frac{|Z|e^2}{r_0 T_i}\right)^{-1} \right] ,$$

where m_e and m_i are the electron and ion masses. Let us rewrite this formula in terms of the reduced cluster charge $x = |Z|e^2/(r_0 T_e) = |Z|e^2/(r_W n^{1/3} T_e)$ (r_W is the Wigner–Seitz radius):

$$x = \left[\ln \sqrt{\left(\frac{m_i T_e}{m_e T_i}\right)} - \ln \left(1 + x \frac{T_e}{T_i}\right) \right] , \quad |Z| = x \frac{r_0 T_e}{e^2} . \tag{6.21}$$

In particular, for an argon plasma and temperatures $T_e = 2\,\text{eV}$ and $T_i = 400\,\text{K}$ of electrons and ions, this formula gives $x = 2.6$, which leads to the particle charge $Z = -4 \times 10^3 e$ for the particle radius $r_0 = 1\,\mu\text{m}$.

Formula (6.21) holds true in the limit $|Z| \gg 1$ if a change of cluster charge by one leads to a small variation of the electron and ion fluxes. In particular, in the case of identical electron and ion temperatures ($T_i = T_e$), this formula gives

$$x = \ln\left(\frac{1}{1+x}\sqrt{\frac{m_i}{m_e}}\right) . \tag{6.22}$$

Table 6.1 gives the solution of this equation for a quasineutral ionized inert gas (and nitrogen) with positive atomic ions (and N_2^+ ions for nitrogen).

We now consider the limit of small cluster size

$$r_0 \ll \frac{e^2}{T}$$

when clusters are neutral or singly charged and are in a quasineutral ionized gas with different temperatures of electrons T_e and ions T_i. The charging processes involving electrons, ions A_+, and clusters M_n^{+Z} consisting of n atoms and having charge Z proceed according to the scheme

$$e + M_n \to M_N^- , \quad e + M_n^+ \to M_N ,$$
$$A^+ + M_n \to M_N^+ + A , \quad A^+ + M_n \to M_N^+ + A .$$

This scheme gives the following set of balance equations for the number density of neutral clusters N_0, singly negatively charged clusters N_-, and singly positively charged clusters N_+:

$$\frac{dN_0}{dt} = -k_e N_e N_0 - k_i N_i N_0 + k_i\left(1 + \frac{e^2}{r_0 T_i}\right) N_i N_- + k_e\left(1 + \frac{e^2}{r_0 T_e}\right) N_e N_e ,$$
$$\frac{dN_+}{dt} = k_i N_i N_0 - k_e\left(1 + \frac{e^2}{r_0 T_e}\right) N_e N_e ,$$
$$\frac{dN_-}{dt} = k_e N_e N_0 - k_i\left(1 + \frac{e^2}{r_0 T_i}\right) N_i N_- .$$

Here k_i is the rate constant for ion attachment to a neutral cluster and k_e is the rate constant for electron attachment to a neutral cluster according to

$$k_i = \pi r_0^2 \sqrt{\frac{8T_i}{\pi m_i}} , \quad k_e = \pi r_0^2 \sqrt{\frac{8T_e}{\pi m_e}} .$$

Table 6.1 The solution of (6.22) [41].

Buffer gas	He	Ne	Ar	Kr	Xe	N_2
x	3.26	3.90	4.17	4.47	4.65	4.03

Correspondingly, the rate constants for neutralization of a singly negatively charged cluster in collisions with ions $k_{-1,0}$ and the rate constants for neutralization of a singly positively charged cluster in collisions with electrons $k_{+1,0}$ are given by

$$k_{-1,0} = k_\mathrm{i}\left(1 + \frac{e^2}{r_0 T_\mathrm{i}}\right), \quad k_{+1,0} = k_\mathrm{e}\left(1 + \frac{e^2}{r_0 T_\mathrm{e}}\right).$$

The rate constant for electron attachment to a negatively charged cluster contains an additional factor, $\exp(-e^2/(r_0 T_\mathrm{e}))$, and the rate constant for ion attachment to a positively charged cluster contains the factor $\exp(-e^2/(r_0 T))$, and therefore these processes are weak. Next, the ratios of the number densities of neutral N_0, negatively charged N_- and positively charged N_+ clusters are equal depending on the plasma parameters:

$$\frac{N_-}{N_0} = \frac{k_\mathrm{e} N_\mathrm{e}}{k_\mathrm{i} N_\mathrm{i}\left(1 + \frac{e^2}{r_0 T_\mathrm{i}}\right)}, \quad \frac{N_+}{N_0} = \frac{k_\mathrm{i} N_\mathrm{i}}{k_\mathrm{e} N_\mathrm{e}\left(1 + \frac{e^2}{r_0 T_\mathrm{e}}\right)},$$

$$\frac{N_+}{N_-} = \left(\frac{k_\mathrm{e} N_\mathrm{e}}{k_\mathrm{i} N_\mathrm{i}}\right)^2 \frac{\left(1 + \frac{e^2}{r_0 T_\mathrm{e}}\right)}{\left(1 + \frac{e^2}{r_0 T_\mathrm{i}}\right)}.$$

6.1.6
Ionization Equilibrium for Large Dielectric Clusters

The character of the ionization equilibrium for a dielectric cluster in an ionized gas is determined by the properties of the cluster surface. Each dielectric particle or cluster has on its surface traps for electrons that we call active centers. Electrons are captured by these centers, and this leads to the formation of negative ions at certain points on the cluster surface. The ionization equilibrium of a dielectric cluster in a plasma corresponds to the equilibrium of these bound negative ions and free plasma electrons. Although bound electrons can transfer between neighboring active centers, this process proceeds slowly. The ionization equilibrium in this case results from detachment of bound negative ions on the cluster surface by electron impact and the capture of electrons by active centers of the cluster surface.

Thus, metal and dielectric clusters are characterized by different charging processes in ionized gases. In the case of a metal cluster, valence electrons are located over the entire cluster volume, whereas they form negative ions on the surface of dielectric clusters. Ionization of metal clusters at high temperatures results from collision of internal electrons that leads to the formation of fast electrons which are capable of leaving the cluster surface. This thermoemission process and attachment of plasma electrons to the cluster surface determine the ionization balance for clusters which have a positive charge at high temperatures and a negative charge at low temperatures. The electron binding energy for a large metal cluster tends to the work function of a bulk metal. Dielectric clusters are negatively charged since their ionization potential greatly exceeds their electron affinity.

Let us consider the diffusion regime of charging of dielectric clusters if the mean free path of ions and electrons in a buffer gas is small compared with the cluster size. Then the ionization equilibrium for a dielectric cluster results from the processes [66]

$$e + A_n^{-Z} \to \left(A_n^{-(Z+1)}\right)^{**} , \quad \left(A_n^{-(Z+1)}\right)^{**} + A \to A_n^{-(Z+1)} + A ,$$
$$B^+ + A_n^{-(Z+1)} \to B + A_n^{-Z} , \tag{6.23}$$

so an autodetaching state $(A_n^{-(Z+1)})^{**}$ is quenched by collisions with surrounding atoms. Because the rate constant for electron attachment to a dielectric cluster greatly exceeds the cluster ionization rate constant for ionization by electron impact, these particles are negatively charged.

In contrast to metal particles, the binding energies for active centers do not depend on particle size, because the action of each center is concentrated in a small region of space. Evidently, the number of such centers is proportional to the area of the particle's surface, and for particles of micrometer size this value is large compared with that occupied by the charges. Hence, we consider the regime of charging of a small dielectric particle far from the saturation of active centers. Then positive and negative charges can exist simultaneously on the particle's surface. They move over the surface and can recombine there. Usually, the binding energy of electrons in negative active centers is in the range $\text{EA} = 2\text{–}4\,\text{eV}$, and the ionization potential for positive active centers is about $J_0 \approx 10\,\text{eV}$. Hence, attachment of electrons is more profitable for electrons of a glow discharge, and a small dielectric particle has a negative charge in a glow gas discharge.

The electric potential on the cluster surface is $\varphi = Ze/r_0$, where Z and r_0 are the cluster charge and cluster radius, respectively. The electron state is stable, $e\varphi < \text{EA}$, and from this we have the critical cluster charge Z_* where the electron state on the cluster surface is stable [66]:

$$Z_* = r_0 \cdot \frac{\text{EA}}{e^2} . \tag{6.24}$$

In particular, for $r_0 = 1\,\mu\text{m}$ and $\text{EA} = 3\,\text{eV}$ for a dielectric cluster, we have $Z_* = 2 \times 10^3$, and the cluster electric potential is 3 V. Under these parameters, a typical distance between neighboring occupied active centers is about $0.3\,\mu\text{m}$, which is one or two orders of magnitude larger than a typical distance between neighboring active centers. If the particle charge exceeds the critical charge Z_* according to (6.24), the bound electron state becomes an autodetaching state, but in this case the barrier size is $R = |Z|e^2/\text{EA}$, where Z is the cluster charge. From this it follows that because this distance exceeds a typical atomic size, the lifetime of the autodetaching state is long enough. Therefore, the cluster charge may exceed greatly the critical value given by (6.24).

6.2
Particle Fields in an Ionized Gas

6.2.1
Self-Consistent Particle Field in a Rare Ionized Gas

If a micrometer-sized particle is inserted in an ionized gas, it creates a large field in some region, so the particle potential energy $U(R)$ in this region exceeds a thermal energy of electrons and ions. Since ions provide the main contribution to screening of the particle field, we define the dimension l of the particle field as

$$|U(l)| \sim T_{\rm i} , \tag{6.25}$$

where $T_{\rm i}$ is the ion temperature. We now consider a rareness gas when this dimension is small compared with the mean free path of ions λ in this gas:

$$l \ll \lambda . \tag{6.26}$$

We assume that the presence of particles in a gas does not influence the ionization balance in the gas, and the number density of electrons and ions far from the particle is equal to the equilibrium number density N_0 that is realized in the absence of particles. Next, the number density of electrons in the particle field is lower than that far from the particle, whereas near the particle the ion number density exceeds the equilibrium one N_0 remarkably. Hence, we ignore below the influence of electrons on particle field screening. In addition, the screening of the particle field influences weakly the particle charge, which follows from the equality of the electron and ion fluxes to the particle surface, and these fluxes originate far from the particle where its field does not act.

In considering the distribution of free ions in the particle field, we use the ergodic theorem [67–69]: the probability $dP_{\rm i}$ for a particle to be in a distance range between R and $R + dR$ from the particle is proportional to the time dt during which an ion is found in this region. We have from the ion motion equation [70]

$$dt = \frac{dR}{v_R} = \frac{dR}{v\sqrt{1-\rho^2/R^2 - U(R)/\varepsilon}} ,$$

where $v_R(R)$ is the normal component of the ion velocity in the particle field at a distance R from the particle center, v is the ion velocity far from the particle, $\varepsilon = m_{\rm i} v^2/2$ is the ion energy far from the particle, and ρ is the impact parameter for ion motion with respect to the particle. From this within the framework of statistical mechanics [71–73] we have that the probability of ion location in a given region is proportional to a time of ion location in this region ($dP_{\rm i} \sim dt$) and is proportional to the number density of ions $N_{\rm i}(R) \sim dP_{\rm i}$. From this we have for the number density of ions

$$N_{\rm i}(R) = C\frac{\int \rho d\rho dP_{\rm i}}{4\pi R^2 dR} = N_0 \int_0^{\rho(R)} \frac{\rho d\rho}{\sqrt{1 - \frac{\rho^2}{R^2} - \frac{U(R)}{\varepsilon}}} , \tag{6.27}$$

where C is the normalization factor, and we use only half of the trajectory when an ion moves far away from the particle. In the case of free ion motion $U(R) = 0$ and $\rho(R) = R$, this formula gives $N_i = N_0$.

If we divide the ion trajectories into two groups and account for the absence of the removed part for $\rho \leq \rho_c$, where ρ_c is the boundary impact parameter for ion capture by this particle, we obtain for the number density of free ions in the particle field [60]

$$N_i(R) = \frac{N_0}{2}\left[\sqrt{1 - \frac{U(R)}{\varepsilon}} + \sqrt{1 - \frac{\rho_c^2}{R^2} - \frac{U(R)}{\varepsilon}}\right], \tag{6.28}$$

where the boundary impact parameter for capture ρ_c is connected to the particle radius r_0 by the relation [70]

$$\rho_c^2 = r_0^2\left[1 - \frac{U(r_0)}{\varepsilon}\right].$$

Let us average the ion number density over the Maxwell velocity distribution function far from the particle, that is,

$$f(\varepsilon) = N_0 \cdot \frac{2\varepsilon^{1/2}}{\sqrt{\pi}T_i^{3/2}} \exp\left(-\frac{\varepsilon}{T_i}\right), \quad \int f(\varepsilon)\varepsilon^{1/2}d\varepsilon = N_0,$$

where T_i is the ion temperature expressed in energy units. In the spatial region $|U(R)| \gg T_i$ (the potential energy $U(R)$ is negative) we have

$$N_i(R) = N_0\left[\sqrt{\frac{|U(R)|}{\pi T_i}} + \sqrt{\frac{|U(R)| - |U(r_0)|r_0^2/R^2}{\pi T_i}}\right]. \tag{6.29}$$

In particular, near the particle, this formula gives

$$N_i(R) = N_0\sqrt{\frac{|U(r_0)|}{\pi T_i}}, \quad R - r_0 \ll r_0; \quad N_i(R) = N_0\sqrt{\frac{4|U(R)|}{\pi T_i}}, \quad R \gg r_0.$$

In the other limiting case, $U(R) \ll T_i$, which corresponds to large ion distances from the particle, $R \to \infty$, we have $N(R) = N_0$. At distances far from the particle, (6.29) takes the form

$$N_i(R) = N_0\sqrt{1 + \frac{4|U(R)|}{\pi T_i}}. \tag{6.30}$$

As is seen, in a region of action of the particle field the number density of ions exceeds the equilibrium number density N_0, whereas the number density of electrons is below the equilibrium number density. Hence, screening of the particle field is determined mostly by ions.

We now consider a nonequilibrium plasma when the electron and ion temperatures (T_e and T_i) are different. Moreover, the energy distribution function for electrons differs from the Maxwell one, and its form is essential for the particle charge. Nevertheless, we introduce the electron temperature such that a real energy distribution of electrons provides the same flux of electrons on the particle surface and, correspondingly, the same particle charge, as takes place for the Maxwell electron distribution with a given electron temperature. Screening of the particle field is determined by ions and therefore is independent of the electron distribution function. Next, by analogy with (6.21) and (6.22), the particle charge in this case is given by [37, 41]

$$|Z| = \frac{r_0 T_e}{2e^2} \ln\left(\frac{T_e m_i}{T_i m_e}\right) . \tag{6.31}$$

Let us introduce the Coulomb field size R_0:

$$|U(R_0)| = \frac{|Z|e^2}{R_0} = T_i , \quad R_0 \gg r_0 ,$$

where

$$R_0 = \frac{|Z|e^2}{T_i} = r_0 X , \quad X = \frac{T_e}{2T_i} \ln\left(\frac{T_e m_i}{T_i m_e}\right) , \quad |Z| = \frac{T_i R_0}{e^2} . \tag{6.32}$$

Let us apply this to a typical gas discharge plasma in argon with $T_e = 1\,\text{eV}$ and $T_i = 400\,\text{K}$ and where the particle radius is $r_0 = 1\,\mu\text{m}$. According to the above formulas, for this plasma we have $|Z| = 5 \times 10^3$, $X = 210$, and $R_0 = 210\,\mu\text{m}$. The criterion for a rare plasma, if this plasma does not screen the particle field, is $N_0 R_0^3 \ll 1$, which gives for these parameters $N_0 \ll 5 \times 10^8\,\text{cm}^{-3}$. As is seen, at typical number densities of electrons and ions in gas discharge this criterion is not fulfilled.

To determine the self-consistent field of the particle and ions in the region of action of this field, we introduce the current charge $z(R)$ inside a sphere of radius R, and according to the Gauss theorem [74, 75], the electric field strength $E(R)$ and the current charge $z(R)$ inside a sphere are

$$E(R) = \frac{z(R)e}{R^2} , \; z(R) = |Z| - \int_{r_0}^{R} N_i(r) \cdot 4\pi r^2 dr , \; \frac{dz(R)}{dR} = -4\pi R^2 N_i(R) . \tag{6.33}$$

Using (6.30) for the ion number density, we have for the potential energy of the particle field

$$U(R) = \int_R^l \frac{z(r)e^2}{r^2} dr = \frac{z(R)e^2}{R^2} + \Delta U , \quad \Delta U = \int_R^l \frac{[z(R) - z(r)]e^2}{r^2} dr , \tag{6.34}$$

where l is a dimension for the region of action of the particle field that is defined as

$$z(l) = 0 . \tag{6.35}$$

Let us restrict ourselves to the first expansion term for the potential energy of the particle field $U(R) = z(R)e^2/R$, which for the current charge gives

$$z = \left(\sqrt{|Z|} - \frac{8\sqrt{\pi}}{5} N_0 R^{5/2} \sqrt{\frac{e^2}{T_i}}\right)^2 = |Z| \left[1 - \left(\frac{R}{l}\right)^{5/2}\right]^2 \tag{6.36}$$

and for the dimension of the self-consistent field gives

$$l = \frac{0.66}{N_0^{2/5}} \left(\frac{|Z| T_i}{e^2}\right)^{1/5} = \frac{0.66 |Z|^{2/5}}{N_0^{2/5} R_0^{1/5}} \,. \tag{6.37}$$

In particular, with the parameters in the above example ($T_e = 1\,\text{eV}$, $T_i = 400\,\text{K}$, the particle radius is $r_0 = 1\,\mu\text{m}$) and for the number density of electrons and ions $N_0 = 10^{10}\,\text{cm}^{-3}$ of an argon plasma, we find from this formula $l = 43\,\mu\text{m}$, and for $N_0 = 10^9\,\text{m}^{-3}$ we obtain $l = 108\,\mu\text{m}$. All these values are less than the size of the Coulomb particle field.

Note that a simplified expression ($\Delta U = 0$) for the particle potential energy according to (6.34) allows us to represent the results in an analytic form, but this simplification leads to an error. Figure 6.2 gives the ratio $\Delta U(R)/U(R)$ at the reduced number density of a plasma where the contribution due to free and trapped ions is comparable. This gives that the error due to the above simplification does not exceed 10%. Another remark relates to the behavior of the self-consistent field near its boundary (6.35). Indeed, the above consideration is valid under the condition $|U(R)| \gg T_i$, and dependence (6.36) is violated near the boundary (6.35). In particular, in the case $T_e \gg T_i$, we have Debye screening in the region $|U(R)| \lesssim T_e$,

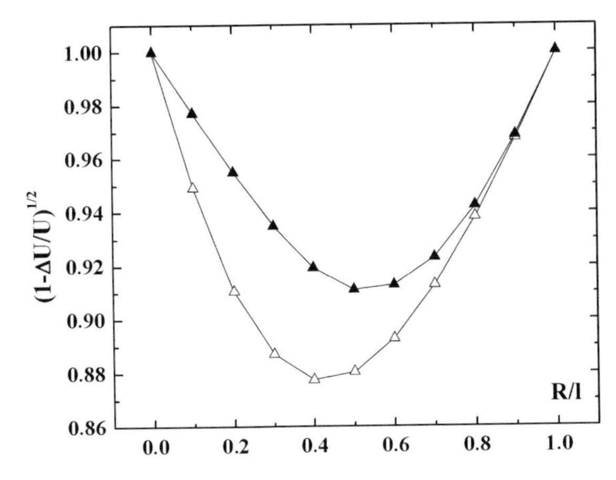

Figure 6.2 The ratio of the correction $\Delta U(R)$ to the particle potential energy in a plasma to the accurate interaction potential $U(R)$ as a function of the distance R from the particle for the versions where free ions (filled triangles) or trapped ions (open triangles) dominate at $N_0 r_0^2 = 100\,\text{cm}^{-1}$. At this reduced number density of plasma particles, the contribution to particle screening due to free and trapped ions is comparable.

which is

$$|U(R)| = \frac{z_0 e^2}{R} \exp\left(-\frac{R}{r_D}\right) ,$$

where the Debye–Hückel radius r_D is given by (1.9), and z_0 follows from combining this formula and (6.36) in an intermediate region.

The criterion for gas rareness has the form

$$N_a R_0 \sigma^* \ll 1 ,$$

where σ^* is the diffusion cross section of ion–atom scattering, which is assumed to be independent of the collision velocity. Because this ion scattering proceeds through the resonant charge exchange process, where an ion and an atom move along straight trajectories, we have $\sigma^* = 2\sigma_{res}$ [179], so the criterion for gas rareness takes the form

$$2N_a R_0 \sigma_{res} \ll 1 . \tag{6.38}$$

In this case the electron and ion currents originate at distances of approximately λ from the particle where interaction of ions with the particle field is negligible. In particular, for the above example we have the cross section of resonant charge exchange involving an argon atom and its ion to be $\sigma_{res} = 83\,\text{Å}^2$ at a collision energy of 0.01 eV [76]. Criterion (6.38) for argon pressure p takes the form

$$p \ll 1\,\text{Torr} .$$

We now consider one more aspect of ion–cluster interactions. If a cluster or a particle is moving in a rareness gas with a low velocity w compared with a thermal velocity of atoms, the frictional force F is determined by atom scattering by this particle, and on the basis of (6.1) we have for this force

$$F = N_a \sigma^* v_T \cdot m w , \quad v_T = \frac{4}{3}\sqrt{\frac{8T}{\pi m}} , \tag{6.39}$$

where N_a is the number density of gas particles, m is the mass of these atoms, the diffusion cross section of atom–particle scattering is $\sigma^* = \pi r_0^2$ (r_0 is the particle radius), and the velocity v_T is of the order of the average atom velocity. We now consider this problem from another standpoint: if a flux of ions passes through an ionized gas containing particles. Then the frictional force for ions due to particles, the drag force [4]), by analogy with (6.39) is given by

$$F = N_p \sigma^* v_T \cdot m w , \tag{6.40}$$

where N_p is the number density of gas atoms and m is the ion mass. If an ion is not captured by the particle, it is scattered elastically, and the diffusion cross section of this scattering is $\sigma^* \approx \pi l^2$ because of strong ion interaction with a self-consistent particle field at distances R from its center, $R < l$. From this it follows that although the number density of particles N_p in an ionized gas is small, the large radius of action of its field ($l \gg r_0$) may compensate for this smallness, and interaction of ions with particles may be of importance for their braking in an ionized gas.

6.2.2 Trapped Ions of Low-Density Plasma in a Particle Field

In an ionized gas of low density along with ion trajectories that correspond to ion capture by the particle surface and free trajectories that correspond to ion removal, capture of ions in closed trajectories is of importance for screening of the particle field [77]. These trajectories are different for the Coulomb (Figure 6.3) and screened Coulomb (Figure 6.4) fields [70, 78]. In the case of the Coulomb field, a captured ion moves along the same elliptic orbit after each period, whereas for the screened Coulomb field the elliptic orbit of the ion rotates after each period. Formation of trapped ions results from the resonant charge exchange process

$$\widetilde{A}^{+} + A \rightarrow \widetilde{A} + A^{+} , \qquad (6.41)$$

where $\widetilde{A}$ represents one of colliding particles. In analyzing of the behavior of trapped ions, it is of importance that the cross section of resonant charge exchange σ_{res} exceeds significantly that of elastic ion–atom scattering, and therefore an ion and an atom involved in this process move along straight trajectories [79, 80]. The parameters that determine the subsequent ion behavior are given in Figure 6.5 and include R, the point where this process takes place, ε, the atom energy at this

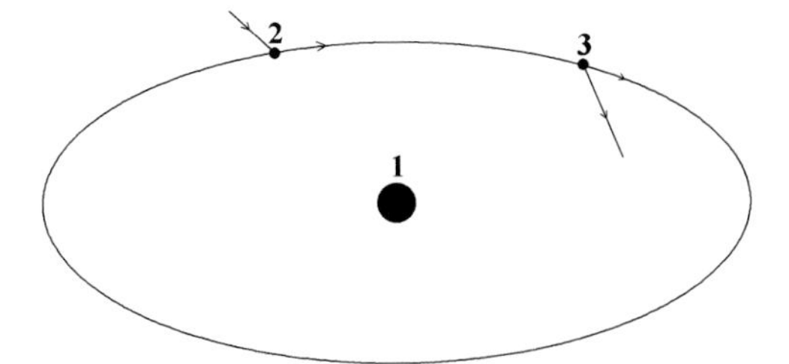

Figure 6.3 Trajectories of motion of a captured ion in the Coulomb field of the particle.

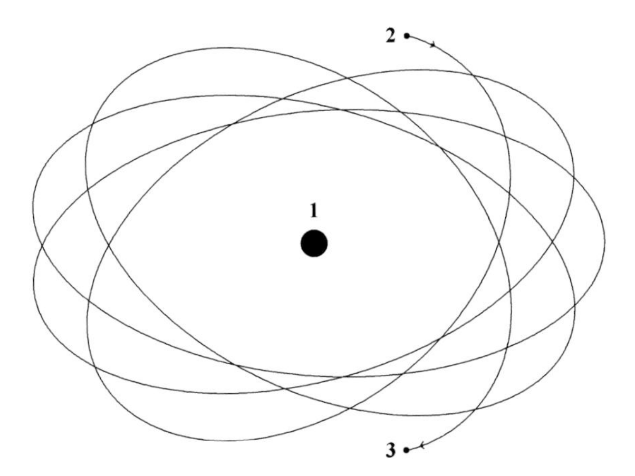

Figure 6.4 Trajectories of motion of a captured ion in a screened particle field.

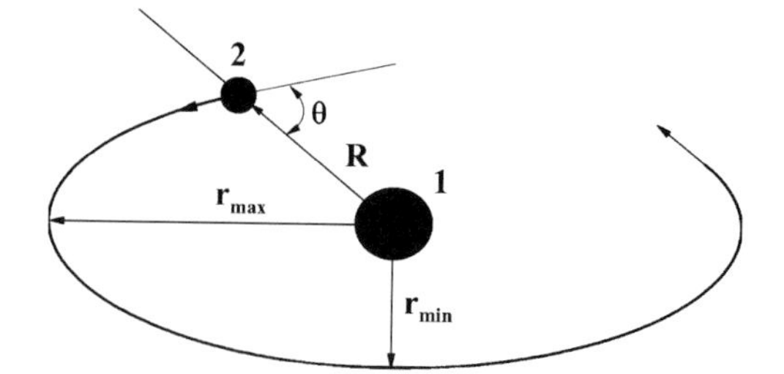

Figure 6.5 Parameters of motion of a trapped ion along a closed trajectory: r_{min} is the minimum distance from the particle, r_{max} is the maximum distance from the particle, **R** is the coordinate of the point where the resonant charge exchange event proceeds, and θ is the angle between the direction of ion motion after charge exchange and vector **R**.

point, which becomes the ion energy, and the angle θ between the atom velocity and vector **R**. We consider below the region of action of the particle field,

$$|U(R)| \gg \varepsilon\,, \tag{6.42}$$

from which a forming ion cannot move from the region of action of the particle field. Therefore, after a subsequent resonant charge exchange event it can transfer to another close trajectory or fall to the particle surface. As a result, a self-consistent field $U(R)$ is created on the basis of the Coulomb field of the negatively charged particle, and on the basis of fields of free and trapped ions.

Let us find the number density of trapped ions N_{tr} that follows from the balance equation

$$N_a \sigma_{res} N_i P_{tr} v_i = N_a \sigma_{res} N_{tr} v_{tr} (1 - p_{tr})\,. \tag{6.43}$$

Here N_i is the number density of free ions, v_i is the relative velocity for a free ion and atom partaking in the resonant charge exchange process, v_{tr} is the relative velocity of a trapped ion and atom, P_{tr} is the probability of forming a trapped ion as a result of resonant charge exchange involving a free ion, and p_{tr} is the probability for a trapped ion to remain in a closed trajectory after the resonant charge exchange. If we assume $v_i \sim v_{tr}$, $P_{tr} \sim 1$, and $p_{tr} < 1$, we obtain from (6.43), the same order of magnitude for the number densities of trapped ions as for free ions in the region of action of the particle field. Then the smallness of the probability of resonant charge exchange for a free ion because of a low density of buffer gas atoms is compensated for by the long lifetime of trapped ions compared with the flight time of free ions in the region of action of the particle field.

In analyzing the balance equation (6.43) under conditions (6.42), we note that the velocity of a free ion is $v_i = \sqrt{2|U(R)|/M}$ in this region. Assuming the particle field to be close to that in the Coulomb field, we obtain on the basis of the virial theorem [81] for the average kinetic energy of trapped ions $|U(R)|/2$, and the velocity is $v_{tr} = \sqrt{2|U(R)|/M}$, which gives $v_i/v_{tr} = \sqrt{2}$. Next, because the parameters of a forming ion are independent of the parameters of an incident ion, we have in the region given in (6.42) $P_{tr} = p_{tr}$, and (6.43) gives for the number density of trapped ions

$$N_{tr}(R) = N_i(R) \cdot \frac{P_{tr}\sqrt{2}}{(1-P_{tr})} \,.$$

Using the conservation laws for the energy and orbital momentum of a captured ion, we find that ion capture in a closed orbit is possible at $R \geq \sqrt{r_0 R_0}$, and the probability P_{tr} of ion capture in a closed orbit is

$$P_{tr}(R,\varepsilon) = p_{tr}(R,\varepsilon) = \int\limits_0^{\cos\theta_0} d\cos\theta = \cos\theta_0 = \sqrt{1-\frac{r_0 R_0}{R^2}}\,, \quad R \geq \sqrt{r_0 R_0}\,,$$

where θ_0 is the boundary angle in Figure 6.5 at which $r_{min} = r_0$. This gives for the number density of trapped ions in the region given in (6.42)

$$N_{tr}(R) = N_i(R)\frac{R^2\sqrt{2}}{r_0 R_0}\sqrt{1-\frac{r_0 R_0}{R^2}}\left(1+\sqrt{1-\frac{r_0 R_0}{R^2}}\right), \quad R \geq \sqrt{R_0 r_0}\,.$$

One can include in this formula the probability for an ion to be removed from the particle field if the interaction potential is relatively small. The probability P_{tr} of capture in a closed orbit is zero at the boundary of the region of the action of the particle field, $R = l$, and we account for this fact by an additional factor $1 - R/l$. As a result, we obtain for the number density of trapped ions [82]

$$N_{tr}(R) = N_i(R)\frac{R^2\sqrt{2}}{r_0 R_0}\sqrt{1-\frac{r_0 R_0}{R^2}}\left(1+\sqrt{1-\frac{r_0 R_0}{R^2}}\right)\cdot\left(1-\frac{l}{R}\right), \quad R \geq \sqrt{R_0 r_0}\,. \tag{6.44}$$

We now analyze the screening of the particle field if it is created by trapped ions. Using like for free ions the current charge $z(R)$ inside a sphere of radius R, replacing in (6.33) the number density of free ions by that of trapped ions, and repeating the operation in deduction (6.36), we obtain for the current charge $z(R)$

$$z(R) = \left[\sqrt{|Z|} - \frac{16\sqrt{\pi}}{9}\frac{N_0 R^{5/2}}{r_0 R_0}\sqrt{\frac{e^2}{T_i}}\left(1-\frac{R}{l}\right)\Phi(R)\right]^2 = |Z|\left[1-\left(\frac{R}{l}\right)^{9/2}\right]^2, \tag{6.45}$$

where

$$\Phi(R) = \frac{1}{2}\sqrt{1-\frac{r_0 R_0}{R^2}}\left(1+\sqrt{1-\frac{r_0 R_0}{R^2}}\right), \quad l = 1.05\left(\frac{|Z| r_0\sqrt{R_0}}{N_0\Phi(9l/11)}\right)^{2/9}.$$

We assume a weak dependence, $\Phi(R)$, and take it at a distance where the integrand has a maximum. One can find a small correction to the result by expansion over a small parameter $1 - \Phi(R)$. Using the last formula for the above example of an argon dusty plasma ($T_e = 1\,\text{eV}$, $T_i = 400\,\text{K}$, $r_0 = 1\,\mu\text{m}$), we obtain $l = 36\,\mu\text{m}$ for the number density of electrons and ions $N_0 = 10^{10}\,\text{cm}^{-3}$, and $l = 59\,\mu\text{m}$ for $N_0 = 10^9\,\text{cm}^{-3}$. Comparing these values with those in Section 6.2.1 for screening by free ions, we conclude that trapped ions dominate in screening the particle field if the number densities of plasma electrons and ions is below $N_0 = 10^{10}\,\text{cm}^{-3}$.

We also can use these formulas for the criterion for plasma rareness if plasma ions do not screen the particle field. As for free ions, this criterion has the form $N_0 \ll N_*$, where $l(N_*) = R_0$. For the parameters in the example under consideration ($T_e = 1\,\text{eV}$, $T_i = 400\,\text{K}$), we have from this

$$N_0 r_0^2 \ll 0.03\,\text{cm}^{-1}\,,$$

which corresponds to $N_* = 3 \times 10^6\,\text{cm}^{-3}$ for particle radius $r_0 = 1\,\mu\text{m}$.

In considering the case where trapped ions dominate in shielding of the particle field, we also give the expressions for the number density of free and trapped ions in the particle field according to (6.30) and (6.44). Using a simplified formula, $U(R) = z(R)e^2/R$, for the potential energy of the particle field and (6.45) for the current charge $z(R)$, we obtain for the number density of free $N_i(R)$ and trapped $N_{tr}(R)$ ions at distance R from the particle [82]

$$N_i(R) = N_0 \sqrt{1 + \frac{4R_0}{\pi R}\left[1 - \left(\frac{R}{l}\right)^{9/2}\right]^2}\,,$$
$$N_{tr}(R) = N_i(R) \cdot \frac{2R^2\sqrt{2}}{r_0 R_0}\Phi(R)\left(1 - \frac{R}{l}\right)\,. \tag{6.46}$$

In the case (6.36) where free ions dominate in screening of the particle field, we have the following expressions for the number density of free and trapped ions on the basis of (6.30) and (6.36) [82]:

$$N_i(R) = N_0 \sqrt{1 + \frac{4R_0}{\pi R}\left[1 - \left(\frac{R}{l}\right)^{5/2}\right]^2}\,,$$
$$N_{tr}(R) = N_i(R) \cdot \frac{2R^2\sqrt{2}}{r_0 R_0}\Phi(R)\left(1 - \frac{R}{l}\right)\,. \tag{6.47}$$

On the basis of these limiting cases one can give a general algorithm for determination of the parameters of a self-consistent field accounting for screening by both free and trapped ions. Indeed, the screening charges due to free Q_i and trapped Q_{tr} ions are given by the relations

$$Q_i = \int_{r_0}^{l} 4\pi N_i(R) R^2 dR\,, \quad Q_{tr} = \int_{\sqrt{r_0 R_0}}^{l} 4\pi N_{tr}(R) R^2 dR\,,$$

and according to the definition of the dimension l of the particle field region we have

$$Q = Q_i + Q_{tr} = |Z| . \tag{6.48}$$

Using this relation as the equation for l, we introduce the part of the screening charge ξ that is created by trapped ions as

$$\xi = \frac{Q_{tr}}{Q_i + Q_{tr}} . \tag{6.49}$$

In evaluating these parameters of the particle field screening we use (6.46) or (6.47) for the number densities of free and trapped ions. Note that in the range of competition of these screening channels, the difference in the results for these versions is not remarkable. In particular, for the above example of the argon dusty plasma ($T_e = 1\,\text{eV}$, $T_i = 400\,\text{K}$, $r_0 = 1\,\mu\text{m}$) at $N_0 = 10^{10}\,\text{cm}^{-3}$ the contribution of trapped ions to the particle screening is $\xi = 0.40$ and $\xi = 0.43$, respectively, if we use (6.46) or (6.47) for the number densities of free and trapped ions, and the size l of the particle field is 30 μm and 31 μm, respectively, for these cases.

The above formulas exhibit the scaling law according to which the parameters of the length dimension are proportional the particle radius, and the dependence on the number density of electrons and ions N_0 far from the particle is through $N_0 r_0^2$. In particular, free and trapped ions give the same contribution to the screening charge if $N_0 r_0^2 = 70\,\text{cm}^{-1}$, which means that at $r_0 = 1\,\mu\text{m}$ the corresponding number density of the plasma is $N_0 = 7 \times 10^9\,\text{cm}^{-3}$ and at $r_0 = 10\,\mu\text{m}$ this number density is $N_0 = 7 \times 10^7\,\text{cm}^{-3}$. Next, trapped ions disappear at $N_0 r_0^2 = 10^3\,\text{cm}^{-1}$, which corresponds to $N_0 = 1 \times 10^{11}\,\text{cm}^{-3}$ for $r_0 = 1\,\mu\text{m}$, and the number density is $N_0 = 1 \times 10^9\,\text{cm}^{-3}$ for $r_0 = 10\,\mu\text{m}$. From the last value it follows that trapped ions are not important for electric probes in a plasma at $r_0 \approx 10\,\mu\text{m}$ but are significant for a rare dusty plasma in the solar system.

Figure 6.6 shows the dependence of the part of the screening charge ξ (6.49) that is created by trapped ions on the reduced number density $N_0 r_0^2$ of a surrounding plasma, and Figure 6.7 gives the dependence of the reduced size of the particle field on this parameter. Figure 6.8 shows the reduced number densities of free $N_i r_0^2$ and trapped $N_{tr} r_0^2$ ions as a function of the reduced distance R/r_0 from the particle. These number densities correspond to the reduced number density of a surrounding plasma $N_0 r_0^2 = 100\,\text{cm}^{-1}$ at which the contribution from trapped ions to the particle screening is approximately 40%. Comparing the results of the two cases in Figures 6.6 and 6.7 where the number densities of ions are determined by (6.46) or (6.47), one can estimate the accuracy of the above operations as approximately 10%.

We note one more aspect of this phenomenon. We consider the case $R_0 \gg r_0$, which allows us to divide the ion trajectories from which an ion attaches to the particle or goes into the surrounding plasma. Next, the nearest elliptic ion orbit is far from the particle. All this according to (6.32) is possible only for a nonequilibrium plasma where $T_e \gg T_i$, as is realized in a gas discharge plasma containing dust

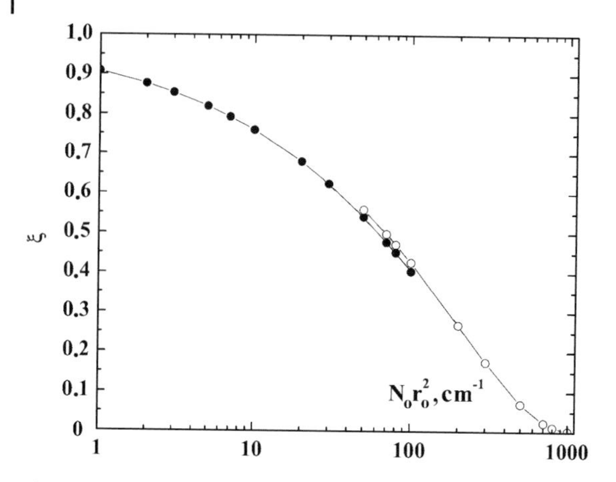

Figure 6.6 The part of the screening charge owing to trapped ions as a function of the reduced number density of ions for a particle in an ionized argon with $T_e = 1$ eV and $T_i = 400$ K. Open circles correspond to the case when free ions dominate, and closed circles relate to the case when trapped ions dominate.

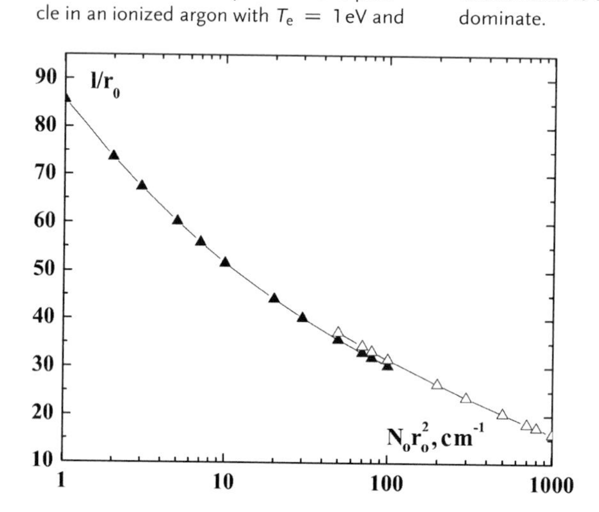

Figure 6.7 The reduced size of the particle field in an ionized argon as a function of the reduced number density of electrons with $T_e = 1$ eV and $T_i = 400$ K. Open triangles correspond to the case when free ions dominate, and closed triangles relate to the case when trapped ions dominate.

particles. In an equilibrium plasma ($T_e = T_i$) we have $R_0/r_0 \equiv X \approx 5$–6, and the above effects are mixed. Thus, trapped ions are of importance for a rare plasma but not for a dense plasma.

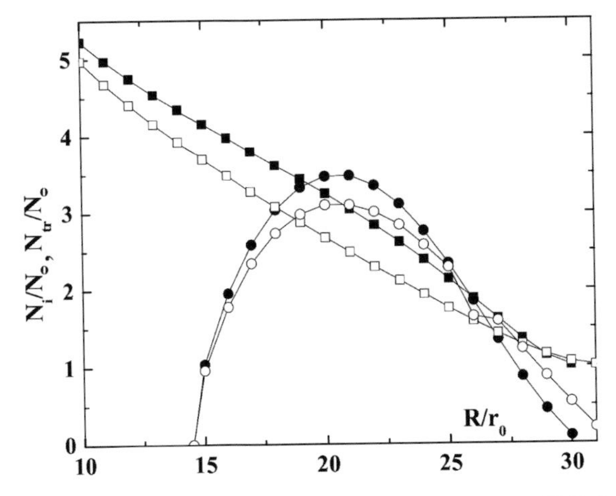

Figure 6.8 The number densities of free (squares) and trapped (circles) ions near a particle in a rare ionized gas as a function of the reduced number density of electrons and ions of a plasma with $T_e = 1$ eV and $T_i = 400$ K. Open symbols correspond to the case when free ions dominate, and closed symbols relate to the case when trapped ions dominate.

6.2.3
Interaction of Dust Particles with the Solar Wind

In analyzing the equilibrium between a particle and surrounding rare ionized gas as a result of attachment of electrons and ions to the particle, we find that trapped ions are of importance for a low number density of plasma electrons and ions. This occurs in a dusty plasma in the solar system [83–85] that is formed as a result of interaction of the solar wind with dust in comet tails and the rings of Saturn, Jupiter, and Uranus. The basis of these dusty plasmas is the solar wind [12, 13, 86], which is a plasma flux from the solar corona. The solar wind consists mostly of electrons and protons and is propagated from the Sun in all directions more or less uniformly. The solar wind accounts for the loss of approximately 2×10^{-14} times the mass of the Sun per year. This plasma flux varies as $1/R^2$ with increasing distance R from the Sun. On the basis of the average parameters of the solar wind on the Earth level, we have that the number densities of electrons and protons are about 7 cm^{-3}, the electron temperature is 2×10^5 K, the proton temperature is 5×10^4 K, and the flux velocity is 4×10^7 cm/s (this velocity corresponds to a proton kinetic energy of about 1 keV). When the solar wind encounters dust particles, a specific dusty plasma is formed [4, 83, 84].

As a demonstration of a cosmic dusty plasma, we consider below two examples of a cosmic plasma containing dust particles. The first example relates to the rings of Saturn (see Figure 6.9 [87]). In particular, the E and F rings of Saturn contain ice particles of size ranging from 0.5 to 10 μm [88, 89] and a typical number density of ice particles is 30 cm^{-3} [88]. The sources of the ice particles are the neighboring satellites: Enceladus for the E ring [24, 90], and Prometheus and Pandora for the F ring [91]. Simultaneous measurements within the framework of the Cassini

Figure 6.9 The rings of Saturn contain dust particles from its satellites and are supported by the action of its magnetic field and the solar wind [87].

project provided evidence for the dust parameters [92, 93] and plasma parameters [26, 94] from Saturn's E ring. The solar wind also influences transport of dust particles from Enceladus [24, 90, 95]. The number density of the plasma that results from the interaction of the magnetic field of Saturn and the solar wind is $N_0 = 30–100\,\mathrm{cm}^{-3}$, the electron temperature is 10–100 eV [96], and the magnetic field influences significantly the properties of this plasma. Ions of various types exist in a ring plasma, and basic types of ions are OH^+ and H_2O^+ with temperatures of the order of 10^3 K.

Let us analyze the parameters of a cosmic dusty plasma on the basis of the above analysis. Taking for definiteness $r_0 = 1\,\mu\mathrm{m}$ and $T_\mathrm{e} = T_\mathrm{i} = 30\,\mathrm{eV}$, we obtain from (6.22) $x = 2.5$, which corresponds to the particle charge $|Z| = 5 \times 10^4$. Next, taking the number density of particles $N_\mathrm{p} = 30\,\mathrm{cm}^{-3}$ [88], we obtain $N_\mathrm{p} R_0^2 \approx 10^{-5}\,\mathrm{cm}^{-1}$, which is small compared with $N_\mathrm{p}^{1/3}$, that is, scattering of electrons and ions of the solar wind by each particle proceeds independently. Note that we obtain an overestimated value for the particle charge because the total number density of particle charge $N_\mathrm{p}|Z| \sim 10^6\,\mathrm{cm}^{-3}$ exceeds significantly the observed value $N_0 \sim 10^2\,\mathrm{cm}^{-3}$ [96]. Moreover, from the observed number density of electrons and ions it follows that $|Z| \sim 3$, and the negative charge of this plasma is connected to particles, whereas ions are located in the space between particles and $N_\mathrm{i} \sim 10^2\,\mathrm{cm}^{-3}$. But this is large compared with the number density of fast electrons and protons in the solar wind on the Saturn level, which is approximately $0.1\,\mathrm{cm}^{-3}$.

Let us consider the problem regarding this plasma from another standpoint. Formation of a dusty plasma of the solar system results from interaction of the solar wind with an interplanetary dust. As a result of mixing, a dusty plasma is formed, and the number density of charged particles in this plasma exceeds that in the solar wind by several orders of magnitude. Above, we verified this for the dusty plasma of the rings of Saturn, and the same can be done for other types of dusty plasma occurring in the solar system. In particular, in the case of comets [27, 28], comet tails result from the interaction of the solar wind with comet dust and comet ions (see Figure 6.10 [97]). Although the magnetic properties of comets are important for the properties of this plasma [29], processes in the field of charged particles

Figure 6.10 As a result of interaction with the solar wind, the comet tails consisting of ions and dust particles are separated from the comet nucleus, but because of the action of magnetic forces the binding between the nucleus and tails is conserved in the course of comet evolution [97].

of this plasma proceed as described above. Correspondingly, the long lifetime of trapped ions greatly increases the number density of the plasma of a comet tail. This number density is 10^3–$10^4\,\mathrm{cm}^{-3}$ [14, 15] and exceeds that in the solar wind, but the electron temperature $T_e \sim 10^4\,\mathrm{K}$ [15–17] corresponds to that in the solar wind. One of the reasons for an increase in the plasma density in a dusty plasma of comet tails is the formation of trapped ions in the field of dust particles with a relatively long lifetime.

Thus, in contrast to a laboratory dusty plasma, where dust particles do not influence the plasma parameters outside of the action of the particle field, dust particles in the solar system change the parameters of a plasma, created by the solar wind. In addition, interaction of a cosmic dusty plasma with the magnetic field is of importance and determines the spatial structure of this plasma.

6.2.4
Screening of the Particle Field in a Dense Ionized Gas

In a dense ionized gas where the mean free path electrons and ions λ is small compared with the particle radius r_0, we use the Fuks formulas (6.7) and (6.9) for the rates of electron and ion attachement to the particle. This gives the following formula for the number density of electrons N_e and ions N_i near the particle surface.

$$N_e(R) = N_0 \frac{\exp(-X) - \exp(-X_0)}{1 - \exp(-X_0)}\,, \quad N_i(R) = N_0 \frac{1 - \exp(X - X_0)}{1 - \exp(-X_0)}\,. \quad (6.50)$$

Here R is the distance from the particle, N_0 is the number density of electrons and ions far from the particle, and $X = U(R)/T$, $X_0 = U(r_0)/T$, where $U(R)$ is the potential energy of electrons at a distance R from the particle and the electron and ion temperatures T are identical. Note that formulas (6.50) follow from the conservation of the rates of intersection of spheres of different radius for electrons and ions. According to the character of electron and ion attachment, the number densities of electrons and ions are zero at the particle surface ($X = X_0$) and tend to the equilibrium number density N_0 of electrons and ions far from the charged particle.

For the difference between the ion and electron number densities, formulas (6.50) give

$$\Delta N \equiv N_\mathrm{i}(R) - N_\mathrm{e}(R) = N_0 \cdot \frac{1 - \exp(X - X_0) + \exp(-X_0) - \exp(-X_0)}{1 - \exp(-X_0)} \,. \tag{6.51}$$

This quantity is symmetric with respect to transformation $X \to X_0 - X$ and has a maximum at $X = X_0/2$, where this quantity is

$$\Delta N = N_0 \cdot \frac{1 - \exp(-X_0/2)}{1 + \exp(-X_0/2)} \,.$$

In addition from equality of electron and ion attachment rates we have $X_0 = \ln(D_-/D_+)$.

The electric potential $\varphi(R) = U(R)/e$ that is created by the particle follows from the Poisson equation:

$$\Delta\varphi \equiv \frac{1}{R}\frac{d^2(R\varphi)}{dR^2} = 4\pi e \Delta N \,. \tag{6.52}$$

If $X = U(R)/T \ll 1$, we have $\Delta N = N_0 X$ and the Poisson equation has the form

$$\frac{1}{R}\frac{d^2(R\varphi)}{dR^2} = \frac{\varphi}{r_\mathrm{D}^2} \,, \qquad r_\mathrm{D} = \sqrt{\frac{T}{4\pi Z N_0 e^2}} \,.$$

Its solution is

$$\varphi(R) = \frac{Ce}{R}\exp\left(-\frac{R}{r_\mathrm{D}}\right) , \tag{6.53}$$

and in the Fuks approach $C = Z$. One can see that the Fuks approach $U(R) = Ze^2/R$ is valid for weak shielding of the particle charge by surrounding electrons and ions, which leads to the criterion

$$N_0 r_0^3 \ll 1 \,. \tag{6.54}$$

In a general case we use expressions (6.51) for the difference ΔN between the number densities of electrons and ions, and the Poisson equation (6.52) has the form

$$\frac{1}{R}\frac{d^2[RX(R)]}{dR^2} = \frac{\Delta N}{N_0 r_\mathrm{D}^2} \,,$$

where $X \le X_0$ and $X_0 = \ln(D_-/D_+)$. From this one can formulate the general character of variation of the particle potential $\varphi(R)$ and the number density difference ΔN for ions and electrons on approaching the particle from infinity. At large distances from the particle surface, the particle potential and the number density difference drop exponentially with distance from the particle because of the Debye screening. At the particle surface the number density difference ΔN is zero and has a maximum at some distance from the particle, whereas the particle electric potential $\varphi(R)$ drops monotonically with distance from the particle. The connection between these quantities is determined by the Poisson equation (6.52).

Let us consider this connection from another standpoint. According to the Gauss theorem, the particle electric field strength is

$$E = \frac{Z(R)e}{R^2} ,$$

and the electric potential at distance R from the particle is

$$\varphi(R) = \int_R^\infty \frac{Z(r)}{r^2} dr .$$

This gives the following connection between the reduced particle electric potential and the number density difference:

$$X = \frac{1}{r_D^2} \int_R^\infty \left(\frac{1}{R} - \frac{1}{r} \right) \frac{\Delta N}{N_0} r^2 dr . \tag{6.55}$$

In reality, this equation is identical to the Poisson equation (6.52). In particular, at small values of X where $\Delta N = N_0 X$ the solution of this equation is given by (6.53), as well as by the Poisson equation (6.52). In addition, multiplication of (6.55) by R and its double differentiation leads to the Poisson equation (6.52), that is, (6.52) and (6.55) are equivalent.

We thus have that the particle charge in a dense ionized gas and the screening of this charge near the particle are determined by the Fuks approach, which is valid at high plasma density (6.5). As the number density of electrons and ions increases and criterion (6.5) is violated, the general concept of the Fuks approach for electron and ion fluxes is conserved and is expressed by (6.50). These formulas are valid at distances from the charged particle where screening of the particle charge is weak, and hence the range of these distances is shortened as the number densities of electrons and ions increase. Nevertheless, (6.50) and (6.51) may be used for evaluation of both the particle charge and its screening by surrounding electrons and ions if the above formulas are used in the right-hand side of (6.55). In this case the particle charge and screening parameters are determined by competition between two dimension parameters, the particle radius r_0 and the Debye–Hückel radius r_D, and the Fuks limit corresponding to the criterion

$$r_D \gg r_0 . \tag{6.56}$$

The inverse ratio is valid in the positive column of gas discharge of not low pressure if the effective particle radius is taken as the radius of curvature for walls. The conditions were given above to generalize the Fuks approach, and they may be used in the course of numerical solutions of the above equations. Unfortunately, in contemporary evaluations (e.g., [98–102]) all these conditions are not fulfilled.

6.2.5
Influence of Particles on the Properties of Ionized Gases

The above analysis shows that if a micrometer-sized particle is inserted in a plasma, it acquires a negative charge owing to attachment of electrons and ions to its surface, and the Coulomb field of this particle is screened by plasma electrons and ions. This screening has a nature other than the Debye screening in a plasma due to attachment of electrons and ions to the particle surface. Hence, plasma electrons and ions determine both the establishment of the equilibrium particle charge and shielding of the particle field by the plasma electrons and ions. We now consider another aspect of interaction of particles with a plasma. Indeed, particles are sinks for electrons and ions, and their presence in a plasma may influence the plasma parameters.

We compare the rate of attachment of electrons and ions to the walls ν_w in a gas discharge tube and to particles ν_p in a plasma. We make this comparison for an argon plasma in the positive column of glow discharge (see Figure 6.11), and this plasma is inside a cylindrical tube of radius R_0 into which particles of radius r_0 and number density N_p are inserted. We have for the flux of electrons and ions to the tube surface in the case of the Schottky regime [103] of the positive column $j = -D_a \partial N/\partial R = 1.25 D_a N_0/R_0$, where N is the number density of electrons and ions near the walls, N_0 is their number density at the axis, and D_a is the coefficient of ambipolar diffusion. From this we obtain for the rate of electron and ion attachment to the walls per unit tube length

$$\nu_w = 2.5\pi D_a N_0 = 2.5\pi \frac{T_e}{T_i} D_i N_0 ,$$

where the attachment rate ν_w is the number of attaching electrons and ions per unit tube length and unit time, D_i is the diffusion coefficient of ions, and T_e and T_i are the electron and ion temperatures.

In determining the rate ν_p of electron and ion attachment to particles we use for definiteness the case of a rare gas when the rate constant k for ion attachment to a negatively charged particle is given by (2.101) and the particle charge is determined by (6.21). Then we have for the total rate ν_p of electron and ion attachment to particles per unit tube length

$$\nu_p = k N_i N_p \pi R_0^2 = \frac{2\sqrt{2\pi}|Z|e^2 r_0}{\sqrt{m_i T_i}} \cdot \pi R_0^2 N_i k N_p ,$$

where N_i is the current number density of ions, Z is the particle charge, and m_i is the ion mass. Averaging the number density of ions over the tube cross section, we obtain from this

$$\nu_p = 6.8\sqrt{\frac{T_i}{m_i}\frac{|Z|e^2}{T_i}}\, r_0 R_0^2 N_0 N_p \,.$$

From this we find for the ratio of the attachment rates

$$\frac{\nu_p}{\nu_w} = \frac{0.87}{D_i}\sqrt{\frac{T_i}{m_i}\frac{|Z|e^2 r_0}{T_e}}\, R_0^2 N_p = \xi \frac{S_p}{S_w}\,, \quad \xi = \frac{0.14}{D_i}\sqrt{\frac{T_i}{m_i}\frac{|Z|e^2 r_0}{T_e}}\, R_0\,, \tag{6.57}$$

where $S_p = 4\pi r_0^2 N_p \cdot \pi R_0^2$ is the total area of the particle surface per unit tube length, and $S_w = 2\pi R_0$ is the wall area per unit tube length. Since $|Z|e^2/r_0 T_e \sim 1$, we have $\xi \sim R_0/\lambda \gg 1$, where λ is the mean free path of ions in an ionized gas. Thus, the particle surface is more effective than the tube walls for recombination of electrons and ions.

$$\frac{\nu_p}{\nu_w} = \xi \frac{S_p}{S_w} = 2\pi r_0^2 R_0 N_p \xi \,.$$

As a demonstration of these results, we consider an example of an argon plasma with ion temperature $T_i = 400$ K and a gas discharge tube of radius $R_0 = 1$ cm. We find the reduced electric field strength in the absence of particles [60], $E/N =$

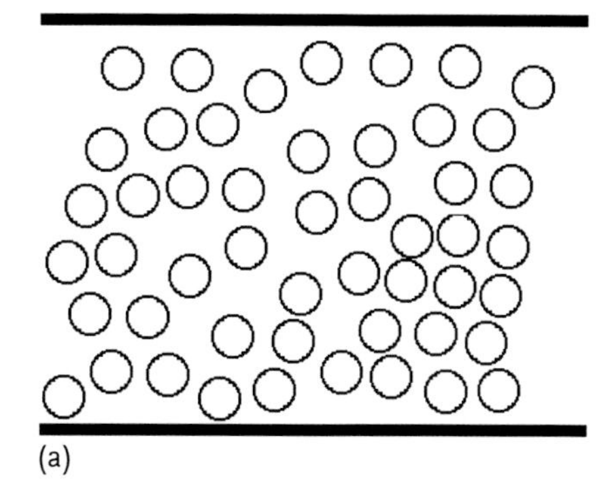

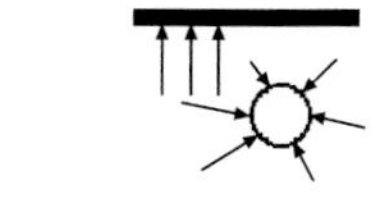

Figure 6.11 Dust particles in the positive column of gas discharge (a) and fluxes of electrons and ions to the walls and particle surface (b). As a result, electrons and ions recombine on the particle surfaces as well as on the surface of the gas discharge tube.

13 Td, using the values of the first Townsend coefficient and the electron drift velocity [104]. If attachment of electrons and ions to particles gives the same contribution to plasma destruction as that due to attachment of electrons and ions to particles, the reduced electric field strength of the positive column becomes $E/N = 16\,\text{Td}$.

Taking the particle radius as $r_0 = 1\,\mu\text{m}$ and the electron temperature as $T_e = 1\,\text{eV}$, we obtain in the case of a rare ionized gas according to (6.21) $|Z|e^2/r_0 T_e = 2.6$ and $Z = -1.8 \times 10^3 e$. Next, for $p = 10\,\text{Torr}$ we have the number density of argon atoms $N_a = 2.4 \times 10^{17}\,\text{cm}^{-3}$, the diffusion coefficient for atomic argon ions at this pressure is $D_i = 5.2\,\text{cm}^2/\text{s}$, and $\sqrt{T_i/m_i} = 2.9 \times 10^4\,\text{cm/s}$. We obtain $\xi = 780$, which is the efficiency of attachment of electrons and ions to particles with respect to their attachment to the walls of the gas discharge tube. In addition, we obtain the number density of particles $N_p = 1/(2\pi r_0^4 R_0 \xi) = 2.0 \times 10^4\,\text{cm}^{-3}$ at which the rates of plasma destruction as a result of attachment to particles and walls are equal, so if these particles form a crystal with close-packed particles that corresponds to the face-centered cubic or hexagonal structure, the distance between nearest particles is 0.4 mm. If the mass density of particles corresponds to a glass, we obtain from this for the mass density ρ of particles in space $\rho_p = 1.8 \times 10^{-7}\,\text{g/cm}^3$. For comparison, we have the mass density of argon atoms under the above conditions, $N_a = 1.6 \times 10^{-5}\,\text{g/cm}^3$. As seen in this example, the particles influence the processes in this gas discharge plasma if the mass density of particles is less than that of gas atoms by two orders of magnitude.

6.2.6
Particle Structures in Dusty Plasma

We have that a micrometer-sized particle in an ionized gas is surrounded by an ion coat, and ions screen the particle charge that is established due to attachment of electrons and ions to the particle surface. Electrons of a rare ionized gas give a small contribution to the particle field shielding. We characterize the region of action of the particle field by size l, the values of which are given in Figure 6.7. The character of interaction of two particles in an ionized gas is represented in Figure 6.12. At large distances between two particles (Figure 6.12) the particles do not interact and they are independent particles. In contrast, at small separations (Figure 6.12b) the particles are repulsed owing to interaction of the partially screened charges. From this it follows that under the given conditions the number density of particles in an ionized gas is restricted. Next, interaction of particles in an ionized gas has a short-range character because at separations $R > 2l$ two particles do not interact. If this separation is below $2l$, the charges of the particles are partially screened, and the particles are strongly repulsed. Note that screening of two interacting particles differs from that for individual particles, and hence it is not correct to apply to the particle screened charges that relate to individual particles.

One can confirm the general character of particle interaction in an ionized gas given in Figure 6.12 and based on measurements of interaction of two particles

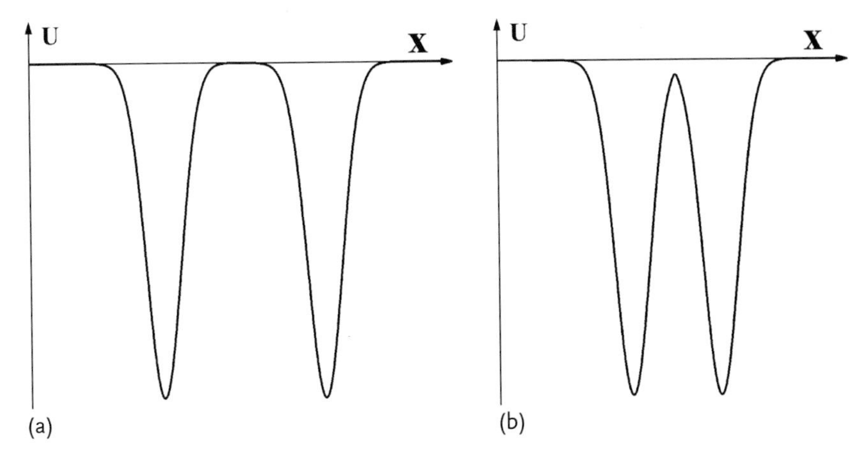

Figure 6.12 The character of interaction of two micrometer-sized particles in an ionized gas. If the distance between particles exceeds twice the size ($2l$) of action of the particle field, these particles do not interact (a), and if the total screening of the particle charges is absent, that is a distance between particles of less than $2l$, the particles are repulsed (b).

which are in an electric trap that is created in the sheath of radio-frequency gas discharge [105]. The interaction potential of two particles may be modeled by a short-range interaction of particles, but the plasma electric potential between particles if the distance between them exceeds $2l$ differs from that far from particles. This means that electron and ion fluxes to particles from distances where the particle fields are shielded differ from those toward one particle. The reason for this is such that a large size is required for the region where the equilibrium number density of electrons and ions is restored. In this way particles influence the potential of an ionized gas in which they are located.

Thus, one can model interacting particles inserted in an ionized gas as a system of hard spheres (2.10) with the radius of screening of the particle Coulomb field l. It is of importance that the radius of action of the particle field l or the radius of the hard sphere model, may be varied by variation of the gas discharge current and other parameters. This allows one to govern what structures can be formed by these particles. The most beautiful property of a dusty plasma formed by insertion of micrometer-sized dust particles into a gas discharge plasma is the possibility to form a crystal structure under some conditions. Variation of gas discharge parameters may lead to melting of such a crystal. This possibility was proposed by Ikezi [106] and was realized in 1994 by several experimental groups [20, 30–32]. Evidently, the so-called plasma crystal, that is, the structure formed by dust particles in a gas discharge plasma, attracted much attention to this problem and resulted in much research in this area (e.g., [1, 4, 35–39]). Note that millimeter distances between nearest neighbors of the structures formed by micrometer-sized particles allow one to use simple visual methods for their observation. For this reason, a dusty plasma is a convenient model for the analysis of both the formation and destruction of ordered structures and various hydrodynamic effects.

We restricted ourselves to structures that dust particles form in gas discharge plasma and give some examples of such structures. Then dust particles are inserted in a plasma and are captured by electric traps in this plasma. Because the distance between nearest neighbors of the structures formed is close to l, the radius of action of an individual particle, the character of the preparation of the particle structure is similar to filling a box with hard balls. Then it is necessary that the trap of the gas discharge creates a vertical force for compensation of the gravitational force acting on the particle. This may be realized in radio-frequency gas discharge where a sheet is created between the electrodes and plasma with a voltage of approximately 10 V. The electric field due to this voltage may prevent particles from falling down, and this force compensates for the gravitational force acting on particles. Figure 6.13 shows examples of structures [107] formed in an electric trap of this discharge.

An electric trap of striations may be used for particle confinement, and Figure 6.14 demonstrates the particle structure [34] formed in glow discharge under conditions of striation existence. Specific particle structures may be formed in an inductively coupled discharge with an inductor of special construction [108].

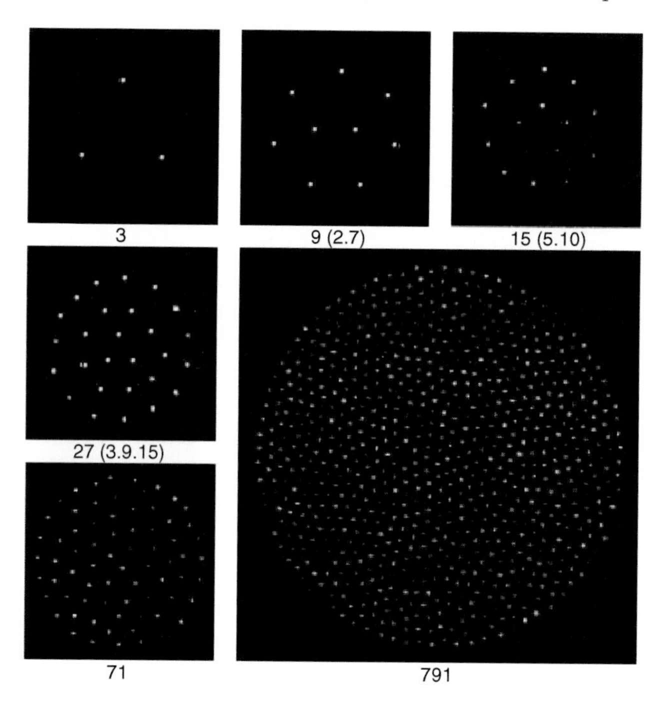

Figure 6.13 Particle structures formed in the sheath of radio-frequency gas discharge near the electrode [107].

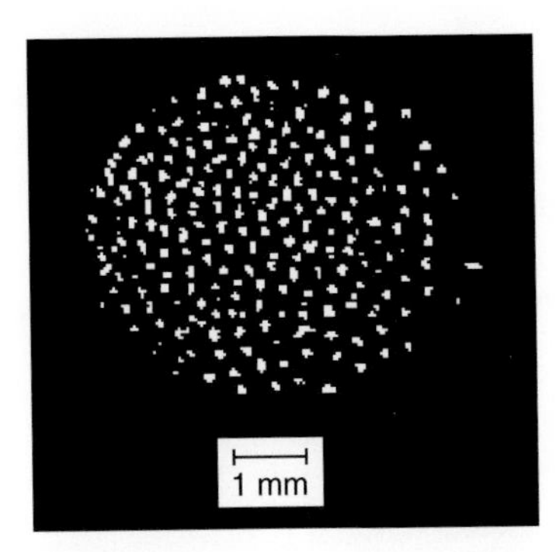

Figure 6.14 A particle structure formed in striations of glow discharge [34].

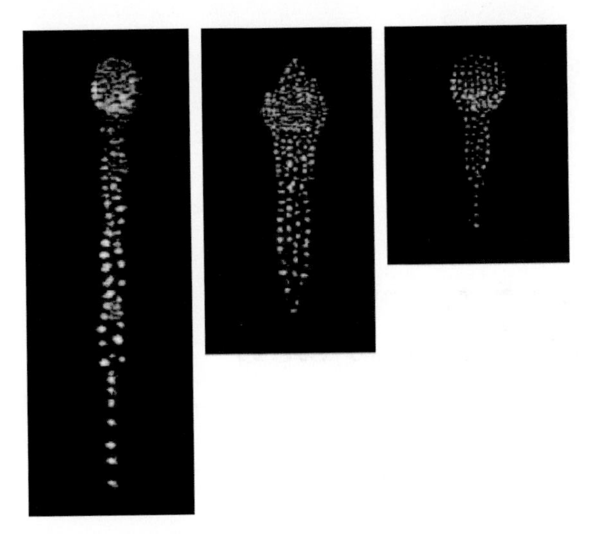

Figure 6.15 Structures of particles formed in an inductively coupled discharge with an inductor of special construction [108].

This construction provides forces on the boundary of a radio-frequency plasma directed upward, which leads to specific structures of dust particles injected in this plasma. Some structures of this type are represented in Figure 6.15. Along with electric forces, thermophoretic forces due to temperature gradients may be used for equalizing the gravitational force. A complex structure of particles formed in stratified glow discharge is represented in Figure 6.16 [109]. Thus, there are various possibilities to have a trap in a gas discharge plasma that allows one to obtain some structures of particles. These structures are a convenient model for study of the solid–liquid phase transition and some hydrodynamic effects in gas and liquid flows [110].

We note that fluxes of electrons and ions to the particle surface determine both the particle charge and screening of the particle Coulomb field by plasma ions.

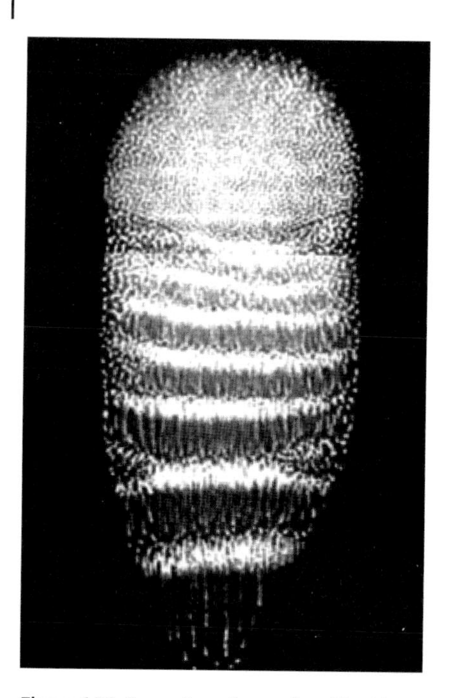

Figure 6.16 Formation of crystal and liquid structures of particles due to thermophoretic forces which are created by the temperature gradient [109].

If two particles are located at a distance of less than $2l$, the particle interaction results from interaction of nonshielding particle charges at a given distance. These charges are close to those for isolated particles if screening is determined by free ions, because the main contribution to screening is from ion trajectories which start outside the action of the other particle field. If screening is determined by trapped ions, the current charge inside a sphere of a given radius is isotropic and identical for an isolated particle and if another particle is present because ions undergo many rotations around the particle in the course of their lifetime in a close orbit. Thus, we have that the shielded charge of the particle in an ionized gas on its interaction with another particle is close to or coincides with that of the other particle. Next, the current charge $z(R)$ at a given distance R from it is determined by (6.36) or (6.45) depending on the role of free and trapped ions in screening of the particle field. These dependencies are closer to the stepwise distance dependence for the current charge rather than to Debye or Yukawa dependencies. Therefore, models based on the exponential dependencies of the distances between particles do not describe the system of particles in an ionized gas.

6.3
Cluster Plasma

6.3.1
Ionization Equilibrium for Metal Clusters in a Cluster Plasma

A cluster plasma is an ionized gas containing clusters [40, 41, 111], and the number density of clusters is small compared with the number density of atoms and molecules. Clusters in a cluster plasma may grow and evaporate during evolution of the cluster plasma. We will be guided by a cluster plasma consisting of a buffer gas and metal clusters. Hence, this cluster plasma includes both metal clusters and metal atoms, and the total number density of metal atoms is small compared with that of gas atoms. Under these conditions, metal clusters and metal atoms are of importance for cluster processes, and a buffer gas has the role of a stabilization component. Below we analyze several such systems which may be used for generation of metal clusters.

There are various methods to generate metal clusters in a cluster plasma. In all cases, metal atoms are injected in a gas or gas flow, and then metal clusters are formed from these atoms. Metal atoms may be injected in a gas as a result of evaporation of metal atoms from a metal surface and erosion of a metal surface by laser radiation or the ion (electron) beam of a gas discharge. The effective method of creation of a cluster plasma is based on injection of some metal compounds in a flow of a dense plasma, where these compounds are destroyed, and metal atoms are joined in metal clusters under appropriate conditions. This scenario uses a large binding energy of atoms in clusters as it takes place for most metals.

If we deal with a cluster plasma including metal clusters, a positive charge of this plasma is connected with metal clusters since their ionization potential is lower than that for buffer gas atoms. We consider below the ionization equilibrium involving metal clusters that is realized according to the scheme

$$M_n^{+Z+1} + e \leftrightarrow M_n^{+Z} \, . \tag{6.58}$$

Here M is a metal atom and M_n^Z is a metal cluster consisting of n atoms and having charge Z. By analogy with the Saha formula, we have for the probability $P_Z(n)$ for a cluster consisting of n atoms to have charge Z

$$\frac{P_Z(n) N_e}{P_{Z+1}(n)} = 2 \left(\frac{m_e T_e}{2\pi\hbar^2} \right)^{3/2} \exp\left[-\frac{I_Z(n)}{T_e} \right] , \tag{6.59}$$

where T_e is the electron temperature, m_e is the electron mass, N_e is the electron number density, and $I_Z(n)$ is the ionization potential of the cluster consisting of n atoms and having charge Z.

Note that the ionization equilibrium (6.58) in an ionized gas results from stepwise ionization of charged clusters in collisions with electrons and three body recombination of charged clusters and electrons. But because the number density of the internal electrons of a metal cluster greatly exceeds that of the plasma electrons, a third particle in the recombination process may be a bound electron. Under equilibrium between internal and plasma electrons, the Saha formula (6.59) is valid, but the electron release process is connected to internal electrons.

The equilibrium (6.58) between free electrons and bound electrons in a metal cluster corresponds to a long mean free path of electrons in a gas compared with the cluster radius in accordance with criterion (6.5) and the ionization potential $I_Z(n)$ of a large cluster for $n \gg 1$. In this case removal of an electron from the cluster surface outside the region of action of the cluster Coulomb field requires that this interaction is overcome, which gives

$$I_Z(n) = \frac{Ze^2}{r_0} + I_0(n) , \tag{6.60}$$

where $I_0(n)$ is the ionization potential of a neutral cluster consisting of n atoms. This relation allows one to transform the Saha formula (6.59) to the form

$$\frac{P_Z}{P_{Z-1}} = A \exp\left(-\frac{Ze^2}{r_0 T_e}\right) , \quad A = \frac{2}{N_e}\left(\frac{m_e T_e}{2\pi\hbar^2}\right)^{3/2} \exp\left(-\frac{I_0}{T_e}\right) . \tag{6.61}$$

From this formula we find the Saha formula (6.59) is reduced to the Gauss form for a large cluster charge $Z \gg 1$:

$$P_Z(n) = P_{\overline{Z}}(n) \exp\left[-\frac{(Z-\overline{Z})^2}{2\Delta^2}\right] ,$$

$$\overline{Z} = \frac{r_0 T_e}{e^2}\left\{\ln\left[\frac{2}{N_e}\left(\frac{m_e T_e}{2\pi\hbar^2}\right)^{3/2}\right] - \frac{I_0(n)}{T_e}\right\} , \quad \Delta^2 = \frac{r_0 T_e}{e^2} . \tag{6.62}$$

One can use a rough model for the cluster ionization potential assuming it to be a monotonic function of the number n of cluster atoms. We represent the ionization potential $I_0(n)$ of a neutral metal as

$$I_0(n) = W + \frac{\text{const}}{n^{1/3}} ,$$

where W is the metal work function, that is, the electron binding energy for a macroscopic cluster. Including in the consideration the atomic ionization potential $I_0(1) = I$ and transferring to the ionization potential of a charged cluster according to (6.60), we have in this model for the cluster ionization potential

$$I_Z(n) = \frac{Ze^2}{r_0} + W + \frac{I - W}{n^{1/3}} . \tag{6.63}$$

Evidently, within the framework of this model we have for the binding energy of a negatively charged cluster

$$I_{-1}(n) = W + \frac{\text{EA} - W}{n^{1/3}} , \tag{6.64}$$

where EA is the electron affinity of an atom.

From the above formulas, we have that clusters that are found in thermodynamic equilibrium with an ionized gas may be positively and negatively charged. In particular, according to the Saha formula we have for the ratio of the number densities for singly positively charged N_+ and singly negatively charged N_- clusters

$$\frac{N_+}{N_-} = \zeta^2 \exp\left(-\frac{\Delta}{T_e n^{1/3}}\right), \quad \zeta = \frac{2}{N_e}\left(\frac{m_e T_e}{2\pi\hbar^2}\right)^{3/2} \exp\left(-\frac{W}{T_e}\right),$$
$$\Delta = I_0(1) + \mathrm{EA} - 2W. \tag{6.65}$$

For example, in the case of copper we have $I_0(1) = 7.73\,\mathrm{eV}$, $\mathrm{EA} = 1.23\,\mathrm{eV}$, and $W = 4.4\,\mathrm{eV}$, which gives $\Delta = 0.16\,\mathrm{eV}$, and the ratio $\Delta/(T_e n^{1/3})$ is small for large clusters and high electron temperatures. Hence, the sign of the mean cluster charge is determined by the sign of the function $\ln \zeta(T_e)$. This is a monotonic function of the electron temperature, so at high electron temperatures clusters are positively charged, and at low electron temperatures they have mean negative charge. The transition electron temperature T_* is determined by the equation $\zeta(T_*) = 1$. In particular, in the case of copper and for large clusters it is $T_* = 2380\,\mathrm{K}$ for $N_e = 1 \times 10^{11}\,\mathrm{cm}^{-3}$, $T_* = 2640\,\mathrm{K}$ for $N_e = 1 \times 10^{12}\,\mathrm{cm}^{-3}$, and $T_* = 2970\,\mathrm{K}$ for $N_e = 1 \times 10^{13}\,\mathrm{cm}^{-3}$. As is seen, the temperature T_* of zero cluster charge increases with increasing number density of plasma electrons.

Thus, a metal cluster is positively charged at high temperatures and negatively charged at low temperatures. As a demonstration of this fact, Figure 6.17 gives the average charge of tungsten clusters in a plasma as a function of the cluster temperature, and also the cluster temperature T_* at which this cluster becomes neutral on average, that is, $\zeta(T_*) = 1$. Table 6.2 contains the parameters in (6.65) for some parameters of a cluster and plasma. In addition, according for (6.62), the cluster charge is proportional to $n^{1/3}$, and Table 6.2 contains the proportionality factor for the parameters considered.

6.3.2
Conversion of an Atomic Vapor into a Gas of Clusters

Being guided by a cluster plasma consisting of a buffer gas and addition of metal in the form of metal clusters, we consider the first stage of formation of a cluster plasma from a supersaturated atomic metal vapor. This allows one to ignore the processes of cluster evaporation, whereas the vapor condensation consists in the formation of nuclei of condensation, and cluster growth results from atom attachment to nuclei of condensation. In the case of metal vapor condensation the nuclei of condensation are diatomic metal molecules which are formed in the three body process. One can assume in the next stage of the growth process that the energy excess in atom attachment to metal molecules is transferred into oscillation degrees of freedom, and then this energy is transferred to the buffer gas atoms in collisions. Therefore, we have the following scheme for the cluster growth process in this case [112, 113]:

$$2M + A \rightarrow M_2 + A \,, \quad M_n + M \longleftrightarrow M_{n+1} \,, \tag{6.66}$$

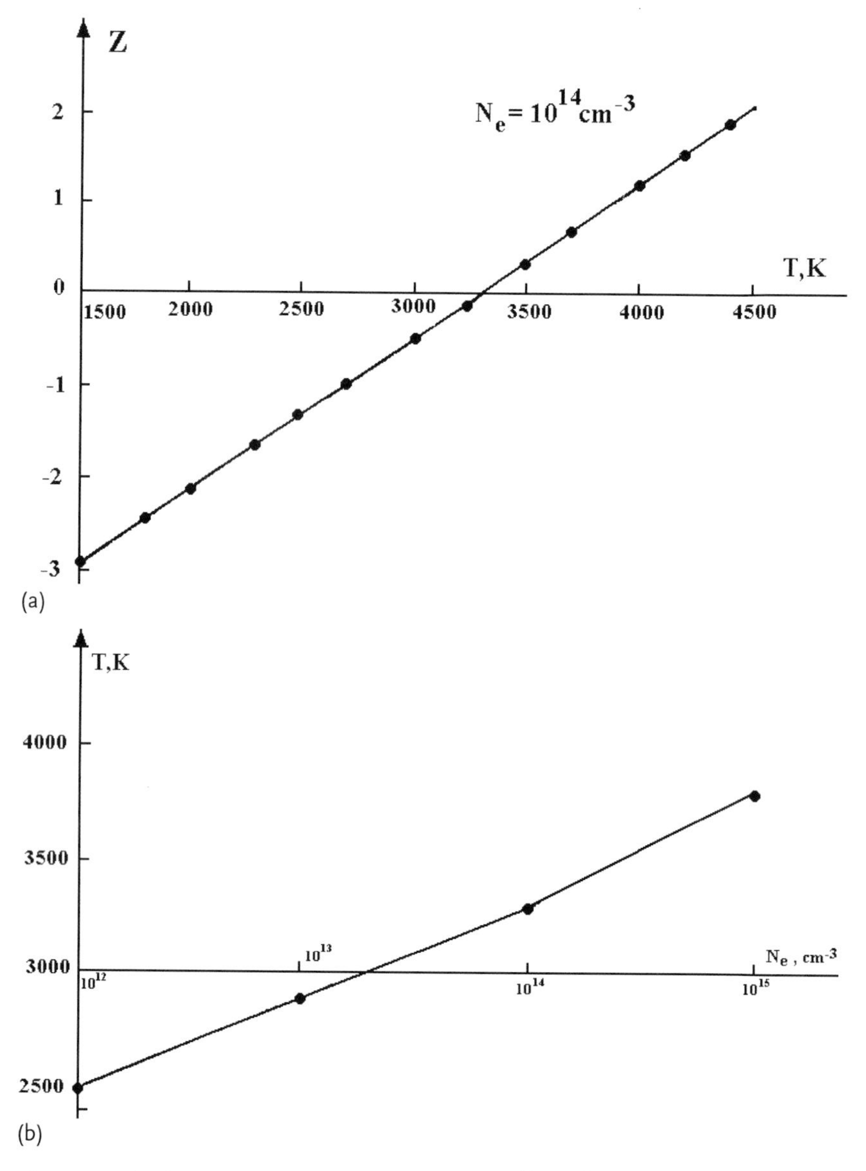

Figure 6.17 The dependence on the electron temperature for the mean charge of a tungsten cluster consisting of $n = 1000$ atoms in a plasma with the number density of electrons $N_e = 10^{14}$ cm^{-3} (a) and the dependence of the electron temperature at which this charge is zero on the electron number density (b).

where M and A are a metal atom and an atom of the buffer gas, respectively, and n is the number of cluster atoms. Assuming each contact of a metal atom with a cluster leads to their sticking together, we have according to formula (2.18) $\sigma = \pi r_0^2$ for the cross section of this process, where r_0 is the cluster radius within the framework of the liquid drop model. Correspondingly, the rate constant for atom attachment to a cluster of n atoms is

$$k_n = \sqrt{\frac{8T}{\pi m_a}} \cdot \pi r_0^2 = k_0 n^{2/3} , \quad k_0 = \sqrt{\frac{8\pi T}{m_a}} \cdot r_W^2 . \tag{6.67}$$

Here m_a is the mass of the metal atom. The rate of the pairwise process in scheme (6.66) exceeds significantly the rate of three body process for a number density of buffer gas atoms that is governed by the parameter

$$G = \frac{k_0}{N_a K} \gg 1 , \tag{6.68}$$

where N_a is the number density of buffer gas atoms, K is the rate constant for formation of a diatomic molecule in three body collisions, and its typical value in

Table 6.2 Parameters of charged metal clusters. I is the atomic ionization potential, W is the metal work function, EA is the atom electron affinity, $\Delta = I + \mathrm{EA} - 2W$, T_* is the temperature where the average cluster charge is zero for a the mean cluster size $n = 1000$ and electron number density $N_e = 10^{13}\ \mathrm{cm}^{-3}$, and Z is the average cluster charge at a temperature of 1000 K for argon as a buffer gas if the cluster charging is determined by processes of attachment of electrons and positive ions to the cluster. The rate constant k_e for electron–cluster collision is given at $T = 1000$ K [41].

Element	I, eV	W, eV	EA, eV	Δ, eV	T_*, 10^3 K	$Z/n^{1/3}$	k_e, 10^{-8} cm^3/s
Ti	6.82	3.92	0.08	−0.96	2.51	0.084	1.7
V	6.74	4.12	0.52	−0.98	2.63	0.078	1.5
Fe	7.90	4.31	0.15	−0.57	2.75	0.074	1.3
Co	7.86	4.41	0.66	−0.30	2.82	0.073	1.3
Ni	7.64	4.50	1.16	−0.20	2.87	0.072	1.3
Zr	6.84	3.9	0.43	−0.53	2.51	0.093	2.1
Nb	6.88	3.99	0.89	−0.21	2.57	0.085	1.7
Mo	7.10	4.3	0.75	−0.75	2.74	0.081	1.6
Rh	7.46	4.75	1.14	−0.90	3.00	0.078	1.5
Pd	8.34	4.8	0.56	−0.70	3.04	0.080	1.5
Ta	7.89	4.12	0.32	−0.03	2.65	0.085	1.7
W	7.98	4.54	0.82	−0.28	2.90	0.081	1.6
Re	7.88	5.0	0.2	−1.92	3.12	0.080	1.5
Os	8.73	4.7	1.1	0.43	3.01	0.078	1.5
Ir	9.05	4.7	1.56	1.21	3.03	0.080	1.5
Pt	8.96	5.32	2.13	0.45	3.38	0.081	1.6
Au	9.23	4.30	2.31	2.94	2.85	0.083	1.9

this case is $K \sim 10^{-32}\,\mathrm{cm^6/s}$, whereas a typical rate constant for the attachment process is $k_0 \sim 3 \times 10^{-11}\,\mathrm{cm^3/s}$ for metal atoms in a buffer gas. Below we consider the regime $G \gg 1$ of cluster growth that takes place at $N_a \ll 3 \times 10^{21}\,\mathrm{cm^{-3}}$,

The number density N_m of free metal atoms, the number density N_{cl} of clusters, the number density N_b of bound atoms in clusters, and the number of cluster atoms n are given by the following set of balance equations according to scheme (6.68):

$$\frac{dN_b}{dt} = -\frac{dN_m}{dt} = \int N_m k_0 n^{2/3} f_n dn + K N_m^2 N_a\,, \qquad \frac{dN_{cl}}{dt} = K N_m^2 N_a\,, \tag{6.69}$$

where f_n is the size distribution function for clusters and is introduced such that $f_n dn$ is the number density of clusters with a size range between n and $n + dn$, and this distribution function is normalized by the condition

$$N_{cl} = \int f_n dn\,, \quad N_b = \int n f_n dn\,.$$

Because the ratio of the first and second terms on the right-hand side of the first balance equation (6.69) is of the order of $G n^{2/3}$, and since a typical number n of cluster atoms and the cluster parameter G are large, one can ignore the second term in the first equation of the balance equations (6.69).

The set of balance equations (6.69) must be supplemented with the equation for cluster growth

$$\frac{dn}{dt} = k_0 n^{2/3} N_m\,, \tag{6.70}$$

and the boundary condition $N_b(\tau) = N_m$ for this model $N_m = \mathrm{const}$, where τ is the time for conversion of an atomic vapor into a gas of clusters.

The solution of the last equation gives for a cluster size $n_{max}(t)$ at time t after the this process has begun

$$n_{max}(t) = \left(\frac{N_m k_0 t}{3}\right)^3\,.$$

Correspondingly, the maximum cluster size $n_{max}(\tau)$ at the end of the growth process is

$$n_{max}(\tau) = \left(\frac{N_m k_0 \tau}{3}\right)^3\,. \tag{6.71}$$

Next, according to the nature of this process, the value $f_n dn$ is proportional to the time range dt of formation of diatomic molecules which are the condensation nuclei for the clusters. This leads to the following size distribution function for clusters:

$$f_n = \frac{C}{n^{2/3}}\,, \quad n < n_{max}(t)\,, \tag{6.72}$$

where C is the normalization constant and $f_n = 0$ if $n > n_{max}(t)$.

We have the following relations for the number density of clusters N_{cl} and the total number density of bound atoms N_b:

$$N_{cl} = \int f_n dn \,, \quad N_b = \int n f_n dn \,.$$

Using (6.72) for the size distribution function for clusters, we get for these parameters

$$N_{cl} = 3Cn_{max}^{1/3} \,, \quad N_b = \frac{3}{4} C n_{max}^{4/3} \,.$$

On the basis of the average cluster size $\overline{n} = N_b/N_{cl}$, we obtain

$$\overline{n} = \frac{n_{max}}{4} \,. \tag{6.73}$$

Thus, within the framework of the model we have at the end of the process from the second balance equation of set (6.69)

$$N_b = N_m = \frac{n_{max}}{4} N_{cl} = \frac{n_{max}}{4} K N_m^2 N_a \tau \,.$$

From this and (6.71) we find for the maximum cluster size n_{max} at the end of this process and the duration τ of the cluster growth process [112, 114]

$$n_{max} = 1.2 G^{3/4} \,, \quad \tau = \frac{3.2 G^{1/4}}{N_m k_0} \,. \tag{6.74}$$

These formulas describe cluster parameters at the end of the cluster growth process. Because $G \gg 1$, a typical cluster size is large, $n \gg 1$, and the reduced nucleation time τ is large compared with a typical time of $(k_0 N_m)^{-1}$ for attachment of one atom.

6.3.3
Cluster Growth in Coagulation and Coalescence

Figure 6.18 shows the mechanisms of clusters growth [111]. Figure 6.18a corresponds to the above case of conversion of a gas or vapor consisting of free atoms into a gas of clusters when the vapor pressure exceeds the saturated vapor pressure. Figure 6.18b corresponds to the coagulation mechanism due to joining of clusters that proceeds according to the scheme

$$M_{n-m} + M_m \to M_n \,, \tag{6.75}$$

and is an important process for the Earth's atmosphere [115]. Note that although a buffer gas does not partake in the coagulation process, it plays a stabilization role. As is seen, the coagulation process proceeds at a small number density N of free

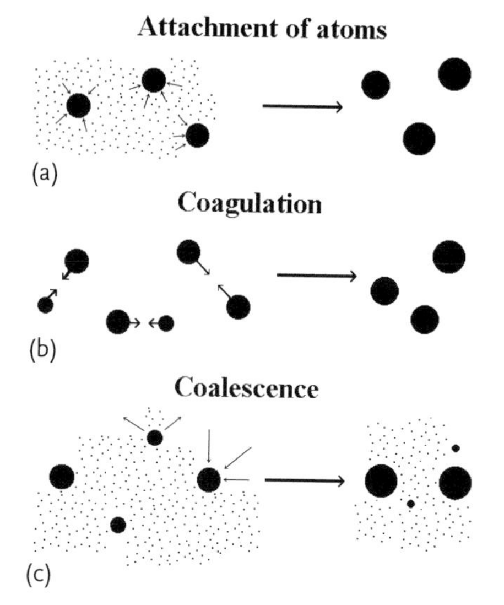

Figure 6.18 The mechanisms of cluster growth: (a) attachment of free atoms to clusters; (b) coagulation resulting in joining of clusters on contact; (c) coalescence that proceeds owing to equilibrium between the cluster and the parent vapor. As a result of atom attachment to clusters and cluster evaporation, small clusters decompose and large clusters grow. As a result, the average cluster size increases.

atoms compared with that of bound atoms N_b, that is, $N \ll N_b$. Another mechanism of cluster growth, given in Figure 6.18c, is known as Ostwald ripening [116] or coalescence and takes place at such temperatures where for small clusters the rate of evaporation exceeds the rate of atom attachment, and for large ones we have the opposite relation between these rates. Then the equilibrium between clusters and their atomic vapor leads to evaporation of small clusters and growth of large clusters. As a result, the average cluster size increases.

The size distribution function f_n for clusters in the coagulation process is described by the Smoluchowski equation [117]:

$$\frac{\partial f_n}{\partial t} = -f_n \int k(n, m) f_m dm + \frac{1}{2} \int k(n-m, m) f_{n-m} f_m dm \,. \qquad (6.76)$$

Here $k(n-m, m)$ is the rate constant for process (6.75), the factor $1/2$ accounts for the fact that collisions of clusters consisting of $n-m$ and m atoms are present in the equation twice, and the distribution function is normalized as

$$\int f_n dn = N_{\text{cl}} \,,$$

where N_{cl} is the number density of clusters. Note that the conservation of the total number density N_b of bound atoms in clusters in the course of the cluster growth process follows from the Smoluchowski equation (6.76). Indeed, multiplication of the the Smoluchowski equation (6.76) by n and integration over dn gives

$$\frac{dN_b}{dt} = -\int nk(n, m) f_n dn f_m dm + \frac{1}{2}\int nk(n-m, m) f_{n-m} f_m dn dm \,,$$

where the number density of bound atoms is $N_b = \int_0^\infty n f_n dn$ and in the second integral we have $n > m$. Changing $n - m$ for n on the right-hand side of this relation, we find the terms to mutually cancel, and $dN_b/dt = 0$. Hence, according to the Smoluchowski equation, the number density N_b of bound atoms is conserved in this process.

We now determine the rate constant for association k_{as} of two liquid clusters of radii r_1 and r_2 in the case of a dense gas if criterion (4.110) holds true. The rate of association of a test cluster (the first cluster) in a gas of clusters of the second kind is given by the Smoluchowski formula (6.8), so the rate constant for association $k_{as} = J_0/N_0$ is given by

$$k_{as} = 4\pi D(r_1 + r_2) = 4\pi(D_1 + D_2)(r_1 + r_2) \,,$$

where we account for the diffusion coefficient D that is responsible for approach of two clusters as $D = D_1 + D_2$. Using (4.110) for the diffusion coefficient of a large cluster, we find

$$k_{as} = \frac{2T}{3\eta}\left[2 + \frac{(r_1 - r_2)^2}{r_1 r_2}\right] \approx \frac{4T}{3\eta} \,, \tag{6.77}$$

where we ignore the second term in parentheses compared with the first one.

As is seen, the rate constant for cluster association in a dense buffer gas is independent of both the size and the material of particles. It is convenient in this case to use a dimensionless time $\tau = N_b k_{as} t$ in the Smoluchowski equation (6.76) and to introduce the concentration of clusters of a given size $c_n = f_n N_b$, where N_b is the total number density of bound atoms in clusters. The normalization condition for the cluster concentration has the form

$$\sum_n n c_n = 1 \,,$$

and the Smoluchowski equation (6.76) in terms of cluster concentrations takes the form

$$\frac{\partial c_n}{\partial \tau} = -c_n \int_0^\infty c_m dm + \frac{1}{2}\int_0^n c_{n-m} c_m dm \,. \tag{6.78}$$

This equation has a simple automodel solution,

$$c_n = \frac{4}{\overline{n}^2}\exp\left(-\frac{2n}{\overline{n}}\right) \,, \quad \overline{n} = \tau \,, \tag{6.79}$$

that holds true at long times when the solution is independent of the initial conditions. This expression satisfies the normalization condition $\int_0^\infty n c_n dn = 1$ and means that the form of the distribution function is conserved in the course of particle evolution, but the average particle size varies in time as

$$\overline{n} = k_s N_b t \,, \tag{6.80}$$

and we assume the total number density of bound atoms to be conserved in time.

This version of coagulation is valid in a dense buffer gas, in particular, for particle growth in an aerosol plasma with micrometer-sized aerosol particles. In the case of nanosized clusters, the case of a gas rareness (4.109) is realized. Assuming each contact of two colliding clusters leads to their joining, we have the cross section of the association process for large clusters, that is, consisting of many atoms, as $\sigma = \pi(r_1 + r_2)^2$, where r_1 and r_2 are the radii of colliding clusters. This leads to a size-dependent expression for the rate constant for cluster joining, which for the liquid drop model of colliding atoms gives

$$k(n, m) = k_0(n^{1/3} + m^{1/3})^2 \sqrt{\frac{n+m}{nm}} \,, \quad k_0 = r_W^2 \sqrt{\frac{8T\pi}{m_a}} \,, \tag{6.81}$$

where r_W is the Wigner–Seitz radius and m_a is the mass of a cluster atom. This leads to a nonlinear dependence of the average cluster size on time that has the form [118, 119]

$$\overline{n} = 6.3(N_b k_0 t)^{1.2} \,. \tag{6.82}$$

Since we used the assumption $n \gg 1$, this formula is valid under the condition

$$k_0 N_b t \gg 1 \,.$$

The coalescence mechanism of cluster growth or Ostwald ripening [116] results from an equilibrium between clusters and their atomic vapor through processes of atom attachment to the cluster surface and cluster vaporization. This mechanism is realized at high temperatures where the pressure of an atomic vapor is high enough. Let the critical cluster size be such that the rates of the attachment and evaporation processes are identical for this size. Then clusters of larger size will grow, whereas clusters of smaller size will evaporate. As a result, the average cluster size will increase.

A strict theory of coalescence [120–124] was elaborated for the case of grain growth in solid solutions. In this case atoms evaporated from the grain surface propagate inside the solid and attach to another grain. As a result, the number of small grains decreases, the number of large grains increases, and the average grain size increases in time. This process is restricted by slow diffusion of atoms inside the solid. In the case of a cluster plasma with small clusters according to (4.109), the processes of atom attachment to the cluster surface involving free atoms proceeds fast by analogy with growth of drops in a supersaturated vapor [125, 126]. Hence, in this case the coalescence process results from equilibrium between gas of clusters and an atomic vapor [41, 43], and other parameters determine the rate of the coalescence process.

We now consider the general character of coalescence in a cluster plasma assuming this plasma to consist of a rareness buffer gas in accordance with criterion (4.109), an atomic (metal) vapor, and a gas of (metal) clusters. The rate constant for attachment of an atom to a cluster consisting of $n \gg 1$ atoms is given by (6.67)

$$k_n = k_0 n^{2/3} ,$$

and the rate of atom evaporation from the cluster surface is determined by the principle of detailed balance and is

$$\nu_n = k_0 n^{2/3} N_{\text{sat}}(T) \exp\left(\frac{\varepsilon_0 - \varepsilon_n}{T}\right) ,$$

where $N_{\text{sat}}(T)$ is the number density of atoms for the saturated vapor pressure at a given temperature and $\varepsilon_0 - \varepsilon_n$ is the difference between the atom binding energies for a bulk particle and a cluster consisting of n atoms. Taking the atom binding energy in a cluster as the sum of the volume and surface parts [127], we obtain the difference given above for the atom binding energies due to the surface cluster energies in the form

$$\varepsilon_0 - \varepsilon_n = \frac{\Delta\varepsilon}{n^{1/3}} ,$$

and then the critical cluster size n_{cr} for which the rates of atom attachment to a cluster and cluster evaporation become identical follows from the relation [40]

$$N_{\text{m}} = N_{\text{sat}}(T) \exp\left(\frac{\Delta\varepsilon}{T n_{\text{cr}}^{1/3}}\right) , \tag{6.83}$$

where N_{m} is the equilibrium number density of free atoms. On the basis of the above rates of cluster growth and evaporation which proceed step by step as a result of addition of one atom to or release of one atom from a cluster, we have the kinetic equation for the size distribution function for clusters f_n in the linear approximation as a flux of atoms in a space of cluster sizes [40]:

$$\frac{\partial f_n}{\partial t} = -\frac{\partial j_n}{\partial n} , \quad j_n = k_0 n^{2/3} f_n \left[N_{\text{m}} - N_{\text{sat}}(T) \exp\left(\frac{\Delta\varepsilon}{T n^{1/3}}\right)\right] . \tag{6.84}$$

The kinetic equation (6.84) describes evolution of the size distribution function for the coagulation character of cluster growth. We note the characteristics of this process. The cluster system may be divided into two groups with sizes above and below the critical size (6.83), and clusters of one group cannot transfer to other group [40]. Therefore, cluster growth in the coalescence process is accompanied by an increase of the cluster critical radius that almost conserves the ratio between the numbers of clusters for the first and second groups. Next, the rate of cluster growth according to (6.84) depends on the parameter

$$a = \frac{\Delta\varepsilon}{T n^{1/3}} . \tag{6.85}$$

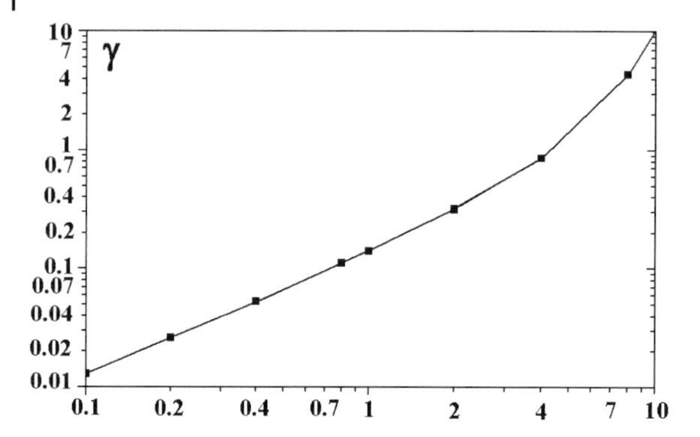

Figure 6.19 Parameter γ in the size growth equation (6.86) as a function of a (6.85) [41, 43].

Let us represent the balance equation for the average cluster size $\overline{n}$ in the coagulation process analogously to (6.70) in the form

$$\frac{d\overline{n}}{dt} = \gamma k_0 N_{\text{sat}}(T)\overline{n}^{2/3} . \tag{6.86}$$

The values of γ depending on a and defined by (6.85) are given in Figure 6.19 [41, 43].

6.3.4
Cluster Growth in a Hot Gas with Metal-Containing Molecules

In considering the cluster plasma, we will be guided by a plasma consisting of a buffer gas with metal clusters. A high number density of bound atoms in metal clusters is realized in the case if a liquid metal-containing compound is inserted in a flow of a dense plasma. This method of cluster generation in a buffer gas allows us to obtain an effective source of light [128–131] and is used for preparation of films of heat-resistant metals as a result of transport of metal clusters and their deposition onto a surface. So, we consider the scheme of cluster generation if metal-containing molecules are injected into a hot gas or plasma and are converted into a gas of metal clusters. The best case for this is if metal-containing molecules contain halogen atoms, that is, these molecules are MX_k, where M is a metal atom and X is a halogen atom.

We first find the conditions under which decomposition of metal-containing molecules in a plasma leads to cluster formation, that is, the process of cluster generation follows the scheme

$$MX_k \longleftrightarrow M + kX , \quad M + M_{n-1} \longleftrightarrow M_n , \tag{6.87}$$

and in a low-temperature region halogen atoms join to form X_2 molecules. The equilibrium between components of process (6.87) is described by the scheme

$$M + \frac{k}{2}X_2 \longleftrightarrow MX_k$$

and the possibility for the compound MX_k to decay into metal clusters and halogen molecules is determined roughly by the criterion [41]

$$\varepsilon_X < \varepsilon_M ,$$

where ε_X is the binding energy per halogen atom in molecule MX_k and ε_M is the binding energy per atom for bulk metal.

From the chemical equilibrium for molecules MX_k it follows that this compound decomposes into atoms at temperatures of the order

$$T_1 = \frac{\varepsilon_X}{\ln(N_0/[X])} , \quad (6.88)$$

where $[X]$ is the total number density of free and bound halogen atoms, and N_0 is of the order of a typical atomic value. In turn, metal clusters are transformed into atomic vapor at temperatures of the order

$$T_2 = \frac{\varepsilon_M}{\ln(N_0/[M])} , \quad (6.89)$$

where $[M]$ is the total number density of free and bound metal atoms. Evidently, clusters remain stable in the temperature range [132, 133]

$$T_1 < T < T_2 . \quad (6.90)$$

Because of the halogen excess, $[X] \gg [M]$, the possibility of the existence of clusters and gas molecules in the system under consideration corresponds to the criterion

$$\varepsilon_X < \varepsilon_M < k\varepsilon_X . \quad (6.91)$$

One can see that these criteria are compatible if the gaseous compound of a heat-resistant metal contains several halogen atoms. The data related to this analysis are given in Table 6.3. This table gives the values of ε_X and ε_M that are obtained on the basis of the Gibbs thermodynamic potential for the compounds under consideration, and the values that follow from the enthalpy values for these compounds are given in parentheses. The temperatures T_1 and T_2 follow from (6.88) and (6.89), and the temperature T_3 coincides with T_2 and results from the saturated vapor pressure of the metal as a function of temperature. The data in Table 6.3 relate to a total number density of halogen atoms of $1 \times 10^{16}\,\mathrm{cm}^{-3}$ and a total number density of metal atoms that is four or six times less depending on the type of metal-containing molecules. If criteria (6.90) are fulfilled, clusters of this metal exist in the temperature range $T_1 < T < T_2$, and such cases are marked in the last column in Table 6.3 by a plus sign.

We now analyze the character of evolution of a plasma flow with inserted metal-containing molecules in the center of the flow. Then we take argon as a buffer gas at a pressure of 1 atm, and metal-containing molecules MX_k are injected into the central part of the flow of radius 1 mm such that the concentration of metal atoms after total destruction of metal-containing molecules is 10% with respect to argon atoms. We compare typical times of various processes which are responsible for evolution of this system. As a result of decomposition of metal-containing molecules, metal atoms are formed and they are transformed into clusters according to scheme (6.66). For simplicity, we assume the binding energies of each halogen atom in the initial molecule MX_k and forming radicals to be identical and equal to ε_X. We take the following rates of decay of molecules and radicals in collisions with buffer gas atoms:

$$\nu_d = N_a k_{gas} \exp\left(-\frac{\varepsilon_X}{T}\right) . \qquad (6.92)$$

Here N_a is the number density of buffer gas atoms, T is a current temperature of buffer gas atoms, and k_{gas} is the gas-kinetic rate constant for collisions of molecules and radicals with atoms of a buffer gas, that is, $k_{gas} = v_T \sigma_{gas}$, where v_T is the thermal velocity of argon atoms, and the gas kinetic cross section of collision is $\sigma_{gas} = 3 \times 10^{-15}\,cm^2$. We assume the rate ν_d to be independent of the number of halogen atoms k in the molecule or radicals.

Let us apply this scheme to metal-containing molecules MoF_6, WF_6, IrF_6, and WCl_6 at the temperature T_* in the center of a flow that is given by $N_m/N_{sat}(T_*) \approx 10$ for MoF_6 and WF_6, and $N_m/N_{sat}(T_*) \approx 1000$ for IrF_6 and WCl_6, where N_m is the total number density of metal atoms in metal-containing molecules, N_{sat} is the number density of metal toms at the saturated vapor pressure for the indicated temperature, and $T_* \ll \varepsilon_X$. Under the above assumption, the time for decompo-

Table 6.3 Parameters of some compounds of heat-resistant metals [41].

Compound	ε_X	ε_M	T_1, 10^3 K	T_2, 10^3 K	T_3, 10^3 K	Cluster existence
$HfCl_4$	(4.2)	6.0 (6.4)	(2.4)	3.0 (3.2)	3.15	+
HfF_4	(6.3)	6.0 (6.4)	(3.5)	3.0 (3.2)	3.15	–
$ThCl_4$	(4.7)	5.8 (6.2)	(2.6)	3.0 (3.1)	3.12	+
ThF_4	6.2 (6.5)	5.8 (6.2)	3.5 (3.6)	3.0 (3.1)	3.12	–
$TiBr_4$	3.0 (3.2)	4.4 (4.9)	1.7 (1.8)	2.2 (2.5)	2.28	+
$TiCl_4$	3.5 (3.8)	4.4 (4.9)	2.0 (2.1)	2.2 (2.5)	2.28	+
UCl_4	3.8 (4.1)	5.1 (5.5)	2.1 (2.3)	2.6 (2.8)	2.73	+
UF_4	5.7 (5.9)	5.1 (5.5)	3.2 (3.3)	2.6 (2.8)	2.73	–
$ZrCl_4$	4.3 (4.8)	5.9 (6.3)	2.4 (2.7)	3.0 (3.2)	3.08	+
ZrF_4	6.5 (6.9)	5.9 (6.3)	3.6 (3.9)	3.0 (3.2)	3.08	–
IrF_6	2.2 (2.5)	6.4 (6.9)	1.2 (1.4)	3.2 (3.5)	3.15	+
MoF_6	3.9 (4.2)	6.3 (6.8)	2.2 (2.4)	3.2 (3.4)	3.19	+
UCl_6	3.0 (3.3)	5.1 (5.5)	1.7 (1.8)	2.6 (2.8)	2.73	+
UF_6	4.7 (5.0)	5.1 (5.5)	2.6 (2.8)	2.6 (2.8)	2.73	–
WCl_6	(3.0)	8.4 (8.8)	(1.7)	4.2 (4.4)	4.04	+
WF_6	4.5 (4.9)	8.4 (8.8)	2.5 (2.7)	4.2 (4.4)	4.04	+

sition of metal-containing molecules is

$$\tau_{\text{chem}} = \frac{6}{\nu_{\text{d}}(T)} = \frac{6}{N_{\text{a}} k_{\text{gas}} \exp\left(-\frac{\varepsilon_X}{T}\right)} . \tag{6.93}$$

We compare this time with the time for conversion of free metal atoms into metal clusters in the case of prompt decomposition of metal-containing molecules that is given by (6.74). Since this time is short compared with the time for flow drift in an aggregation tube, the coagulation mechanism may determine the subsequent cluster growth, and the average cluster size is given by (6.80) or by (6.82).

Table 6.4 contains some results for this regime of cluster generation to represent the real character of the cluster generation process. We assume metal-containing molecules to be injected into a cylinder of radius $\rho_0 = 1$ mm in the flow center, and this radius is small compared with the tube radius; the argon pressure is $p = 1$ atm, and the concentration of metal-containing molecules is $c_M = 1\%$ in the central part with respect to argon atoms. For definiteness we take the flow velocity $w = 300$ cm/s and the tube length $l = 30$ cm, that is, the time for the flow being located inside the tube is 0.1 s.

Let us give some comments regarding the data in Table 6.4. Here ρ is the density of metal-containing compounds at room temperature, T_{m} and T_{b} are their melting and boiling temperatures, respectively, ε_X is the binding energy per halogen atom in this compound, ε_M is the binding energy per metal atom in the bulk metal, the temperatures T_1 and T_2 are given by (6.88) and (6.89), $c_M \delta T$ is the temperature increase resulting from process (6.87) if the extracted energy is consumed on heating of the buffer gas, T_* is the temperature at the tube center, which is chosen as indicated above, and N_{a} is the number density of argon atoms at a pressure of 1 atm, so the total number density of metal atoms (free and bound) is $N_{\text{m}} = 0.01 N_{\text{a}}$.

To estimate the character of consumption of metal-containing compound, we give in Table 6.4 the rate dM/dt of the compound flux, that is, the compound mass inserted in the argon flux, and the rate of deposition dl/dt, that is, the rate of film growth dl/dt if the flux of metal clusters is deposited on a substratum. These values are given by formulas

$$\frac{dM}{dt} = \pi \rho^2 w N_{\text{m}} m_c \, , \quad \frac{dl}{dt} = \frac{4 r_{\text{W}}^3}{3 m \rho^2} \frac{dM}{dt} \, ,$$

where M_{c} and m are the masses of a metal-contained molecule and a metal atom, respectively.

We now determine the kinetic parameters of this system under the above conditions. The time τ_{chem} for destruction of a metal-containing molecule follows from (6.93), and a typical time for heat transport is given by

$$\tau_{\text{heat}} = \frac{\rho_0^2}{4\chi} \, ,$$

where χ is the thermal diffusivity coefficient of argon. As is seen, usually heat transport proceeds faster than release of free metal atoms, and heating of argon due

Table 6.4 Parameters of evolution of metal-containing molecules in hot argon ($p = 1$ atm, $c_M = 10\%$, $\rho_0 = 1$ mm).

Compound	MoF_6	IrF_6	WF_6	WCl_6
ρ, g/cm^3	2.6	6.0	3.4	3.5
T_m, K	290	317	276	548
T_b, K	310	326	291	620
ε_X, eV	4.3	2.5	4.9	3.6
ε_M, eV	6.3	6.5	8.4	8.4
T_1, K	2200	1200	2500	1700
T_2, K	4100	4000	5200	5200
δT, 10^3 K	12	3.7	13	4.8
T_*, K	3600	2900	4600	4200
N_m, 10^{16} cm^{-3}	2.0	2.5	1.6	1.8
dM/dt, μg/s	0.66	1.2	0.74	1.1
dl/dt, nm/s	23	20	13	20
τ_{chem}, 10^{-3} s	5	0.1	1	0.1
τ_{heat}, 10^{-4} s	2.0	2.8	1.3	1.5
$\overline{n}$	1.6×10^7	1.9×10^7	1.5×10^7	1.5×10^7
r, nm	40	42	40	40
N_{cl}, 10^9 cm^{-3}	1.2	1.3	1.1	1.1
D_0, cm^2/s	25	18	36	31
$\sqrt{\overline{\rho^2}}$, μm	88	71	110	100

to the transport process is not strong. Released metal atoms join in clusters, and for the basic time of residence of this mixture in the tube $\tau_{dr} = 0.1$ s the cluster growth process proceeds through coagulation. We give in Table 6.4 the average cluster size $\overline{n}$ according to (6.82) at the end of the coagulation process together with its radius r, and the number density of clusters N_{cl}.

Let us evaluate the broadening of the region occupied by metal clusters as a result of diffusion in the course of the cluster growth process. We have for the mean square displacement during this process

$$\overline{\rho^2} = 4\int D\,dt = 4D_0\int \frac{dt}{\overline{n}^{2/3}} = \frac{20\,D_0\tau_{dr}}{\overline{n}^{2/3}}\,. \tag{6.94}$$

We used above the case of cluster diffusion in a rare gas (4.110) and (4.114) for the diffusion coefficient. As is seen, the relative displacement of clusters in the course of their growth is small.

6.3.5 Passage of Cluster Plasma Flow through an Orifice

There are various methods for generation of metal cluster beams. The first stage of this process is formation of an atomic vapor, and then the total conversion of metal atoms into metal clusters proceeds in a flow of a buffer gas. We analyze below the next stage of generation of a cluster beam, that is, its extraction from the aggregation chamber by passage of this flow through an orifice. We will be guided by the geometry of the aggregation chamber near the orifice according to Figure 6.1 that provides a laminar flow of a buffer gas. But reduction of the cross section of the aggregation chamber near the exit orifice can lead to cluster attachment to walls near the orifice, and this process determines the efficiency of the cluster beam generation. The subsequent separation of the buffer gas flow and cluster beam is made in a simple manner by pumping. Since the momentum of a buffer gas atom is small compared with the momentum of a cluster, buffer gas atoms are removed from the flow by pumping, and this flow is transformed into a cluster beam. In the following stage, clusters may be charged by a weak crossed electron beam, and then a beam of charged clusters is governed by electric optics.

In considering the behavior of clusters in a buffer gas flow near the exit orifice, we assume the orifice radius ρ_0 to be large compared with the mean free path of buffer gas atoms, $\rho_0 \gg \lambda$, which corresponds to the gas-dynamic character of the flow. When a gas reaches the orifice, its velocity increases up to the sound speed and increases sharply near the orifice. Clusters are moving far from the orifice with the flow velocity, but their drift velocity w_0 may differ from the flow velocity near the orifice and is given by (6.2), $w_0 = c_s(\nu\tau)^{2/3}$, where c_s is the sound speed. The cluster drift velocity is affected by the small parameter $\nu\tau$, where $1/\nu$ is the time for relaxation of the cluster momentum and τ is the time for variation of the flow velocity. Table 6.5 contains examples of this under the flow conditions in Table 6.4, and w_0 is the drift velocity of clusters at the orifice under these conditions when the equilibrium between the buffer gas flow and clusters is violated.

We also give simple estimations for attachment of clusters to the walls of the aggregation chamber near the orifice. When clusters are located in all the cross section of the tube or its conic part, the dependence of the cluster flux j to the walls on the tube and cluster parameters is $j \sim D_{cl}/\rho^2$, where D_{cl} is the diffusion coefficient for clusters in a buffer gas and ρ is the radius of the current cross section.

Table 6.5 Character of relaxation of the cluster momentum near an orifice of radius $\rho_0 = 1$ mm for the parameters of the argon flow with metal clusters under the conditions in Table 6.4.

Compound	MoF_6	IrF_6	WF_6	WCl_6
$\nu\tau$	0.13	0.087	0.051	0.06
w_0/c_s	0.70	0.53	0.37	0.41
w_0, 10^4 cm/s	7.8	5.3	4.7	5.0

For the portion of clusters ξ which attach to the walls near the orifice this gives

$$\xi \sim \frac{D_{\mathrm{cl}}}{w_0 \rho_0 \tan \alpha} ,$$

where w_0 is the drift velocity of clusters near the orifice, ρ_0 is the orifice radius, and α is the cone angle (Figure 6.1). If we use (4.114) for the diffusion coefficient and (6.2) for the drift velocity of clusters, we obtain the following dependence for the potion of attachment clusters on a typical cluster size n, the number density N_{a} of buffer gas atoms, and the orifice radius:

$$\xi \sim \frac{1}{n^{4/9} N_{\mathrm{a}}^{5/3} \rho_0^{5/3}} .$$

From this it follows that attachment of clusters to the walls near the orifice becomes important at low pressures of the buffer gas, a small radius of the orifice, and not large cluster sizes.

Let us consider one more plasma effect that can influence transport of clusters to the walls. If a plasma consists of electrons and positive ions, clusters are negatively charged because of the higher mobility of electrons. As this plasma relaxes in the course of its drift toward the orifice, electrons and ions go to the walls. As a result, this plasma becomes charged, and its charge may be connected with clusters. This charge creates an electric field that accelerates attachment of clusters to the walls. This effect may be of importance for a certain cluster plasma [60].

6.4 Plasma Processes in the Earth's Atmosphere

6.4.1 Processes in Atmospheric Plasmas

In the foregoing discussion, we examined general principles and concepts relating to plasmas. We shall now apply general plasma principles and concepts to atmospheric plasmas, the properties of which depend strongly on the altitude above the Earth's surface and have specifics for each altitude. Plasmas in the upper atmosphere are generated by the absorption of solar radiation that ionizes oxygen atoms and nitrogen molecules. Plasmas in the lower layers of the atmosphere have properties depending on transport processes of charged and excited atoms, and also on the heat balance of these low-lying atmospheric layers.

Atmospheric plasmas can be used to illustrate the elementary collision processes that can occur in a weakly ionized plasma. Table 6.6 contains a list of basic atmospheric plasma processes, and we will be referring to this in the analysis of atmospheric phenomena. These processes are an explicit illustration of the data in Tables 2.16–2.19. We can use photoprocesses as the first example. Processes 1–3 in Table 6.6 are responsible for formation of charged particles in the upper atmosphere, and process 4 is associated with the production of atomic oxygen. The process that is the inverse of process 5 gives the main contribution to glow in the night sky, and processes 7–10 determine auroral radiation.

As an introduction to the study of atmospheric plasmas, we consider general properties of the Earth's atmosphere. The heat balance of the Earth is represented in Table 6.7. The source of the Earth's heat balance is solar radiation. Solar radiation can be approximated by radiation of a blackbody with an effective temperature of 5800 K, corresponding to a radiation flux of $6.4\,kW/cm^2$ from its surface, mostly in the visible region of the spectrum. But in contrast to a blackbody, solar radiation contains a remarkable flux in UV and vacuum UV (VUV) spectra that is responsible for ionization processes in the Earth's atmosphere. The radiative flux is $0.137\,W/cm^2$ at the distance of the Earth from the Sun, which translates to 1.74×10^{14} kW of solar power reaching the Earth's atmosphere. The same amount of power then emanates from the Earth in the form of emission from the Earth's surface and atmosphere as infrared radiation, and reflection by the Earth's surface and atmosphere. The surface of the Earth receives and returns 2.5×10^{14} kW of power, whereas the Earth including its atmosphere receives and returns 2.69×10^{14} kW of power. If the Earth is regarded as an ideal blackbody, the emitted infrared radiation corresponds to an effective surface temperature of 291 K. Absorption of solar UV radiation by molecules and atoms of the Earth's atmosphere determines ionization and chemical processes in upper atmosphere layers. The optical density of the Earth's atmosphere for this radiation is high, and therefore radiation of short wavelengths is absorbed at high altitudes. Figure 6.20 shows the altitudes at which solar radiation of a given wavelength is absorbed [134].

The Earth's atmosphere at sea level consists mostly of molecular nitrogen (78%), molecular oxygen (21%), and argon (about 1%); the total number density of molecules and atoms is $2.7 \times 10^{19}\,cm^{-3}$ (at a pressure of 1 atm). The number density of molecules decreases with increasing altitude h as given by the barometric

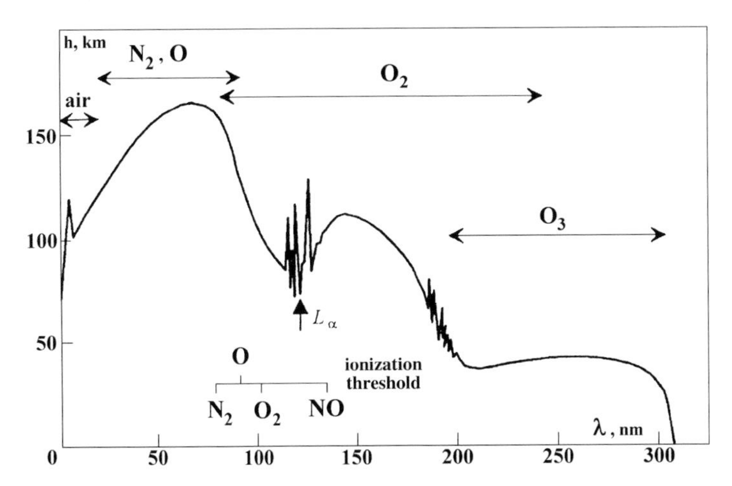

Figure 6.20 The boundary altitudes for absorption of solar UV radiation of a given wavelength [134].

Table 6.6 Elementary collision processes in the Earth's atmosphere. The wavelength (λ) of photons is expressed in nanometers, the lifetime (τ) is given in seconds, the rate constant (k) for pairwise processes is in cubic centimeters per second, and the rate constant (K) for three body processes is expressed in centimeters to the sixth power per second. All rate constants correspond to room temperature [90].

Type	Number	Typical process	Parameters
Photo-processes	1	$\hbar\omega + O \to O^+ + e$	$\nu = 3 \times 10^{-7}\,s^{-1}$ [a]
	2	$\hbar\omega + O_2 \to O_2^+ + e$	$\nu = 4 \times 10^{-7}\,s^{-1}$ [a]
	3	$\hbar\omega + N_2 \to N_2^+ + e$	$\nu = 3 \times 10^{-7}\,s^{-1}$ [a]
	4	$\hbar\omega + O_2 \to O(^1D) + O(^3P)$	$\sigma_{max} \sim 10^{-17}\,cm^2$
	5	$\hbar\omega + O^- \to O + e$	$\nu = 1.4\,s^{-1}$ [a]
	6	$\hbar\omega + O_2^- \to O_2 + e$	$\nu = 0.3\,s^{-1}$ [a]
	7	$O(^1S) \to O(^1D) + \hbar\omega$	$\tau = 0.8\,s$, $\lambda = 558\,nm$
	8	$O(^1D) \to O(^3P) + \hbar\omega$	$\tau = 140\,s$, $\lambda = 630\,nm$
	9	$N(^2D_{5/2}) \to N(^4S) + \hbar\omega$	$\tau = 1.4 \times 10^5\,s$, $\lambda = 520\,nm$
	10	$N(^2D_{3/2}) \to N(^4S) + \hbar\omega$	$\tau = 6 \times 10^4\,s$, $\lambda = 520\,nm$
Ionization	11	$e + O_2 \to O_2^+ + e$	
	12	$e + N_2 \to N_2^+ + e$	
Recombination	13	$e + N_2^+ \to N + N$	$k = 2 \times 10^{-7}\,cm^3/s$
	14	$e + O_2^+ \to O + O$	$k = 2 \times 10^{-7}\,cm^3/s$
	15	$e + NO^+ \to N + O$	$k = 4 \times 10^{-7}\,cm^3/s$
	16	$e + N_4^+ \to N_2 + N_2$	$k = 2 \times 10^{-6}\,cm^3/s$
	17	$O^- + O_2^+ + N_2 \to O + O_2 + N_2$	$k_{ef} = 2 \times 10^{-6}\,cm^3/s$ [b]
	18	$O_2^- + O_2^+ + O_2 \to 3O_2$	$K = 1.6 \times 10^{-25}\,cm^6/s$
	19	$NO_2^- + NO^+ + N_2 \to NO + NO_2 + N_2$	$K = 1 \times 10^{-25}\,cm^6/s$
Electron attachment	20	$e + 2O_2 \to O_2^- + O_2$	$K = 3 \times 10^{-30}\,cm^6/s$
	21	$e + O_2 + N_2 \to O_2^- + N_2$	$K = 1 \times 10^{-31}\,cm^6/s$
	22	$e + O_2 \to O^- + O$	$k = 2 \times 10^{-10}\,cm^3/s$, $\varepsilon = 6.5\,eV$
Negative ion detachment	23	$O^- + O_3 \to e + 2O_2 + 2.6\,eV$	$k = 3 \times 10^{-10}\,cm^3/s$
	24	$O_2^- + O_2 \to e + O_3 + 0.62\,eV$	$k = 3 \times 10^{-10}\,cm^3/s$
	25	$O^- + O_2 \to e + O_3 - 0.42\,eV$	$k < 1 \times 10^{-12}\,cm^3/s$
Dissociation	26	$e + O_2 \to 2O + e$	$k = 2 \times 10^{-10}\,cm^3/s$, $\varepsilon > 6\,eV$

Table 6.6 Continued.

Type	Number	Typical process	Parameters
Ion reactions	27	$O^+ + N_2 \rightarrow NO^+ + N + 1.1\,eV$	$k = 6 \times 10^{-13}\,cm^3/s$
	28	$O^+ + O_2 \rightarrow O_2^+ + O + 1.5\,eV$	$k = 2 \times 10^{-11}\,cm^3/s$
	29	$N^+ + O_2 \rightarrow O^+ + NO + 2.3\,eV$	$k = 2 \times 10^{-11}\,cm^3/s$
	30	$N_2^+ + O \rightarrow NO^+ + N + 3.2\,eV$	$k = 1 \times 10^{-10}\,cm^3/s$
Quenching	31	$2O_2(^1\Delta_g) \rightarrow O_2 + O_2(^1\Sigma_g^+)$	$k = 2 \times 10^{-17}\,cm^3/s$
	32	$O_2(^1\Delta_g) + O_2 \rightarrow 2O_2$	$k = 2 \times 10^{-18}\,cm^3/s$
	33	$O_2(^1\Sigma_g^+) + N_2 \rightarrow O_2 + N_2$	$k = 2 \times 10^{-17}\,cm^3/s$
	34	$N_2(A^3\Sigma^+) + O_2 \rightarrow N_2 + O_2$	$k = 4 \times 10^{-12}\,cm^3/s$
	35	$O(^1D) + O_2 \rightarrow O + O_2$	$k = 5 \times 10^{-11}\,cm^3/s$
	36	$O(^1D) + N_2 \rightarrow O + N_2$	$k = 6 \times 10^{-11}\,cm^3/s$
	37	$O(^1S) + O_2 \rightarrow O + O_2$	$k = 3 \times 10^{-13}\,cm^3/s$
	38	$O(^1S) + O \rightarrow O + O$	$k = 7 \times 10^{-12}\,cm^3/s$
Chemical reaction	39	$O + O_3 \rightarrow 2O_2$	$k = 7 \times 10^{-14}\,cm^3/s$
Three body	40	$O + 2O_2 \rightarrow O_3 + O_2$	$K = 7 \times 10^{-34}\,cm^6/s$
association	41	$O + O_2 + N_2 \rightarrow O_3 + N_2$	$K = 6 \times 10^{-34}\,cm^6/s$

a Per atom, molecule, or ion for daytime atmosphere at zero zenith angle.
b At a pressure of 1 atm.

formula (1.48)

$$N(h) = N(0)\exp\left(-\int_0^h \frac{mgdz}{T}\right), \tag{6.95}$$

where m is the average molecular mass, g is the free-fall acceleration, and $T(z)$ is the temperature at altitude z. For air temperature $T = 300$ K the barometric formula (6.95) gives that the number density of molecules decreases twice at an altitude of 6 km. Note that because of convection the concentration of the air components does not vary up to altitudes of the order of 100 km, where the ionization processes start and the atmosphere content varies.

To determine the change of temperature with altitude, one can ignore heat transfer processes, and employ the adiabatic law $TN^{1-\gamma} = \text{const}$, where $\gamma = 1.4$ is the adiabatic exponent for air at the temperatures considered. This leads to the equation $dT/T = (\gamma - 1)dN/N$, and from the barometric formula (6.95) we have $dN/N = -mgdz/T$. Hence, the temperature gradient in the atmosphere is $-dT/dz = mg(\gamma - 1) = 14$ K/km. The real value (5–10 K/km) is lower than this because of water vaporization in the atmosphere and other heat transport processes. Nevertheless, near the Earth's surface the temperature of the atmosphere drops in accordance with the adiabatic model up to the tropopause at an altitude of (12 ± 4) km, where the temperature is 200 ± 20 K. Note that various atmospheric parameters vary depending on latitude and time of day, and we now deal with mean

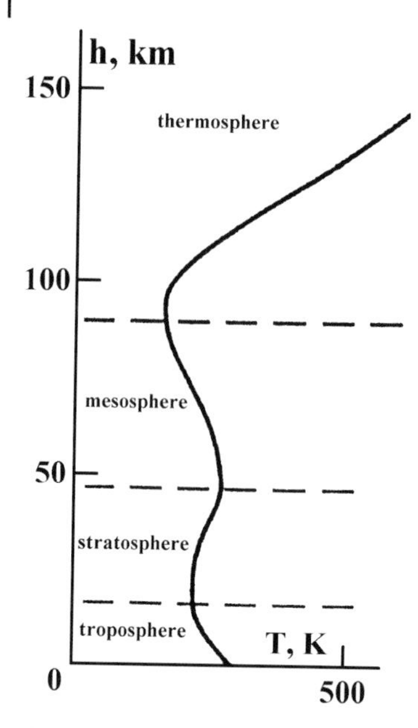

Figure 6.21 Classification of atmospheric layers depending on their altitude [135].

values. In particular, the tropopause temperature is higher near the equator and lower at the poles. The stratosphere lies above the tropopause up to the stratopause at an altitude of (50 ± 5) km, and the temperature of the stratosphere increases with altitude and reaches 270 ± 20 K in the stratopause. In the mesosphere, which starts from the stratopause, the temperature drops up to 160–190 K at an altitude of 85–100 km. At higher altitudes the temperature of the atmosphere increases with increasing altitude. Figure 6.21 gives definitions of some atmospheric layers depending on their altitude [135].

This behavior of the temperature of the atmosphere results from the heat balance of the atmosphere at these altitudes. In particular, most of the ozone is in the stratosphere, and absorption of VUV solar radiation by ozone leads to an increase of the temperature of the stratosphere compared with neighboring regions. On the basis of the temperature gradient at low altitudes near the Earth's surface one can find the contribution of the heat flux q due to the thermal conductivity of air to the Earth's heat balance. Our estimate for the heat flux resulting from thermal conductivity is $q = \kappa |dT/dz| = 3 \times 10^{-8}\,\mathrm{W/cm^2}$, whereas according to the data in Table 6.7, convection is responsible for an average heat flux of $2 \times 10^{-3}\,\mathrm{W/cm^2}$, which is significantly more.

Table 6.7 Heat balance of the Earth. Convection of air and evaporation of water from the Earth's surface leads to transport of energy from the surface to the atmosphere.

Form of energy	Process	Power, 10^{13} kW
Solar radiation	Reaching the Earth's atmosphere	17.3
	Reflected by the Earth	0.7
	Absorbed by the Earth	8.3
	Reflected by the atmosphere	5.4
	Absorbed by the atmosphere	2.9
Infrared radiation	Atmospheric emission to space	10.2
	Atmospheric emission to the Earth	16.7
	Absorbed by the atmosphere	18.7
	Absorbed by the Earth	16.7
	Emitted by the Earth	19.7
	Transmitted through the atmosphere	1.0
Convection		1.3
Evaporation		4.0

6.4.2
Aerosol Plasma

According to the definition, an aerosol plasma is weakly ionized atmospheric air located at various altitudes with an admixture of aerosol particles – solid and liquid particles. Aerosol particles or aerosols include water drops, ice crystallites, and dusty particles of size of the order of 0.01–100 μm that exist in atmospheric air. The kind and size of aerosols in the Earth's atmosphere depend both on the altitude and on local conditions there. In particular, water aerosols are located close the the Earth's surface, and a tropospheric aerosol has as a base SO_2, NH_3, and NO_x molecules, mineral salts, and other compounds that result from natural processes on the Earth's surface and may reach high altitudes.

The smallest aerosols, Aitken particles [8, 9] are formed as a result of photochemical reactions at altitudes above the cloud boundary. They have a size of 10–100 nm and a number density of 10^2–10^4 cm^{-3} [10, 11]. Aitken particles scatter the solar radiation and are responsible for the blue color of the sky. Aitken particles partake in atmospheric chemical reactions and charge transfer from initially formed electrons and simple ions to large aerosols.

The processes involving aerosols are different at different altitudes and are responsible for interaction of atmospheric air with solar radiation [6, 7]. Having general properties of small particles in neutral and ionized gases [136–138], water aerosols located at altitudes of a few kilometers determine electric phenomena in atmospheric air, including lightning [6, 7, 139]. The process of Earth charging that is the basis of atmospheric electric processes is complicated and includes many stages. But falling of charged water aerosols under the action of gravitational forces is the only way to create an electric current that charges the Earth's surface. In turn, the Earth's electric field is of importance for charging of aerosols, and such processes will be considered below. Aerosol particles in the Earth's atmosphere may be formed by natural processes and as a result of human activity [140].

When solid clusters join as a result of contact between them in atmospheric air, porous structures can be formed in which solid clusters conserve their individuality. These structures are fractal aggregates if their joining takes place in the absence of external fields. This name results from the dependence of the matter density in fractal aggregates on their size. The specifics of growth of fractal aggregates in a buffer gas are connected to the diffusion character of their motion. This can lead to a nonlinear dependence of the nucleation rate on the aggregate number density. Because of the micrometer size of fractal aggregates, they interact effectively with external fields. Even small electric fields can cause an effective interaction between fractal aggregates through the interaction of induced charges. As a result of nucleation in an external electric field, so-called fractal fibers are formed that are elongated fractal structures.

6.4.3 The Ionosphere as a Mirror for Electromagnetic Waves

The history of exploration of the Earth's ionosphere starts from Marconi's experiment in 1901 when he tried to establish radio contact between two continents. The transmitter had been set up in Europe on the Cornwall peninsula in England, and the receiver was located in Canada, on the Newfoundland peninsula. From the standpoint of wave propagation theory, this experiment seemed to be hopeless. According to the laws of geometrical optics, radio waves should propagate at such distances with rectilinear beams, and radio connection for these distances seemed to be precluded by the spherical form of the Earth's surface. The experiment led to a surprising result: the signal was detected by the receiver and its intensity exceeded estimates by many orders of magnitude.

The only explanation for this discrepancy was the existence of a radio mirror in the Earth's atmosphere that reflects radio waves. The radio mirror model assumes that waves follow rectilinear paths and repeatedly reflect between the surface of the Earth and this upper-atmosphere radio mirror. In 1902, O. Heaviside and A. Kennelly assumed that the role of this radio mirror is played by an ionized layer in the atmosphere. This was confirmed in 1924–1925 by an experiment conducted by a group of English physicists from Cambridge University ([180, 181]). Placing the receiver at a distance of 400 m from the transmitter, they determined by measurement of the time delay for the reflected signal to arrive that the reflector is at an altitude of 100–120 km. This reflecting layer, which was referred to earlier as the Heaviside layer, is now called the E layer (based on the designation of the electric field vector for radio waves). Subsequently, the existence of ionized gas was discovered at other altitudes. The part of the atmosphere that contains ionized gases is called the ionosphere [141].

In reality, the Earth's ionosphere is not a uniform plasma because of different compositions of the atmosphere and different degrees of ionization for different altitudes. Hence, the ionosphere is divided into layers, as given in Table 6.8.

The presence of electrons in the ionosphere prevents long electromagnetic waves from propagating therein if their frequencies are less than the frequency of plasma oscillations. Hence, radio waves reflect from the ionosphere and return to the Earth's surface [142, 143]. Table 6.9 gives limiting electron number densities at which the plasma frequency coincides with the frequency of plasma oscillations, so longer radio waves cannot propagate in such a medium. The maximum electron number density of the order of $10^6\,cm^{-3}$ may be reached at an altitude of about 200 km. Therefore, the best reflection conditions are fulfilled for short waves within the range of 30–100 m. Because the electron number density at any particular altitude depends on season, time of day, and other factors, the quality of radio communications at a given wavelength varies continually. Long waves are reflected at low altitudes, where the atmospheric density is relatively high, and reflection of the waves is accompanied by damping.

It has been established that several ionized layers differ in their properties, and different processes are of importance for each layer [144–147]. The lower D layer of the atmosphere occupies the altitude region from 50 to 90 km. A typical number density of charged particles therein is of the order of $10^3\,cm^{-3}$. The negative charge of the D layer of the Earth's atmosphere arises primarily from the presence of negative ions, and a great variety of both negative and positive ions reside there. In particular, the most widespread positive ion is the $H_3O^+ \cdot H_2O$ cluster ion. Figure 6.22 shows the average distribution of the electron and ion number density with altitude. These number densities vary significantly depending on the condi-

Table 6.8 Layers of the ionosphere.

Layer	Altitude, km
D	60–90
E	90–140
F_1	140–300
F_2	300–1000

Table 6.9 Critical number densities of plasma electrons (N_{cr}) for propagation of radio waves.

Type of radio waves	Wavelength, m	N_{cr}, cm^{-3}
Long	10 000–1 000	$10–10^3$
Medium	1 000–100	$10^3–10^5$
Short	100–10	$10^5–10^7$

tions, and Figure 6.22 gives the daytime and nighttime distributions for the number densities of electrons and basic ionospheric ions according to evaluations [148].

The next higher layer of the ionosphere, called the E-layer, is at an altitude from 90 to 140 km. Charged particles of the E layer are formed by photoionization of air caused by solar UV radiation. These charged particles drift to lower layers of the atmosphere and are the source of plasma in the D layer. The electron number density in the E layer of the ionosphere is of the order of $10^5\,\text{cm}^{-3}$. Negative ions are nearly nonexistent in this layer, and the basic types of positive ions are NO^+ and O_2^+. The decay of charged particles in the E layer is due to dissociative recombination of electrons and molecular ions, or to transport of charged particles to lower layers of the atmosphere.

One can estimate a background altitude between D and E layers from the relation

$$\alpha N_e^2 \sim K N_e [O_2]^2 ,$$

where α is the dominant recombination process from processes 13–15 in Table 6.6, and K is the rate constant for three body process 21 in Table 6.6. This estimate means that in the E layer electrons are lost as a result of dissociative recombination, whereas in the D layer they attach to oxygen molecules, and then negative ions of various kinds determine the negative charge of the ionosphere. Using the values of the rate constants given in Table 6.6 and using in the above formula $N_e \sim 10^5\,\text{cm}^{-3}$, we find the boundary between the D and E layers corresponds to the number density of oxygen molecules $[O_2] \sim 10^{14}\,\text{cm}^{-3}$, which corresponds to an altitude of 90 km.

A higher ionospheric layer (the F_1 layer) is located at an altitude of 140–200 km. Above it (up to an altitude of about 400 km) is the F_2 layer of the ionosphere. The electron number density in these F layers is 10^5–$10^6\,\text{cm}^{-3}$. The basic type of positive ions is O^+. Charged particles of the F layers of the ionosphere are formed by photoionization of atmospheric oxygen (the basic component of the atmosphere at these altitudes) by solar radiation. Loss of electrons from these layers is caused by

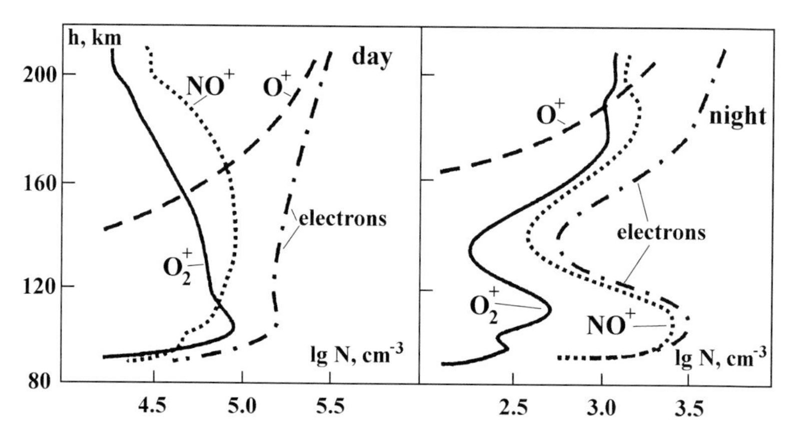

Figure 6.22 The average altitude distributions of electrons and basic positive ions in the ionosphere according to evaluations [148].

photorecombination of electrons and oxygen ions, photoattachment of electrons to oxygen atoms, and by transport of electrons to lower atmospheric layers.

The E and F layers of the ionosphere, as layers with a high number density of electrons, are responsible for reflection of radio signals, so these layers are the radio mirror of the atmosphere for radio waves. But their role is broader than this function. The altitudes at which they occur are the most convenient for operation of artificial satellites. Also, the interesting physical phenomenon, the aurora, occurs at these altitudes. An aurora occurs when a flow of solar protons penetrates the atmosphere. Charge exchange of these protons and their deceleration takes place in the E and F layers of the atmosphere. The subsequent chain of elementary processes leads to formation of excited atoms whose radiation is observed from the Earth as an aurora.

One of the characteristics of auroras is that their radiation is created by forbidden transitions of atoms and ions that cannot be modelled in the laboratory. In particular, processes 7–10 in Table 6.6 usually give the principal contributions to the radiation of auroras. Under laboratory conditions, these excited states of atoms and ions are quenched by collisions with air molecules. The probabilities for an excited atom to radiate and to be quenched are equal at a number density of quenching particles given by $(k\tau)^{-1}$, where τ is the radiative lifetime of the excited atom, and k is the rate constant for collision quenching. The excited oxygen atom $O(^1S)$ is quenched in the upper atmosphere mainly by atomic oxygen (process 38 in Table 6.6), and the threshold number density of oxygen atoms is $2 \times 10^{11}\,cm^{-3}$. In the case of $O(^1D)$ the main quenching mechanism in the ionosphere is process 36 in Table 6.6. This gives the threshold number density of nitrogen molecules as $10^8\,cm^{-3}$. If the number density of quenching particles is lower than the values given above, radiation of the corresponding excited atoms may be significant.

6.4.4 Atomic Oxygen in the Upper Atmosphere

Atomic oxygen is one of the basic components of the upper atmosphere, so we shall examine those processes that determine its relative abundance in the atmosphere. The formation of atomic oxygen results from photodissociation of molecular oxygen (process 4 in Table 6.6) by solar radiation in the spectral range from 132 to 176 nm (corresponding to the photon energy range from 6 to 10.3 eV). This absorption range is called the Schumann–Runge continuum and is characterized by cross sections of 10^{-19}–$10^{-17}\,cm^2$. The Schumann–Runge continuum is the determining factor in the generation of atomic oxygen at altitudes higher than 120 km. At lower altitudes, up to $h \approx 80$ km, the generation of atomic oxygen is accomplished mostly by dissociation of oxygen molecules due to the weak Herzberg continuum in the wavelengths range from 140 to 175 nm, where the absorption cross section is less than $2 \times 10^{-24}\,cm^2$.

The recombination rate of two oxygen atoms in three body collisions decreases as the altitude increases, and hence the concentration of atomic oxygen increases. The number density of atomic oxygen becomes equal to the number density of molecular oxygen at altitudes from 100 to 120 km, and the number density of atomic oxygen is compared with that of molecular nitrogen at altitudes from 150 to 200 km. For comparison, we note that the average total number densities of nitrogen and oxygen molecules range from 7×10^{12} to $3 \times 10^{9}\,\mathrm{cm}^{-3}$, and the number density of atomic oxygen is in the range from 6×10^{11} to $4 \times 10^{9}\,\mathrm{cm}^{-3}$ as the altitude varies from 100 up to 200 km. Thus, atomic oxygen is one of the basic components of the upper atmosphere at high altitudes.

To analyze the absorption of short-wave solar radiation as a result of photodissociation of molecular oxygen, we write the balance equation for the intensity I_ω of solar radiation of a given frequency in the atmosphere in the form

$$\frac{dI_\omega}{dh} = -I_\omega \sigma_\omega [O_2]\,, \tag{6.96}$$

where h is the altitude, $[O_2]$ is the number density of molecular oxygen, and σ_ω is the photodissociation cross section. We assume that incident solar radiation is perpendicular to the Earth's surface, and the number density of molecular oxygen varies according to the barometric formula (6.95), $[O_2] = N_0 \exp(-h/L)$, where $L = T/mg \approx 10\,\mathrm{km}$. Then the solution of (6.96) is

$$I_\omega(h) = I_\omega(\infty) \exp\left[-\exp\left(-\frac{h-h_0}{L}\right)\right]\,, \tag{6.97}$$

where $I_\omega(\infty)$ is the intensity of solar radiation above the atmosphere, and the altitude h_0 is determined by the relation $\sigma_\omega L[O_2](h_0) = 1$. Formula (6.97) shows that the principal portion of the solar radiation of this part of the spectrum is absorbed near h_0. Because the photodissociation cross section is in the range of $\sigma_\omega \sim 10^{-19}$–$10^{-17}\,\mathrm{cm}^2$, this absorption takes place at altitudes where $[O_2] \sim 10^{11}$–$10^{13}\,\mathrm{cm}^{-3}$.

It is instructive to compare the decay time for the photodissociation of molecular oxygen and a typical time for transport of molecular oxygen to the altitudes at which absorption occurs. A typical dissociation time is

$$\tau_{\mathrm{dis}} \sim \left[\frac{1}{4}\int \frac{dI_\omega \sigma_\omega}{\hbar\omega}\right]^{-1} \sim 2 \times 10^{6}\,\mathrm{s}\,,$$

where the factor 1/4 results from averaging the radiation flux over the entire surface of the Earth, and $\hbar\omega = 6$–$10\,\mathrm{eV}$ is the photon energy. The drift velocity of an atom or molecule as a result of its own weight is, according to the Einstein relation (4.38),

$$w = \frac{Dmg}{T} \sim g\left[N\sigma_{\mathrm{g}}\sqrt{\frac{m}{T}}\right]^{-1} \sim \frac{3 \times 10^{13}\,\mathrm{cm}^{-2}\,\mathrm{s}^{-1}}{N}\,, \tag{6.98}$$

where we use the estimate (4.53) for the diffusion coefficient D of oxygen atoms and $\overline{\sigma} \approx \sigma_{\mathrm{g}} \approx 3 \times 10^{-15}\,\mathrm{cm}^2$ is of the order of the gas-kinetic cross section for atmospheric molecules and atoms. In this expression, N is the total number density of atoms and molecules at a given altitude of the atmosphere. Assuming molecular nitrogen to be the primary component of the atmosphere at altitudes of maximum photodissociation, we have $[O_2] = N/4$, where the concentration of molecular oxygen is $[O_2] \sim 10^{11}\,cm^{-3}$, and a typical transport time is $\tau_{dr} \sim L/w \sim 10^4\,s$. This gives the inequality $\tau_{dr} \ll \tau_{dis}$, so the photodissociation process does not disrupt the barometric distribution of molecular oxygen at altitudes where photoabsorption of solar radiation takes place.

To estimate the number density of atomic oxygen, we can compare the flux $w[O]$ of oxygen atoms and the rate of their formation, which is of the order of the flux of solar photons causing photodissociation. This flux is $I \sim 3 \times 10^{12}\,cm^{-2}\,s^{-1}$ at optimal altitudes. The number density of atomic oxygen is

$$[O] \sim \frac{I}{w} \sim 0.1N \tag{6.99}$$

at altitudes where it is formed, and (6.98) has been used for the atom drift velocity. From this it follows that the concentration of atomic oxygen (approximately 10%) does not vary with altitude until the mechanisms for loss of atomic oxygen become weak. Decay of atomic oxygen is determined by the three body process for recombination of atomic oxygen $2O + M \rightarrow O_2 + M$, where M is N_2, O, or O_2. The maximum number density of atomic oxygen $[O]_{max}$ is observed at altitudes where a typical time for the three body association process is equal to the time for atomic transport to lower layers, that is,

$$\frac{w[O]_{max}}{L} \sim K[O]^2_{max}N\,.$$

Because the value of the three body rate constant is $K \sim 10^{-33}\,cm^6/s$, we obtain $[O]_{max} \sim 10^{13}\,cm^{-3}$. Part of the atomic oxygen penetrating to lower layers in the atmosphere is transformed into ozone by processes 40 and 41 in Table 6.6. Thus, the photodissociation process in the upper atmosphere leads to a high ozone concentration (up to 10^{-4}) at altitudes from 40 to 80 km.

6.4.5
Ions in the Upper Atmosphere

Ions in the upper atmosphere such as N_2^+, O_2^+, O^+, and N^+ are formed by photoionization of the corresponding neutral species (see Figure 6.22) The spectrum of solar radiation in the VUV range that is responsible for ionization of atmospheric atomic particles is created by the solar corona. Hence, the intensity of this radiation can vary over a wide range. The average flux of photons with the wavelength below 100 nm is about $2.4 \times 10^{10}\,cm^{-2}\,s^{-1}$. This part of the spectrum causes Lyman-series transitions of atomic hydrogen, including the Rydberg spectrum and the continuum. The contribution in the spectral range from 84 to 103 nm averages about $1.3 \times 10^{10}\,cm^{-2}\,s^{-1}$. Among the more notable processes are the CIII transition at 99.1 nm ($9 \times 10^8\,cm^{-2}\,s^{-1}$), the CIII transition at 97.7 nm ($4.4 \times 10^9\,cm^{-2}\,s^{-1}$), the OV transition at 63 nm ($1.3 \times 10^9\,cm^{-2}\,s^{-1}$), the HeI transition at 58.4 nm ($1.3 \times 10^9\,cm^{-2}\,s^{-1}$), and the Lyman HeII transition at 30.4 nm ($7.7 \times 10^9\,cm^{-2}\,s^{-1}$). The maximum rate constant for the generation of electrons as a result of photoionization processes occurs at an altitude of about 160 km and is $4 \times 10^{15}\,\mathrm{cm^{-3}\,s^{-1}}$.

One can obtain rough estimates of the number density of charged particles at medium altitudes, where the photoionization rate constant is a maximum. The photoionization cross section is $\sigma_{\mathrm{ion}} \sim 10^{-18}$–$10^{-17}\,\mathrm{cm}^2$, so the number density of molecules is $N_{\mathrm{m}} \sim (\sigma_{\mathrm{ion}} L)^{-1} \sim 10^{11}$–$10^{12}\,\mathrm{cm}^{-3}$ at altitudes where photoionization occurs. The number density of molecular ions N_{i} at these altitudes follows from the balance equation $\alpha N_{\mathrm{e}} N_{\mathrm{i}} \sim I_{\mathrm{ion}}/L$, where α is the dissociative recombination coefficient for processes 13–15 in Table 6.6 and $I_{\mathrm{ion}} \sim 2 \times 10^{10}\,\mathrm{cm^{-2}\,s^{-1}}$ is the flux of photons whose absorption leads to photoionization. From this it follows that

$$N_{\mathrm{i}} \sim \sqrt{\frac{I_{\mathrm{ion}}}{\alpha L}} \sim 5 \times 10^5\,\mathrm{cm}^{-3}\,. \tag{6.100}$$

This corresponds to the maximum number density of ions in the atmosphere. The number density of molecular ions in the lower layers of the atmosphere follows from the balance equation

$$\frac{w N_{\mathrm{i}}}{L} \sim \alpha N_{\mathrm{i}}^2\,.$$

This gives the number density relation $N_{\mathrm{i}} N_{\mathrm{m}} \sim 10^{15}\,\mathrm{cm}^{-6}$, where N_{m} is the number density of nitrogen molecules.

A typical time for establishment of the equilibrium for molecular ions that follows from the estimate (6.100) is $\tau_{\mathrm{rec}} \sim (\alpha N_{\mathrm{e}})^{-1} \sim 20\,\mathrm{s}$, whereas a typical ion drift time for these altitudes is $\tau_{\mathrm{dr}} \sim 10^4\,\mathrm{s}$. Therefore, local equilibrium for molecular ions results from the competing ionization and recombination processes. Because of the short time for establishment of this equilibrium, the number density of charged particles for daytime and for nighttime atmospheres is different. The above estimate refers to the daytime atmosphere. The ion number density of the nighttime atmosphere follows from the relation $\alpha N_{\mathrm{i}} t \sim 1$, where t is the duration of the night. This gives the estimate $N_{\mathrm{i}} \sim 10^2$–$10^3\,\mathrm{cm}^{-3}$ for the nighttime atmosphere.

Atomic ions formed as a result of the photoionization process participate in the ion–molecule reactions listed as processes 27–30 in Table 6.6. A typical time for these processes is $\tau \sim (k N)^{-1} \sim 0.01$–$10\,\mathrm{s}$, and is small compared with a typical recombination time. This explains why ions in the ionosphere are molecular ions (see Figure 6.22). In addition, it shows the origin of NO^+ molecular ions. These ions cannot result from photoionization because of the small number density of NO molecules.

Atomic ions are removed from high altitudes because of a short transport time. One can estimate the maximum number density of atomic oxygen ions by comparing a typical drift time L/w and the characteristic time for the ion–molecule reactions listed in Table 6.6 and estimated as by $(k[N_2])^{-1}$. Assuming the basic component of the atmosphere at these altitudes to be atomic oxygen, we find that the maximum number density of O^+ atomic ions occurs at altitudes where $[N_2][O] \sim$

$3 \times 10^{19}\,\mathrm{cm}^{-6}$. This corresponds to altitudes of approximately 200 km. The maximum number density of atomic ions N_i follows from the balance equation

$$\int \sigma_{ion} d\, I_{ion} \sim k[N_2]N_i \,.$$

For photoionization of atomic oxygen, we have $\int \sigma_{ion} d\, I_{ion} = 2 \times 10^{-7}\,\mathrm{s}^{-1}$, so the maximum number density is

$$N_i \sim 2 \times 10^5\,\mathrm{cm}^{-3} \frac{[O]}{[N_2]} \sim 10^6\,\mathrm{cm}^{-3} \,. \tag{6.101}$$

In higher layers of the atmosphere, the number density of atomic ions is determined by the barometric formula, because it is proportional to the number density of primary atoms, and declines with increasing altitude.

At altitudes where photoionization occurs, the negative charge of the atmospheric plasma comes from electrons. In the D-layer of the ionosphere, electrons attach to oxygen molecules in accordance with processes 20–22 in Table 6.6, and it is negative ions that govern the negative charge of the atmosphere. At the altitudes where this transition takes place, the balance equation $w N_e/L \sim K N_e[O_2]^2$ is appropriate, where $K \sim 10^{-31}\,\mathrm{cm}^6/\mathrm{s}$ is the rate constant for processes 20 and 21 in Table 6.6. From this it follows that formation of negative ions occurs at altitudes where $N[O_2]^2 \sim 3 \times 10^{39}\,\mathrm{cm}^{-9}$ or $[O_2] \sim 10^{13}\,\mathrm{cm}^{-3}$. We account for the coefficient of ambipolar diffusion of ions being of the order of the diffusion coefficient for atoms. In the D layer of the ionosphere, recombination proceeds according to the scheme $A^- + B^+ \to A + B$, and is characterized by the rate constant $\alpha \sim 10^{-9}\,\mathrm{cm}^3/\mathrm{s}$. This leads to the relation

$$N_i N \sim 10^{17}\,\mathrm{cm}^{-6} \tag{6.102}$$

for the number density of ions N_i. Thus, the properties of the middle and the upper atmosphere are established by processes in excited and dissociated air involving ions, excited atoms, and excited molecules.

6.5 Electric Machine of the Earth's Atmosphere

6.5.1 The Earth as an Electric System

Processes in the lower atmosphere lead to charging and discharging of the Earth, and we consider below the Earth as an electric system and the processes which govern its electric properties. The Earth is negatively charged, and the average electric field strength near the Earth's surface is $E = 4\pi\sigma = 130\,\mathrm{V/m}$ (σ is the average surface charge density) [149, 150]. This gives the total Earth charge $Q = 5.8 \times 10^5$ C. In addition, the electric potential of the Earth with respect to the environment is

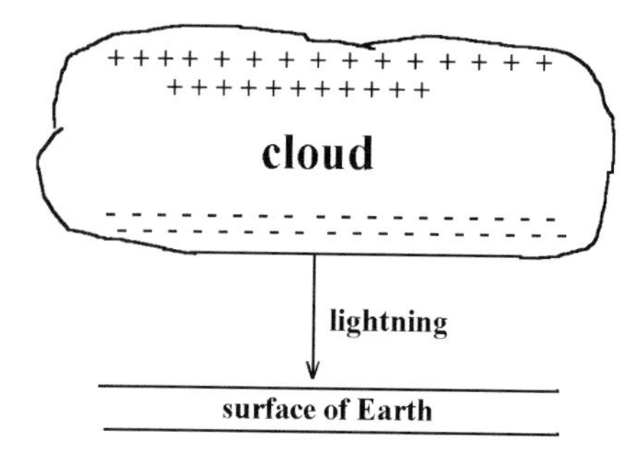

Figure 6.23 Character of charging of the Earth's surface through a cloud when charge separation results from the falling of negatively charged water drops and charge transfer from a cloud to the Earth's surface proceeds through lightning [153, 167].

$U = 300\,\text{kV}$) if we model the Earth as a spherical capacitor with one electrode as the Earth's surface and the other electrode is a sphere at a distance of 6 km from the Earth's surface. From this it follows that all the electric processes of the Earth's atmosphere take place in low atmospheric layers at altitudes below approximately 10 km.

Our goal is to construct the scheme of processes which create the electric machine of the Earth. Because the Earth's atmosphere contains charged particles, its conductivity determines the electric currents. The average current density for a quiet atmosphere over land is $2.4 \times 10^{-12}\,\text{A/m}^2$, and over the ocean it is $3.7 \times 10^{-12}\,\text{A/m}^2$. This gives a total Earth current of approximately $I = 1700\,\text{A}$, which leads to the Earth discharging in a time of typically $\tau = Q/I = 340\,\text{s}$, that is, the time for the Earth to discharge owing to the conductivity of the Earth is about 6 min. If we take the average mobility of atmospheric ions to be $K = 2.3\,\text{cm}^2/(\text{V cm})$, we find the average ion number density to be $N_\text{i} = 300\,\text{cm}^{-3}$.

We use the following simple scheme of the electric machine of the Earth's atmosphere [151]. The discharge of the Earth results from currents of positive and negative atmospheric ions which are formed under the action of both cosmic radiation and radioactivity of the Earth's surface. Charging of the Earth's surface begins in clouds, where the positive and negative charges are separated, and the lower cloud edge has a high electric potential with respect to the Earth's surface of 20–100 MV [152]. Subsequent lightning between the lower cloud edge and the Earth's surface leads to charging of the Earth (see Figure 6.23 [153]).

Note that the character of charge separation in clouds as a result of the falling of water aerosols depends on the sort of atmospheric ions, which may be different depending on small admixtures to atmospheric air in a given region. Therefore, the lower cloud edge may be charged both negatively and positively. Nevertheless, the negative charge of the lower cloud edge is preferable. In reality, the number of lightning discharges from clouds to the ground with transfer of the negative charge to the Earth's surface exceeds that for transfer of positive charge by 2.1 ± 0.5 times on average [154]. The ratio of the amount of negative charge to positive charge transferred in this way is 3.2 ± 1.2 on average [154]. In addition, there are 100 negative flashes per second over the entire globe on average [152].

Thus, although it may seem strange at first glance, the charging of the Earth occurs as a result of lightning in the atmosphere. Charge transfer processes in clouds that are connected with charging and precipitation of small drops or aerosols lead to a redistribution of the charges therein. The lower part of a cloud is usually negatively charged and the upper part is usually positively charged. The electric potential of a cloud can reach hundreds of millions of Volt, causing breakdown of the air. The discharge of a cloud to the Earth's surface in the form of a lightning stroke is accompanied by the transfer of electric charge to the Earth's surface. About 10% of thunderstorms transfer positive charge to the Earth's surface, and 90% transfer negative charge. The net result is that the Earth acquires a negative charge.

We can estimate the timescales for the electric processes of the Earth. Positive and negative ions recombine under atmospheric conditions in three body collisions, with an effective ion recombination coefficient of $\alpha = 2 \times 10^{-6}\,\mathrm{cm^3/s}$ for recombination according to processes 17–19 in Table 6.6. We note that electrons are practically absent in the atmosphere near the Earth's surface although electron formation is the first stage of ionization processes in the lower Earth atmosphere. But according to processes 20 and 21 in Table 6.6, conversion of electrons into negative ions O_2^- near the Earth's surface proceeds during a time of approximately 10^{-8} s, which is short compared with the lifetime of charged particles in the atmosphere. Hence, we assume the negative charge in the lower atmosphere to be connected partially to the negative molecular ions. One can estimate from this a time for recombination of molecular ions as about $\tau = (\alpha N_{\mathrm{i}})^{-1} = 0.5\,\mathrm{h}$ for $N_{\mathrm{i}} = 10^3\,\mathrm{cm^{-3}}$. This contrasts with the much shorter characteristic time of $q/I = 6\,\mathrm{min}$ for discharging of the Earth, where q is the Earth's charge and I is the average atmospheric current.

One can estimate a typical distance an ion travels during its lifetime τ, which is $s = KE\tau$, where K is the ion mobility and E is the electric field strength near the Earth. Table 6.10 contains the measured mobilities of some ions in nitrogen. These ions may be formed in a humid atmosphere and are characterized by high electron binding energies. Indeed, the electron affinity is 2.3 eV for the NO_2, 3.7 eV for NO_3, and 2.8 eV for CO_3. For comparison, the electron affinity for O_2 is 0.43 eV and for the oxygen atom is 1.5 eV. These stable negative ions can form complex ions in a humid atmosphere. Using the data in Table 6.10, we take the ion mobility $K \approx 2\,\mathrm{cm^2/(V\,s)}$, which gives for the distance traveled by an ion during its lifetime as $s \approx 50\,\mathrm{m}$. This distance is short compared with the size of the atmosphere, which is evidence for continuous formation of charged particles in the atmosphere.

One can estimate the average number density of molecular ions taking into account, as follows from measurements, the average current density over land of $2.4 \times 10^{-16}\,\mathrm{A/cm^2}$ and over the ocean of $3.7 \times 10^{-16}\,\mathrm{A/cm^2}$ [149]. Let us assume the mobility of negative and positive molecular ions to be close to $2\,\mathrm{cm^2/(V\,s)}$ as follows from Table 6.10. Taking the number densities N_{i} of negative and positive molecular ions to be identical, on the basis of the electric field strength $E = 1.3\,\mathrm{V/cm}$ we obtain that the indicated current density over land is provided by a ion

Table 6.10 The mobilities of some negative and positive ions in nitrogen [155–158]. The mobility K is reduced to normal conditions (the nitrogen number density $N = 2.69 \times 10^{19}\,\mathrm{cm}^{-3}$ corresponds to the temperature $T = 273\,\mathrm{K}$ and a pressure of 1 atm) and is expressed in square centimeters per volt per second.

Ion	K
NO_2^-	2.5
NO_3^-	2.3
CO_3^-	2.4
$NO_2^- \cdot H_2O$	2.4
$NO_3^- \cdot H_2O$	2.2
$CO_3^- \cdot H_2O$	2.1
N_2^+	1.9
CO_2^+	2.2
N_2H^+	2.1
N_3^+	2.3
N_2O^+	2.3
N_4^+	2.3
$H^+ \cdot H_2O$	2.8
$H^+ \cdot (H_2O)_2$	2.3
$H^+ \cdot (H_2O)_3$	2.1

number density of $N_i = 300\,\mathrm{cm}^{-3}$ and over the ocean we have $N_i = 500\,\mathrm{cm}^{-3}$. Note that molecular ions carry only a part of the atmospheric charge.

Let us consider the ionization balance in the lower atmosphere. There are two mechanisms for the ionization process. Near the Earth they result from radioactivity of the Earth's surface, but these processes proceed in a thick layer. Another mechanism is determined by galactic and solar cosmic rays [159–161]. Maximum ionization is observed at altitudes between 11 and 15 km, and is characterized by ionization rates of [161] $30–40\,\mathrm{cm}^{-3}\,\mathrm{s}^{-1}$. This leads to an ion number density of about $6 \times 10^3\,\mathrm{cm}^{-3}$ if we use the recombination coefficient of about $10^{-6}\,\mathrm{cm}^3/\mathrm{s}$ at this altitude. This number density exceeds that at low atmosphere altitudes.

Thus, one can compose the following character of electric processes for the Earth that is negatively charged owing to the falling of water drops in clouds. The charge transfer from lower cloud edges to the Earth's surface results from electric breakdown in the atmosphere that is realized by lightning.

6.5.2
Lightning

Let us consider the characteristics of lightning as an atmospheric phenomenon. Lightning is a powerful electric breakdown between a cloud and the Earth's surface, between two clouds, or within one cloud. The length of the lightning channel is measured in kilometers. Lightning is the most widespread electric phenomenon in the atmosphere. The action of lightning and the descent of charged raindrops to the Earth's surface leads to charging of the Earth. The average charge carried by a single lightning stroke is about 25 C. If we assume that all charging of the Earth is accomplished by lightning, we conclude that it is necessary to have about 70 lightning strokes per second (i.e., about six million lightning strokes per day) for maintenance of the observed charging current. If we assume each lightning event is observable at a distance up to 10 km, then one can observe an average of three or four lightning strokes every day. In actuality, the frequency depends on the season and geographical location, but the above estimate shows that this phenomenon is widespread.

We shall examine the distinguishing features of lightning as an electric discharge. The charge separation in a cloud results from the charging of drops (aerosols) that subsequently descend. This leads to separation of charge in a cloud, and creates an electric potential in the lower part of the cloud that amounts to hundreds of millions of Volt. The subsequent discharging of the cloud in the form of a lightning stroke leads to the transfer of this charge to the Earth's surface. The breakdown electric field strength of dry air is about 25 kV/cm, which exceeds by an order of magnitude the average electric field strength between the Earth's surface and a cloud during a thunderstorm. Hence, lightning as a gas breakdown has a streamer nature. Random inhomogeneities of the atmosphere, dust, aerosols, and various admixtures decrease the breakdown voltage. The first stage of lightning is creation of the discharge channel. This stage is called the stepwise leader. The stepwise leader is a weakly luminous breakdown that propagates along a broken line with a segment length of tens of meters. A typical propagation velocity of the stepwise leader is of the order of 10^7 cm/s, and that, in turn, is of the order of the electron drift velocity in air in the fields under study.

After creation of the conductive channel, the electric current begins to flow through it, and its luminosity increases sharply. This stage is called the recurrent stroke and is characterized by a propagation velocity of up to 5×10^9 cm/s, corresponding to the velocity of propagation of the electric field front in conductors. The recurrent stroke is relatively short. Its first phase (the peak-current phase) lasts some microseconds and the discharge is complete in less than 10^{-3} s. During this time the current channel does not expand, so the energy released is used to heat the channel and for ionization of the air in it. To estimate the temperature of the channel, we take the charge transferred to be $q = 2\,\mathrm{C}$, with an electric field strength E in the channel of 1 kV/cm. The energy released per unit length of the channel is qE, and a typical increase of the air temperature in the channel is

$$\Delta T \sim \frac{qE}{c_p \rho S}, \qquad (6.103)$$

where $c_p \sim 1\,\mathrm{J/g}$ is the heat capacity of air, $\rho \sim 10^{-3}\,\mathrm{g/cm^3}$ is the air density, and $S \sim 100\,\mathrm{cm^2}$ is the channel cross section (its radius is approximately 10 cm). From this it follows that $\Delta T \sim 2 \times 10^4$ K. This rough estimate makes clear that the air in the lightning channel is highly ionized. At such air temperatures, radiative processes restrict the subsequent increase of temperature. Usually, the maximum temperature of the lightning channel is about 3×10^4 K.

The recurrent stroke creates a hot lightning channel. Its supersonic expansion creates an acoustic wave – thunder. Equilibrium between the conductive channel and the surrounding air is established, and then the principal part of the atmospheric charge is carried through the channel. The lightning current in this stage of the process falls drastically in time, and the total duration of the recurrent stroke is about 10^{-3} s. Then the conductive channel disintegrates. If during its decay a redistribution of charges in the cloud has occurred, then the lightning flash can recur through this channel. This can take place if the time interval from the previous flash is no more than about 0.1 s. Then the new flash begins from the so-called arrowlike leader, which is similar to the stepwise leader, but is distinguished from it in that it travels the distance along the existing channel continuously, without delay at each step as in the case of the stepwise leader. After the passage of the arrowlike leader, a recurrent stroke occurs as in the first breakdown. After some delay, breakdown may again occur by passage of an electric current through the existing channel. As a rule, one lightning stroke contains several charge pulses through the same channel.

Lightning is of interest as an intense source of radiation. To find the maximum efficacy of the transformation of the electric energy into radiation, we shall assume that the conductive channel is a blackbody of temperature 30 000 K. Then the radiation flux of the conductive channel is $j = \sigma T^4 \sim 5 \times 10^5\,\mathrm{W/cm^2}$, and the radiation flux per unit length of the channel is $2\pi R j \sim 10^7\,\mathrm{W/cm}$, where the channel radius is taken to be $R = 10\,\mathrm{cm}$. Since the above temperature is maintained for $\tau \sim 10^{-4}\,\mathrm{s}$, we have an estimated upper limit for the specific radiation energy of $\varepsilon \sim 10^3\,\mathrm{J/cm}$. The electric energy per unit length of the lightning channel is $QE = 2\,\mathrm{kJ/cm}$. This rough estimate leads to the conclusion that the electric energy of lightning is transformed very efficiently into radiated energy.

Another distinguishing feature of radiation emitted by the lightning channel relates to the role of UV radiation. According to Wien's law, the maximum radiation energy for a blackbody with a temperature of 30 000 K occurs at wavelengths of about 100 nm. This estimate shows the possibility for lightning to emit hard UV radiation, although in reality, owing to the limited transparency of an atmospheric plasma of the above temperature for UV and VUV radiation, visible radiation constitutes the principal portion of the radiation due to lightning.

6.5.3
Electric Processes in Clouds

We thus consider the basis of the electric machine of the Earth to be the result of charge separation in clouds due to the falling of negatively charged water, and below we analyze the reality of this scheme. Then we consider the action of the Earth's electric machine as a secondary process of water circulation in the Earth's atmosphere. Indeed, every year 4×10^{14} t (metric tons) of evaporated water pass through the atmosphere, corresponding to 13 million tonnes of water per second, which corresponds to a power of 4×10^{13} kW. The power associated with passage of electric current through the atmosphere is $UI = 5 \times 10^5$ kW, where $U = 300$ kV is the Earth's potential and $I = 1700$ A is the atmospheric current. Since the cloud potential during a thunderstorm exceeds the potential of the Earth by a factor of 10^3, we find that the power expended in the charging of the Earth is three orders of magnitude greater than the power of discharging of the Earth, but five orders of magnitude smaller than the power consumed in the evaporation of water. Therefore, the power expended on electric processes during the circulation of water in the atmosphere is relatively small. To create the observed charging current, it is necessary that the charge transfer should be 1.4×10^{-10} C/g of water. We will use this value for subsequent estimates.

Being guided by micrometer-sized water drops, we consider the case (6.5) where the drop radius r_0 is large compared with the mean free path of ions in air, which gives $r_0 \gg 0.1\ \mu$m for atmospheric air. Then the average charge Z of a drop (an aerosol particle) that is established by attachment of negative and positive ions to an aerosol is given by (6.12)

$$\overline{|Z|} = \frac{r_0 T}{e^2} \ln \frac{K_-}{K_+} , \tag{6.104}$$

where K_- and K_+ are the mobilities of the negative and positive ions in atmospheric air. The mobility of various ions in nitrogen, the basic component of air, are given in Table 6.10. Comparison of the mobilities for positive and negative ions shows that they are nearby. From this it follows that in principle aerosol particles in the Earth's atmosphere may be charged both negatively and positively, although the negative charge of aerosol particles is preferable. On the basis of the charge of clouds, one can conclude that the negative charge of aerosol particles is realized in roughly 90% of cases.

Let us consider a cloud where the mobility of negative molecular ions is higher than that of positive ions, and we take for definiteness $K_-/K_+ = 1.1$. We assume that water drops have a spherical shape with radius r_0 and extract the dependence of various parameters on the drop radius, and reduce various parameters to the drop radius $a = 1\ \mu$m. Then we have for the mass of the drop m expressed in grams, the fall velocity w of a water drop expressed in cm/s, and the average drop charge $\overline{|Z|}$ at temperature $T = 300$ K given in electron charges

$$m = 4.2 \times 10^{-12} \left(\frac{r_0}{a}\right)^3 , \quad w = 0.012 \left(\frac{r_0}{a}\right)^2 , \quad \overline{|Z|} = 1.7 \left(\frac{r_0}{a}\right) .$$

These formulas give for the ratio of the drop charge to its mass (expressed in C/g)

$$\frac{\overline{|Z|}}{m} = 6.6 \times 10^{-8} \left(\frac{r_0}{a}\right)^2 .$$

Since the ratio of the total electric charge transferred in the Earth's atmosphere to the total water mass is 1.4×10^{-10} C/g, we obtain from this that the aerosol size is restricted by

$$r_0 < 22\,\mu\text{m}\,. \tag{6.105}$$

We note also that the charge distribution function (6.15) for drops (aerosol particles) is wide. In particular, on the basis of the charge distribution function we have for the width of this distribution function

$$\overline{(Z - \overline{Z})^2} = 18\left(\frac{r_0}{a}\right)\,.$$

One can see that for $r_0 \approx 20\,\mu\text{m}$ the width of the distribution function is close to the average drop charge.

We assume in this case an equilibrium between molecular ions in atmospheric air and charged aerosols, and the formation of charged aerosols does not influence the number density of molecular ions. Let us determine a typical time for aerosol charging under these conditions. The charging time, that is, the time for establishment of charge equilibrium between aerosols and molecular ions, is [60, 65]

$$\tau_{\text{rec}} = \frac{T}{4\pi N_{\text{i}} e^2 D_{\text{i}}} = \frac{1}{4\pi N_{\text{i}} e K_{\text{i}}}\,, \tag{6.106}$$

where N_{i} is the number density of molecular ions, D_{i} and K_{i} are the diffusion coefficient and the mobility for positive molecular ions, and this time is independent of aerosol size. In accordance with the data in Table 6.10, we take $K_{\text{i}} \sim 2\,\text{cm}^2/(\text{V s})$, which gives $\tau_{\text{rec}} \sim 10^3$ s at the number density of molecular ions $N_{\text{i}} \sim 300\,\text{cm}^{-3}$. This time is less than or comparable to the time for cloud formation.

We now analyze charging of clouds from another standpoint. Let us find the value of the total mass of water drops ρ per unit volume that provides the above specific charge of drops $|Z|e/m = 1.3 \times 10^{-10}\,\text{C/g}$ and the current density of falling charged drops $3.7 \times 10^{-16}\,\text{A/cm}^2$ as is observed on average over the ocean [149]. We obtain on the basis of the above formulas that this value is independent of the drop radius and is $\rho = 4.6 \times 10^{-7}\,\text{g/cm}^3$, which corresponds to a water content in the atmosphere of 0.35 g/kg of air, whereas the average content of water in the Earth's atmosphere is 7 g/kg of air [150], which relates to the temperature of 10 °C at which the saturated vapor pressure corresponds to the above water content. These estimates show that a small part of atmospheric water is used for charging the Earth.

For this water content we obtain the number density of water drops N_{dr} per cubic centimeter:

$$N_{\text{dr}} = 8.4 \times 10^4 \left(\frac{a}{r_0}\right)^3\,.$$

The charge density $N_Z = |Z| N_{\text{dr}}$ per cubic centimeter is

$$N_Z = 1.4 \times 10^5 \left(\frac{a}{r_0}\right)^2\,.$$

If we require this charge density not to exceed a typical value of approximately $10^3\,\text{cm}^{-3}$, we find the drop radius must be not small:

$$r_0 > 10\,\mu\text{m}\,. \tag{6.107}$$

One more estimate relates to the electric field strength on cloud edges if it is created by a noncompensated charge of falling water drops. From the Poisson equation we have the following estimate for the electric field strength E and the electric potential U of a cloud:

$$E \approx 2\pi N_Z L\,, \quad U = 2\pi e N_Z L^2\,,$$

where we take a typical charge density of $N_Z \sim 10^3\,\text{cm}^{-3}$ and the cloud depth $L \sim 1\,\text{km}$. We obtain $E \sim 100\,\text{V/cm}$ and $U \sim 10\,\text{MV}$, which correspond to observational values [152]. Thus, assuming that charge separation in clouds is determined by falling of charged water drops, we obtain a noncontradictory picture of this phenomena, but the radii of falling drops lie in a narrow range: $r_0 = 10$–$20\,\mu\text{m}$.

6.5.4 Characteristics of Earth Charging

In considering Earth charging in atmospheric processes as a result of lightning phenomena [151] which lead to charge transfer from clouds to the Earth's surface, we find that the charge separation in clouds results from falling of charged water drops under the action of the gravitational force. The origin of this process is evaporation of water from the Earth's surface, and condensation of water proceeds at altitudes of several kilometers, where the temperature is lower that that at the Earth's surface. Therefore, the atmospheric electric phenomena are secondary phenomena with respect to water circulation in the Earth's atmosphere. The above estimates show that the main contribution to the charge separation in clouds follows from water drops with radius $r_0 = 10$–$20\,\mu\text{m}$. We now consider these processes in detail to convince ourselves of the reality of this picture of atmospheric electric processes.

In the above analysis we assumed that charging of water drops results from attachment of positive and negative molecular ions, and this process does not influence the number density of molecular ions, which is $N_\text{i} \sim 10^3\,\text{cm}^{-3}$. Cosmic rays are of importance for ionization of the Earth's atmosphere [159, 162, 163]. Indeed, penetrating the Earth atmosphere up to altitudes of several kilometers and ionizing atmospheric molecules, cosmic rays (mostly protons) lead to the formation of molecular ions. Let us determine the number density of molecular ions if they are formed under the action of cosmic rays and are destroyed as a result of attachment to water drops. We assume the mobilities of positive and negative ions to be similar and according to the data in Table 6.10 the mobility is $K = 2\,\text{cm}^2/(\text{V s})$ at atmospheric pressure and the ion diffusion coefficient for $T = 300\,\text{K}$ is $D = 0.05\,\text{cm}^2/\text{s}$. From this for the average water content in atmospheric air of 7 g/kg of air, we obtain for the rate of destruction of molecular ions as a result of attachment to water drops

$$\frac{dN_\text{i}}{dt} = -4\pi D r_0 N_\text{i} N_\text{dr} = -\frac{N_\text{i}}{\tau}\left(\frac{a}{r_0}\right)^2 ,$$

where $a = 1\,\mu\text{m}$ as earlier and $\tau = 0.2\,\text{s}$. Equalizing this rate to the rate of ion formation under the action of cosmic rays for optimal altitudes $h = 11-15\,\text{km}$, where this rate is $Q = 30\,\text{cm}^{-3}\,\text{s}^{-1}$ [161], we obtain for the equilibrium number density of molecular ions

$$N_\text{i} = 6\,\text{cm}^{-3} \cdot \left(\frac{r_0}{a}\right)^2 .$$

In particular, for the optimal drop size $r_0 = 20\,\mu\text{m}$ this formula gives $N_\text{i} \sim 2 \times 10^3\,\text{cm}^{-3}$. Therefore, ignoring the influence of the processes of attachment of molecular ions to water drops does not change the character of the charging processes under consideration.

We also note that in the above estimations we assumed the atmosphere to be uniform, that is, these parameters do not depend on the altitude. But this is not correct because the charge separation leads to a nonuniform distribution of charged particles, including molecular ions, whose number density may depend on the altitude, and this dependence may be different for positive and negative molecular ions. Nevertheless, this complexity does not violate the general character of charge separation in the Earth's atmosphere.

We have the character of charge separation is determined by molecular ions in the atmosphere, which in turn are connected to corresponding chemical processes and may be different depending on certain atmospheric conditions [164]. In the first stage of these processes, ionization of air under the action of cosmic rays leads to the formation of O^- or O_2^- ions, and these negative ions in turn partake in chemical processes which result in the formation more stable negative ions. In the same manner, positive ions, which start from simple O^+, O_2^+, N^+, and N_2^+ ions, are transformed into other ions depending on some admixtures to air molecules in the atmosphere. Subsequently, more stable positive and negative ions attach to aerosol particles which are charged negatively if the diffusion coefficient or the mobility of negative ions is higher than that for positive ions, and the aerosol particles are charged positively for the inverse relation between ion mobilities. Therefore, small additions to atmospheric air may change the sign of the charge of falling aerosol particles, which is evidence of the complexity of charge separation in the Earth's atmosphere.

Above we considered charging of water drops in atmospheric air as a result of attachment of molecular ions to the drop surface, and this process proceeds in clouds where the water concentration is relatively high [165–169]. But there are other mechanisms of charge transfer to aerosol particles in the Earth's atmosphere. The characteristic feature of the condensation process for water is that it proceeds at low temperatures close to the meting point, and then ice particles may be formed simultaneously with liquid drops. These conditions are typical for clouds which are formed at altitudes of several kilometers where the atmospheric temperature is close to 0 °C. Interaction of ice particles with water drops may be of importance for charge separation in the atmosphere [170–176].

Taking ice aerosols with needle shape and liquid aerosols as spherical drops, we give in Figures 6.24 and 6.25 the mechanism of charging in an external electric field. Then the needle particle is polarized such that charges are located at its ends, and contact with a spherical particle leads to charge transfer to the spherical particle and correspondingly to charging of the needle particle, which is neutral at the beginning. But charging by collision of ice and liquid aerosols also occurs in the absence of an external electric field. Indeed, contact of these aerosols creates an interface potential that leads to charge transfer between aerosols. Assuming this interface potential to be large compared with the thermal energy of aerosols, one can estimate a typical charge Z of each aerosol such that the potential of their attraction after charge transfer does not exceed the thermal aerosol energy T, that is,

$$\frac{Z^2 e^2}{r_0} \sim T \, .$$

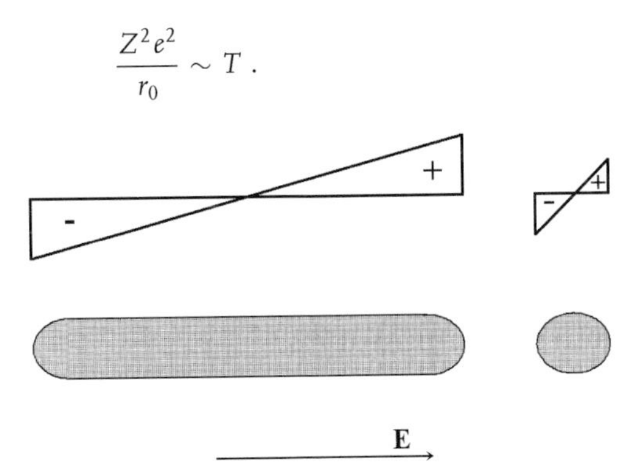

Figure 6.24 Distribution of charges induced under the action of a uniform electric field in joining of cylindrical and spherical particles of identical radius.

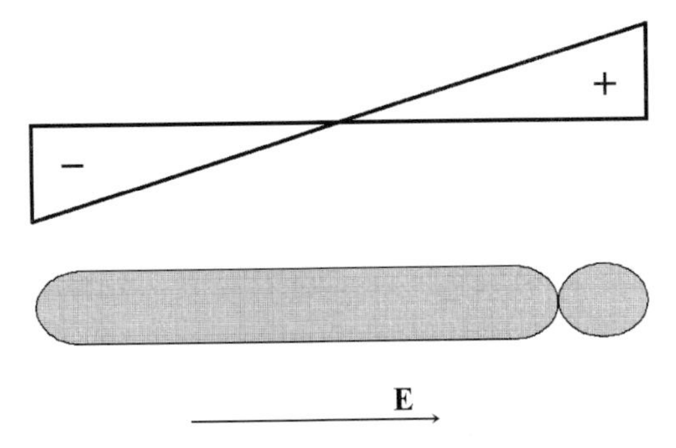

Figure 6.25 Redistribution of charges as a result of joining of cylindrical and spherical particles.

This leads to the following estimate for a typical charge of aerosol particles after collision of ice and liquid water aerosols:

$$Z \approx \sqrt{\frac{r_0 T}{e^2}} = 4 \cdot \sqrt{\frac{r_0}{a}} .$$

This charge corresponds to the width of the charge distribution function (6.15) for particles if their charging results from attachment of molecular ions to their surface.

We now compare the rate of this mechanism with that due to molecular ions. We take a typical time for this process on the basis of the Smoluchowski formula (6.8) and this time coincides with the time for the coagulation process:

$$\frac{1}{\tau_{col}} = 4\pi D_{dr} N_{dr} r_0 . \tag{6.108}$$

According to (4.110), the diffusion coefficient of aerosols is [60, 65]

$$D_{dr} = \frac{T}{4\pi r_0 \eta} = D_0 \cdot \left(\frac{a}{r_0}\right) ,$$

where η is the air viscosity, and in air at room temperature and atmospheric pressure we have $D_0 = 1.2 \times 10^{-7}\,\mathrm{cm^3/s}$. Next, taking the average water content in the Earth's atmosphere to be 7 g/kg of air, that is, the water density in drops is $\rho = 9.1 \times 10^{-5}\,\mathrm{g/cm^{-3}}$, and assuming the drop radius to be identical for all drops, we obtain for the number density of drops (expressing it in per cubic centimeter)

$$N_{dr} = 2.2 \times 10^6 \left(\frac{a}{r_0}\right)^3 .$$

As a result, we obtain a typical time for collision of a test water drop with other drops, assuming collision as contact of two drop surfaces:

$$\frac{1}{\tau_{col}} = \frac{1}{\tau_0} \left(\frac{a}{r_0}\right)^3 ,$$

where $\tau_0 = 3.1 \times 10^3$ s. It is seen that the time for collision is large compared with the lifetime of neutral drops in a cloud for typical drop sizes. From this it follows that the mechanism of drop charging as a result of collision of ice particles and liquid drops is not important for electric processes in the Earth's atmosphere.

A typical time (6.108) for collision of two drops characterizes also the time for growth of neutral drops in atmospheric air. But because water drops are charged by attachment of molecular ions, the process of drop growth results from collision of charged drops of the opposite sign. We determine below a typical time for drop recombination. Let us divide drops into two parts, positively and negatively charged ones, and because the mobilities of positive and negative ions are similar, we take in this estimate the average drop charge to be zero, and the number densities of positively and negatively charged drops to be $N_{dr}/2$, where N_{dr} is the total number density of drops. The recombination coefficient of two oppositely charged drops in a gas is given by the Langevin formula (6.10) [64]:

$$k_{\rm rec} = |Z_+ Z_-| e(K_+ + K_-) \,, \tag{6.109}$$

where Z_+ and Z_- are the charges of positive and negative drops and K_+ and K_- are their mobilities ($K_+ = K_- = K_{\rm dr}$). On the basis of the charge distribution function for drops (6.15), we have for the average positive $|\overline{Z_+}|$ and negative $|\overline{Z_-}|$ drop charges

$$|\overline{Z_+}| = |\overline{Z_+}| = \sqrt{\frac{r_0 T}{\pi e^2}} \,.$$

Using the expression for the mobility of drops in a gas for case (6.5),

$$k_{\rm dr} = \frac{e^2}{4\pi r_0 \eta} \,,$$

where $\eta = 1.85 \times 10^{-4}$ g/(cm s) is the air viscosity at room temperature, we obtain for the average rate constant for recombination for a test charged drop

$$\overline{k_{\rm rec}} = \frac{T}{2\pi^2 \eta} \,. \tag{6.110}$$

As is seen, the rate constant for recombination for a test charged drop is independent of the drop size (we assume all the drops have identical radius r_0) and is

$$\overline{k_{\rm rec}} = 1.1 \times 10^{-11}\,{\rm cm}^3/{\rm s} \,.$$

We now find an optimal drop size by equalizing the typical time $\tau_{\rm rec} = 2/N_{\rm dr} k_{\rm dr}$ and the time $\tau_{\rm fal} = L/w$ for falling of drops in a cloud, where $L \approx 2$ km is a typical cloud size. Taking the average water content in the atmosphere (7 g/kg of air), we have

$$\frac{1}{\tau_{\rm rec}} = \frac{2}{N_{\rm dr} k_{\rm rec}} = 1.2 \times 10^{-5}\,{\rm s}^{-1} \left(\frac{a}{r_0}\right)^3 \,,$$
$$\frac{1}{\tau_{\rm fal}} = \frac{w}{L} = 6 \times 10^{-8}\,{\rm s}^{-1} \left(\frac{r_0}{a}\right)^3 \,.$$

From this we obtain for the size of charged water drops when they reach the lower edge of a cloud $r_0 \approx 3\,\mu$m.

We note that in considering electric phenomena in the atmosphere, we simplified the problem and extracted the principal mechanism of Earth charging where separation of charges in the atmosphere results from falling of charged water drops. In reality, this process is more complex. In particular, convective motion of atmospheric air leads to mixing of atmospheric layers [177], and a charged layer may be transported to other altitudes up to 100 km. This can cause breakdown and lightning at high altitudes.

6.5.5
Charged Particles in an Aerosol Plasma

Ions in an aerosol plasma can attach to aerosol particles, and this is of importance for atmospheric plasmas. We now consider a plasma regime wherein charged atomic particles are created in the plasma by an external source, and their decay occurs through ion–ion recombination. The presence of aerosol particles in such a plasma can change its properties. The balance equations for the number densities N_+ of positive ions and N_- of negative ions in the plasma have the form

$$\frac{dN_+}{dt} = G - \alpha N_+ N_- - \frac{I_+}{N_p} e \,, \quad \frac{dN_-}{dt} = G - \alpha N_+ N_- - \frac{I_-}{N_p} e \,, \quad (6.111)$$

where G is the rate of ionization by an external source, α is the ion–ion recombination coefficient, N_p is the number density of particles, and I_+ and I_- are the ion currents to a particle as given by (6.7) and (6.9).

For simplicity we consider the case $Z < r_0 T/e^2$ when the ion current is described by the Smoluchowski formula (6.8). Then $N_+ = N_- = N_i$, and the balance equation in the stationary case has the form

$$G - \alpha N_i^2 - k N_i N_p = 0 \,,$$

where $k = 4\pi D r_0$ and $D = (D_+ + D_-)/2$ is the average diffusion coefficient for ions in a gas. This equation shows that recombination of ions involving aerosols is stronger than that in an aerosol-free space when

$$N_p \gg \frac{\sqrt{G\alpha}}{k} \,, \quad (6.112)$$

which leads to the number density of ions

$$N_i = \frac{G}{k N_p} \,. \quad (6.113)$$

To make numerical estimates for an atmospheric plasma, we take $G \sim 10$ $\text{cm}^{-3}\text{s}^{-1}$ and $\alpha \sim 10^{-6}\,\text{cm}^3/\text{s}$, so for ions with mobility $K = 1\,\text{cm}^2/(\text{V s})$ we have $k/r_0 = 0.3\,\text{cm}^2/\text{s}$, and criterion (6.112) gives $N_p r_0 \gg 0.01\,\text{cm}^{-2}$. In particular, for particles of size $r_0 \sim 10\,\mu\text{m}$ this gives $N_i \gg 100\,\text{cm}^{-3}$ (corresponding to density $\rho \gg 4 \times 10^{-7}\,\text{g/cm}^3$ for water vapor), so (6.113) leads to the estimate $N_i N_p \sim 3 \times 10^4\,\text{cm}^{-6}$. Because the maximum rate of ionization in the low atmosphere due to cosmic rays is 30–$40\,\text{cm}^{-3}\,\text{s}^{-1}$ [161], from this estimation it follows that aerosols of size $r_0 \sim 10\,\mu\text{m}$ determine the ionization balance in atmospheric air if $N_p \gg 100\,\text{cm}^{-3}$.

We now consider another aspect of charged aerosol particles in the Earth's atmosphere when their falling under the gravitational force leads to charge separation in the atmosphere and creates there the electric field of strength E which is established by currents of atmospheric ions, which leads to the balance equation

$$Z e N_p w = e E (K_+ N_+ + K_- N_-) \,,$$

where N_p is the number density of aerosol particles, Ze is their mean charge, w is the velocity of its falling, which is given by (6.3), N_+ and N_- are the number densities of positive and negative ions, and K_+ and K_- are their mobilities. Taking the regime when aerosol particles give a small contribution to the ionization balance, that is, $N_+ = N_- = N$, and the ion mobilities are nearby $K_+ = K_- = K$, we obtain from this balance the simple relation for the electric field strength:

$$E = \frac{Z N_p w}{2KN} .$$

In particular, from this it follows that since for large aerosol particles $Z \sim r_0$ according to (6.11) and $w \sim r_0^2$ according to (6.3), the electric field strength at a given mass density of water in the atmosphere is independent of a typical size of aerosol particles.

6.5.6
Prebreakdown Phenomena in the Atmosphere

We now analyze one more aspect of electric phenomena in the atmosphere. As noted above, the electric field strength of a cloud during lightning is less by one or two orders of magnitude than the breakdown strength at atmospheric pressure of about $E = 25\,\mathrm{kV/cm}$ ($E/N = 90\,\mathrm{Td}$) between plane electrodes, or the breakdown strength due to propagation of a positive streamer (about 5 kV/cm). We shall consider some reasons for this that extend our understanding of the nature of processes that accompany electric phenomena in atmospheric air.

Consider a chain of processes that can lead to breakdown. We shall consider only one of several mechanisms in order to demonstrate the character of these processes [178]. Assume that atmospheric air is subjected to an electric field of strength E, and the elementary processes 11, 12, 17, 20, 23, 26, 40, and 41 in Table 6.6 are the determining factors. These processes lead to balance equations for the number densities of electrons N_e, negative oxygen ions N_-, and ozone $[O_3]$ that can be stated as

$$\frac{dN_e}{dt} = (\nu_{ion} - \nu_{at}) N_e + N_-[O_3]k_{23} ,$$
$$\frac{dN_-}{dt} = \nu_{at} N_e - k_{17} N_-^2 - N_-[O_3]k_{23} ,$$
$$\frac{d[O_3]}{dt} = M - \frac{[O_3]}{\tau} + 2\nu_{26} N_e . \qquad (6.114)$$

We employ here the quantities $\nu_{ion} = k_{11}[O_2] + k_{12}[N_2]$, $\nu_{at} = k_{20}[O_2]$, and $\nu_{26} = k_{26}[O_2]$. We assume that process 17 in Table 6.6 gives the main contribution to recombination of positive and negative charges and that the number density of ozone molecules is relatively small, so the dissociation of an oxygen molecule leads automatically to the formation of two ozone molecules as a result of processes 40 and 41. The rate of ozone formation is denoted by M, and τ is the lifetime of ozone molecules.

Solving the set of two equations (6.114) under the assumption that $N_e \ll N_-$, we have for the number densities of charged particles

$$N_e = \frac{k_{23}^2}{k_{17}} \frac{\nu_{ion}}{(\nu_{at} - \nu_{ion})^2} [O_3]^2 , \quad N_- = \frac{k_{23}}{k_{17}} \frac{\nu_{ion}}{\nu_{at} - \nu_{ion}} [O_3] . \tag{6.115}$$

The subscripts in the rate constants correspond to the process numbers in Table 6.6. Expressions (6.115) show that a reverse connection between processes restores electrons. As a result, there are stable currents in atmospheric air subjected to an electric field. At $\nu_{at} = \nu_{ion}$, an instability develops that leads to breakdown in pure air. This will occur at $E = 25\,\text{kV/cm}$, which is the breakdown strength of dry air in the case of uniform currents. However, the sequence of processes specified above leads to breakdown at smaller electric fields. An increase of the ozone number density leads to an increase of the electron number density, which, in turn, causes an increase of the ozone number density. Then at certain sets of parameter values an explosive instability can develop, meaning that breakdown has occurred. We now seek the threshold for this instability.

For this purpose, we analyze the third equation in (6.114). Substituting it for the electron number density in (6.115), we have

$$\frac{d[O_3]}{dt} = \frac{N_0}{\tau} - \frac{[O_3]}{\tau} + k_{ef}[O_3]^2 , \tag{6.116}$$

where $N_0 = M\tau$ is the ozone number density in the absence of the electric field. The effective rate constant determined by

$$k_{ef} = \frac{2k_{23}^2 \nu_{ion} \nu_{26}}{k_{17}(\nu_{at} - \nu_{ion})^2}$$

is the combination of the rate constants for the corresponding processes. Under typical atmospheric conditions we have $N_0 \sim 10^{12}\,\text{cm}^{-3}$. The effective rate constant k_{ef} depends strongly on the electric field strength. It has the values $8 \times 10^{-13}\,\text{cm}^3/\text{s}$ at $E/N = 30\,\text{Td}$, $2 \times 10^{-11}\,\text{cm}^3/\text{s}$ at $E/N = 50\,\text{Td}$, and $6 \times 10^{-10}\,\text{cm}^3/\text{s}$ at $E/N = 80\,\text{Td}$.

In the stationary case, (6.116) has two solutions, of which one is stable. The instability corresponds to imaginary values for the solutions. Hence, the instability that leads to breakdown corresponds to the condition

$$k_{ef} \geq (4N_0\tau)^{-1} . \tag{6.117}$$

The real lifetime of ozone molecules in the atmosphere is rather long ($\tau \sim 10^6\,\text{s}$) because ozone is produced at high altitudes (40–80 km) and its loss results from chemical reactions (ozone cycles) or from transport to the Earth's surface. The parameter τ in condition (6.117) refers to the time during which an ozone molecule is located in a region with a large electric field strength. If we take the size of such a region as the cloud size $L \sim 1\,\text{km}$, and the velocity of transport from this region to be the wind velocity $v \sim 1\,\text{m/s}$, then we have $\tau \sim 10^3\,\text{s}$, and condition (6.117) gives

$k_{ef} \gg 2 \times 10^{-16}$ cm^3/s. The above values of k_{ef} show that the chain of processes we are considering leads to reduction of the breakdown electric field strength by several times.

We consider the ozone version of breakdown reduction to be merely illustrative, being guided by breakdown in homogeneous air, whereas breakdown in the atmosphere at large distances between electrodes develops in the form of a streamer, a nonuniform ionization wave, which requires lower threshold fields than uniform breakdown. Nevertheless, the above analysis illustrates important reasons for reduction of the electric field required to cause breakdown in natural air. Other reasons for a reduction in the breakdown threshold are heating of some regions of the atmosphere, the presence of particles, drops, or chemical compounds in the atmosphere, inhomogeneities of the electric field, and so on. This listing exemplifies the wide variety of processes that can influence atmospheric electric properties.

References

1 Fortov, V. *et al.* (2005) *Phys. Rep.*, **421**, 1.

2 Vladimirov, S.V., Ostrikov, K., and Samarian, A.A. (2005) *Physics and Applications of Complex Plasmas*, Imperial College, London.

3 Tsytovich, V.N., Morfill, G.E., Vladimirov, S.V., and Thomas, H. (2008) *Elementary Physics of Complex Plasmas*, Springer, Berlin.

4 Fortov, V.E. and Morfill, G.E. (eds) (2010) *Complex and Dusty Plasmas*, CRC Press, London.

5 Bonitz, M., Honig, N., and Ludwig, P. (eds) (2010) *Introduction to Complex Plasmas*, Springer, Berlin.

6 Flleagle, R.G. and Businger, J.A. (1980) *An Introduction to Atmospheric Physics*, Academic Press, San Diego.

7 Salby, M.L. (1996) *Fundamentals of Atmospheric Physics*, Academic Press, San Diego.

8 Aitken, J. (1881) *Nature*, **23**, 195.

9 Aitken, J. (1892) *Proc. R. Soc. A*, **51**, 408.

10 Harris, S.H. (1990) in *Encyclopedia of Physics*, (eds. R.G. Lerner and G.L. Trigg), Wiley & Sons Inc., New York, p. 30.

11 Harris, S.H. (2005) in *Encyclopedia of Physics*, (eds. R.G. Lerner and G.L. Trigg), Wiley-VCH Verlag GmbH, p. 61.

12 Somov, B.V. (2000) *Cosmic Plasma Physics*, Kluwer Academic Publishing, Dordrecht.

13 Meyer-Vernet, N. (2007) *Basics of Solar Wind*, Cambridge University Press, Cambridge.

14 Wright, C.C. and Nelson, G.J. (1979) *Icarus*, **38**, 123.

15 Meyer-Vernet, N., Steinberg, J.L., Strauss, M.A. *et al.* (1987) *Astron. J.*, **93**, 474.

16 Vourlidas, A., Davis, C.J., Eyles, C.J. *et al.* (2007) *Astrophys. J.*, **668**, L79.

17 Buffington, A., Bisi, M.M., Clover, J.M. *et al.* (2008) *Astrophys. J.*, **677**, 798.

18 Axford, W.I. and Mendis, D.A. (1974) *Annu. Rev. Earth Planet. Sci.*, **2**, 419.

19 Goertz, C.K. (1989) *Rev. Geophys.*, **27**, 721.

20 Mendis, D.A. and Rosenberg, M. (1994) *Annu. Rev. Astron. Astrophys.*, **32**, 419.

21 Horanyi, M. (1996) *Annu. Rev. Astron. Astrophys.*, **34**, 383.

22 Tsytovich, V.N. (1997) *Phys. Usp.*, **40**, 53.

23 Morfill, G.E., Grün, E., and Johnson, T.V. (1980) *Planet. Space Sci.*, **28**, 1087.

24 Morfill, G.E., Grün, E., Johnson, T.V. (1983) *J. Geophys. Res.*, **88**, 5573.

25 Horanyi, M. (2000) *Phys. Plasmas*, **7**, 3847.

26 Kurth, W.S. *et al.* (2006) *Planet. Space Sci.*, **54**, 988.

27 Krishna Swamy, K.S. (1997) *Physics of Comets*, World Scientific, Singapore.

28 Brandt, J.C. and Chapman, R.D. (2004) *Introduction to Comets*, Cambridge University Press, Cambridge.

29 Mendis, D.A. (1988) *Ann. Rev. Astron. Astrophys.*, **26**, 11.

30 Chu, J.H. and Lin, I. (1994) *Phys. Rev. Lett.*, **72**, 4009.

31 Thomas, H. *et al.* (1994) *Phys. Rev. Lett.*, **73**, 652.

32 Hayashi, Y. and Tachibana, S. (1994) *Jpn. J. Appl. Phys.*, **33**, L804.

33 Melzer, A., Trottenberg, T., and Piel, A. (1994) *Phys. Lett. A*, **191**, 301.

34 Nefedov, A.P., Petrov, O.F., and Fortov, V.E. (1997) *Phys. Usp.*, **40**, 1163.

35 Shukla, P.K. and Mamun, A.A. (2001) *Introduction to Dusty Plasma Physics*, IOP Publishing, Bristol.

36 Fortov, V.E. *et al.* (2004) *Phys. Usp.*, **47** 447.

37 Fortov, V.E., Jakubov, I.T., and Khrapak, A.G. (2006) *Physics of Strongly Coupled Plasma*, Oxford University Press, Oxford.

38 Melzer, A. and Goree, J. (2008) Fundamentals of dusty plasmas, in *Low Temperature Plasmas*, (eds R. Hippler *et al.*), vol. 1, John Wiley & Sons, Inc., Hoboken, p. 129.

39 Hippler, R. and Kersten, H. (2008) Applications of dusty plasmas, in *Low Temperature Plasmas*, (eds R. Hippler *et al.*), vol. 2, John Wiley & Sons, Inc., Hoboken, p. 787.

40 Smirnov, B.M. (1997) *Phys. Usp.*, **40**, 1117.

41 Smirnov, B.M. (2000) *Phys. Usp.*, **43**, 453.

42 Smirnov, B.M. (2004) *Contrib. Plasma Phys.*, **44**, 558.

43 Smirnov, B.M. (2006) *Principles of Statistical Physics*, John Wiley & Sons, Inc., Hoboken.

44 Echt, O., Sattler, K., and Recknagel, E. (1981) *Phys. Rev. Lett.*, **94**, 54.

45 Echt, O. *et al.* (1982) *Ber. Bunsenges. Phys. Chem.*, **86**, 860.

46 Ding, A. and Hesslich, J. (1983) *Chem. Phys. Lett.*, **94**, 54.

47 Knight, W.D. *et al.* (1984) *Phys. Rev. Lett.*, **52**, 2141.

48 Harris, I.A., Kidwell, R.S., and Northby, J.A. (1984) *Phys. Rev. Lett.*, **53**, 2390.

49 Phillips, J.C. (1986) *Chem. Rev.*, **86**, 619.

50 Harris, I.A., Norman, K.A., Mulkern, R.V., and Northby, J.A. (1986) *Chem. Phys. Lett.*, **130**, 316.

51 Miehle, W., Kandler, O., Leisner, T., and Echt, O. (1989) *J. Chem. Phys.*, **91**, 5940.

52 Mott-Smith, H.M. and Langmuir, I. (1926) *Phys. Rev.*, **28**, 727.

53 Druyvesteyn, M.J. and Penning, F.M. (1940) *Rev. Mod. Phys.*, **12**, 87.

54 Baum, E. and Chapkis, R.L. (1970) *AIAA J.*, **8**, 1073.

55 Allen, J.E. (1992) *Phys. Scr.*, **45**, 497.

56 Shun'ko, E.V. (2009) *Langmuir Probe in Theory and Practice*, Brown Walker Press, Boca Raton.

57 Al'pert, Y.L., Gurevich, A.V., and Pitaevsky, L.P. (1965) *Space Physics with Arti-facial Satellites*, Consultants Bureau, New York.

58 Wigner, E.P. and Seitz, F. (1934) *Phys. Rev.*, **46**, 509.

59 Wigner, E.P. (1934) *Phys. Rev.*, **46**, 1002.

60 Smirnov, B.M. (2010) *Cluster Processes in Gases and Plasmas*, John Wiley & Sons, Inc., Hoboken.

61 Smirnov, B.M., Shyjumon, I., and Hippler, R. (2007) *Phys. Rev. E*, **77**, 066402.

62 Fuks, N.A. (1964) *Mechanics of Aerosols*, Macmillan, New York.

63 Smoluchowski, M.V. (1916) *Z. Phys.*, **17**, 585.

64 Langevin, P. (1905) *Ann. Chem. Phys.*, **8**, 245.

65 Smirnov, B.M. (1999) *Clusters and Small Particles in Gases and Plasmas*, Springer, New York.

66 Illenberger, E. and Smirnov, B.M. (1998) *Phys. Usp.*, **41**, 651.

67 Birkhoff, G.D. (1931) *Proc. Natl. Acad. Sci. USA*, **17**, 656.

68 Neumann, J. (1932) *Proc. Natl. Acad. Sci. USA*, **18**, 70, 263.

69 Landau, L.D. and Lifshitz, E.M. (1980) *Statistical Physics*, vol. 1, Pergamon Press, Oxford.

70 Landau, L.D. and Lifshitz, E.M. (1980) *Mechanics*, Pergamon Press, London.

71 Currie, I.G. (1974) *Fundamental Mechanics of Fluids*, McGraw-Hill, London.

72 Huang, K. (1987) *Statistical Mechanics*, John Wiley & Sons, Inc., New York.

73 Acheson, D.J. (1990) *Elementary Fluid Dynamics*, Oxford University Press, Oxford.

74 Smythe, W.R. (1950) *Static and Dynamic Electricity*, McGraw-Hill, New York.

75 Jackson, J.D. (1998) *Classical Electrodynamics*, John Wiley & Sons, Inc., New York.

76 Smirnov, B.M. (2008) *Reference Data on Atomic Physics and Atomic Processes*, Springer, Heidelberg.

77 Bernstein, I.B. and Rabinovitz, I.N. (1959) *Phys. Fluids*, **2**, 112.

78 Zaslavskii, G.M. and Sagdeev, R.Z. (1988) *Introduction to Nonlinear Physics. From Pendulum to Turbulence and Chaos*, Nauka, Moscow.

79 Sena, L.A. (1946) *Zh. Eksp. Teor Fiz.*, **16**, 734.

80 Sena, L.A. (1948) *Collisions of Electrons and Ions with Atoms*, Gostekhizdat, Leningrad.

81 Hirschfelder, J.O., Curtiss, Ch.F., and Bird, R.B. (1954) *Molecular Theory of Gases and Liquids*, John Wiley & Sons, Inc., New York.

82 Smirnov, B.M. (2010) *JETP*, **110**, 1042.

83 Bliokh, P.V., Sinitsin, V., and Yaroshenko, V. (1995) *Dusty and Self-Gravitational Plasmas in Space*, Kluwer Academic Publishing, Dordrecht.

84 Shukla, P.K. and Mamun, A.A. (2002) *Introduction to Dusty Plasma Physics*, Institute of Physics Publishing, Bristol.

85 Horanyi, M., Havnes, O., and Morfill, G.E. (2010) in *Complex and Dusty Plasmas*, (eds. V.E. Fortov and G.E. Morfill), CRC Press, London, p. 291.

86 Hundhausen, A.J. (1996) *Introduction to Space Physics*, Cambridge University Press, Cambridge.

87 Hamilton, C.J. (1995) *Solar Views*, http://www.solarviews.com/browse/pia/PIA06193.jpg (accessed 10 August 2011).

88 Mendis, D.A. and Rosenberg, M. (1996) *Annu. Rev. Astron. Astrophys.*, **32**, 419.

89 Bosh, A.S., Olkin, C.B., French, R.G., and Nicholson, P.D. (2002) *Icarus*, **157**, 57.

90 Dikarev, V.V. (1999) *Astron. Astropys.*, **346**, 1011.

91 Matthews, L.S., Hyde, T.W. (2004) *Adv. Space Res.*, **33**, 2292.

92 Srama, R. *et al.* (2006) *Planet. Space Sci.* 54, 967.

93 Porco, C.C. *et al.* (2006) *Science*, **311**, 1393.

94 Wang, Z. *et al.* (2006) *Planet. Space Sci.*, **54**, 957.

95 Havnes, O., Morfill, G., Melandsø, F. (1992) *Icarus*, **98**, 141.

96 Gan-Baruch, Z., Eviator, A., Richardson, J.D. (1994) *J. Geophys. Res.*, **99**, 11063.

97 Hamilton, C.J. (1995) *Views of the Solar System*, http://www.solarviews.com/cap/comet/west.htm (accessed 10 August 2011).

98 Bryant, P. (2003) *J. Phys. D*, **36**, 2859.

99 Khrapak, S.A. *et al.* (2005) *Phys. Rev. E*, **72**, 016406.

100 Khrapak, S.A., Morfill, G.E., Khrapak, A.G., and D'yachkov, L.G. (2006) *Phys. Plasmas*, **13**, 052114.

101 D'yachkov, L.G., Khrapak, A.G., Khrapak, S.A., and Morfill, G.E. (2007) *Phys. Plasmas*, **14**, 042102.

102 Gatti, M. and Kortshagen, U. (2008) *Phys. Rev. E*, **78**, 046402.

103 Schottky, V.W. (1924) *Phys. Z.*, **25**, 342.

104 Dutton, J. (1975) *J. Phys. Chem. Ref. Data*, **4**, 577.

105 Konopka, U., Morfill, G.E., and Rathke, L. (2000) *Phys. Rev. Lett.*, **84**, 891.

106 Ikezi, H. (1986) *Phys. Fluids*, **29**, 1764.

107 Juan, W.T. *et al.* (1998) *Phys. Rev. E*, **58**, R6947.

108 Fortov, V.E. *et al.* (2000) *Phys. Lett. A*, **267**, 179.

109 Karasev, V.Yu. *et al.* (2008) *JETP*, **106**, 399.

110 Fortov, V.E. *et al.* (2010) in *Complex and Dusty Plasmas*, (eds. V.E. Fortov, G.E. Morfill), CRC Press, London, p. 1.

111 Smirnov, B.M. (2003) *Phys. Usp.*, **46**, 589.

112 Smirnov, B.M. (1994) *Phys. Usp.*, **37**, 621.

113 Smirnov, B.M. and Strizhev, A.Yu. (1994) *Phys. Scr.*, **49**, 615.

114 Smirnov, B.M., Shyjumon, I., and Hippler, R. (2006) *Phys. Scr.*, **73**, 288.

115 Voloshchuk, V.M. (1984) *Kinetic Theory of Coagulation*, Gidrometeoizdat, Leningrad.

116 Ostwald, W. (1900) *Z. Phys. Chem.*, **34**, 495.

117 Smoluchowski, M.V. (1918) *Z. Phys. Chem.*, **92**, 129.

118 Rao, B.K. and Smirnov, B.M. (1997) *Phys. Scr.*, **56**, 588.

119 Rao, B.K. and Smirnov, B.M. (2002) *Mater. Phys. Mech.*, **5**, 1.

120 Lifshitz, I.M. and Slezov, V.V. (1958) *Sov. Phys. JETP*, **35** 331.

121 Lifshitz, I.M. and Slezov, V.V. (1959) *Fiz. Tverd. Tela*, **1**, 1401.

122 Lifshitz, I.M. and Slezov, V.V. (1961) *J. Phys. Chem. Sol.*, **19**, 35.

123 Slezov, V.V. and Sagalovich, V.V. (1987) *Sov. Phys. Usp.*, **30**, 23.

124 Lifshitz, E.M. and Pitaevskii, L.P. (1981) *Physical Kinetics*, Pergamon Press, Oxford.

125 Zhukhovizkii, D.I., Khrapak, A.G., and Jakubov, I.T. (1983) *High Temp.*, **21**, 1197.

126 Zhukhovizkii, D.I. (1989) *High Temp.*, **27**, 515.

127 Ino, S. (1969) *J. Phys. Soc. Jpn.*, **27**, 941.

128 Weber, B. and Scholl, R. (1992) *J. Illum. Eng. Soc.*, Summer **1992**, 93.

129 Scholl, R. and Weber, B. (1992) in *Physics and Chemistry of Finite Systems: From Clusters to Crystals*, (eds. P. Jena, S.N. Khana, and B.K. Rao), vol. 2, Kluwer Academic Press, Dordrecht, p. 1275.

130 Weber, B. and Scholl, R. (1993) *J. Appl. Phys.*, **74**, 607.

131 Scholl, R. and Natour, G. (1996) in *Phenomena in Ionized Gases*, (eds. K.H. Becker, W.E. Carr, E.E. Kunhardt), AIP Press, Woodbury, p. 373.

132 Smirnov, B.M. (1998) *Phys. Scr.*, **58**, 363.

133 Smirnov, B.M. (2000) *JETP Lett.*, **71**, 588.

134 Chamberlain, J. (1978) *Theory of Planetary Atmospheres*, Academic Press, London.

135 Deminov, M.G. (2008) in *Encyclopedia of Low Temperature Plasma*, (ed V.E. Fortov), ser. B, vol. 1, Yanus, Moscow, p. 72, in Russian.

136 Reist, P.C. (1984) *Introduction to Aerosol Science*, Macmillan, New York.

137 Green, H.L. and Lane, W.R. (1964) *Particulate Clouds: Dust, Smokes and Mists*, Van Nostrand, Princeton.

138 Friendlander, S.K. (2008) *Smoke, Dust and Haze: Fundamentals of Aerosol Dynamics*, Oxford University Press, Oxford.

139 Uman, M.A. (1986) *About Lightning*, Dover, New York.

140 Hinds, W.C. (1999) *Aerosol Technology: Properties, Behavior and Measurement of Airborne Particles*, John Wiley & Sons, Inc., New York.

141 Kelley, M.C. (1989) *The Earth's Ionosphere: Plasma Physics and Electrodynamics*, Academic Press, San Diego.

142 Alpert, Ja.L. (1972) *Propagation of Radiowaves in Atmosphere*, Nauka, Moscow.

143 Davies, K. (1990) *Ionospheric Radio*, Peter Peregrinus Ltd, London.

144 McIven, M. and Philips, L. (1978) *Chemistry of Atmospheres*, Mir, Moscow.

145 Brunelli, B.E. and Laminadze, A.A. (1988) *Physics of Ionosphere*, Nauka, Moscow.

146 Rees, M.H. (1989) *Physics and Chemistry of Upper Atmosphere*, Cambridge University Press, Cambridge.

147 Shunk, R.W. and Nagy, A.E. (2000) *Ionospheres: Physics, Plasma Physics, and Chemistry*, Cambridge University Press, Cambridge.

148 Bilitza, D. (2002) *Radio Science*, **36**, 261.

149 Israel, H. (1973) *Atmospheric Electricity*, Keter Press Binding, Jerusalem.

150 Moore, C.B. and Vonnegut, B. (1977) in *Lightning*, (ed. R.H. Golde), Academic Press, London, p. 51.

151 Feynman, R.P., Leighton, R.B., and Sands, M. (1964) *The Feynman Lectures on Physics*, vol. 2, Addison-Wesley, Reading.

152 Berger, K. (1977) in *Lightning*, (ed. R.H. Golde), Academic Press, London, p. 119.

153 Fowler, R.G. (1982) in *Applied Atomic Collision Physics*, (eds H.S.W. Massey, E.W. McDaniel, and B. Bederson), vol. 5, Academic Press, New York, p. 35.

154 Latham, J. and Stromberg, I.M. (1977) in *Lightning*, (ed. R.H. Golde), Academic Press, London, p. 99.

155 Ellis, H.W., Pai, R.Y., McDaniel, E.W., Mason, E.A., and Viehland, L.A. (1976) *Atom. Data Nucl. Data Tables*, **17**, 177.

156 Ellis, H.W., McDaniel, E.W., Albritton, D.L., Viehland, L.A., Lin, S.L., and Mason, E.A. (1978) *Atom, Data Nucl. Data Tables*, **22**, 179.

157 Ellis, H.W., Trackston, M.G., McDaniel, E.W., and Mason, E.A. (1984) *Atom. Data Nucl. Data Tables*, **31**, 113.

158 Viehland, L.A. and Mason, E.A. (1995) *Atom. Data Nucl. Data Tables*, **60**, 37.

159 Wilson, C.T.R. (1925) *Proc. R. Soc.*, **37**, 32D.

160 Neher, H.V. (1971) *J. Geophys. Res.*, **76**, 1637.

161 Nicolet, M. (1975) *Planet. Space Sci.*, **23**, 637.

162 Schouland, B. (1930) *Proc. R. Soc.*, **A130**, 37.

163 Appelton, E. and Bowen, E. (1933) *Nature*, **132**, 965.

164 Seinfeld, J.H. and Pandis, S.N. (1998) *Atmospheric Chemistry and Physics*, John Wiley & Sons, Inc., New York.

165 Byers, H.R. (1965) *Elements of Cloud Physics*, University of Chicago Press, Chicago.

166 Fletcher, N.H. (1969) *The Physics of Rainclouds*, Cambridge University Press, London.

167 Mason, B.J. (1971) *The Physics of Clouds*, Clarendon Press, Oxford.

168 Twomey, S. (1977) *Atmospheric Aerosols*, Elsevier, Amsterdam.

169 Proppacher, H. and Klett, J. (1978) *Microphysics of Clouds and Precipitation*, Reidel, London.

170 Javaratne, E.R., Saunders, C.P.R., and Hallett, J. (1983) *Q. J. R. Meteorol. Soc.*, **109**, 609.

171 Williams, E.R., Zhang, R., and Rydock, J. (1991) *J. Atmos. Sci.*, **48**, 2195.

172 Dong, Y. and Hallett, J. (1992) *J. Geophys.Res.*, **97**, 20361.

173 Mason, B.L. and Dash, J.G. (2000) *J. Geophys.Res.*, **105**, 10185.

174 Dash, J.G., Mason, B.L., and Wettlaufer, J.S. (2001) *J. Geophys.Res.*, **106**, 20395.

175 Berdeklis, P. and List, R. (2001) *J. Atmos.Sci.*, **58**, 2751.

176 Nelson, J. and Baker, M. (2003) *Atmos. Chem. Phys. Discuss.*, **3**, 41.

177 MacGorman, D.R. and Rust, W.D. (1998) *Electrical Nature of Storms*, Oxford University Press, New York.

178 Eletskii, A.V. and Smirnov, B.M. (1991) *J. Phys. D*, **24**, 2175.

179 Holstein, T. (1952) *J. Chem. Phys.*, **56**, 832.

180 Appleton, E.V. and Barnett, M.A.F. (1925) *Nature*, **115**, 333.

181 Appleton, E.V. and Barnett, M.A.F. (1925) *Proc. Roy. Soc. A*, **109**, 621.

7
Conclusion – Plasmas in Nature and the Laboratory

Summarizing the analysis of properties of ionized gases or a weakly ionized plasma, we consider briefly the problems in our environment where we encounter a plasma. A terrestrial plasma occurs in various forms in the Earth's atmosphere and the solar system, and these forms are different depending on the location of this plasma. At low altitudes above the Earth's surface the plasma is maintained by ionization of air under the action of cosmic rays and atmospheric electric fields, whereas near the Earth's surface, some of the ionization of air arises from the decay of radioactive elements in the Earth's crust. A plasma in lower layers of the atmosphere is characterized by a low density of charged particles. The presence of a plasma is limited by the tendency of electrons to attach to oxygen molecules fast, with formation of negative ions. Therefore, plasmas in the lower atmosphere acquire a negative charge in the form of negative ions. Plasma processes are responsible for the action of the electric machine of the Earth (Section 6.5) and accompany the circulation of water in the Earth through its atmosphere and create electric fields in the atmosphere. This leads to various electric phenomena in the Earth's atmosphere, such as lightning (Section 6.5.2) – electric breakdown under the action of electric fields in the atmosphere, corona discharge in a quiet atmosphere due to passage of electric currents near conductors, and Saint Elmo's fire – a glow in the vicinity of protruding objects. Electric discharges in atmospheric air are often associated with energetic natural phenomena of a nonelectric nature, such as volcanic eruptions, earthquakes, sand storms, and water spouts.

The Earth's ionosphere (Section 6.4.3) is a typical plasma consisting of electrons and ions. It results from photoionization of atomic oxygen or molecular nitrogen under the action of the hard UV radiation from the Sun. The ionosphere reflects radio signals from the Earth, that is, it is a mirror for electromagnetic waves (Section 6.4.3), and allows the propagation of radio waves over large distances. The aurora or aurora borealis are observed in the regions of the geographical poles where the magnetic lines of force due to the Earth's magnetic field are directed perpendicular to the Earth's surface. Electrons and protons of the solar wind move along the magnetic field lines of the Earth, and the more energetic particles proceed to the vicinity of the Earth's magnetic poles, where they are braked as a result of collisions with atmospheric atoms or molecules, and these protons ionize and excite atoms in the upper atmosphere. Radiation of metastable oxygen and nitrogen atoms formed

Fundamentals of Ionized Gases, First Edition. Boris M. Smirnov.

by these collisions creates the observed glow of the aurora. In addition, electrons and protons may be captured by the magnetic field of the Earth at a distance of several Earth radii, and radiation belts are formed in this manner (see Figure 7.1). Thus, the upper atmosphere of the Earth contains different types of plasmas whose properties are determined by interaction of solar radiation and the solar wind with the Earth's fields and flows.

The properties of a solar plasma depend on where in the Sun it is located. The Sun's core contains a plasma with a temperature of about 17 million Kelvin and a number density of the order of 10^{24} cm^{-3}. As one moves from the center of the Sun, the number density of charged particles in the plasma and the temperature decrease. The Sun's photosphere is of principal importance both for Earth processes and for the solar energy balance. The photosphere is a thin layer of the Sun whose thickness (about 1000 km) is small compared with the Sun's radius, but this layer, with a temperature of about 6000 K, emits most of the energy radiated by the Sun owing to electron photoattachment to hydrogen atoms (Section 3.3.7) and determines the yellow color of the Sun.

The next layer of the solar atmosphere is the chromosphere, whose name originates from the red color observed during a total eclipse of the Sun by the Moon. The temperature of the chromosphere decreases at first and then increases with an increase of the distance from the Sun's center, whereas the density of the solar atmosphere decreases with increase of distance from the Sun's center. The chromosphere is transparent because it is rarefied. The next higher, more rarefied region of the Sun's atmosphere is called the solar corona. This beautiful name is suggested by the glow in this region that is observed during a total eclipse of the Sun. The

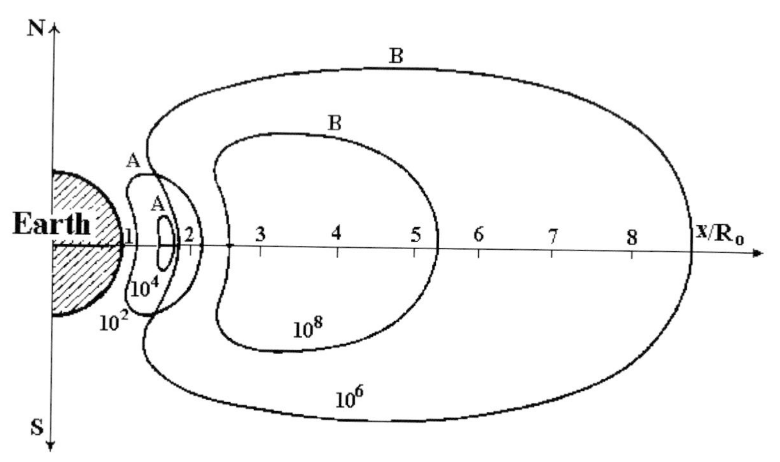

Figure 7.1 Radiation belts near the Earth. Curves *A* correspond to regions of capture of fast protons with energy of more 30 MeV, curves *B* correspond to capture of slow electrons with energy of 0.1–5 MeV. The mean fluxes of protons are expressed in units of per square centimeter per second, R_0 is the Earth's radius, x is the distance from the Earth.

characteristic feature of the solar corona is its high temperature (approximately 10^6 K), and therefore the solar corona emits hard UV radiation.

The solar corona is not stable because of its high temperature and not low temperature gradient. Expansion of the solar corona into space creates a plasma flow, the solar wind, directed outward from the Sun. The solar wind creates interplanetary plasma and leads to various phenomena in the upper terrestrial atmosphere. The solar wind interacts with dust from Saturn's and Jupiter's satellites and the magnetic fields of these planets create Saturn's and Jupiter's rings (Section 6.2.3). Interaction of the solar wind with comet dust involving solar radiation and magnetic fields determines the properties of the comet tail (Section 6.2.3).

In the Sun there are a number of phenomena that result from interaction of the solar plasma with its magnetic fields. Spicules – separated columns of moving plasma – are formed in the chromosphere and create chromospheric bushes. Under typical solar conditions, approximately 10^6 spicules are observed simultaneously. Spicules result from reconnection of magnetic lines of force (Section 4.5.9). An increase in the magnetic field strength in some zones of the solar atmosphere leads to scattering of gas in these zones, and causes the formation of cylindrical structures in the plasma motion. This mechanism is the basis for the generation of spicules.

Another structure of the solar plasma observed in the solar corona is a prominence. Prominences are clots of dense plasma of different forms with temperatures of the order of 10^4 K. They are generated as a result of interaction of plasma streams with magnetic fields and propagate in the corona from the chromosphere. Solar flares are observed in the upper chromosphere or in the corona. They are a result of development of plasma instabilities, in particular, reconnection of magnetic lines of force (Section 4.5.9). Solar flares last several minutes. A sharp increase of the plasma temperature in flares causes an increase in emission of short-wave radiation. Solar flares generate short intense fluxes of solar plasma and can strongly enhance the solar wind. Such signals reach the Earth in 1.5–2 days, and cause magnetic storms and auroras in the terrestrial atmosphere. In summary, there are several varieties of terrestrial plasma and this plasma is responsible for various phenomena which we observe.

In considering briefly plasma applications, we note the variety of plasma forms, which also create the variety of plasma applications. Hence, there are a great number of applications, and this number is increasing with time. The most widespread is a gas discharge plasma that is produced in the simplest way, and the main plasma application corresponds to light sources on the basis of gas discharge. There are many types of gas discharge lamps depending on their use. In particular, fluorescent gas discharge lamps are used now instead of incandescent lamps and provide an efficacy of up to 100 lm/W, whereas sodium lamps of yellow color are used for outdoor lighting. A plasma may be used for determination of additions in gases or vapors of low concentrations of admixtures up to 10^{-10}–10^{-9} g/g within the framework of the optogalvanic method [1, 2]. In this determination, gas discharge is burned, and a laser signal is tuned to a resonant line of a given element, which leads to variation of the gas discharge current because of the subsequent ionization processes.

Gas discharge is a simple method to create a plasma, and because gas discharge has many regimes and forms which are increasing with time, the number of applications of a gas discharge plasma is growing. New applications with old ideas are being developed when the technology allows these ideas to be used. An example of this are plasma displays[1], the concept of which for television was described in 1936 by Kalman Tihanyi as a system of single transmission points of a grid of cells arranged in a thin panel display, where these points are excited to different levels by varying the voltages applied to these point. Nevertheless, production of plasma displays started in the 1980s when they could compete with electron-beam tubes or liquid crystal displays. Roughly, a plasma display includes two parallel planes with parallel bus bars in each plane. Bus bars of each plane are surrounded by dielectric planes and bus bars of different planes are in perpendicular directions. As a result, the space inside the bus-bar plane is divided into some cells, and each cell relates to one intersection of bus bars. The space is filled by a mixture of nitrogen and neon (or other inert gases), and gas discharges occur in each cell almost independently. Behind the cathode each cell is divided into three parts covered by colored luminophors, so after absorption radiation of local gas discharge, one of subcells gives a red color, the second one gives a green color, and the third subcell transforms radiation of gas discharge into a blue color. These three colors are joined in the overall color of this element, which depends on the intensity of local discharge, and this intensity in turn is created through bus bars. Competing with other types of displays, plasma displays or plasma panels are favorable for screens of large size and their applications are determined by the possibilities of new technology.

Various laboratory devices and systems contain a plasma, where it is divided into a low temperature plasma and a hot plasma, depending on the concentration of electrons and ions in the plasma or the temperature of the charged particles. In a hot plasma, a thermal energy of the atomic particles exceeds a characteristic atomic value (which may be the ionization potential of the plasma atoms), and is much less than this characteristic value in a low temperature plasma. Correspondingly, plasma devices containing hot plasmas or low temperature plasmas are different in principle. An example of a hot plasma is a thermonuclear fusion plasma, that is, a plasma for a controlled thermonuclear reaction [3]. This reaction proceeds with participation of nuclei of deuterium or of tritium – isotopes of hydrogen. To achieve this reaction, it is necessary that during the time of plasma confinement, that is, during the time when ions of deuterium or tritium are present in the reaction zone, these ions have a chance to participate in a thermonuclear reaction. There must be both a high ion temperature (about 10 keV) and a high number density of ions for the thermonuclear reaction to proceed. The threshold number density of ions N_{i} for the thermonuclear reaction depends on the plasma lifetime τ such that the product of these values $N_{\mathrm{i}}\tau$ must exceed a certain value. This condition, the Lawson criterion, at ion temperatures of several thousand electronvolts corresponds to an

1) http://ru.wikipedia.org/wiki/ (accessed 10 August 2011)

onset value of $N_i\tau = 10^{16}\,\text{cm}^{-3}\,\text{s}$ for a deuterium plasma and an onset value of $10^{14}\,\text{cm}^{-3}\,\text{s}$ for a tritium plasma. These values are reached in contemporary plasma fusion devices, where a hot plasma is created by fast beams of hydrogen atoms, but it is necessary to solve a circle of engineering and technological problems to construct an effective fusion power station.

Another method to solve the problem of thermonuclear fusion uses pulsed systems. In this case a deuterium pellet is irradiated over its entire surface by a laser pulse or a fast ion beam. When the pulse impinges on the pellet, the pellet is heated and compressed by a factor of 10^2–10^3. The heating and compression is intended to promote a thermonuclear fusion reaction. Note that the dense hot plasma that is formed during compression of the pellet is a special state of matter which does not have an analogue in an ordinary laboratory setting.

Various concepts are used in plasma devices for special use. The magnetohydrodynamic (MHD) generator works on the basis of the Hall effect (Section 4.5.2), where a stream of hot, weakly ionized gas flows in a transverse magnetic field, and the electric current is induced in a direction perpendicular to the directions of the stream and magnetic field (Figure 7.2a). As a result, the flow energy is converted into electric energy. According to thermodynamic laws, the efficiency of this transformation is quite high because a flowing gas is hot.

The principles of action of MHD generators and plasma engines may have an analogy, as shown in Figure 7.2. In plasma engines, which are used in the cosmos as rocket engines of space vehicles [4–8], a flux of ions causes the motion of a system in the opposite direction according to Newton's third law. There are two types of ion engines. In Hall thrusters, electrons are magnetized by a strong magnetic field, and ions leave the plasma region under the action of the electric field. In ion thrusters, a magnetic field is used to prevent electron attachment to the anode. In both cases ions leave the plasma region with a typical velocity of 10^7 cm/s, which

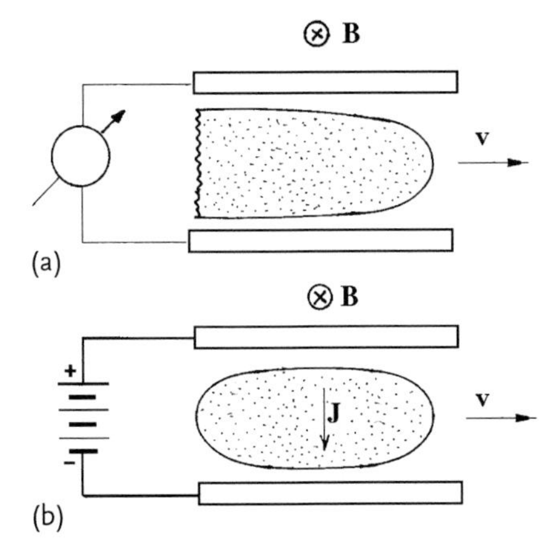

Figure 7.2 A magnetohydrodynamic generator (a) and an ion engine (b). In the first case, the electric field occurs as a result of motion of a plasma in a magnetic field; in the second case, electrons and ions obtain their velocity owing to crossing electric and magnetic fields.

exceeds the velocity of the products of the chemical reaction in power engines by an order of magnitude, so ion propulsion is effective for cosmic engines, where the problem of the fuel weight must be overcome. Ion and Hall thrusters in the cosmos exhibit the properties of a long lifetime and high reliability.

Along with stationary MHD generators, which now do not have a suitable place in contemporary energetics, there are pulsed MHD generators, which are analogous to guns with a weakly ionized plasma instead of shells. The plasma is formed as a result of combustion of gunpowder or some equivalent source of propellant gas. When this plasma passes through a region with an external magnetic field, electric power is created. To estimate the possibilities of this system, let us perform a simple calculation. We assume that the plasma velocity equals to the velocity which a bullet acquires in the gun. We take the efficiency of energy transformation to be of the order of 50%, and the length of the region with a magnetic field to be 1 m. Then the specific power of this generator is of the order of 10^8 W/g and the pulse duration is approximately 7×10^{-4} s. Thus, the power output of a pulsed MHD generator, for this brief instant, corresponds to the total power output of all the electric power plants in the world if the mass of the powder used is of the order of 10 g. The pulsed MHD generator transforms the chemical energy of the powder into electric energy. Owing to the simplicity and yield parameters of these systems, they are convenient for special applications as autonomous pulsed sources of electric energy.

The other system where a plasma is used for generation of electric energy is the thermoemission converter. It contains two parallel metal plates with different work functions (the work function is the binding energy of an electron at a surface). One of these plates is heated and emits electrons, which reach the other plate, so an electric current is created. Connection of the plates through a load leads to release of the electric energy in the load. It is evident that a plasma is not the underlying basis for this device. Nevertheless, the use of a plasma in the gap between the plates makes it possible to overcome an important difficulty attendant on this system. If there is no plasma in the gap, electron charge is accumulated in this region and creates an electric potential between the plates that increases with the electron number density in the gap, and is opposite in direction to the potential that initiated the phenomenon. Beyond a certain level, the counterpotential will stop generation of the electric energy. For typical energy fluxes in these systems (approximately 1 W/cm^2) the distance between plates must be less than 10 μm. It is difficult to combine this condition with the high temperature of the heated plate (approximately 2000 K). Introduction of a plasma into the gap between plates provides a means of overcoming this difficulty. It is interesting that the principles of operation of MHD generators, thermoemission converters, and plasmotrones were suggested as early as the end of the nineteenth century. Now, with the advantage of contemporary materials and technology, these devices have acquired a new life.

In various applications of a plasma with an intense input of energy, the plasma is generated in a moving medium. The plasma generator, or plasmotrone, is usually an arc discharge established in a flowing gas or vapor that is an equilibrium thermal plasma [9]. Such plasma generators produce plasma torches which have wide applications in various technological areas, including incineration of waste and special medical uses. Many plasma applications are based on the possibility of introducing a high level of electric energy into the plasma. This leads to the creation and maintenance of an ionized gas containing active atomic particles: electrons, ions, excited atoms, radicals. These particles can be analyzed by a variety of techniques, and therefore a plasma can used not only in energetic systems, but also in measuring instruments. In particular, plasma-based methods of spectral analysis are widely practiced in metallurgy. In these methods a small amount of metal in the form of a solution or powder is introduced into a flowing discharge plasma, and spectral analysis of the plasma makes it possible to determine the metal composition. The accuracy of spectral determination of admixture concentration with respect to a primary component is of the order of 0.01–0.001%. On the basis of the optogalvanic method [1, 2], one can increase this limit significantly.

The number of plasma applications is increasing with time and these applications are included in new aspects of human activity. Plasma applications in medicine started several decades ago and are based on the mechanical or chemical action of a plasma on a living object. For example, a plasma knife, which is a thick plasma flux, is used similarly to a knife in surgery. The action of a plasma as an active medium may be of various types because of the existence of many plasma forms. In particular, destruction of medical waste, including bandages, medical cotton, and other disposable materials, by using an arc plasma proceeds similarly to combustion of these materials in an oven, but the plasma setup is more compact and excludes dangerous products, but it is also more expensive. Nevertheless, plasma methods are more favorable in practice than combustion of medical materials in an oven. A not so powerful plasma is used for decontamination and sterilization of medical tools. Because a plasma contains various active particles, such as electrons, ions, metastable atoms and molecules, and radicals, it is applied in wound healing, dermatology, and dentistry, where the plasma kills microbes and does not destroy living tissue. Of course, a suitable plasma form and a certain current regime must be used for each case and follows from the corresponding study. In addition, plasma technology is used for production of specific medical materials and devices.

Plasma processing for environmental applications is developing in two directions. The first is decomposition of toxic substances, explosive materials, and other hazardous wastes, that can be decomposed in a plasma into their simple chemical constituents. The second is improvement of air quality. A corona discharge of low power is used for this purpose. This gas discharge generates active atomic particles such as oxygen atoms. These atoms have an affinity for active chemical compounds in air, and react with them. Such discharges also destroy microbes, but present no hazards to humans because of low concentrations of these particles.

Plasmas are currently widely employed in industry, and their range of application is increasing continuously [10–12, 15]. The usefulness of plasmas in technology can be ascribed to two qualities: Plasmas make available much higher temperatures than can be achieved with chemical fuel torches; and a large variety of ions, radicals, and other chemically active particles are generated therein. Therefore, either directly within a plasma or with its help, one can conduct chemical processes of practical importance. Another advantage of a plasma follows from the possibility of introducing large specific energies in a simple fashion.

The oldest applications of a plasma as a heat-transfer agent [9, 13] are in the welding or cutting of metals. Since the maximum temperature of chemical torches is about 3000 K, they cannot be used for some materials. The arc discharge (electric arc) makes it possible to increase this temperature by a factor of three, so melting or evaporation of any material is possible. Therefore, an electric discharge is used, starting a century ago, for welding and cutting of metals. Presently, plasma torches with a power of up to 10 MW are used for iron melting in cupolas, for scarp melting, for production of steel alloys, and for steel reheating in tundishes and landlies. Plasma processing is used for extraction of metals from ores. In some cases, plasma methods compete with traditional ones based on chemical heating. Comparing the plasma methods with chemical methods in cases when either can be used, one can conclude that plasma methods provide a higher specific output, a higher quality of product, a smaller amount of waste, but require greater energy expenditure and more expensive equipment.

Another application of a plasma as a heat-transfer agent relates to fuel energetics. The introduction of a plasma into the burning zone of low-grade coals leads to improvements in the efficiency of the burning, with an accompanying reduction in particulate emissions, in spite of a relatively low energy input from the plasma. Plasmas are also used also for pyrolysis and other methods of processing and cleaning of fuel. As a powerful heat-transfer agent, a plasma is used for treatment and destruction of waste [14].

Plasmas have been employed extensively for the processing and treatment of surfaces or for the preparation of new materials in the form of films [15–20]. The good heat-transfer capabilities of a plasma are useful for treatment of surfaces. During plasma processing of surfaces, the chemical composition of the surface does not change, but its physical parameters may be improved. Another aspect of the processing of a surface by a plasma refers to the case when active particles of the plasma react chemically with the surface. The upper layer of the surface can acquire a chemical composition different from that of the substrate. For example, plasma hardening of a metallic surface occurs when metal nitrides or carbides are formed in the surface layer. These compounds are generated when ions or active atoms in the plasma penetrate the surface layer. A third mechanism for the plasma action on a surface is realized when the surface material does not itself participate in the chemical process, but material from the plasma is deposited on the surface in the form of a thin film. This film can have special mechanical, thermal, electric, optical, and chemical properties depending on specific problems and requirements. It is convenient to use for this purpose plasma beams that flow from jets. Beam methods for deposition of micrometer-thickness films are widespread in the manufacture of microelectronics, mirrors, and special surfaces.

An important area of plasma applications is plasma chemistry [11, 15, 17, 21, 22], which relates to the production of chemical compounds. The first industrial plasmachemical process was employed for ammonia production at the beginning of the twentieth century. It was later replaced by a cheaper method that produced ammonia from nitrogen and hydrogen in a high-temperature, high-pressure reactor with a platinum catalyst. Another plasmachemical process involves production of ozone in a barrier discharge. This method has been used for several decades. Large-scale development of plasma chemistry was long retarded by the required high power intensity. When other criteria became the limiting factors, new plasmachemical processes were mastered. Presently, the array of chemical compounds produced industrially by plasmachemical methods includes C_2H_2, HCN, TiO_2, Al_2O_3, SiC, XeF_6, KrF_2, O_2F_2, and many others. Although the number of such compounds constitutes a rather small fraction of the products of the chemical industry, this fraction is steadily increasing. The products of a plasmachemical process may exist either in the form of a gas or in the form of condensed particles. Plasmachemical industrial production of ceramic-compound powders such as SiC, Si_3N_4, or powders of metals and metal oxides leads to a yield product of high quality. Plasmachemical processes with participation of organic compounds are used as well as those with involvement of inorganic materials. Organic-compound applications include the production of polymers and polymeric membranes, processes of fine organic synthesis in a cold plasma [11, 15, 16], and so on. In a qualitative assessment of the technological applications of plasmas, we conclude that plasma technologies have a sound basis, and have promising prospects for important further improvements.

Etching [15, 16, 23, 24] is a typical example of plasma processing. It is one stage in the fabrication of microelectronics. The process consists in the replication of the desired pattern on an element of an integrated microelectronic scheme. The replicator is a sheet of glass or quartz patterned with a thin film of a metal or metal oxide that absorbs UV radiation. This sheet, or wafer, is a substrate on which several layers of different materials are deposited. The upper layer, with thickness of the order of 1 μm, is a photoresist – an organic substance which absorbs UV radiation that causes volatilization at low temperatures. Other deposited materials are exemplified by Si, SiO_2, S_3N_4, and Al, which can play the role of a dielectric, a semiconductor, or a metal in an integrated manufacturing scheme.

A sequence in the fabrication of patterns is illustrated in Figure 7.3. The first step of the replication (lithography) process is to establish the replicator pattern on the sample. Then UV radiation is directed to the sample through the replicator. The transmitted radiation evaporates a photoresist and transfers the pattern to the sample in this way (Figure 7.3b). The following stage is the etching process, in which an underlying film is removed at points where a photoresist has been removed. This may be accomplished with electron beams, ion beams, X-rays, or by chemical or plasma methods. The etching process must be anisotropic in that material must be removed in a vertical direction only. At the same time, the process must be selective, so that it acts only on the removed material. These requirements seem to be incompatible. For example, the ion beam method is anisotropic because ions are directed perpendicular to the surface. But it is not selective because ions destroy the etching layer and the photoresist to an almost equal degree. Discharge methods that use chemically active particles (O, F, Cl) are selective, as chemical

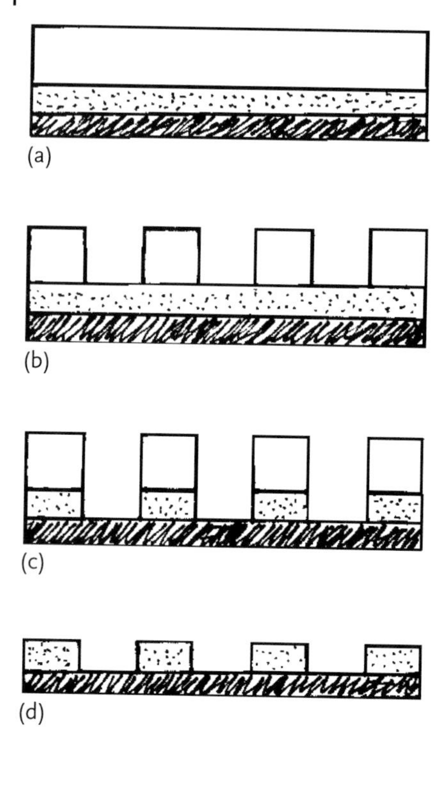

Figure 7.3 Stages of the microfabrication process: (a) the initial view of a sample, the upper layer is a photoresist; (b) irradiation of a sample and removal of part of the photoresist according to patterns of a replicate; (c) the etching process; (d) removal of the photoresist; (e) deposition of another material.

methods, but they are not anisotropic. Chemically active particles may not act on a photoresist, but they do act on the total surface of a removal layer, and the structure thus formed does not have vertical boundaries. The best result for the etching process comes from a combination of plasma methods that provides a selective and anisotropic etching process.

As a demonstration of this possibility, we consider the etching of Al film. This film may be removed by reaction with chlorine atoms or chlorine molecules that are formed in a gas discharge. The product of these reactions is $AlCl_3$ gas, which then leaves the system. Because Cl and Cl_2 do not react with a photoresist, such a process is selective, but it is not anisotropic. Addition of CCl_4 or $CHCl_3$ chlorocarbons to chlorine in the discharge gas leads to the formation of a polymer that covers the sample with a thin film. The film thus formed hampers the penetration of Cl and

Cl_2 to the Al, and slows the chemical process. But as a result of ion bombardment, the chlorocarbon protective film is destroyed, and chemical reactions between Cl or Cl_2 and Al proceed at such locations until a chlorocarbon film again forms there. This provides an anisotropy of the etching process due to the vertical direction of the ions. Thus, this process combines selectivity and anisotropy. Using a radio-frequency discharge in low-density chlorine with an admixture of chlorocarbons, one can obtain simultaneously an ion beam that is formed on the plasma boundary due to the plasma sheath, chemically active Cl and Cl_2 particles, and material for a polymer film.

Etching is one of the stages of the microfabrication process (Figure 7.3). In addition, the photoresist must be removed, and structures may be deposited in the etched areas by other materials. Plasma processes may be useful here also. In particular, a simple technique for photoresist removal is based on its reaction with oxygen atoms that can be obtained from a gas discharge plasma. The analysis of the etching process shows the complexity of plasma application methods. Because of the variety of possible plasma systems and methods for any particular applied problem, the choice of the optimal scheme, method, or regime depends on certain requirements and conditions. In conclusion, we note that this description of plasma applications has a schematic character and cannot give details of this plasma area.

References

1 Ochkin, V.N., Preobrazhensky, N.G. and Shaparev, N.Ya. (1998) *Optohalvanic Effect in Ionized Gas*, Gordon and Breach, London.

2 Ochkin, V.N. (2009) *Spectroscopy of Low Temperature Plasma*, Wiley-VCH Verlag GmbH.

3 Chen, F.F. (1984) *Introduction to Plasma Physics and Controlled Fusion*, Plenum, New York.

4 Stuhlinger, F. (1964) *Ion Propulsion for Space Flight*, McGraw-Hill, New York.

5 Jahn, R.G. (1968) *Physics of Electric Propulsion*, McGraw-Hill, New York.

6 Brewer, G.R. (1970) *Ion Propulsion Technology and Applications*, Gordon and Breach, New York.

7 Grishin, S.D. and Leskov, L.V. (1989) *Electrical Rocket Engines of Space Vehicles*, Mashinostroenie, Moscow, in Russian.

8 Goebel, D.M. and Katz, I. (2008) *Fundamentals of Electric Propulsion*, John Wiley & Sons, Inc., Hoboken.

9 Boulos, M.I., Fauchais, P., and Pfender, E. (1994) *Thermal Plasmas*, Plenum, New York.

10 Rossnagel, S.W., Cuomo, J.J., and Westwood, W.D. (1990) *Handbook of Plasma Processing Technology*, Noyes Publications, Park Ridge.

11 National Academy (1995) *Plasma Science: from Fundamental Research to Technological Applications*, National Academy Press, Washington.

12 Chen, F.F. and Chang, J.P. (2003) *Lectures Notes on Principles of Plasma Processing*, Kluwer, New York.

13 Nemchinsky, V.A. and Severance, W.S. (2006) *J. Phys.*, **39**, R423.

14 Gomez, E. *et al.* (2009) *J. Hazard. Mater.*, **161**, 604.

15 Lieberman, M.A. and Lichtenberger, A.J. (2005) *Principles of Plasma Discharge and Materials Processing*, John Wiley & Sons, Inc., Hoboken.

16 D'Agostino, R. (1990) *Plasma Deposition, Treatment, and Etching of Polymers*, Academic Press, New York.

17 Proud, J. *et al.* (1991) *Plasma Processes of Materials: Scientific Opportunities and Technological Challenges*, National Academy Press, Washington.

18 Rickbery, D.S. and Mathews, A. (1991) *Surface Engineering: Processes, Characterization and Applications*, Blackie and Son, Glasgow.

19 Howlett, S.R., Timothy, S.R., and Vaughan, D.A. (1992) *Industrial Plasmas: Focusing UK Skills on Global Opportunities*, CEST, London.

20 Grill, A. (1994) *Cold Plasma in Material Fabrication. From Fundamentals to Applications*, IEEE Press, New York.

21 Holohan, J.R. and Bell, A.T. (1974) *Techniques and Applications of Plasma Chemistry*, John Wiley & Sons, Inc., New York.

22 Fridman, A.A. (2008) *Plasma Chemistry*, Cambridge University Press, Cambridge.

23 Chapman, B.N. (1980) *Glow Discharge Processes – Sputtering and Plasma Etching*, John Wiley & Sons, Inc., New York.

24 Manos, D.M. and Flamm, D.L. (1989) *Plasma Etching. An Introduction*, Academic Press, San Diego.

Appendix A
Physical Constants and Units

Table A.1 Physical constants.

Electron mass	$m_e = 9.10939 \times 10^{-28}$ g
Proton mass	$m_p = 1.67262 \times 10^{-24}$ g
Atomic mass unit	$m_a = \frac{1}{12} m(^{12}\text{C}) = 1.66054 \times 10^{-24}$ g
Ratio of proton and electron masses	$m_p/m_e = 1836.15$
Ratio of atomic and electron masses	$m_a/m_e = 1822.89$
Electron charge	$e = 1.602177 \times 10^{-19}\,\text{C} = 4.8032 \times 10^{-10}$ esu
	$e^2 = 2.3071 \times 10^{-19}$ erg cm
Planck constant	$h = 6.62619 \times 10^{-27}$ erg s
	$\hbar = 1.05457 \times 10^{-27}$ erg s
Light velocity	$c = 2.99792 \times 10^{10}$ cm/s
Fine-structure constant	$\alpha = e^2/(\hbar c) = 0.07295$
Inverse fine-structure constant	$1/\alpha = \hbar c/e^2 = 137.03599$
Bohr radius	$a_0 = \hbar^2/(m_e e^2) = 0.529177$ Å
Rydberg constant	$R = m_e e^4/(2\hbar^2) = 13.6057\,\text{eV} = 2.17987 \times 10^{-18}$ J
Bohr magneton	$\mu_B = e\hbar/(2m_e c)$
	$= 9.27402 \times 10^{-24}$ J/T
	$= 9.27402 \times 10^{-21}$ erg/G
Avogadro number	$N_A = 6.02214 \times 10^{23}\,\text{mol}^{-1}$
Stefan–Boltzmann constant	$\sigma = \pi^2/(60\hbar^3 c^2) = 5.669 \times 10^{-12}\,\text{W}/(\text{cm}^2\,\text{K}^4)$
Molar volume	$R = 22.414$ l/mol
Loschmidt number	$L = N_A/R = 2.6867 \times 10^{19}\,\text{cm}^{-3}$
Faraday constant	$F = N_A e = 96485.3$ C/mol

Fundamentals of Ionized Gases, First Edition. Boris M. Smirnov.

Table A.2 Conversion factors for energy units.

	1 J	1 erg	1 eV	1 K	1 cm^{-1}	1 MHz	1 kcal/mol
1 J	1	10^7	6.2415×10^{18}	7.2429×10^{22}	5.0341×10^{22}	1.5092×10^{27}	1.4393×10^{20}
1 erg	10^{-7}	1	6.2415×10^{11}	7.2429×10^{15}	5.0341×10^{15}	1.5092×10^{20}	1.4393×10^{13}
1 eV	1.6022×10^{-19}	1.6022×10^{-12}	1	11604	8065.5	2.4180×10^8	23.045
1 K	1.3807×10^{-23}	1.3807×10^{-16}	8.6174×10^{-5}	1	0.69504	2.0837×10^4	1.9872×10^{-3}
1 cm^{-1}	1.9864×10^{-23}	1.9864×10^{-16}	1.2398×10^{-4}	1.4388	1	2.9979×10^4	2.8591×10^{-3}
1 MHz	6.6261×10^{-28}	6.6261×10^{-21}	4.1357×10^{-9}	4.7992×10^{-5}	3.3356×10^{-5}	1	9.5371×10^{-9}
1 kcal/mol	6.9477×10^{-21}	6.9477×10^{-28}	4.3364×10^{-2}	503.22	349.76	1.0485×10^7	1
1 kJ/mol	1.6605×10^{-21}	1.6605×10^{-28}	1.0364×10^{-2}	120.27	83.594	2.5061×10^6	0.23901

The data in Tables A.3– A.6 were taken from [1].

Table A.3 Conversion factors in formulas used in general physics involving atomic particles.

Number	Formula	C	Units
1	$v = C\sqrt{\varepsilon/m}$	5.931×10^7 cm/s	ε in eV, m in emu[a]
		1.389×10^6 cm/s	ε in eV, m in amu[b]
		5.506×10^5 cm/s	ε in K, m in emu
		1.289×10^4 cm/s	ε in K, m in amu
2	$v = C\sqrt{T/m}$	1.567×10^6 cm/s	T in eV, m in amu
		1.455×10^4 cm/s	ε in K, m in amu
3	$\varepsilon = Cv^2$	3.299×10^{-12} K	v in cm/s, m in emu
		6.014×10^{-9} K	v in cm/s, m in amu
		2.843×10^{-16} eV	v in cm/s, m in emu
		5.182×10^{-13} eV	v in cm/s, m in amu
4	$\omega_H = CH/m$	1.759×10^7 s^{-1}	H in G, m in emu
		9655 s^{-1}	H in G, m in amu
5	$r_H = C\sqrt{\varepsilon m}/H$	3.372 cm	ε in eV, m in emu, H in G
		143.9 cm	ε in eV, m in amu, H in G
		3.128×10^{-2} cm	ε in K, m in emu, H in G
		1.336 cm	ε in K, m in amu, H in G
6	$p = CH^2$	4.000×10^{-3} Pa = 0.04 erg/cm^3	H in G
7	$r_W = Cm/\rho^{1/3}$	0.7346 Å	m in m_a, ρ in g/cm^3
	$n = C(r/r_W)^3$	4.189	r and r_W in Å
8	$k_0 = Cr_W^2\sqrt{T/m}$	4.5714×10^{-12} cm^3/s	r_W in Å, T in K, m in amu
9	$v = C\rho r_0^2/\eta$	0.2179 cm/s	r_0 in µm, ρ in g/cm^3, η in 10^{-5} g/(cm s)
	$v = Cr_0^2/\eta$	0.01178 cm/s	r_0 in µm, ρ in g/cm^3, η for air at $p = 1$ atm, $T = 300$ K

a emu is the electron mass unit ($m_e = 9.108 \times 10^{-28}$ g).
b amu is the atomic mass unit ($m_a = 1.6605 \times 10^{-24}$ g).

1 The particle velocity is $v = \sqrt{2\varepsilon/m}$, where ε is the energy and m is the particle mass.
2 The average particle velocity is $v = \sqrt{8T/(\pi m)}$ with a Maxwell distribution function for velocities of particles; T is the temperature expressed in energy units and m is the particle mass.
3 The particle energy is $\varepsilon = mv^2/2$, where m is the particle mass and v is the particle velocity.
4 The Larmor frequency is $\omega_H = eH/(mc)$ for a charged particle of mass m in a magnetic field of a strength H.
5 The Larmor radius of a charged particle is $r_H = \sqrt{2\varepsilon/m}/\omega_H$, where ε is the energy of a charged particle, m is its mass, and ω_H is the Larmor frequency.
6 The magnetic pressure (4.162), $p_m = H^2/(8\pi)$.
7 The Wigner–Seitz radius (2.17) and the number of atoms in a cluster (2.16).
8 The reduced rate constant for atom attachment to a cluster according to (2.19).
9 The free-fall velocity for a cluster of radius r that is given by $v = 2\rho g r^2/(9\eta)$, where g is the free-fall acceleration, ρ is the cluster density, and η is the viscosity of matter through which a cluster moves.

Table A.4 Conversion factors in formulas for radiative transitions between atomic states.

Number	Formula	C	Units
1	$\varepsilon = C\omega$	4.1347×10^{-15} eV	ω in s^{-1}
		6.6261×10^{-34} J	ω in s^{-1}
2	$\omega = C\varepsilon$	$1.519 \times 10^{15}\ s^{-1}$	ε in eV
		$1.309 \times 10^{11}\ s^{-1}$	ε in K
3	$\omega = C/\lambda$	$1.884 \times 10^{15}\ s^{-1}$	ε in eV
4	$\varepsilon = C/\lambda$	1.2398 eV	λ in μm
5	$f_{0*} = C\omega d^2 g_*$	1.6126×10^{-17}	ω in s^{-1}, d in D[a]
		0.02450	$\Delta\varepsilon = \hbar\omega$ in eV, d in D[a]
6	$f_{0*} = Cd^2 g_*/\lambda$	0.03038	λ in μm, d in D[a]
7	$1/\tau_{*0} = C\omega^3 d^2 g_0$	$3.0316 \times 10^{-40}\ s^{-1}$	ω in s^{-1}, d in D[a]
		$1.06312 \times 10^{6}\ s^{-1}$	$\Delta\varepsilon = \hbar\omega$ in eV, d in D[a]
8	$1/\tau_{*0} = Cd^2 g_0/\lambda^3$	$2.0261 \times 10^{6}\ s^{-1}$	λ in μm, d in D[a]
9	$1/\tau_{*0} = C\omega^2 g_0 f_{0*}/g_*$	$1.8799 \times 10^{-23}\ s^{-1}$	ω in s^{-1}, d in D[a]
		$4.3393 \times 10^{7}\ s^{-1}$	$\Delta\varepsilon = \hbar\omega$ in eV, d in D[a]
10	$1/\tau_{*0} = C f_{0*} g_0/(g_* \lambda^2)$	$6.6703 \times 10^{7}\ s^{-1}$	λ in μm, d in D[a]

a D is debye, 1 D $= ea_0 = 2.5418 \times 10^{-18}$ CGSE.

1 The photon energy $\varepsilon = \hbar\omega$, where ω is the photon frequency.

2 The photon frequency is $\omega = \varepsilon/\hbar$.

3 The photon frequency is $\omega = 2\pi c/\lambda$, where λ is the wavelength and c is the light speed.

4 The photon energy is $\varepsilon = 2\pi\hbar c/\lambda$.

5 The oscillator strength for a radiative transition from the lower 0 to the upper $*$ state of an atomic particle that is averaged over lower states 0 and is summed over upper states of the group $*$ according to (2.113) is $f_{0*} = 2m_e\omega/(3\hbar e^2)d^2 g_*$, where $\mathbf{d} = \langle 0|\mathbf{D}|*\rangle$ is the matrix element for the operator of the dipole moment of an atomic particle taken between transition states. Here m_e and $\hbar$ are atomic parameters, g_* is the statistical weight of the upper state, and $\omega = (\varepsilon_* - \varepsilon_0)/\hbar$ is the transition frequency, where ε_0 and ε_* are the energies of the transition states.

6 The oscillator strength for radiative transition is given by [2–5] $f_{0*} = 4\pi c m_e/(3\hbar e^2\lambda)d^2 g_*$. Here λ is the transition wavelength; other notation is the same as above.

7 The rate of the radiative transition (2.112) is $1/\tau_{*0} = B_{*0} = 4\omega^3/(3\hbar c^3)d^2 g_0$, where B is the Einstein coefficient; other notation is as above.

8 The rate of radiative transition is given by $1/\tau_{*0} = B_{*0} = 32\pi^3/(\hbar\lambda^3)d^2 g_0$, where λ is the wavelength of this transition; other notation is the same as above.

9 The rate of radiative transition according to (2.112) is $1/\tau_{*0} = 2\omega^2 e^2 g_0/\left(m_e c^3\right) g_* f_{0*}$.

10 The rate of radiative transition (2.112) is given by $1/\tau_{*0} = 8\pi^2 g_0/\left(\hbar g_* \lambda^2 c\right) f_{0*}$.

Table A.5 Conversion factors in formulas for transport coefficients.

Number	Formula	C	Units
1	$D = CKT$	$8.617 \times 10^{-5}\,\text{cm}^2/\text{s}$	K in $\text{cm}^2/(\text{V s})$, T in K
		$1\,\text{cm}^2/\text{s}$	K in $\text{cm}^2/(\text{V s})$, T in eV
2	$K = CD/T$	$11\,604\,\text{cm}^2/(\text{V s})$	D in cm^2/s, T in K
		$1\,\text{cm}^2/(\text{V s})$,	D in cm^2/s, T in eV
3	$D = C\sqrt{T/\mu}/(N\overline{\sigma_1})$	$4.617 \times 10^{21}\,\text{cm}^2/\text{s}$	$\overline{\sigma_1}$ in Å^2, N in cm^{-3}, T in K, μ in amu.
		$1.595\,\text{cm}^2/\text{s}$	$\overline{\sigma_1}$ in Å^2, $N = 2.687 \times 10^{19}\,\text{cm}^{-3}$, T in K, μ in amu
		$171.8\,\text{cm}^2/\text{s}$	$\overline{\sigma_1}$ in Å^2, $N = 2.687 \times 10^{19}\,\text{cm}^{-3}$, T in eV, μ in amu
		$68.1115\,\text{cm}^2/\text{s}$	$\overline{\sigma_1}$ in Å^2, $N = 2.687 \times 10^{19}\,\text{cm}^{-3}$, T, μ in emu
		$7338\,\text{cm}^2/\text{s}$	$\overline{\sigma_1}$ in Å^2, $N = 2.687 \times 10^{19}\,\text{cm}^{-3}$, T in eV, μ in emu
4	$K = C(\sqrt{T\mu}N\overline{\sigma_1})^{-1}$	$1.851 \times 10^4\,\text{cm}^2/(\text{V s})$	$\overline{\sigma_1}$ in Å^2, $N = 2.687 \times 10^{19}\,\text{cm}^{-3}$, T in K, μ in amu
		$171.8\,\text{cm}^2/(\text{V s})$	$\overline{\sigma_1}$ in Å^2, $N = 2.687 \times 10^{19}\,\text{cm}^{-3}$, T in eV, μ in amu
		$7.904 \times 10^5\,\text{cm}^2/(\text{V s})$	$\overline{\sigma_1}$ in Å^2, $N = 2.687 \times 10^{19}\,\text{cm}^{-3}$, T in K, μ in emu
		$7338\,\text{cm}^2/(\text{V s})$	$\overline{\sigma_1}$ in Å^2, $N = 2.687 \times 10^{19}\text{cm}^{-3}$, T in eV, μ in emu
5	$\kappa = C\sqrt{T/m}/\overline{\sigma_2}$	$1.743 \times 10^4\,\text{W}/(\text{cm K})$	T in K, m in amu, $\overline{\sigma_2}$ in Å^2
		$7.443 \times 10^5\,\text{W}/(\text{cm K})$	T in K, m in emu, $\overline{\sigma_2}$ in Å^2
6	$\eta = C\sqrt{Tm}/\overline{\sigma_2}$	$5.591 \times 10^{-5}\,\text{g}/(\text{cm s})$	T in K, m in amu, $\overline{\sigma_2}$ in Å^2

1 The Einstein relation (4.38) for the diffusion coefficient of a charged particle in a gas $D = KT/e$, where D and K are the diffusion coefficient and mobility of a charged particle and T is the gas temperature.
2 The Einstein relation (4.38) for the mobility of a charged particle in a gas $K = eD/T$.
3 The diffusion coefficient of an atomic particle in a gas in the first Chapman–Enskog approximation (4.53), $D = 3\sqrt{2\pi T/\mu}/(16N\overline{\sigma_1})$, where T is the gas temperature, N is the number density of gas atoms or molecules, μ is the reduced mass of a colliding particle and a gas atom or molecule, and $\overline{\sigma_1}$ is the average cross section of collision.
4 The mobility of a charged particle in a gas in the first Chapman–Enskog approximation (4.21), $K = 3e\sqrt{2\pi/(T\mu)}/(16N\overline{\sigma_1})$; the notation is the same as above.
5 The gas thermal conductivity in the first Chapman–Enskog approximation (4.54), $\kappa = 25\sqrt{\pi T}/(32\sqrt{m}\overline{\sigma_2})$, where m is the mass of the atom or molecule and $\overline{\sigma_2}$ is the average cross section of collision between gas atoms or molecules; other notation is the same as above.
6 The gas viscosity in the first Chapman–Enskog approximation (4.56), $\eta = 5\sqrt{\pi Tm}/(24\overline{\sigma_2})$; the notation is the same as above.

Table A.5 Continued.

Number	Formula	C	Units
7	$\xi = CE/(TN\sigma)$	1.160×10^{20}	E in V/cm, T in K, σ in Å^2, N in cm^{-3}
		1×10^{16}	E in V/cm, T in eV, σ in Å^2, N in cm^{-3}
8	$D_0 = C\sqrt{T/m}/(Nr_W^2)$	1.469×10^{21} cm^2/s	r_W in Å, N in cm^{-3}, T in K, m in amu
		0.508 cm^2/s	r_W in Å, $N = 2.687 \times 10^{19}$ cm^{-3}, T in K, m in amu
		54.69 cm^2/s	r_W in Å, $N = 2.687 \times 10^{19}$ cm^{-3}, T in eV, m in amu
9	$K_0 = C(\sqrt{Tm}Nr_W^2)^{-1}$	1.364×10^{19} cm^2/(V s)	r_W in Å, N in cm^{-3}, T in K, m in amu
		0.508 cm^2/(V s)	r_W in Å, $N = 2.687 \times 10^{19}$ cm^{-3}, T in K, m in amu
		54.69 cm^2/(V s)	r_W in Å, $N = 2.687 \times 10^{19}$ cm^{-3}, T in K, m in amu
10	$D_0 = CT/(r_W\eta)$	7.32×10^{-5} cm^2/s	r_W in Å, T in K, η in 10^{-5} g/(cm s)
11	$K_0 = C/(r_W\eta)$	0.085 cm^2/(V s)	r_W in Å, η in 10^{-5} g/(cm s)

7 The ion drift parameter in a gas in a constant electric field $\xi = eE/(TN\sigma)$, where E is the electric field strength, T is the gas temperature, N is the number density of atoms or molecules, and σ is the cross section of collision.

8 The reduced diffusion coefficient for a spherical cluster of small size in the first Chapman–Enskog approximation (4.114), $D_0 = 3\sqrt{2T/\pi m}/(16Nr_W^2)$, where T is the gas temperature, N is the number density of gas atoms, m is the mass of a cluster atom, and r_W is the Wigner–Seitz radius.

9 The zero-field reduced mobility for a spherical cluster of small size in the first Chapman–Enskog approximation (4.113), $K_0 = 3e/(8Nr_W^2\sqrt{2\pi mT})$, where the notation is given above.

10 The reduced diffusion coefficient for a spherical cluster of large size (4.110) in a gas, $D_0 = T/(6\pi r_W\eta)$, where η is the gas viscosity; other notation is as above.

11 The zero-field reduced mobility for a spherical cluster of large size (4.110), $K_0 = e/(6\pi r_W\eta)$, where the notation is given above.

Table A.6 Conversion factors in formulas used in gas and plasma physics.

Number	Formula	C	Units
1	$N = Cp/T$	$7.339 \times 10^{21}\ \mathrm{cm}^{-3}$	p in atm, T in K
		$9.657 \times 10^{18}\ \mathrm{cm}^{-3}$	p in Torr, T in K
2	$p = CNT$	1.036×10^{-19} Torr	N in cm^{-3}, T in K
3	$\gamma = CN_e/T^3$	2.998×10^{-21}	N_e in cm^{-3}, T in eV
		4.685×10^{-9}	N_e in cm^{-3}, T in K
4	$\xi = Cm^{3/2}T^{3/2}$	$2.415 \times 10^{15}\ \mathrm{cm}^{-3}$	T, m in emu
		$3.019 \times 10^{21}\ \mathrm{cm}^{-3}$	T in eV, m in emu
		$1.879 \times 10^{20}\ \mathrm{cm}^{-3}$	T in K, m in amu
		$2.375 \times 10^{26}\ \mathrm{cm}^{-3}$	T in eV, m in amu
5	$K = C/T_e^{9/2}$	$2.406 \times 10^{-31}\ \mathrm{cm}^6/\mathrm{s}$	T_e in 1000 K
6	$\omega_p = C\sqrt{N_e/m}$	$5.642 \times 10^4\ \mathrm{s}^{-1}$	N_e in cm^{-3}, m in emu
		$1322\ \mathrm{s}^{-1}$	N_e in cm^{-3}, m in amu
7	$r_D = C\sqrt{T/N_e}$	525.3 cm	N_e in cm^{-3}, T in eV
		4.876 cm	N_e in cm^{-3}, T in K
8	$i = CU^{3/2}m^{-1/2}l^{-2}$	$5.447 \times 10^{-8}\ \mathrm{A/cm}^2$	m in amu, l in cm, U in V
		$2.326 \times 10^{-6}\ \mathrm{A}^2$	m in emu, l in cm, U in V
9	$U = Cm^{1/3}l^{4/3}i^{-2/3}$	69 594	m in amu, l in cm, i in $\mathrm{A/cm}^2$
		5697 V	m in emu, l in cm, i in $\mathrm{A/cm}^2$
10	$l = Cm^{-1/4}U^{3/4}i^{-1/2}$	2.3338×10^{-4} cm	m in amu, U in V, i in $\mathrm{A/cm}^2$
		1.525×10^{-3} cm	m in emu, U in V, i in $\mathrm{A/cm}^2$

1 The gas state equation (4.12).
2 The gas state equation (4.12).
3 The ideality plasma parameter (1.3), $\gamma = N_e e^6/T_e^3$.
4 Preexponent of the Saha formula (1.52) $\xi = [mT/(2\pi\hbar^2)]^{3/2}$.
5 The rate constant of three body electron–ion recombination (2.71), $K = 1.5e^{10}/(m_e^{1/2}T_e^{9/2})$.
6 The plasma frequency (1.13), $\omega_p = \sqrt{4\pi N_e e^2/m_e}$.
7 The Debye–Hückel radius (1.7), $r_D = \sqrt{T/(8\pi N_e e^2)}$.
8 The three-halves law (1.31), $i = 2/(9\pi)\sqrt{e/(2m)}U^{3/2}/l^2$, where l is the distance between electrodes, U is the difference in electrode voltages, i is the maximum current density, and e and m are the particle charge and mass.
9 The three-halves law in the form $U = (9\pi i/2)^{2/3}(2m/e)^{1/3}l^{4/3}$.
10 The three-halves law in the form $l = (2/(9\pi))^{1/2}(e/(2m))^{1/4}U^{3/4}/i^{1/2}$.

References

1 Smirnov, B.M. (2008) *Reference Data on Atomic Physics and Atomic Processes*, Springer, Heidelberg.

2 Condon, E.U. and Shortley, G.H. (1967) *Theory of Atomic Spectra*, Cambridge University Press, Cambridge.

3 Sobelman, I.I. (1979) *Radiative Spectra and Radiative Transitions*, Springer-Verlag, Berlin.

4 Bethe, H. and Salpeter, E. (1975) *Quantum Mechanics of One- and Two-Electron Atoms*, Springer-Verlag, Berlin.

5 Krainov, V.P., Reiss, H., and Smirnov, B.M. (1997) *Radiative Transitions in Atomic Physics*, John Wiley & Sons, Inc., New York.

Appendix B
Parameters of Atoms and Ions

Table B.1 Ionization potential (J) of atoms in the ground state.

Atom	J, eV	Atom	J, eV	Atom	J, eV	Atom	J, eV
$H(^2S_{1/2})$	13.598	$K(^2S_{1/2})$	4.341	$Rb(^2S_{1/2})$	4.177	$Cs(^2S_{1/2})$	3.894
$He(^1S_0)$	24.586	$Ca(^1S_0)$	6.113	$Sr(^1S_0)$	5.695	$Ba(^1S_0)$	5.212
$Li(^2S_{1/2})$	5.392	$Sc(^2D_{3/2})$	6.562	$Y(^2D_{3/2})$	6.217	$La(^2D_{3/2})$	5.577
$Be(^1S_0)$	9.323	$Ti(^3F_2)$	6.82	$Zr(^3F_2)$	6.837	$Ce(^1G_4)$	5.539
$B(^2P_{1/2})$	8.298	$V(^4F_{3/2})$	6.74	$Nb(^6D_{1/2})$	6.88	$Ta(^4F_{3/2})$	7.89
$C(^3P_0)$	11.260	$Cr(^7S_3)$	6.766	$Mo(^7S_3)$	7.099	$W(^5D_0)$	7.98
$N(^4S_{3/2})$	14.534	$Mn(^6S_{5/2})$	7.434	$Tc(^6S_{5/2})$	7.28	$Re(^6S_{5/2})$	7.88
$O(^3P_2)$	13.618	$Fe(^5D_4)$	7.902	$Ru(^5F_5)$	7.366	$Os(^5D_4)$	8.73
$F(^2P_{3/2})$	17.423	$Co(^4F_{9/2})$	7.86	$Rh(^4F_{9/2})$	7.46	$Ir(^4F_{9/2})$	9.05
$Ne(^1S_0)$	21.565	$Ni(^3F_4)$	7.637	$Pd(^1S_0)$	8.336	$Pt(^3D_3)$	8.96
$Na(^2S_{1/2})$	5.139	$Cu(^2S_{1/2})$	7.726	$Ag(^2S_{1/2})$	7.576	$Au(^2S_{1/2})$	9.226
$Mg(^1S_0)$	7.646	$Zn(^1S_0)$	9.394	$Cd(^1S_0)$	8.994	$Hg(^1S_0)$	10.438
$Al(^2P_{1/2})$	5.986	$Ga(^2P_{1/2})$	5.999	$In(^2P_{1/2})$	5.786	$Tl(^2P_{1/2})$	6.108
$Si(^3P_0)$	8.152	$Ge(^3P_0)$	7.900	$Sn(^3P_0)$	7.344	$Pb(^3P_0)$	7.417
$P(^4S_{3/2})$	10.487	$As(^4S_{3/2})$	9.789	$Sb(^4S_{3/2})$	8.609	$Bi(^4S_{3/2})$	7.286
$S(^3P_2)$	10.360	$Se(^3P_2)$	9.752	$Te(^3P_2)$	9.010	$Rn(^1S_0)$	10.75
$Cl(^2P_{3/2})$	12.968	$Br(^2P_{3/2})$	11.814	$I(^2P_{3/2})$	10.451	$Ra(^1S_0)$	5.278
$Ar(^1S_0)$	15.760	$Kr(^1S_0)$	14.000	$Xe(^1S_0)$	12.130	$U(^5L_6)$	6.194

Fundamentals of Ionized Gases, First Edition. Boris M. Smirnov.

Table B.2 Electron affinities (EA) of atoms. Here EA is the electron binding energy of the negative ion, whose electron state is marked; "not" means that the the electron affinity of the atom is not a positive value.

Ion	EA, eV	Ion	EA, eV	Ion	EA, eV	Ion	EA, eV
$H^-(^1S)$	0.75416	$S^-(^2P)$	2.0771	$Se^-(^2P)$	2.0207	$I^-(^1S)$	3.0590
He^-	Not	$Cl^-(^1S)$	3.6127	$Br^-(^1S)$	3.3636	Xe^-	Not
$Li^-(^1S)$	0.618	Ar^-	Not	Kr^-	Not	$Cs^-(^1S)$	0.4716
Be^-	Not	$K^-(^1S)$	0.5015	$Rb^-(^1S)$	0.4859	Ba^-	Not
$B^-(^3P)$	0.28	$Ca^-(^2P)$	0.018	$Sr^-(^2P)$	0.026	$La^-(^3F)$	0.5
$C^-(^4S)$	1.2629	$Sc^-(^1D)$	0.19	$Zr^-(^4F)$	0.43	Hf^-	Not
$C^-(^2D)$	0.035	$Ti^-(^4F)$	0.08	$Nb^-(^5D)$	0.89	$Ta^-(^5D)$	0.32
N^-	Not	$V^-(^5D)$	0.53	$Mo^-(^6S)$	0.75	$W^-(^6S)$	0.816
$O^-(^2P)$	1.4611	$Cr^-(^6S)$	0.67	$Tc^-(^5D)$	1.0	$Re^-(^5D)$	0.15
$F^-(^1S)$	3.4012	Mn^-	Not	$Ru^-(^4F)$	1.5	$Os^-(^4F)$	1.4
Ne^-	Not	$Fe^-(^4F)$	0.151	$Rh^-(^3F)$	1.14	$Ir^-(^3F)$	1.57
$Na^-(^1S)$	0.5479	$Co^-(^3F)$	0.662	$Pd^-(^2D)$	0.56	$Pt^-(^2D)$	2.128
Mg^-	Not	$Ni^-(^2D)$	1.15	$Ag^-(^1S)$	1.30	$Au^-(^1S)$	2.3086
$Al^-(^3P)$	0.441	$Cu^-(^1S)$	1.23	Cd^-	Not	Hg^-	Not
$Si^-(^4S)$	1.394	Zn^-	Not	$In^-(^3P)$	0.3	$Tl^-(^3P)$	0.5
$Si^-(^2D)$	0.526	$Ga^-(^2P)$	0.5	$Sn^-(^4S)$	1.112	$Pb^-(^4S)$	0.364
$Si^-(^2P)$	0.034	$Ge^-(^4S)$	1.233	$Sb^-(^3P)$	1.05	$Bi^-(^3P)$	0.95
$P^-(^3P)$	0.7465	$As^-(^3P)$	0.80	$Te^-(^2P)$	1.9708	$Po^-(^2P)$	1.6

Index

Fundamentals of Ionized Gases, First Edition. Boris M. Smirnov.